Richard,

Another Christmas
another book!

With all my love,
Carole.

Christmas 1987

TWENTIETH-CENTURY
WARRIORS

TWENTIETH-CENTURY
WARRIORS

*The Development of the Armed Forces of the Major Military Nations
in the Twentieth Century*

Field Marshal Lord Carver

WEIDENFELD AND NICOLSON · LONDON

First published in Great Britain by
George Weidenfeld & Nicolson Limited
91 Clapham High Street, London SW4 7TA

ISBN 0 297 79160 5

Printed in Great Britain at The Bath Press, Avon

Contents

Maps   vii

Acknowledgements   ix

Introduction   1

1 Britain   3

2 France   40

3 Germany   90

4 Russia   171

5 The Soviet Union   207

6 The United States of America   260

7 Japan   331

8 China   379

9 Conclusion   431

Select Bibliography   447

Notes   451

Index   453

Maps

1 The Western Front 1914–18   15
2 The Eastern Front 1914–18   102
3 The Russo-Japanese War 1904–5   176
4 The Eastern Front 1941–45   226
5 The Western Pacific 1941–45   292
6 China   381

Acknowledgements

I am deeply grateful to the following for reading and commenting on the chapters on which they are expert: Professors Louis Allen, John Erickson, Sir Michael Howard, Douglas Johnson and Dennis Showalter, Dr Gerald Segal and David Steeds. Any errors and omissions that may be detected are entirely my responsibility.

I also wish to acknowledge the courtesy and help I have received, once again, from the Librarians and staff of the Prince Consort's Army Library at Aldershot, and of the Ministry of Defence Libraries and Historical Branches in London; also from the United States National Archives, the US Naval Historical Center and the US Air Force Historical Research Center.

Finally, I owe a huge debt to Judy Rous for her accurate, skilful and rapid typing. I had previously thought that the feat of deciphering my script was one which only I, not without difficulty, could perform. In triumphantly proving me wrong, she has lifted a great burden from my fingers.

Michael Carver
1987

Introduction

In this book I have set out to trace the development of the armed forces of the major military nations in this century. I have restricted my study to Britain, France, Germany, Russia and the Soviet Union, the United States of America, Japan and China. Armed forces of other nations have played important parts in the history of the century, but none of them have exerted the same influence throughout the period. The Indian Army is perhaps an exception but, until 1947, its development and activity is covered by those of Britain. Since 1948 the development and action of the Israeli armed forces have been of considerable significance, and those of her principal Arab opponents have been impressive in scale, but neither of them has exercised the degree of influence over the whole period as those I have considered.

Military histories tend to look at events from the point of view of the nationality of the writer. That produces a serious distortion. I have tried to avoid that bias by tracing the development of a nation's armed forces, and describing the campaigns in which they took part, from that nation's point of view from the beginning to the end of the period, at the expense of some repetition. I hope that the reader, as I have done, will find interesting and valuable this illustration of the military history of the same period in seven different pictures, painted from different viewpoints.

I have tried to arrive at some conclusions from drawing these pictures together. One clear one emerges: that the wars in which all were engaged were none of them inevitable. They were not the product of forces beyond human control, nor were they the inescapable result of the development of the armed forces themselves and the trade in arms which supplied them, although both influenced the policies which governments pursued in support of what they believed to be the interests of their nations. It was those policies which were themselves the causes of war, as Professor Sir Michael Howard has so eloquently reminded us, heeding the advice of Clausewitz. In the event, the policies which some nations, notably Germany, Russia and Japan, pursued during the period, although designed to further their national interests, proved to be counterproductive because of the opposition they aroused by conflict with the interests of others. They should have heeded the advice of the Chinese writer on military affairs in the first century BC, Sun Tzu, who taught that one should not forget that after a war one has to live peaceably with one's neighbours. Expectations for the future must rest, under the shadow of the awe-inspiring effects of the nuclear weapon, on the hope that communities of differing racial origins and cultures, trusting to different ways of organizing their societies, and competing for resources in a world where the distribution of population is changing dramatically, will find some better way, as large groups of them already have, of adjusting their changing relationships than killing each other.

1

BRITAIN

1900–18

Queen Victoria's Diamond Jubilee in 1897 appeared to mark the apogee of British power and influence. Since the Battle of Trafalgar in 1805 her Royal Navy had dominated the oceans almost unchallenged and, since that of Waterloo ten years later, her army had been free of any commitment on the continent of Europe, apart from the limited one in the Crimea from 1854 to 1856. This happy state of affairs, combined with her lead in the Industrial Revolution, had made it possible for her to expand and secure an empire scattered all over the world. The primary task of her armed forces was seen as its defence. This is made clear by the Stanhope memorandum, a paper signed by Edward Stanhope, Secretary of State for War, in 1891, 'laying down the Requirements from our Army'. It stated:

Her Majesty's Government have carefully considered the question of the general objects for which our Army is maintained. It has been considered in connection with the programme of the Admiralty, and with knowledge of the assistance which the Navy is capable of rendering in the various contingencies which appear to be reasonably probable, and they decide that the general basis of the requirements from our Army may be correctly laid down by stating that the objects of our military organization are:

(a) The effective support of the civil power in all parts of the United Kingdom.

(b) To find the number of men for India, which has been fixed by arrangement with the Government of India.

(c) To find garrisons for all our fortresses and coaling stations, at home and abroad, according to a scale now laid down, and to maintain these garrisons at all times at the strength fixed for a peace or war footing.

(d) After providing for these requirements, to be able to mobilize rapidly for home defence two Army Corps of Regular troops and one partly composed of Regulars and partly of Militia, and to organize the Auxiliary Forces, not allotted to Army Corps or garrisons, for the defence of London and for the defensible positions in advance, and for the defence of mercantile ports.

(e) Subject to the foregoing considerations and to their financial obligations, to aim at being able, in case of necessity, to send abroad two complete Army Corps, with Cavalry Division and Line of Communication. But it will be distinctly understood that the probability of the employment of an Army Corps in the field in any European war is sufficiently improbable to make it the primary duty of the military authorities to organize our forces efficiently for the defence of this country.

The Jubilee Naval Review at Spithead, in which 165 warships had taken part, including twenty-one first-class battleships and fifty-four cruisers, appeared to reinforce the British public's view that naval supremacy would make it possible for the Prime Minister, Lord Salisbury, to continue the policy of 'splendid isolation' which he had pursued for so long, but there were dark clouds on the horizon which caused anxiety to those responsible for defence. The principal one was the alliance between France and Russia, signed in 1894, brought about primarily by their concern at the threat from Germany, since Wilhelm II had dismissed Bismarck in 1890. Britain, however, saw it in a different light. She was already sensitive to the fact that other nations were catching her up in the industrial race, which was reflected in their ability to challenge her naval power, seen as the guarantee of her commercial wealth. The threat was a dual one: to her naval supremacy and, through that, to her commerce, on which, with the great increase in her population, she was so dependent. Those were the days when nations still assumed that one had to exercise sovereignty over an area in order to preserve one's trade with it, although Britain herself had been trying to move away from that in the direction of 'Free Trade'. The emerging industrial nations, Germany, France, Russia, the USA and Japan, were jealous of Britain's privileged position and determined to obtain a greater share themselves. Naval power, as expounded by the American Captain Alfred Mahan, in his influential book *The Influence of Sea Power upon History* published in 1890, was seen as the path to achieve this. These fears are well illustrated by a comparison of the battleship strength of the countries concerned in 1883 and 1897.[1]

Country	Battleships in 1883	Battleships in 1897 (plus those building)
Britain	38	62
France	19	36
Germany	11	12
Russia	3	18
Italy	7	12
USA	0	11
Japan	0	7

There had been a series of alarms about the naval situation ever since the Crimean War, principally over the threat from France. In the 1850s this had led to the construction of the Palmerston forts to protect naval bases on the Channel, and when relations with France deteriorated after Britain's intervention in Egypt in 1882, similar fears were aroused and an extensive programme of fortifications covering the approaches to London was embarked upon in 1889, the year in which the Naval Defence Act was passed, initiating the construction of the *Resolution* class of battleships. Some wits dubbed this alarm about the possibility of invasion as 'The Blue Funk School'.

It was the problem of how the army was to meet the demands of home defence that had led to the Stanhope memorandum. In 1898 the regular army consisted of 233,560 men.* The infantry was almost equally split between foreign and home service, the former including fifty-two battalions in India, the latter twenty in Ireland. The Cardwell reforms, prompted both by the Crimean War and the Indian Mutiny, established this pattern of linking battalions on home service to sister battalions of the same regiment serving overseas. In practice it subordinated the needs of the army as a whole to the requirements of British troops in India, increased as a result of the Mutiny. This applied not only to numbers and organization, but also to equipment, training and general military and social habits, an influence which persisted up to 1940, reinforcing objections to a continental commitment.

To back up the regular army there were three different forces of part-time soldiers: the Militia, the Yeomanry and the Volunteers. The first had its origins in the power of Lord-Lieutenants of counties to raise forces in emergency for home defence by compulsory recruiting by ballot; but it had become a volunteer force, largely used as a stepping-stone towards joining the regular army, either because the recruit was not up to regular physical standards, or as a trial period, being easier and cheaper to get out of, once committed. It did little training, was poorly officered and was not an efficient force. In 1898 it was about 100,000 strong. A more efficient force was the Volunteers, in 1898 about 250,000 strong. It was a keener body of men, of a much more middle-class character than the Militia, owing its origin largely to the wave of patriotic feeling at the time of the invasion alarm in 1858, which had led to the Palmerston forts. The War Office and the Militia treated it with reserve, as it was seen as undermining the latter

---

* Of these 150,376 were infantry, in 157 battalions; 19,267 cavalry in 31 regiments; 40,152 artillery in 125 batteries and 99 companies of garrison artillery; 8,100 engineers; 9,100 in the supporting services, and 6,565 in colonial corps.

which the regular army had long looked on as its natural source of reserve manpower. The third body was the Yeomanry, a volunteer force of cavalry which up to 1888 had been liable only to service in its own area, and primarily regarded as an internal security force, Peterloo being one of its most notorious engagements. It was naturally drawn from landowners and the agricultural community. Its standard of training was low, and there was little organization higher than the regiment, or even the squadron. In 1899 there were thirty-eight regiments, totalling in theory nearly 12,000, in fact about 9,000 men.

This was the army which, on 9 October 1899, found itself suddenly once more at war with the Boer Republic of the Transvaal in South Africa, where there were only 10,000 British troops, although an equal number, mostly from India, was on its way there, and an expeditionary force of 47,000 men from England had been authorized a few days before. It was expected that this force would be sufficient to bring the recalcitrant Boers to heel, but the defeats suffered in the 'Black Week' in mid December came as a shock. The army faced much the same problems as it had in the Crimea. Almost everything had to be improvised. By March 1900 nearly 100,000 men had been sent to South Africa, two-thirds of them from Britain, those sent from garrisons like Malta and Gibraltar being replaced by Militia and Volunteers who had volunteered for foreign service.

As the war entered its second phase, and Kitchener had taken over from the aged Field Marshal Lord Roberts in October 1900, the War Office faced increasing difficulty in keeping up the strength of his army; and, as soon as it was over in 1902, a Royal Commission, chaired by Lord Elgin, was established to 'inquire into the Military Preparations for the War in South Africa, and into the supply of Men, Ammunition, Equipment and Transport by Sea and Land etc.'. A member of the commission, who was to wield a significant influence on the country's defence, particularly on the army, in the next ten years, was Lord Esher. However, before the commission had even begun its inquiry, the Secretary of State for War, Mr St John Brodrick, had proposed a major reorganization of the army to provide six corps (each of three divisions of two brigades each) based in six territorial districts (one in Ireland), three of the corps consisting largely of Volunteers and Militia, both of which for the first time would man artillery. The three regular corps would be organized as an expeditionary force.

Brodrick's proposal was motivated not only by the desire to remedy the faults in organization which the Boer War had revealed, but also by the realization that Britain's isolation was no longer splendid. She had no friends. Germany had sympathized with and supplied the Boer Republic. Both Wilhelm ii and his ambitious admiral, Tirpitz, saw it as a revelation of Britain's weakness: that her much-vaunted empire was like 'some huge giant sprawling over the globe, with gouty fingers and toes stretched in every direction, which cannot be approached without eliciting a scream'.[2] Germany's First Navy Law of 1898 provided for the construction of nineteen battleships in furtherance of Tirpitz's aim that Germany should achieve 'world political freedom' by establishing a fleet in the North Sea which would not only defend her coasts, but threaten Britain's world-wide naval supremacy. Its prime function would be as a political lever. Britain, unable owing to her imperial commitments to concentrate superior strength in the North Sea, would become more amenable to Germany's ambition to stride the world stage. Britain's involvement in South Africa seemed to afford an ideal opportunity to further this policy, and the Second Navy Law of 1900 doubled the target figure of battleships to thirty-eight. Hitherto the Royal Navy had based its future plans on a Two-Power Standard; that is that it had to be capable of defeating a combination of two major naval powers, which it had assumed were France and Russia. While they remained potentially hostile, the German threat, on the same reasoning, would force Britain to adopt a Three-Power Standard. Not only would the cost of this be astronomical, but it was doubtful if it could be manned by voluntary recruitment.

Faced by the apparent need to strengthen both the navy and the army, Arthur Balfour, who succeeded his uncle, Lord Salisbury, as Prime Minister in 1902, converted the former Colonial Defence Committee of the Cabinet into a wide-based Committee of Imperial Defence and, most significantly, gave it a permanent secretariat, which was to be the forerunner both of the Chiefs of Staff Committee and of the Cabinet Secretariat. Major Hankey of the Royal Marines was to wield great influence within that secretariat over a long period.

The committee faced a number of interlocking strategic conundrums. The first was the direct defence of the British Isles. Would the navy any longer be able to guarantee the country against invasion either by France or Germany? If not, what army was needed to defeat it, and how should it be organized and recruited? Against professional armies of the strength of the German or the French, reliance could not be placed solely on Volunteers, Militia and Yeomanry; but, if the regular army was to be held at home for that purpose, it would be impossible to reinforce India or the areas which controlled the sea routes to India, such as Malta, Egypt and South Africa, against a Russian or Franco-Russian threat. Russian expansion into Turkestan, based on the development of railways leading into Afghanistan, was causing grave concern. India's bill for reinforcement in the event of a Russian move into that country was for 100,000 men; and although the Royal Navy's largest fleet was in the Mediterranean, a combination of the French, Russian and possibly Austro-Hungarian fleets, with Turkey coming increasingly under German influence, cast doubt on its ability to ensure safe passage through that sea, as it did also on the ability of the small garrison in Egypt to hold that country and secure the Suez Canal.

These large potential commitments, particularly that of home defence, led Roberts and others to press for conscription, which was recommended in 1904 by the Norfolk Commission, set up to inquire into the future of the Militia and Volunteers.

Two steps to reduce the apparently overwhelming burden of commitments, both primarily affecting the navy, were taken at this time. The first was to accept that the USA had become a power on the American continent and the seas surrounding it, which could not be challenged, and that a Pax Americana had to be accepted all the way from Canada to Cape Horn. The other was the acceptance of the same sort of realism in respect of Japan, and the signature of the Anglo-Japanese alliance in January 1902. This paved the way for the major naval reorganization which was to be implemented by Admiral Sir John Fisher when he became First Sea Lord in 1904. Not only did he launch the new series of battleships with *Dreadnought* and of battle-cruisers with *Invincible*, scrap 154 out-of-date vessels and shake up the whole training and efficiency of the Royal Navy, but he set about concentrating the fleet nearer home waters, reducing strength in the Mediterranean and withdrawing all battleships from elsewhere. The Foreign Office viewed with concern the effect reduction in the Mediterranean might have on uncommitted powers like Italy, Austria-Hungary and Turkey, while the Colonial Office, the India Office, the Dominions and the War Office were alarmed at the concept, derived from Mahan, that all depended on bringing the enemy's main battle-fleet to action and defeating it. Once that were accomplished, there would be no threat either to the homeland or to its overseas dependencies, other than the overland threat to India and, possibly, to Egypt.

Doubts about Fisher's policy were to a large extent dispelled, and the whole orientation of Britain's strategy changed, by two events. The first was the Russo-Japanese War of 1904–5 and Russia's defeat, notably the loss of her fleet at the Battle of Tsushima. For the present, at least, the Russian threat had been removed. The second was the removal also of the French threat and its conversion into the Entente in 1904, strengthened by the Tangier incident in 1905. Henceforth the potential enemy was more and more clearly Germany, until the Triple Entente of Britain, France and Russia faced the Triple Alliance of Germany, Austria-Hungary and Italy.

Not only did this simplify the navy's problem, but it clarified the issue for the army which had been subjected to a major reorganization and reform. Brodrick's plan had been superseded by a similar but different one, initiated by his successor Hugh Arnold-Forster, who had tried to divide the army into two elements, a long-service one for service overseas and a short-service one, which could therefore expand the reserves, for home defence. But it was his successor Lord Haldane, Secretary of State for War from 1905 until 1912, who, acting on the recommendations of the powerful Esher Committee, brought the army's organization into the twentieth century. The post of Commander-in-Chief was abolished; an Army Council established; a General Staff formed, the Chief of which was the senior military member of the Council; the Yeomanry and Volunteers combined into one Territorial Army of all arms, organized into divisions; and the Militia abolished, its reserve converted into one of individuals to supplement the regular army, which was reorganized to provide an expeditionary force of six divisions, provided with full logistic support. The only real argument which still continued was where the army was to fight. India, Egypt and home defence all staked their claims. After the war Haldane claimed that he had consistently supported a commitment to fight alongside the French; but at the time he was careful not to offend the Liberal Party's isolationist policy by giving open support to the preference of most of the senior officers of the General Staff, led by the Director of Military Operations, Henry Wilson, for a firm commitment to the continent. The secretariat of the Committee of Imperial Defence (CID), headed by Sir George Clarke, in which Hankey was influential, took a navalist view.

In the early days of the Liberal administration in January 1906 Haldane had received the permission of the Prime Minister, Sir Henry Campbell-Bannerman, with the support of Sir Edward Grey, the Foreign Secretary, for the General Staff to open discussions with their French and Belgian equivalents as to how to implement the suggestion made by Grey's predecessor, Lord Lansdowne, to the French ambassador, Paul Cambon, at the time of the Tangier crisis, caused by Kaiser Wilhelm's support of Moroccan independence in 1905, that the two governments 'should, so far as possible, discuss in advance any contingencies by which they might in the course of events find themselves confronted'.[3] From that beginning Haldane never wavered from his firm conviction that the aim of the expeditionary force, which he was to create, was to help the Belgians and the French to prevent the Germans from reaching the Channel ports. Its destination was therefore on the far side of the Channel. As staff talks progressed with the French, and were refused by the Belgians, this developed into deployment on the left flank of the French army and, when Henry Wilson became Director of Military Operations in 1910, detailed plans to implement this were pressed on with vigour.

However there were opposing schools of thought, both in the navy and in the army. Fisher, and his successor Arthur Wilson, strongly opposed a continental commitment which they feared, correctly, would become open-ended; but the alternatives they proposed were so manifestly ridiculous in military terms – landings on the Baltic coast of Germany or in the Heligoland Bight, even towing barges laden with soldiers up the Rhine – that the army had no difficulty in persuading the CID to reject them. The navy's enthusiasm for such landings contradicted their assurance that the Germans could not carry out similar operations against Britain. The argument continued about what the army needed to keep at home to deal with that. A CID subcommittee in 1906 had recommended that 70,000 men were needed to repel raids and compel the enemy to consider using such a large force that it could not fail to be intercepted by the navy. Two of the six divisions of the expeditionary force should be retained in Britain for this purpose. A third school, led by Esher, the CID secretariat and elements of the army, favoured another version of navalist strategy: that imperial defence, especially that of India and Egypt, should have priority over a continental commitment. This argument was not helped by the navy insisting that several months would pass before they had established sufficient command of the sea to be able to provide escorts to convoys transporting the force to its destination, even across the Channel. The defence of Egypt had caused concern in 1906, when Turkey occupied Tabah in the Gulf of Aqaba, claimed by Egypt. Various schemes for counteroffensives against the Turks were considered, culminating in examination of an operation against the Dardanelles. It was rejected, the navy insisting that ships could not force a passage unless the army landed to subdue the Turkish forts, and the army that the navy should guarantee landing it unopposed in an area large enough in which 'to form up for battle' against a force which, in a period of tension, could have been built up to a strength of 100,000. The plan was rejected and arguments about how to deal with the Turkish threat to the Suez Canal continued until 1914.

The crucial meeting which decided Britain's strategy took place in the CID on 23 August 1911. Henry Wilson argued that Britain must mobilize on the same day as France; that all six regular divisions should be sent to fight on the left flank of the French army and that plans had to be made to keep this force up to strength as long as the war lasted. The navy put forward all their objections and their alternative proposals for a strategy of blockade and raids. The effect of these would be to force the enemy's main fleet to come out and face battle, the result of which would then decide the war. Although strongly supported by his chief, General Sir William Nicholson, Henry Wilson did not achieve all he asked for, the Prime Minister summing up by saying that the question for decision at the time would be whether or not to send all six divisions to France. The General Staff continued to assume in their planning that they would.

The navy, meanwhile, had continued the process of concentrating in home waters. Although slow to initiate discussions with the French, the Agadir crisis of 1911, the appointment of Winston Churchill as

First Lord of the Admiralty and German naval plans spurred them into action. Tirpitz prepared to increase the active German fleet from seventeen battleships and four battle-cruisers to twenty-five and eight respectively. Britain normally had twenty-two capital ships in full commission in home waters, including six at Gibraltar. The only way to meet the German threat was to transfer ships from the Mediterranean Fleet and Churchill saw clearly that this demanded naval talks with the French in order that they should assume the responsibility for dealing with the threat of the Italian, Austro-Hungarian and Turkish fleets. The decision to reduce the Mediterranean Fleet and to enter into naval talks with the French met strong opposition in many quarters, and Churchill was forced to modify the proposal into one which maintained a fleet, based at Malta, 'equal to a one-power standard, excluding France'. This reduced the capital ship strength there to two and brought that in home waters up to thirty-three.

Although by 1912 it appeared that Britain's strategy, and the organization of its army and navy to implement it, had been adjusted to meet the threat which now came clearly from Germany, no decision or commitment had been made as to whether Britain should go to the help of France if she became involved in war with Germany, nor, if she did, what form that would take. A month had passed after the incident at Sarajevo on 28 June 1914 before the British government finally faced up to the urgent need to decide on the answers to these questions. Until the Germans presented their demand to the Belgian government on 2 August to agree to the passage of their forces through its territory, Herbert Asquith's Cabinet was divided as to whether Britain should fulfil its moral obligations to France or remain neutral. A guarantee had been given on 1 August that the Royal Navy, which Churchill had mobilized without Cabinet authority, would not permit the German fleet to enter the Channel, and Grey made much of the fact that this was in Britain's own interests, and an obligation to France in return for her naval support in the Mediterranean, when he addressed the House of Commons on 3 August. He insisted that Britain was not bound, as France was, to support Russia, to whom Germany had delivered an ultimatum on 31 July, but concentrated on the issue of Belgian neutrality, saying,

> I ask the House from the point of view of British interests to consider what may be at stake. If France is beaten to her knees ... if Belgium fell under the same dominating influence and then Holland and then Denmark ... if in a crisis like this, we run away from these obligations of honour and interest as regards the Belgian Treaty ... I do not believe for a moment that, at the end of this war, even if we stand aside, we should be able to undo what had happened, in the course of the war, to prevent the whole of the West of Europe opposite us from falling under the domination of a single person ... and we should, I believe,

sacrifice our respect and good name and reputation before the world and should not escape the most serious and grave economic consequences.

After the speech he sent an ultimatum to Germany to stop their invasion of Belgium within twenty-four hours. Next day Britain was at war.

Since the Curragh incident in March, Asquith had himself held the seals of office as Secretary of State for War, and had called in Haldane, then Lord Chancellor, to act for him during the previous week. The latter summoned a meeting on 5 August, technically of the Army Council, but enlarged to include the Prime Minister, Foreign Secretary, First Lord and First Sea Lord, and Sir John French, who was to command the British Expeditionary Force (BEF), his Chief of Staff and two corps commanders, Sir John Grierson and Sir Douglas Haig, Sir Ian Hamilton and the two veteran field marshals, Lord Roberts and Lord Kitchener. To Henry Wilson's dismay there was no automatic agreement to implementing the plans so long prepared. At least there was unanimity that home defence could be entrusted to the fourteen divisions of Haldane's new Territorial Army (TA). Kitchener, who foresaw a long war, wished to keep the regular army at home to train a wartime army of seventy divisions, large enough to have a real influence on the continent. Haig went some of the way with him, wishing to delay deployment of the BEF for two or three months while the TA was trained up to a standard to fight alongside it. Only with a force of some twenty divisions would Britain be able to influence the course of events. French wanted to take his force to Antwerp. Nobody but Wilson much liked entirely entrusting the fate of the BEF to the French, which was bound to happen if such a comparatively small force was tacked on to theirs, and there was a strong feeling that to concentrate it at Maubeuge, as had been planned, was too far forward. If it had to go, Amiens was safer and committed it less firmly. But Wilson's argument that it was too late to change the plans, including the laboriously prepared railway time-tables, prevailed, and the decision was finally taken to implement the plan to send all six divisions – only to be changed next day when it was decided to hold two back until the naval situation became clearer. That, in turn, was reversed after a visit by Kitchener to France on 12 August, at which the realities of the situation were brought home to him by the French. He agreed to send the other two to Maubeuge as the concentration area and abandoned his attempt to insist that the BEF should act independently of the French Commander-in-Chief General Joseph Joffre.

All Haldane's efforts had come to fruition and the best organized and equipped expeditionary force that Britain had ever despatched overseas at the outbreak of a war marched confidently and cheerfully up to Mons. But Kitchener was right in his prognostication of a long war, needing seventy divisions, and that 'We must be prepared to put armies of millions in the field and maintain them for several years',[4] just as Fisher had been right in fearing that a continental commitment would become open-ended; but, as Michael Howard has written,

> What alternatives were really open at the time? ... It is far from clear that in August 1914 any of the alternatives that sea power made possible were really preferable to intervention at what was certainly the point of greatest logistical simplicity and, arguably, that of greatest military effect. No one could foretell, either, that the commitment in France was going to grow quite so inexorably. (And no one can be sure that a commitment anywhere else would not have grown just as inexorably and, in the hands of the same commanders, have produced any more effective results with any less loss of life.)[5]

In contrast to the proposals the navy had made before the war for offensive action against German coasts, it observed a prudent strategy of distant blockade as soon as it started, blocking the Channel with nineteen pre-dreadnoughts based at Portland, and the exit from the North Sea between Scotland and Norway with Admiral Sir John Jellicoe's Grand Fleet of twenty-one dreadnoughts, eight pre-dreadnoughts and four battle-cruisers, based initially at Loch Ewe and, after the first few months, once he was satisfied with the anti-submarine defences, at Scapa Flow. The threat of submarines, torpedoes and minefields made him wary of venturing anywhere near the Heligoland Bight, although Admiral Sir David Beatty's battle-cruisers had a successful engagement there on 28 August, sinking three German cruisers and a destroyer without loss to themselves, and another at the Dogger Bank on 24 January 1915. There was no way by which the German High Seas Fleet, to which Tirpitz had devoted so much effort, could have any effect on the war, unless it could catch a detached portion of Jellicoe's fleet unawares, and the latter was determined not to allow that to happen. If he were not 'to lose the war in an afternoon' he had to pursue a prudent defensive strategy, only risking his battleships by steaming out into the North Sea if he knew that his opponent Admiral Reinhard Scheer had done the same. It was far removed from the tradition of Nelson, and both the Royal Navy itself and the British public were disappointed, expecting something more dramatic. That moment did not come until 31 May 1916; and the Battle of Jutland which resulted seemed an anti-climax, Jellicoe losing three battle-cruisers and three armoured cruisers with 6,000 men, to Scheer's one pre-dreadnought and one battle-cruiser with 2,500. Scheer came out again in August, but the two fleets never met, Jellicoe's fear of submarines being intensified by a near-miss to his own flagship. By that time it was becoming clear that the real naval war was not between the battle fleets, but a *guerre de course* against merchant shipping by

German submarines, a higher priority for them now than acting as scouts for the High Seas Fleet. Deprived of them, Scheer henceforth refused to put to sea, while Jellicoe, for fear of submarine attack, refused to let his fleet go south of a line east from Newcastle without an escort of destroyers, which were diverted to meet the threat to merchant shipping. The sortie in August, in which not a shot was fired between the battle-fleets, was in fact one of the decisive naval engagements of the war, very different from what had been expected.

Just as the war at sea had turned out to be different from what the admirals had expected, so had the war from the prognostications of the generals.

When the war of manœuvre, envisaged by the prophets, had come to an end, as far as the British were concerned, in the muddy trenches round Ypres in November 1914, Kitchener expected a period of stalemate until his new army of seventy divisions would be ready to take the field in 1916. Meanwhile it was clear that there was no real threat of invasion and the TA could be used partly to relieve regular units in overseas garrisons, so that they could come home to train Kitchener's army, and partly to reinforce the BEF. By the time of the First Battle of Ypres on 20 October 1914 it had been expanded to four infantry and one cavalry corps, a total of 250,000 men, the strength of the whole regular army at the outbreak of war. (The TA had also been of that number.)

As the fighting died down in mid winter the question had to be faced: what was the army to do in 1915? Joffre was pressing for an offensive in the spring. Unless pressure were exerted from the west, there was a danger that Germany, now allied to Turkey, would deal a fatal blow to Russia, after which she could transfer all her strength against France. If pressure were applied in the west, Russia might be able to exploit her success in Galicia against Austria-Hungary which, with Italy still sitting on the fence, but with every reason to wish to take advantage of Austro-Hungarian weakness, might collapse. Sir John French backed Joffre but had to fight against the navalist school. Fisher, who had been recalled from retirement to replace Arthur Wilson as First Sea Lord, proposed the totally impracticable idea of a landing in the Baltic. Churchill, like Joffre concerned to help Russia keep up the pressure, proposed forcing the passage of the Dardanelles, French being fobbed off with a reinforcement by two Territorial divisions. Australian and New Zealand troops had already been sent to Egypt to counter the threat of a Turkish attack, and had been joined by a Territorial division.

The story of Gallipoli is well known. It did nothing to help the Russians, hard pressed by the divisions which, as Joffre feared, General Erich von Falkenhayn had transferred from the west. The offensive on the Western Front petered out inconclusively, the German intro-

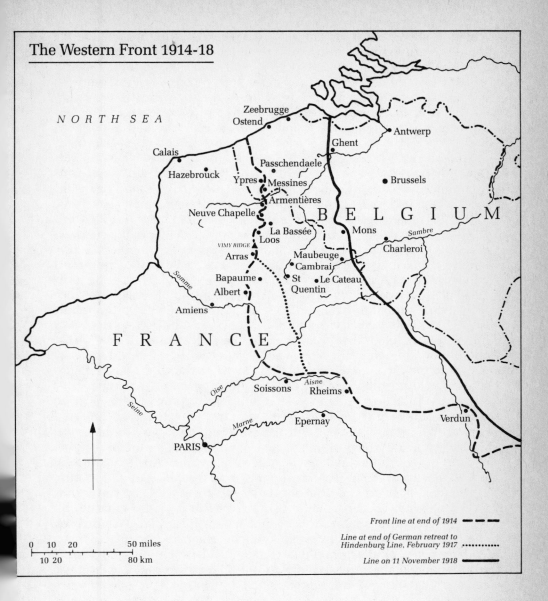

The Western Front 1914-18

NORTH SEA

Zeebrugge
Ostend
Antwerp
Ghent
Calais
Passchendaele
Hazebrouck
Ypres
Messines
Brussels
Armentières
Neuve Chapelle
BELGIUM
La Bassée
Mons
Sambre
VIMY RIDGE
Loos
Charleroi
Arras
Maubeuge
Cambrai
Bapaume
St
Le Cateau
Albert
Quentin
Somme
Amiens
FRANCE
Oise
Aisne
Soissons
Rheims
Seine
Verdun
Marne
Epernay
PARIS

Front line at end of 1914 – – –

Line at end of German retreat to
Hindenburg Line, February 1917 ···········

Line on 11 November 1918 ▬▬▬

0 10 20   50 miles
  10 20   80 km

duction of gas having had an initially sharp effect but not a decisive
one. Italy joined the Entente but was soon seen to be almost as much
of a strategic liability as an asset. In the Middle East the Turks, commanded by a German, had advanced to the Suez Canal but got no
further. In order to protect the oilfield at the head of the Persian Gulf
an expeditionary force from India (where, on the outbreak of war,
75,000 British soldiers were associated with 160,000 Indian) had

foolishly attempted to advance up the Tigris at the hottest time of the year with inadequate logistic preparations, and by the end of the year was besieged in Kut. 1915 had not been a good year for the army, which had suffered half a million casualties since the war had started, 200,000 of them dead or missing. Kitchener's new army had had to be sent over to France in dribs and drabs and confidence in him, in the Prime Minister, Asquith, and in Churchill, who had quarrelled with Fisher, was at a low ebb by the time Gallipoli was evacuated in January 1916. The troops, including the Australian and New Zealand Corps, moved to Egypt. The failure of Gallipoli, Churchill's removal from the Admiralty, the installation of General Sir William Robertson as Chief of Imperial General Staff (CIGS) and Haig's replacement of French in command of the BEF ensured that henceforth the Western Front would receive priority, and that, at any rate as far as the soldiers were concerned, 'side-shows' would be severely discouraged. Some hope of being able to break the deadlock in France lay in an innovation which owed its origin partly to Colonel Swinton, who had been sent from the CID Secretariat to the BEF to handle public relations, and had since returned to his desk in Spring Gardens, partly to Churchill at the Admiralty. This was the tank. However, although some went into action in 1916, they did not show their promise until nearly the end of the following year. The attacks that Haig launched, principally to relieve the pressure which the Germans were exerting on the French round Verdun, were made without their support, massive and prolonged artillery bombardment being relied on to pulverize the enemy defences and help the heavily laden infantry through and beyond the barbed-wire entanglement, while the cavalry hung about behind, waiting for the breakthrough which never came. One of the principal causes of failure was the breakdown of communications. Artillery fire cut telephone lines, with the result that from the moment that battle started commanders had little or no idea what was going on and could not employ or direct reserves, even if they had any idea what to do with them. Realizing that this would be so, their plans were rigid and made little provision for subordinate commanders to use their initiative or exploit a favourable situation. When the four-and-a-half-month-long Somme offensive drew to an end in November, it had added some 420,000 British casualties to the 180,000 which Haig's forces had suffered in the first half of the year. 1916 had proved a worse year for the army than its predecessor.

It had been a bad year for the navy too since Jutland, merchant shipping losses from submarine attacks increasing month by month.* Jellicoe handed over the Grand Fleet to Beatty in November and, as

---

* In September 104,572 tons, October 176,248, November 168,809 and December 182,292 tons.

First Sea Lord, had to face this threat to the whole war effort. He instituted no change in the navy's reliance on patrolling dangerous areas, arming and strictly routeing merchantmen. Only troopships were sailed in convoy, the Admiralty's argument being that escorts could never be provided if all shipping sailed in convoy, of which in any case merchantmen were not capable. It would lead to fewer and slower sailings, congestion in ports and therefore slower turn-round. It was not until the Americans joined in the war in April 1917, with the promise of more escort vessels, that David Lloyd George, who had succeeded Asquith as Prime Minister in December 1916, managed to force Jellicoe to change the Admiralty's policy. It did not transform the situation, which remained serious well into 1918. As late as June of that year Jellicoe was still insisting that the situation would not improve significantly until the U-boat bases on the Belgian coast had been captured.

Jellicoe's view, which attributed an exaggerated importance to the Belgian ports, reinforced Haig's preference for Flanders as the area in which he should make his main effort, rather than alongside the French, part of whose front he had had to take over as his strength built up and theirs diminished: they had suffered 1,675,000 casualties since the war had started. General Robert Nivelle, who succeeded Joffre in December 1916, was as insistent as his predecessor had been that Haig, who now had fifty-six divisions under his command, should not only take over more of the front, but also attack in the Arras–Bapaume sector, while the French attacked between the Somme and the Oise. Unless the Allies kept up the offensive, there was a danger that Russia would collapse or make a separate peace.

Lloyd George had been opposed to any offensive, suggesting a diversion of force to Italy, believing that a blow against Austria-Hungary might have better prospect of success. 'Knocking away the props' was his preferred strategy. However, he was won over by the equally glib Nivelle, but a new situation was created by the German withdrawal to the Hindenburg Line. Haig had been planning an operation from the Ypres salient to exploit the situation created by the threat to the German position of Nivelle's combined Anglo-French offensive further south. It might now pose a less serious threat, for General Erich Ludendorff's shortening of his line provided him with reserves to meet either thrust. But Nivelle remained confident and his plan went ahead. Unfortunately it failed, just as the Russian revolution broke out and the Americans joined the war – both in April. Haig's simultaneous attack went on until mid May and was little more successful, costing him 158,660 casualties. The fact that there was no longer any threat to divert Ludendorff's attention did not deter Haig from wishing to proceed immediately with his planned Flanders offensive, which, it was hoped, would clear the Belgian coast. Lloyd George and Robertson both opposed this but, concerned at the state of the French army as revealed to him by Pétain who had succeeded Nivelle, Haig preferred to continue with his offensive plans, rather than hand the initiative to Ludendorff. Haig now had sixty infantry divisions, of which four were Canadian, five Australian and one New Zealand, and five cavalry divisions. The attack was launched by General Sir Hubert Gough's Fifth Army of ten infantry and two cavalry divisions, supported by 2,174 guns and 136 tanks, the French attacking on his left with six and General Sir Herbert Plumer's Second Army on his right with five. From 31 July when it started, to 10 November when it ended on the Passchendaele ridge, it was dogged by bad weather and cost Haig 250,000 casualties for little gain. 1917 looked like being as depressing a year for the army as its predecessors, but hopes were raised a week later when General Sir Julian Byng's Third Army launched an offensive, originally planned only as a raid, in the comparatively undamaged terrain leading to Cambrai. Six infantry divisions, supported by 375 tanks, 1,000 guns and all five cavalry divisions, broke through the German defences, but the cavalry failed to exploit the success of the infantry and tanks, 179 of which had fallen out for one reason or another on the first day. The Germans counterattacked at the end of the month, regaining all they had lost. Fortunately this was not allowed to obscure the demonstration that, used in sufficient numbers on suitable terrain, the tank had restored mobility to the battlefield. That it also showed that the day of horsed cavalry was over was unfortunately not generally accepted, partly, if not largely, due to the apparent success of that obsolescent arm by General Sir Edmund Allenby against the Turks in Palestine. In the Third Battle of Gaza at the beginning of November and in the subsequent advance to Jerusalem, reached on 9 November, the Desert Mounted Corps played a decisive part, as they were to do in the final stages of the campaign almost a year later.

Lloyd George was determined that 1918 should not be a fourth year of large casualty lists for little gain. Unable to get rid of Haig, he tried to hamstring him by ordering him to send five divisions to Italy, to put heart into the Italians after their defeat at Caporetto in October, by keeping 120,000 soldiers in Britain, and by agreeing that Haig should take over a further twenty-eight miles of the French front in September. Haig's concern was no longer the offensive but, with the Russians effectively out of the war, he feared a German one. By mid February 1918 Haig's fifty-nine divisions faced eighty-one German. A month later they had 201 to face 159 Allied, ninety-nine of them French and one American.

His fears were well founded, as on 21 March Ludendorff launched his offensive designed to envelop the British and cut them off from the Channel ports on which they depended, employing new tactics of infiltration and exploiting both the use of gas and the early morning mist. The left arm of the thrust was initially successful and drove Gough's Fifth Army, transferred to Haig's right, almost to Amiens, for the defence of which Haig had to throw in his reserve of eight infantry and three cavalry divisions. It was a test of the new Allied command arrangements, headed by General Ferdinand Foch, who sent reinforcements which helped to ensure that Ludendorff's second thrust in Flanders was held before it reached any important objective. By the end of April the situation had been stabilized, Haig having lost 260,000 men and Ludendorff 100,000 more. The tide now turned, Foch and Haig determined to exploit the situation with a counter-offensive, while their political leaders, even at the cost of prolonging the war, would have preferred to wait until 1919, when General John J. Pershing's American army would be strong enough to assume a major share of the burden. Fortunately the soldiers won the argument and carried out a series of well-planned and well-executed alternating thrusts which drove the Germans back until, with his allies anticipating his decision and his own people on the verge of revolt, Kaiser Wilhelm II gave up the struggle on 9 November 1918 and the armistice was signed two days later.

The 'Great' War was over. Britain's military effort had been stupendous. The Empire as a whole had provided over eight and a half million men in uniform, 5,700,000 from the British Isles (12.4 per cent of the population), 1,500,000 from India (0.3 per cent), 1,300,000 from the White Dominions (8–11.6 per cent), and 134,000 from the Colonies. Of this number almost a million had died and two million had been wounded, the majority of the casualties occurring in the army on the Western Front, while the Indian Army, mostly serving in the Middle East, suffered fewer both in total and proportionately. At the end of the war on the Western Front Haig had fifty British infantry and three cavalry divisions, five Australian, one New Zealand and four Canadian infantry divisions, totalling nearly two and a half million men, with 420,000 animals. They were supported by 6,146 guns, 1,058 aircraft and 600 tanks. The total number of aircraft supporting both the army and the navy world-wide was over 20,000. Naval strength was fifty-eight capital ships, 103 cruisers, 12 aircraft-carriers, 456 destroyers and 122 submarines. Not only had almost all this equipment, and the ammunition to be fired by it, been manufactured in Britain, but equipment and supplies had also been provided to her allies, even the USA. The war had involved the whole nation and, as the elation and relief of victory faded, people began to question whether this immense effort could not have been better used. Could not Germany have been defeated at a lower cost, human and material, or could not the war have been brought to an end earlier, not by victory but by making peace?

Material to the answers to both questions was a third. What were Britain's war aims? Throughout the war the Allies had been reluctant to discuss the matter, for fear of causing dissension between each other and within their own political sphere, the union sacrée in France, the coalition in Britain. Woodrow Wilson's Fourteen Points had been guardedly accepted as the price of American support. Britain had been reluctant to commit itself too definitely to France's recovery of Alsace and Lorraine, but had no difficulty in insisting on the preservation of an independent Belgium, ensuring that Germany was kept away from the Channel ports, nor that Serbia must at least not be subordinate to Austria-Hungary. When tentative peace-feelers were extended in 1917 by Woodrow Wilson, supported by Lord Lansdowne in Britain, Germany felt confident about her military situation, both on land and at sea, and her terms were clearly not acceptable. They included not only continued absorption of Alsace-Lorraine, but military occupation of Belgium and of fortified towns in eastern France. They would have been the same in 1915 or 1916. Leaving aside Britain's moral commitment to France, German occupation of these areas would have constituted a continuing direct threat to Britain and her naval supremacy. There was no political or diplomatic solution which could have made a satisfactory peace without the defeat of Germany's armed forces.

Could that defeat have been engineered by a less direct method than that favoured and pursued by the Western Front school, represented so forcibly by Robertson and Haig? Professor Michael Howard's verdict has already been quoted[6] and needs no further comment; but there was a difference between the politicians' and the soldiers' view of what was required. Robertson, Haig and Henry Wilson shared the view that the need was to prove to the German military that they could not succeed by military means in establishing Germany as the dominant power not only in Europe but also in the world beyond. The politicians had a wider aim: that the domination of Germany by the military must be eliminated, if a stable balance of power both in Europe and in the wider world were to be maintained. This demanded a much more total defeat, including the detachment of Germany's political allies, than that sought by the generals and the admirals. The attempt, successful as it proved, by the politicians to blame the generals (the admirals getting off scot-free) for the apparently fruitless losses of the war, epitomized in Lloyd George's attack on Haig in his memoirs, was dishonest.

What was the key to victory? Basil Liddell Hart, in the epilogue of his History of the First World War, first published in 1930, when he was intent on proving that the methods employed by Haig had been disastrous, laid great emphasis on the effectiveness of naval blockade, part of 'The British Way in Warfare'. That view dominated historical interpretation of the war in the 1930s. It maintained that it was the political and social collapse behind the front, brought about by the discontent caused by the privations which the civilian population had suffered, accompanied by the defection of her allies, rather than defeat of her armies on the Western Front, which had led to the Kaiser's abdication and the demand for an armistice. That explanation of how victory had been achieved, which was not surprisingly supported by senior German army officers, encouraged those who, after the war, believed that a strategy aimed directly at civilian morale would be more effective and less expensive than one aimed, on Clausewitzian lines, at the destruction of his armed forces. The experience of the Second World War was to restore balance to the historical verdict. It became clear that it was the concentration of all the pressures, of which, in the First World War, the demands of the Western Front were the greatest, which finally brought about Germany's defeat. The idea that the peripheral strategy, relying heavily on naval blockade, could be effective against a major continental land-power, was a fallacy. It was perpetuated by the fact that the stranglehold exerted by the Allies over supplies to Germany after the armistice played an important part in forcing the Germans reluctantly to accept the terms of the Treaty of Versailles.

But it was not only the navy which supported an alternative strategy. A new element had been introduced, an independent air force. The Royal Air Force (RAF) had been formed on 1 April 1918, the impetus coming from the need to co-ordinate procurement of aircraft for the Royal Naval Air Service (RNAS) and the army's Royal Flying Corps (RFC), and as a result of the concern caused by the German air raids on London on 13 June and 7 July 1917, in which 219 people were killed and 625 injured. In November 1911 a sub-committee of the CID headed by Haldane had recommended a 'Flying Corps' with naval and military wings, served by a Central Flying School, the military wing being a conversion of the Air Battalion Royal Engineers which had developed from experiments with balloons in 1878; but the air arms of the two services soon diverged. Their differing demands on the industry became increasingly serious as aircraft production grew into a major element of war industry. Over 14,000 aircraft were produced in 1917, nearly twice as many as in the three previous years, and production in 1918 ran at 2,000 a month, Major-General Hugh Trenchard's RFC, supporting Haig, having 280 squadrons by that time. The high rate of production would have produced more aircraft than were needed for the direct support of the BEF. Out of this surplus was born the concept of a truly independent air force, which would act independently of the operations of the navy or the army and strike directly at targets in the enemy's homeland. The political panic caused by the air raids on London led its sponsors to imagine that an independent air force could create an equal, or even greater, panic in the other camp.

Trenchard was a convinced believer in the offensive as the best means of dealing with the enemy: that attacking the enemy's air force, in the air or on the ground, was a more effective means of frustrating its activity than defensive measures by aircraft or anti-aircraft guns. The air defence of Great Britain had, in any case, not initially been the responsibility of the RFC but of the RNAS, which, not surprisingly, had given it a low priority that amounted virtually to total neglect. This lack of any proper organization for air defence of the home country was a potent factor in influencing the South African General Jan Smuts and his committee to recommend the creation of a separate Royal Air Force. The fact that Trenchard was its first Chief of Staff (although General Sir David Henderson was its real father), and that an independent bombing force (which Trenchard, while he was supporting Haig, had opposed) had been conceived out of the prospective surplus of aircraft, led to a confusion between the concepts of creating a force which would be independent of the navy and the army, although its principal task would be to support them, and of one which would operate on its own, partly to undermine the political will of the potential enemy, partly to frustrate his capability to deliver air attacks on the home country. The exponents of the latter concept were strongly reinforced by the popular revulsion against a repetition of the long-drawn-out trench war of the Western Front, so costly in human and material terms. It was condemned as a war of attrition, the result dependent on a balance of enormous losses. On neither side had the generals conceived it in those terms. They had always been hoping for a breakthrough which could be exploited. General Sir Bernard Montgomery, in Normandy in 1944, was to find himself, on a much smaller scale, faced with the same problem as Haig had been.

1919–45

With Britain's potential enemies gravely weakened and faced with a severe world economic crisis, it was not surprising that the government expected stringent economies in defence expenditure in the 1920s. Servicing the National Debt and social service payments were now major competitors for government funds, the latter demanding £234 million in 1921–2, compared to £41½ million in 1913–14, when government expenditure had totalled £197 million, of which £50 million had gone to the navy and £35 million to the army. The admiralty planned a post-war navy of thirty-three battleships, eight battle-cruisers, sixty cruisers and 352 destroyers, of which eight battleships, all the battle-cruisers, four aircraft-carriers, ten cruisers and forty-three destroyers would be based at a newly constructed base at Singapore, the Dominions being expected to pay for 29 per cent of the cost of the Far East Fleet. This plan would have involved an annual expenditure of £171 million, but by 1922 the naval estimates were down to £56 million. This sharp reduction, even from the £84 million of the previous year, was influenced by the 1921–2 Washington Naval Conference which imposed limits on the number and size of capital ships, aircraft-carriers and cruisers, resulting in a total tonnage of 525,000 each for the USA and Britain, 315,000 for Japan and 175,000 each for France and Italy, maintained by a ten-year ban on new construction which would then be restricted in size. The treaty which concluded it was accompanied by the abolition of the Anglo-Japanese treaty.

The navy's plan for a Far East Fleet based at Singapore was scrapped in favour of acceptance that, to balance the growth of Japanese naval power, Britain and the three white Dominions bordering the Pacific would have to rely on the strength of her other naval rival, the USA. To choose the other path, active rivalry with the USA, possibly involving co-operation with Japan, was out of the question on both political and economic terms. The Royal Navy in the 1920s was a much more modest force than the Admiralty had planned: an Atlantic fleet, originally of fourteen, soon reduced to six, battleships, and a Mediterranean fleet also of six, each with cruiser, destroyer and submarine squadrons, cruiser squadrons also being based in the West Indies, North America, South America, Cape, East Indies and China stations – a reversion to the pre-Fisher flag-showing policy. When the time came for new construction, argument raged about the form it should take. Two schools of thought held that battleships were out of date, the submariners and the air enthusiasts, the latter claiming both that the battleship could not survive air attack, and that the offensive element of the future navy should be the aircraft-carrier. They lost their case, only one aircraft-carrier, designed as such, being constructed, the five others being converted battle-cruisers, while, in addition to the *Nelson*, *Rodney* and *Hood*, a start was made on the *King George V* class of battleship, the construction of five being authorized in 1936 to be followed by four, with larger guns, later. By 1939 the effective strength of the Royal Navy, including the Australian, Canadian and New Zealand ships, was twelve battleships and battle-cruisers, six aircraft-carriers, fifty-eight cruisers, 100 destroyers, 101 smaller escorts and thirty-eight submarines.

Severe as the restrictions on naval strength had been, the army fared worse, in spite of the fact that its commitments had been significantly increased, largely as a result of the destruction of the Ottoman Empire. They included the need to keep garrisons to maintain order in countries which saw no reason to welcome a British imperial yoke in place of a Turkish one, complicated by the decision to establish a national home for the Jews in Palestine; to provide a military presence to ensure compliance, both by Turkey and by Germany, with the armistice terms and the treaties which followed it; to maintain order in India and in the colonies in the face of nationalist independence movements, and to defend India's frontier against incursions from Afghanistan and infiltration of communist agents from the newly formed Soviet Union. To all these was added the age-old task of dealing with insurrection in Ireland, which by 1923 absorbed a force of 80,000 men. To meet all these commitments with voluntary service placed an impossible strain on the army, which was relieved of major commitments in Iraq and Arabia by the RAF, the latter employing a policy, known as air control, which involved the punitive bombing of the villages of recalcitrant tribes. A committee, set up in 1919 to plan the army's future, cognizant of the fact that Haldane's expeditionary force of six regular, backed by fourteen Territorial, divisions had proved inadequate, recommended a plan for twenty-nine divisions, each of which would throw off another on mobilization. The combination of imperial commitments and the Treasury's demand for retrenchment – reducing defence expenditure in 1920 from £502 million in the previous year to £135 million, of which £75 million was to be shared between the army and the air force, backed by the notorious Ten Year Rule – put paid to any such idea. The 'Rule' was, in fact, a reversion to the 1891 Stanhope memorandum; although it was a reasonable enough assumption in 1920, its prolongation every year up to 1932 was to inflict nearly irreparable harm on Britain's defence capability, and thereby to hamstring the nation's ability to pursue a positive foreign policy.

The army's reversion to the habits of the pre-Haldane era gravely affected its transition from the horsed to the mechanical age. When even such an enthusiast for mechanical mobility as Liddell Hart argued against a continental commitment to support France in facing the revised threat from Germany in the 1930s, it is not altogether surprising that the substitution of the motor vehicle for the horse proceeded hesitantly and indecisively. The principal argument lay between those who saw mechanized forces, of which the tank would be the principal arm, as totally replacing the traditional combat units of cavalry, infantry, artillery and engineers, and those who supported a gradual and limited transformation of the traditional arms by providing them, in areas where it was appropriate, with motor vehicles, wheeled or tracked, some of which might be armoured. As long as the army's principal commitment remained imperial defence in undeveloped countries, mobility by horse and foot would, the latter maintained, remain both more appropriate and less expensive. When, in 1934, a rearmament programme was embarked upon, not only were the army's needs subordinated to those of the navy and the air force, but a large share of the £20 million granted to remedy its deficiencies up to 1939 had to be allocated to the anti-aircraft defence of Britain, to which the government attached a high priority.

It was the fear of air attack which had brought the RAF to the front of the stage, encouraged by two exaggerations. The first was fostered by the doctrine which Trenchard, Chief of the Air Staff until 1929, had expounded: that an offensive bomber force could bring the enemy to its knees without the intervention of armies and navies. The second was about the nature, strength and effectiveness of the air force which Hitler, who came to power in 1933, was building up. The air staff's devotion to the bomber offensive was rebounding on their own heads, as politicians, press and public contemplated its application by the *Luftwaffe*. In 1919 the army and the navy had expected and hoped that the infant RAF would disappear, and that they would each retain their own aviation. They were to be disappointed, largely because of the threat of air attack against Britain, which in the early 1920s was feared from France, who had retained a force of 600 aircraft. In 1922 twenty-three squadrons were authorized to meet this threat, fourteen bomber and nine fighter. In 1923 this was increased to fifty-two, two-thirds of which would be bombers, at which figure it remained, although not fully implemented, until the rearmament programme of the mid 1930s, when it was raised to seventy-five, thirty-three of which would be for Home Defence. The Salisbury Committee, which recommended it, also recommended that the RAF should become responsible for air support of the army and the navy, including manning aircraft carried on ships, a decision against which the navy never ceased to fight until it was partially reversed in 1937. The air staff was saved from the consequences of its devotion to an offensive strategy by the invention of radar, a product of the establishment by the Air Ministry of the Committee for the Scientific Survey of Air Defence, chaired by Sir Henry Tizard, which first met in January 1935. Combined with a planned increase in fighter squadrons from thirty-five to forty-nine and the development of the Spitfire and the Hurricane, it provided the foundation for Fighter Command's victory in the Battle of Britain in 1940, although only thirty-five squadrons were in existence in September 1939, sixteen of which were equipped with Hurricanes and ten with Spitfires.

The priority given to the RAF left little over for the army, partly because of uncertainty about its role. Arguments about that bore a close resemblance to those which had raged before 1914. The air staff was strongly opposed to the idea of an expeditionary force on the continent. They argued that it would drag the nation into an unlimited commitment to land warfare, as had happened in 1914. It would not be decisive and would divert resources from the air force, which would bring victory by bombing the enemy's homeland. Their views were modified when they realized that they could not reach targets in Germany unless they used airfields in the Low Countries, and that it was important to deny their use to the enemy. An expedition strictly

limited to that was acceptable. It should not need more than two
divisions, which was all that the army could produce in 1936, and
even they would be ill-equipped to fight enemy tanks or aircraft. The
General Staff showed no enthusiasm for that concept, preferring the
Middle East as the destination of any expeditionary force which could
be produced, a concept supported by Liddell Hart. He suggested an
Imperial Strategic Reserve of four regular and two Territorial armoured
divisions, and eight regular and eight Territorial infantry divisions,
smaller than the standard. The Treasury, led by Neville Chamberlain,
opposed anything approaching that. As late as the end of 1937 the
government's priorities were: first, the security of the UK, especially
from air attack; second, the protection of imperial communications;
third, defence of British imperial possessions; and last, 'co-operation
in defence of territories of any allies Britain might have in war'. No
money was to be spent on the last, if it prejudiced the other three.

However, military talks with the French began in May 1938, but
were limited to discussions, without commitment, of what was needed
to transfer two divisions to France within sixteen days of the outbreak
of war. After the Munich crisis, in September 1938, the naval staff
swung to the support of the army's plan to equip one mobile and
four regular, and as many as possible of its twelve Territorial infantry
divisions for continental warfare; but the opposition of the Treasury
limited it to equipment of the five regular and four Territorial divisions,
while stipulating that the one mobile and two regular infantry divisions, which it was planned would follow the initial two infantry
to France, need not arrive there until forty to sixty days after mobilization.

The parlous state of all these armed forces had led the Chiefs of
Staff to give strong support to Chamberlain's appeasement of Hitler
at the time of the Munich crisis. At the end of 1937 they had warned
the government that they could not 'foresee the time when our defence
forces will be strong enough to safeguard our trade, territory and vital
interests against Germany, Italy and Japan at the same time' and had
stressed the importance 'of any political or international action which
could be taken to reduce the number of our potential enemies and
to gain the support of potential allies'.[7] In March 1938 they repeated
that

> No pressure that we and our possible allies can bring to bear, either
> by sea, or land or in the air, could prevent Germany from invading and
> overrunning Bohemia and from inflicting a decisive defeat on the Cze-
> choslovak Army. We should then be faced with the necessity of undertak-
> ing a war against Germany for the purpose of restoring Czechoslovakia's
> lost integrity and this object would only be achieved by the defeat of
> Germany and as the outcome of a prolonged struggle.[8]

They said much the same a year later before Chamberlain, in a very different mood, gave Britain's guarantee to Poland and Romania.

> Neither Great Britain nor France could afford Poland and Roumania direct support by sea, on land or in the air to help them to resist a German invasion. Furthermore, in the present state of British and French armament production, neither Great Britain nor France could supply any armaments to Poland and Roumania. This emphasises the importance in this respect of assistance from the USSR.[9]

It appears that the Cabinet's decision to announce the guarantee on 31 March, accompanied by the meaningless gesture of doubling the Territorial Army, was made in ignorance of this report.

The war which began with the implementation of this guarantee, Britain's ultimatum to Germany on 2 September 1939, was nothing like what anybody had expected. The well-known military writer, Major-General J.F.C.(Boney)Fuller, had predicted that

> in future warfare, great cities such as London, will be attacked from the air, and a fleet of 500 aeroplanes each carrying 500 ten-pound bombs of, let us suppose, mustard gas, might cause 200,000 minor casualties and throw the whole city into panic within half-an-hour of arrival ... London for several days will be one vast raving Bedlam, the hospitals will be stormed, traffic will cease, the homeless will shriek for help, the city will be in pandemonium ... the government ... will be swept away by an avalanche of terror.

The actuality could not have been more different. The British and French refrained from bombing Germany for fear of retaliation, and the German air force, designed primarily to support the army, concentrated its effort on that in Poland. The first sign that Britain's armed forces were as ill-prepared, ill-organized and ill-equipped to meet the Germans as the Chiefs of Staff had predicted, came in the abortive operation in Norway in April 1940. The immediate reaction of both the navy and the army to that débâcle was that their defeat was due to German air supremacy; but the cause lay deeper than that, as the whole course of the war was to demonstrate. British failure was due to a general lack of professionalism; to a failure to achieve an appropriate balance of weapon systems to deal with new developments in warfare; to the absence of an appropriate organization to co-ordinate the action, not only of the three armed services involved, but the different arms within the army itself. The years devoted to imperial defence, which seldom involved co-operation between all arms, and almost never of the three services, and the neglect of the study and preparation for major warfare on and over the land, took their toll. The ensuing débâcle in Belgium and France in May, culminating in the evacuation of the BEF from Dunkirk, leaving almost all of the army's inventory of equipment behind, provided no clear message other than the inappropriateness, faced by a force strong in tanks and aircraft, of an army consisting overwhelmingly of immobile infantry, supported by a handful of tanks, an apparently ineffective air force, and anti-tank and anti-aircraft weapons derisorily inadequate both in quantity and quality. In fact, General Lord Gort's BEF had engaged in very little fighting, spending most of its time withdrawing, as its Allies on right and left had already done.

If the RAF had been unable to affect the course of operations on the continent, it came into its own both in protecting the evacuation from Dunkirk and in the subsequent Battle of Britain, the German attempt to gain air supremacy over southern England as a preliminary to invasion. The battle lasted from 10 July to 31 October, its decisive phase being in mid September. The fifty-two squadrons of Air Chief Marshal Sir Hugh Dowding's Fighter Command (nineteen of Spitfires and twenty-five of Hurricanes, two of Defiants and six of Blenheims) could operate 644 aircraft, for which 1,259 pilots were available. They faced the threat of 1,480 bombers, to escort which Hermann Goering's *Luftflotten* 2 and 3 could provide 980 fighters. By the end of the battle, when Goering abandoned the attempt to attack airfields and concentrated on night bombing of cities, he had lost 1,733 aircraft to Dowding's 915, to which losses in Bomber Command of 118 and Coastal Command of 130 should be added.[10]

Dowding's gallant pilots had disproved Trenchard's theory, but it was not abandoned by the air staff, led by Air Chief Marshal Sir Charles Portal, or by his successor at Bomber Command, Air Marshal Sir Arthur Harris. While the only other active operations, except at sea, were those of General Sir Archibald Wavell's exiguous forces in the Middle East against the Italians, some means of retaliation against Germany had to be found, if only to sustain the morale of those subjected to nightly aerial bombardment. While Portal was still at Bomber Command in July 1940, he had written:

> We have the one directly offensive weapon in the whole of our armoury, the one means by which we can undermine the morale of a large part of the enemy people, shake their faith in the Nazi regime, and at the same time and with the very same bombs, dislocate the major part of their heavy industry and a good part of their oil products.[11]

That was a gross exaggeration of the capability of Bomber Command in 1940, and did not even become true four years later when Harris was able to despatch fleets of 800 aircraft, carrying far more and heavier bombs, as far as Leipzig and Berlin. German aircraft production rose from 15,288 in 1942 to 25,094 in 1943 and 39,275 in 1944, while monthly tank production increased from 760 to 1,229 during 1943 and to 1,669 in July 1944. By then neither the morale of the German people nor their faith in the Nazi regime had been undermined by air attack. That was not achieved until Russian armies from the east and Anglo-American ones from the west converged on Germany itself.

While the RAF had effectively removed the threat of invasion, in previous ages the principal task of the Royal Navy, the latter had to cope with a dual task: ensuring that supplies, military and other, reached Britain's ports, and that military forces and the equipment for them could leave Britain to reinforce the slender garrisons of the Middle and Far East. The threat came both from the German submarine fleet and, in less distant waters, from land-based aircraft. German occupation of almost the whole coastline of Western Europe, except for Portugal and Spain, and her alliance with Italy, made this task a much more daunting one than in the First World War. Only against the Italian navy in the Mediterranean, before the German air force made its appearance there in 1941, did actions between opposing fleets of large surface ships occur.

The neglect of the air arm between the wars was sorely felt, the fate of the *Prince of Wales* and *Repulse* at the hands of shore-based Japanese aircraft off Malaya in December 1941 bringing it sharply home. Convoy escorts to Russia and all over the world was to be the main activity of the Royal Navy throughout the war, until launching major amphibious operations in the final stages added a more offensive role. The tide did not turn in the war against the submarines until April 1943, when shipping losses fell from an average of 445,000 tons a month to 50,000, and then to 40,000 in 1944 as German submarine losses increased. The key to success had been the development of detection devices carried in aircraft and the allocation of new long-range aircraft to the task. Aircraft, in fact, accounted for the destruction of more submarines than did surface ships: 288 to 246. But it was the aircraft, particularly when land-based, which posed the greatest threat to ships of all kinds, isolating Malta and severely limiting naval action in the Mediterranean until Montgomery's and General Eisenhower's armies had cleared its southern shore, allowing the air forces to operate over the sea. By the end of the war the Royal Navy had 1,065 warships,* 2,907 minor war vessels and 5,477 landing-craft, manned by 863,500 men; but it had never been able to employ its traditional weapon of blockade. Germany's and Japan's territorial conquests had frustrated that. Indeed, until the Battle of the Atlantic had turned in Anglo-American favour, it was Britain that tended to suffer from a form of blockade. Losses of life amounted to 51,000 sailors of the Royal Navy and 35,000 of the Merchant Navy.

After the humiliation of Dunkirk, the army had a brief moment of

* 15 battleships and battle-cruisers, 7 fleet, 4 light fleet and 41 escort aircraft-carriers, 62 cruisers, 131 submarines, 108 fleet destroyers and 383 escort destroyers and frigates, and 314 smaller craft.

glory in its easy victories over the Italians in Libya and Abyssinia in the early months of 1941, but after that nothing seemed to go right for it. The ill-advised expedition to Greece bore a strong resemblance to the brief campaign in France and Belgium. The soldiers, mostly from Australia and New Zealand, saw little fighting before they were withdrawn to conform to the action or inaction of their Greek allies, and both there and subsequently in Crete were to experience the frustration of attempting to move, by land or sea, and fight when the skies were dominated by the enemy air force. Air forces had less influence in the desert battles which swung to and fro in Libya, after the Germans, led by General Erwin Rommel, came to the help of their Italian allies, both because the air forces on each side were more nearly matched and also because targets were not so concentrated or vulnerable as they were at sea or in terrain in which movement was restricted to a limited number of roads which passed through defiles. In the desert, the tank held sway. The victory of British tanks against the Italians had appeared to confirm the theories of Fuller and Liddell Hart, seeking the indirect approach 'either swerving round resistance or piercing it at a weakened spot', but in fact they had never been seriously challenged. Fighting the German Panzer divisions proved to be altogether a more difficult task. Not only the quality of their tanks, but more significantly the quantity and quality of their anti-tank guns, and their general professionalism and co-ordination of all arms in the battlefield, resulted almost invariably in tactical success, which generally prevailed over superior British numbers. Nevertheless, Rommel was flung back to his starting-point between Benghazi and Tripoli at the end of 1941, having lost almost all the 174 German and 146 Italian tanks with which he had begun the battle in November. Resilient (and disobedient) as ever, he caught the British Eighth Army under General Neil Ritchie off balance in January 1942, and in an operation which bore a close resemblance to that with which he had made his debut in Africa a year before, following Liddell Hart's precepts to a nicety, he forced it to withdraw to a position covering Tobruk. Five months later, pre-empting an offensive forced on a reluctant General Sir Claude Auchinleck by Churchill, whose eyes were on saving Malta, Rommel inflicted a resounding defeat on a greatly superior army, which he forced back in confusion to within fifty miles of Alexandria. His initial plan had been grossly over-optimistic, but he recovered and then made fewer mistakes than in the previous November. Ritchie's divisions, armoured and infantry, had generally been slow to react and ineffective in attack. They still did not form a well-knit professional army, partly because they were composed of different elements, British, Indian, South African and some others, such as Poles and French (the New Zealanders were involved in the final stage, and the Australians not at all); but principally because the British army had not been trained in mobile operations involving all arms, nor formed into a cohesive organization for that purpose. Between the wars, except for those specifically allocated to the close support of infantry, tanks had been trained on their own and expected to be able to fight that way. Concentrated tank versus tank battles, in situations of great uncertainty, in clouds of dust, were something which nobody had previously experienced and few had imagined realistically.

Although Auchinleck succeeded in holding Rommel on the El Alamein line, confidence that the latter could be defeated was not restored until Montgomery arrived, held Rommel's major attack in September and then thoroughly trounced him at El Alamein a month later, the methods of battle resembling those which Montgomery had experienced in Haig's successful offensive between August and November 1918; but even he had difficulty in co-ordinating the action of tanks, artillery, infantry and engineers.

Further east the army had fared no better: worse, in fact. When Japan attacked Pearl Harbor, Hong Kong and Malaya on 8 December 1941, the loss of Hong Kong was inevitable; but the collapse in Malaya, followed by the fall of Singapore and the subsequent defeat in Burma, were not. Risks had been taken in leaving the area gravely deficient in naval and air force resources, both in quantity and quality, and the army, mainly drawn from the Indian Army, was mentally, physically and materially ill-prepared to face the experienced Japanese army in a country of few roads, covered in jungle or thick cultivation, in operations over great distances. Operations to dislodge the Japanese from Arakan, and behind their lines by Major-General Orde Wingate's Chindits in the following winter, were no more successful; and it was not until after Admiral Lord Louis Mountbatten's arrival as Supreme Commander and the appointment of General Sir William Slim to command the 14th Army, backed by Auchinleck at Delhi, that the army's self-confidence and skill were improved enough to turn the tables on the Japanese, who, by 1944, had been weakened by the war in the Pacific. The principal novel feature of the campaign was the extensive use of transport aircraft to keep the army supplied, both when cut off by the Japanese and when advancing into Burma, made possible by the establishment of local air supremacy.

The use of airborne forces, landing by parachute or glider, had been tried on a small scale in support of General Dwight D. Eisenhower's forces in North Africa, and on a larger one in the invasion of Sicily which followed it. In the latter there were serious errors in execution which were not encouraging for further operations. But the intervention of American paratroops at the crisis of the Salerno landings in Italy in September 1943, and that of both American and British as part of the landings in Normandy in June 1944, restored faith that they were worth the great effort involved in maintaining them, particularly in air transport. The failure of the operation to secure the bridge at Arnhem in September of that year, overshadowing the successful but less ambitious operations of the two American airborne divisions in the same operation, reinforced the view of those who believed that the impact of this new method of delivering soldiers to the battlefield was doubtfully worth the effort devoted to it. Their use as part of Montgomery's operation to cross the Rhine in March 1945 was not essential to its success.

The operations in which the army was engaged in the final stages of the North African campaign, in Sicily and in Italy, were not basically very different from those of the First World War. The support of large numbers of tanks and an overwhelmingly superior air force did not seem to make it much easier for the soldiers to turn the enemy out of his well-prepared defences, protected by anti-personnel and anti-tank mines, than it had been for their fathers thirty years before. The fighting in Italy was every bit as tough and conditions as unpleasant as it had been in Flanders in 1917.

The 'finest hour' of the British armed forces in the Second World War was undoubtedly the landing on the coast of Normandy on 6 June 1944. For all three services it was a well-planned and well-executed operation, in which the British played the major part. The only disappointment lay in the failure to penetrate as far as Caen on the first day. Success owed a great deal to the preparatory work carried out by Mountbatten's Combined Operations Headquarters, using techniques employed by the US Marine Corps in the Pacific and inventing others; to the 'specialized' armoured vehicles developed by Major-General Sir Percy Hobart's 79th Armoured Division, and to the close co-operation between army and tactical air forces developed in the Middle East. Thereafter disappointment set in. The battlefield solidified on the pattern already familiar in Italy and became one of attrition, as it had been at El Alamein, and the British army found itself short of infantry with which to keep up the pressure, while its considerable number of tanks champed at the bit, unable to engage in the mobile operations which had been envisaged and for which it was now well trained. But, as in that case, the advantage in such a struggle lay with the Allies, as the full strength of the US forces was deployed on land and in the air. When the break came, at the end of July, it was the Americans who were, as had been planned, best placed to exploit it, and their armoured forces, led by General George S. Patton, were able to follow Liddell Hart's precept of an 'expanding torrent' against little opposition, until they reached the eastern borders of France. The British were able to do the same, until they had entered Holland. Thereafter the balance between British and American forces under Eisenhower's command swung even further in favour of the latter, until, as the war ended in May 1945, Montgomery's divisions numbered nine British, five Canadian and one Polish, to seventy-two American.

The British army, that is the element raised from the United Kingdom, had been significantly smaller than that raised in the First World War, reaching a total of forty-eight divisions (eleven armoured, thirty-five infantry and two airborne) a large proportion of which were disbanded before they ever left the shores of Britain; and of 3,643,000 men and 300,000 women enlisted to maintain a strength which at its height was 2,766,000 men and 207,500 women 144,000 were killed. One of the principal reasons for its smaller size, in comparison with Haig's day, was the demand of the air force for men. Its strength was maintained at just over 1,000,000 from 1942 until the end of the war, of whom 70,000 were killed or missing. The Royal Navy's demands had not been much less – 865,300, of whom 51,680 were killed or missing.* An immense effort had been directed to war industry – 112,120 aircraft had been produced by the end of 1944 – but it was not enough to meet all demands, and throughout the war the British armed forces of all kinds were significantly dependent on American and Canadian production. The result of the nation's immense war effort, greatly dependent on imports of all kinds, was a weakening of its economic position even greater than that after the First World War.

What brought about victory? Could it, or peace, have been achieved more quickly and at less cost? The answers are very similar to those given to the same questions posed about the First World War. Victory was brought about by combined and continuous pressure, primarily that exerted on Germany's army by that of the Soviet Union. The prolonged air attack on Germany forced her to devote a considerable effort to air defence of her homeland, which affected the air support she could provide for her own forces; but neither that nor blockade had any significant effect on her material or psychological ability to continue fighting. It was not until military operations to the east, west and south deprived Germany of the resources gained from conquests that any serious shortage of food or materials was felt by her population and industry. The war could have been brought to an end after the fall of France in 1940, but at a political and eventually, no doubt, an economic and military price, which was not acceptable to Britain. Nevertheless, if Hitler had not attacked the Soviet Union and Japan provoked war with the United States in 1941, that price might have to have been paid. The British Commonwealth could not have defeated Germany on its own, and it is doubtful if it could have done so, even with the help of the USA, if the Soviet Union had stood aside.

---

* These figures include Dominion and Colonial personnel. The Merchant Navy lost 34,902 killed or missing.

1945–86

In 1945 Britain's weakness as a military power, relative to the USA and the USSR, was obscured by the even greater weakness of other European countries, notably France, Germany and Italy, and by the collapse of Japan, and her imperial commitments had been increased by the growth of new threats: from internal nationalist movements, by the pressure of Jewish immigration into Palestine, and the threat of communist infiltration, backed by the Soviet Union and the victorious Mao Tse-tung in China. The atom bomb appeared to be the key to future military power, making practicable the claims of the strategic air offensive school, which reality (although not recognized for some time) had proved so exaggerated. But it was irrelevant to imperial problems, which demanded large numbers of men on the ground in many different places. Granting independence to India ended one major potential military involvement, while also removing the Indian Army as a source of military manpower to meet commitments outside the sub-continent. It did not, as it should have done, provoke a major reconsideration of the politico-military policy which was linked to the need to guard the sea and air routes to India. The desire to re-establish British authority east of India, undermined by Japanese occupation, and fear of Soviet encroachment into the oil-rich Middle East, led Britain to maintain an army which could only be sustained by compulsory service. It was not finally ended until 1964, by which time the army's strength had been reduced to 189,000 from the 437,000 at which it had stood in 1955. In addition to the load of imperial commitments, the army faced a new one in maintaining a permanent expeditionary force on the continent of Europe. The uncompromising attitude of the Soviet Union over the future of Germany, and of Eastern Europe generally, combined with its aggressive posture world-wide, led both the USA and Britain (the former impossible without the latter) to station forces on the continent, largely as a guarantee that a rearmed Germany, without which a balance of military power between Western and Eastern Europe could not be established, could never again threaten its neighbours. The main factor in persuading the latter to accept German rearmament, which itself was the key to American participation, was Britain's agreement to station four divisions (later specified as 55,000 men) and a tactical air force on the continent. In accepting that undertaking on top of her ex-imperial ones, she overburdened herself; but it was undoubtedly of a higher priority for her own security than any other. Following closely on it, the Suez fiasco of 1956 persuaded Harold Macmillan, who became Prime Minister because of it, to bow to the 'wind of change' blowing through Africa and Asia and to shed imperial commitments by a combination of trying to create and support successor regimes which would be friendly to Britain, and by associating other nations, notably the USA, whose aim was the preservation of stability and opposition to the influence of communism in the area. This policy, which involved military operations on a small scale in many cases, was by and large successful, Malaysia, East Africa, the Persian Gulf and Cyprus ending up on the credit side of the balance sheet, only Aden being clearly on the debit side. It is too early to judge on which side the conversion of Rhodesia to Zimbabwe will fall.

This disengagement from the imperial role was achieved at low cost in military, financial and political terms, in stark contrast to that of France from Indo-China and Algeria. By the 1980s the army's only major commitment in this field was its oldest one, in Ireland. By then the economics of modern armies had brought its strength down to 160,000. The other aspect of this policy led to the establishment of the Central and South-East Asia Treaty Organizations, which achieved little, other than perhaps a certain bolstering of local confidence, before they faded away as a result of adverse political developments in their respective areas.

The same economic pressures had been at work on the other services, as the cost of their equipment soared. One of the lessons of the Second World War appeared to be the prime importance of air power, both as a strategic weapon aimed directly at the enemy's homeland and in support of armies and navies. The RAF looked forward to a future in which its position would be unchallenged and the priority of its needs recognized. In the first decade after the war this was so, and it received some excellent new aircraft which it was to continue to fly for much longer than had been originally intended: the Hunter, as a fighter-bomber, the Javelin as a night-fighter, the Canberra light-bomber, and the V-bombers, Valiant, Victor and Vulcan, for the strategic role (the latter including the delivery of Britain's own atomic bomb, first tested in 1952), and the Shackleton for long-range maritime operations. The pre-war structure of Fighter, Bomber and Coastal Command, and tactical air forces overseas, seemed set to continue; but, even before the V-bombers came fully into service, this confident assumption was threatened by the development of rocket-propelled missiles, both as offensive and defensive weapons.

The first to be affected was Fighter Command. If the threat of air attack on Britain was the delivery of atom bombs by ballistic missiles, there was little point in devoting considerable effort to anti-aircraft defence by fighters (the army had already ceased to provide anti-aircraft guns for this purpose). In the mid 1950s the function of Fighter Command was reduced to that of identifying hostile aircraft in British airspace, although it was to regain its previous role in the 1970s, when it was argued that non-nuclear attack by aircraft was a possible threat, either because a war could at least have an initial non-nuclear phase or because Britain's possession of its own nuclear weapons protected its homeland from nuclear, but not necessarily conventional, attack. Bomber Command tried to prolong its strategic role, including that of acting as Britain's 'independent deterrent' by marrying a missile to its aircraft, which would then not have to penetrate the enemy's airspace, defended by anti-aircraft missiles. This was to be the American Skybolt, but when the Americans themselves, in 1961, decided not to proceed with its development, the RAF was forced to hand over its strategic nuclear weapon delivery role to the Royal Navy, which built nuclear-powered submarines in which to install American Polaris missiles to deliver British warheads.

In compensation for these blows to its two prestigious commands, the air force switched its effort to a role which it had previously neglected, in spite of continued pressure from the army: air transport. After Suez, Macmillan's government conceived a strategy which would no longer depend on bases which were expensive to maintain, partly because of threats to their security by anti-imperialist movements. Instead, a strategic army reserve would be held in Britain, which could be delivered to where it was needed, and maintained there, by air transport. Transport Command was built up with long, medium, and short-haul fixed-wing aircraft and, late in the day, with helicopters. Island bases were acquired or transformed to act as staging airfields, but could not eliminate the need to seek diplomatic clearance for overflight of other countries. This, and the fact that there were never sufficient aircraft to lift and maintain more than a very small force, meant that bases continued in existence, particularly at Aden and Singapore, until the commitments they were meant to serve were abandoned, Aden being evacuated in 1967 and Singapore and the Persian Gulf in the early 1970s. That withdrawal was accompanied by an assumption by the RAF of a greater responsibility for the air support of the Royal Navy, arising from the decision not to construct any more aircraft-carriers. The Fleet Air Arm's Buccaneer strike aircraft were transferred to the RAF and the latter, with shore-based Phantom fighters, assumed responsibility for fighter defence of the fleet. The remorseless facts of the cost of new aircraft, of the weapons and equipment for them, and of manpower, forced the size of the force down and down, until all combat aircraft in Britain were concentrated in Strike Command and all transport in Support Command, and the number of airfields severely reduced. The principal innovations in the last decade have been the introduction of the Harrier vertical take-off aircraft for the close support of the army and of in-flight refuelling, initially of fighters, but later of all types, the latter coming into prominence in the Falkland Islands operation in 1982, which would not have been possible without the use of the US Air Force's airfield on Ascension Island. By then RAF aircraft squadrons had been reduced to eleven strike/attack (ex-bomber), five offensive support (ex-fighter-bomber), three reconnaissance, nine fighter, four maritime patrol, one airborne early warning, six fixed-wing transport, three tanker, six heli-copter and two air-sea rescue, the whole force employing 93,000 men and women.

If the RAF was confident about its future in the years immediately following the end of the Second World War, the Royal Navy was not. The USA was now undoubtedly the major naval power, overshadowing all others. The fact that the potential enemy was a vast land nation, not susceptible to the application of any form of sea-power; the advent of the atom bomb, which appeared to mean that wars would be short and sharp, and the threat of missile attack, degraded the importance of the navy both in its offensive and defensive roles. The aircraft-carrier had become the capital ship of a modern navy, but the maintenance of any significant number of them, and their demand for escorts, was a very expensive business. Until the mid 1950s the navy managed to coast along, keeping in service the most modern of the ships with which it had finished the war, justifying its role partly on the possibi-lity of a 'broken-backed war' continuing with conventional forces after both sides had used up the comparatively small numbers of atom bombs in existence at that time, partly on flag-flying to support the country's world-wide interests. The moment of truth came in the late 1950s, when the severe reduction in defence expenditure imposed by Macmillan's Conservative administration coincided with the need to build new ships and equip them with missiles instead of guns. Mount-batten's period as First Sea Lord, from 1955 to 1959, was as decisive for the future of the Royal Navy as had been Fisher's before the First World War. It included the development of nuclear-powered sub-marines, both for ballistic-missile and torpedo firing, the abolition of the cruiser, the introduction of the helicopter for a variety of purposes, of anti-ship and anti-aircraft missiles and, in order to give the navy a place in the imperial role, which was then the principal activity of the army and the air force, the conversion of three old light-fleet-carriers to transport Royal Marine Commandos and helicopters to move and supply them. The heart of the fleet remained the strike-carrier, for which the robust Buccaneer aircraft was developed. But Mountbat-ten's 'Way Ahead' fleet pushed naval expenditure to first place in defence expenditure as a whole, and when Harold Wilson's Labour government in 1968 faced economic crisis and decided to reduce this to £2,000 million from the £2,232 million at which it was then running, the navy, like the other two services, had to make drastic cuts. After prolonged and bitter argument the decision was taken to phase out its strike-carriers on the assumption that major military operations east of Suez would no longer be required, and that the fleet could rely for air support on a combination of shore-based aircraft of the

RAF and of the formidable aircraft-carrier fleet of the US navy, with which it would be operating if in the Mediterranean or the Atlantic.

By this time the emphasis of naval operations had changed. The Soviet Union had not only copied the American example in developing nuclear-powered submarines both with nuclear-warheaded ballistic missiles and with torpedoes (and later anti-ship missiles), but was building a large general purpose ocean-going navy which, although its bases were widely separated and their access to the oceans restricted, posed a serious threat to the US and allied navies. The Royal Navy adopted anti-submarine warfare as its highest priority. The RAF's long-range maritime patrol aircraft was probably the most important element in this, but the navy saw the helicopter as an essential adjunct, working in co-operation with frigates firing depth-charges. They had the added advantage of providing a method of saving the Fleet Air Arm. The naval staff concluded that the most efficient method of basing anti-submarine helicopters was in a large so-called 'through-deck cruiser', which could also act as a command ship. The result was remarkably like a small aircraft-carrier, similar to that developed by the Soviet navy, probably for the same purpose. Having slipped that past a vigilant Treasury, they proposed to add to its complement a naval version of the RAF's vertical take-off Harrier, on the argument that it was needed to shoot down the potential enemy's long-range maritime reconnaissance aircraft. In 1975 Wilson's Labour government accepted the requirement, and by 1982, when the Falkland Islands operation took place, one of them was in service and the last of the strike-carriers had been converted to that role. The Fleet Air Arm had reinstated itself in the fixed-wing business.

The Falklands came as a lifebelt to the navy. Ever since Mountbatten's day an argument had raged between those who maintained that the future of maritime warfare lay in a combination of submarines and shore-based aircraft, supplemented by cheap missile-carrying craft of limited range, and those who wished to retain a 'balanced' fleet which could undertake a wide variety of roles in an equally wide variety of circumstances anywhere in the world, independently if need be. The decision had just been made by Margaret Thatcher's Conservative government in favour, in general terms, of the former school, when the Falklands crisis erupted. Since 1945 the navy and the air force, except on a very limited scale in the Korean War, had not had to engage in operations against enemy aircraft or ships. In 1956 the Egyptian navy and air force had offered no challenge.

In the event, although the RAF's air transport and tankers were essential to the operation, and the army provided the majority of the land forces, it was primarily a naval war, fought in a precarious air situation. It brought home to the sailors the vulnerability of surface ships, whether combat or transport, to attack by aircraft equipped with missiles, as it also proved the significance of the threat of nuclear-powered submarines to surface ships. It was the first real operational test of the unified inter-service command, centred on an integrated Ministry of Defence, introduced when Mountbatten was Chief of the Defence Staff (CDS) in 1963. It appeared to work well, but was not a true test. It was primarily a naval operation: the CDS of the time was a sailor: he operated through the naval headquarters of the C-in-C Fleet, whose local subordinate was an admiral commanding a task force, the land force commander of which was a Royal Marine major-general, although most of his troops and all their logistic support were provided by the army. The credit for the fact that it all worked generally smoothly was due primarily to one man who, tragically, was by then dead at the hand of the old enemy of the English, the Irish nationalist: Mountbatten. Ever since he had been appointed Chief of Combined Operations in 1941 he had cajoled, persuaded and forced the three armed forces to act together until they were able to do so pragmatically, regardless of the general organization, to a degree unsurpassed by the armed forces of any other nation. His influence on Britain's armed forces in the second half of the twentieth century was unrivalled.

2

FRANCE

1900–18

The state of the French armed forces at the turn of the century was the product of the political events of the previous thirty years. In spite of its humiliating defeat in 1870, the army had become highly popular. This was partly due to the desire for an instrument of revenge for the loss of Alsace-Lorraine, partly because both those who had wielded political power and those who had tried to seize it on the barricades had been discredited, and partly, in the absence of a respected monarchy, stemming from the desire for an institution to represent the idea of the nation. Conscription for five years' service (reduced to three in 1889) had been introduced in 1872 and an organization created which could mobilize the manpower of the nation into an army designed to fight a major war on the continent. Military matters were to be taken seriously, as they were in Germany, and a General Staff created, to be manned by the graduates of the newly created *Ecole Supérieure de Guerre*; but no amount of study, organization, attention to detail or brilliance of concept could counterbalance the inescapable fact of demography, that Germany could mobilize more men than France. A war to regain the lost provinces was not on the cards. The army's plans concentrated on defence, its forces to be concentrated well behind the frontier, which was backed by fortresses at Toul, Verdun and Epinal. Realization of this awkward fact had led France into an alliance with Russia, signed in 1894, which both nevertheless viewed with caution. Both wanted the help of the other in case either were attacked by Germany, but France was chary of becoming embroiled on account of Russian ambitions in the Balkans and Russia on account of France's desire to regain Alsace and Lorraine.

In spite of the episode of General Georges Boulanger, who, as Minister for War in 1886, appeared to be aiming for personal power, the army in the last decades of the century remained firmly loyal to the Republic. Nevertheless, republicans viewed it with suspicion, and this was accentuated by the change taking place in the character of its officer corps.

The proportion of sons of the traditional aristocracy, and of the conservative and Catholic bourgeoisie, grew, as they found themselves excluded from political life, thrust aside by republicans and socialists. This tendency was accentuated by the new system of promotion, introduced in 1889, which increased the influence of the existing senior officers. An officer's attitude to his religion was an important factor. While the army was becoming more conservative and Catholic, the political and intellectual life of the nation was moving in the opposite direction. The first socialists were elected in 1893 and, from 1891 onwards, the army was increasingly employed to suppress strikes and demonstrations by workers, thus widening the gulf between the army and the political left even further.

It was into this growing divergence between the army, which had been so popular in the 1880s, and the general climate of opinion in the country that the Dreyfus affair broke in 1894. The army was dragged by it into the centre of the political arena. All the suspicions which had been harboured about its loyalty to republican sentiment appeared to have been justified, and not only the army itself, but the whole country was bitterly divided by the issue. It was a tragedy for the army, which had greatly improved its efficiency and professionalism in the previous quarter of a century. Charles Freycinet, as Minister of War, had contributed a great deal, as had a succession of generals holding the newly created post of Chief of the General Staff, notably de Miribel and de Boisdeffre. Their efforts resulted in a mobilization plan which provided 1,400,000 men in the field, 400,000 in fortresses, backed by a further 750,000 men in reserve. Important new items of equipment for them were in the pipeline: the Lebel rifle, the 220-mm mortar, the famous 75-mm artillery piece, backed by howitzers of 120- and 155-mm calibre.

The immediate aftermath of the Dreyfus affair saw a serious drop in the morale of the French army officer. Before it, his meagre pay and poor living conditions had been compensated for by the respect in which he was held and by the feeling that he represented the ideal of the nation. The Dreyfus affair shattered that, especially when the republican and anti-clerical General André became Minister for War and instituted a witch-hunt of the officer corps, designed to weed out fervent Catholics. The result was a dramatic fall in the applications for entry to the cadet schools at St Cyr and the Polytechnique, and a diversion of talent from combat arms to the administrative services, where promotion and pay were better. The demoralization of the officer corps was accompanied by a similar decline in the morale of the rank and file, their increasing use as strike-breakers fuelling the mounting anti-military sentiments among the public from which they themselves were drawn. Desertions and failure to report for reserve training increased, as did the number of courts martial.

A further cause of division within the army itself lay between those who preferred service in France and those who sought adventure overseas. Initially the army had shown little interest in the great wave of imperialism which swept Europe in the second half of the nineteenth century. It had been left to the navy, as it had been in previous times, to concern itself with such affairs and, when soldiers were needed, the navy employed its own, the *infanterie marine*. But Marshal Thomas-Robert Bugeaud's conquest of Algeria, starting in 1830, fired the imagination of French officers bored with the routine of garrison life at home. Some joined the *infanterie marine*, but by the end of the century the army was taking a leading part in colonial expansion, using locally recruited troops. A French colonial expert is quoted as saying, 'While the British Empire was built by businessmen wanting to make money, the French Empire was built by bored officers looking for excitement.' A series of trans-Saharan expeditions, starting in 1879, were aimed at extending French sovereignty into the area to the south, and eastward to Abyssinia, more in a feeling of rivalry with Britain, and of the need to assert France's importance as a nation, both in Europe and in the world, than for any commercial benefit, although fantastic projects were proposed, including not only a trans-Saharan railway but also one from Senegal in the west to the Abyssinian coast of the Red Sea in the east – a dream that led to the Fashoda incident of 1898. The leading figure in this field at the turn of the century was General Joseph-Simon Gallieni. After service in Réunion, West Africa and the Caribbean he undertook the conquest of the Red River delta of Tonkin from 1892 to 1896, an extension of the French military involvement in Indo-China which had started in 1847. He then repeated the performance in Madagascar. One of his subordinates was Colonel Hubert Lyautey, whose life-time task, not completed until his death in 1934, was the conquest and pacification of Morocco. These enterprising officers, gaining direct experience of active service, albeit of a primitive kind, included several who were to hold high command in very different circumstances between 1914 and 1918. They included Joffre, Louis Franchet d'Esperey, Henri Gouraud and Charles Mangin, and, of those who reached prominence in the Second World War, Georges Catroux, Henri Giraud and Alphonse Juin.

While the activities of this colonialist military school were to have important repercussions on the international scene, their concepts of how to conduct operations were the very antithesis of those being put forward at the *Ecole Supérieure de Guerre*, where Foch entered as a student in 1885, the year in which the first lecture was delivered on the theories of Major-General Karl Maria von Clausewitz, whose *Vom Kriege* had been published posthumously between 1832 and 1834. At that time the school was strongly influenced by the ideas of Colonel Ardant de Picq, who, after service in the Crimea, had died leading his regiment near Metz in 1870, having just published privately a study of the history of battle. A public edition appeared in 1880, and a complete version, *Etudes sur le Combat*, in 1902. He had been influenced by Bugeaud and rejected the mechanistic concepts of Baron Henri Jomini and the importance attached to achieving a superiority in numbers. He laid prime emphasis on the importance of morale: of inducing fear into the enemy's hearts and expelling it from one's own ranks by a combination of discipline, founded on enthusiasm rather than slavish obedience, and of a determination to take the offensive, relying on manoeuvre.

When Foch returned to the school as a professor in 1895 and began compiling his famous books, *The Principles of War* (1903) and *The Conduct of War* (1905), he combined the ideas of du Picq with those of Clausewitz, of whom he was a great admirer. To the former's emphasis on the prime importance of morale and of the offensive spirit, and the latter's on the importance of concentrating the maximum force at the decisive point to defeat the enemy's forces, Foch added the importance of the will of the commander. General André Beaufre, a professor of the school from a later generation, wrote:

> True battle, Foch maintained, was the battle of manœuvre prepared, initiated and conducted by the commander and culminating in the decisive attack which was an expression of his will-power and which alone could bring victory. As the supreme act in warfare, a battle must be fought to the limit, with no looking over the shoulder. Every man must play his part in it with all his might and all his resources. ... Indomitable determination to win, however, and confidence in victory on the part of the commander would be of no avail, Foch considered, unless all his men were imbued with the same ideas to the very roots of their being.[1]

This concept lay behind the doctrine of *offensive à outrance*, which, in much more rigid form than du Picq or Foch had conceived it, was translated into the plans which the head of the *Troisième Bureau* of the French General Staff, the influential Colonel de Grandmaison, was to prepare between 1900 and 1914.

The concept took little account of the effect on the battlefield of the weapons which the French themselves had taken a lead in developing: self-loading rifles, machine-guns and breech-loading artillery. The Russo-Japanese War of 1904–5 provided a clear demonstration of the vulnerability of both cavalry and infantry to these modern weapons and of the need for the latter to dig themselves into the ground for protection. One of Foch's contemporaries, an instructor in infantry tactics at the Ecole at that time, Pétain, had absorbed that lesson. He never ceased to emphasize the importance of accuracy and strength of firepower: '*le feu tue*' was his favourite phrase, and in one of his lectures he made the comment: 'The fundamental fact to be gathered

from the 1870 war is the bringing to light of the considerable importance that has been acquired by firepower; this importance has
exceeded all forecasts. The progress in arms will impose on infantry
new combat procedures.' These views were unpopular. Foch did not
disregard the developments in firepower, but believed that they could
be used to the advantage of the offensive, notably by the use of
manœuvre. He did, however, recognize that there was an alternative
concept, which he described as 'the parallel battle' or 'battle of lines',
'in which one goes into action at all points, and in which the commander-in-chief expects a favourable circumstance, or a happy inspiration ... to let him know the place and time when he must act', but
he regarded it as a much inferior strategy, in which the commander-in-
chief had less opportunity to exercise his will, and which was to be
avoided if possible by manœuvre.

While the probability of these theories of war being put into effect
increased as Germany flexed her muscles from the 1905 Tangier crisis
to the Agadir incident in 1911, the means of doing so had been adversely affected by the internal problems of the army, already described,
and the reduction of conscript service from three to two years in 1905,
reducing the strength of the army from 615,000 to 504,000 men. With
Russia's defeat by Japan, France had to look elsewhere for friends.
Britain, smarting from her isolation during the Boer War and apprehen-
sive about Germany's naval and imperial ambitions, felt the same need.
The Tangier crisis, aided by Edward VII's liking for things French,
brought them together, their colonial rivalries being put aside.

Informal contacts between the French military attaché in London,
Major Huguet, and the Director of Military Operations at the War
Office, Major-General Grierson, aided and abetted by the military correspondent of *The Times*, Colonel Repington, had elicited the suggestion that 100,000 to 120,000 men, organized in three corps, might
be transported to the continent within one month of mobilization,
preferably to Belgium, but alternatively to Calais where they could
'unite with the French forces, of whom it would, for example, form
the left wing'. Grierson warned Huguet that, although the General
Staff had made provisional plans on these lines, they could not 'prejudice the decision that the government would take at a given moment'.
When, as happened shortly afterwards, contact between the two
general staffs was given official blessing, that qualification was never
removed and led the French to discount British help in preparing their
own plans. Responsibility for that was divided between the Generalissimo, who was also Vice-President of the War Council and became
C-in-C in war, and the Chief of Staff (*Chef d'Etat-Major*), who was responsible for the day-to-day operation of the army and for the execution
of plans prepared by the Generalissimo. Acting on their information
about the plans which General Alfred, Graf von Schlieffen, Chief of

the German General Staff, was then preparing, the General Staff, at the end of 1904, proposed a change to Plan xv, which called for a major offensive into Lorraine. They feared that, while the main German force of sixteen corps in Alsace-Lorraine parried the French thrust, a reserve of nine corps, concentrated round Aachen, would move through southern Belgium to threaten the French left flank. They therefore proposed the formation, from reserve divisions, of an *Armée du Nord* to counter this. The Generalissimo, General Brugère, rejected both the proposals of the Chief of Staff, General Jean-Marie Pendezec, and, even more forcibly, that of General Duchesne, who favoured abandonment of the planned offensive into Lorraine. In the year that followed a more realistic attitude was adopted by his successors, and in 1909 Plan xvi was adopted, which strengthened the left flank of the French armies, notably by the creation of a Sixth Army in reserve round Chalons and a redeployment of the Fifth Army to the north of it; but it still made no provision for any deployment north of Laon and Vouziers on the Aisne. In 1910, the new Generalissimo, General Michel, having reviewed the draft of Plan xvi concluded that he needed more troops to balance the forty-two corps which the *Deuxième Bureau* then estimated that the German army could deploy against them, an increase of fourteen over their estimate in 1904. This was to be achieved by adding a reserve regiment to every regular one on mobilization, which would make it possible for him to deploy an army of eleven corps (680,000 men) along the French frontier from Lille to Mézières, to which any British contingent that might result from the staff talks would be added. He would take the offensive with this army, reducing that allotted to the Alsace-Lorraine front, which would become a subsidiary sector, reinforced from North Africa. His proposals were rejected by the War Board, and Plan xvi remained in effect.

When Joffre succeeded General Michel in 1911, the posts of Generalissimo and Chief of Staff were amalgamated in that of *Chef d'Etat-Major Général*, the General Staff thus coming under the direct authority of the C-in-C designate. He himself had not attended the *Ecole Supérieure de Guerre* or commanded an army in the annual war games. He was a firm believer in the offensive and had no truck with the devotees of the counteroffensive who wished to await the enemy's first blow before embarking on it. He was dissatisfied with Plan xvi which he considered did not make sufficient provision for the clear signs that a considerable German force would advance through Belgium, while the French were attacking further south. He wished to pre-empt this by launching an offensive through Belgium himself, whatever the attitude of the Belgians. He proposed to move the Fifth Army northward to Mézières and support it with three cavalry corps, the xix Corps from North Africa, and the British Expeditionary Force, which would concentrate round Maubeuge. He paid more attention to the potential value of allies than had his predecessors, sending a mission to Russia to secure assurances that they would take the offensive by the sixteenth day after mobilization, which should contain at least five or six German corps on the eastern front, and adopting a more forthcoming attitude towards the British, now represented by the francophile Major General Wilson as Director of Military Operations. But the government, headed by Raymond Poincaré, backed by the Quai d'Orsay, was adamant that a French breach of Belgian neutrality, in advance of one by Germany, would prejudice a British decision to participate, a view that Wilson endorsed. The latter had entertained hopes of persuading Belgium to commit itself to the Entente in advance, or at least of admitting British troops, but the Agadir crisis, which lasted from July to November 1911, only strengthened its government's insistence on strict neutrality.

Joffre therefore had to forego his hopes of an offensive through Belgium, and directed the General Staff to prepare a new plan. This, Plan XVII, was completed in April 1913 and took account of the increase in conscript service to three years which became law in July. It provided for a field army of 710,000 men in forty-six divisions, twenty-five reserve divisions being allocated to rear area and other secondary duties. It would face a German one, estimated at a total of 880,000, most of which it was expected would be deployed against France. They realized that the German plan would involve an advance through Belgium, but the *Deuxième Bureau* assumed that it would be through the Ardennes, south of the Sambre and the Meuse. Plan XVII, which concentrated the Fifth Army between Mézières, Montmédy and Vouziers, assisted it was hoped by the BEF, would, Joffre believed, be able to deal with this, reinforced if necessary by the Fourth, held in reserve round Bar-le-Duc and St Dizier, behind the offensive to be launched by the Second and Third Armies into Lorraine, while the First stood ready to enter Alsace. Although the general strategy of the plan assumed an initial offensive, it did not, like its predecessors, plan one in detail. It provided for attacks north and south of the Metz-Thionville area or, employing the Fourth and Fifth Armies, through the southern Ardennes. A programme of re-equipment had been put in hand in November 1911, intended to cover five years, costed at 246 million francs. This was reduced in January 1912, but increased at the end of that year to 450 million, and again in 1913 (when an increase of pay was also authorized) in three steps from 450 to 1,400 million. In spite of this, the army only had 2,500 machine-guns to face the German 4,500 and 3,800 of its 75-mm field guns to oppose the German 6,000 77-mm guns. Its principal lack lay in heavier artillery, of which it had 300 pieces, mostly 120-mm howitzers dating from the 1880s, not to be compared, either in quality or in quantity, with the German 150-, 210- and 280-mm guns.

Another field in which Germany was superior was aviation, at least as far as numbers were concerned. As with other items of equipment the French had led the way technically. In 1892 de Freycinet had signed a contract for a military aircraft with M.Ader, who, ten years before, had succeeded in getting his steam-powered machine off the ground. It was to carry two men, 75 kg of explosives and be able to fly for six hours at a height of several hundred metres. When his prototype crashed on its first flight in 1897, no further action was taken until a mission was sent to the USA in 1906 to obtain a licence to manufacture the Wright Brothers' biplane. It was not until 1908 that this was obtained, and modifications made to increase its engine-power and fit wheels instead of skids. General Roques, of the Corps of Engineers, was placed in charge of aviation and appointed Inspector-General of Military Aviation in 1909. A competition was held in August to choose suitable aircraft, and two Farmans, two Wrights and one Blériot, similar to that which had crossed the Channel two weeks before, were ordered, the pilots being trained by the manufacturers. In 1912 three independent aviation groups were formed, consisting of five squadrons of six aircraft each, and a special dark blue uniform introduced. A major expansion was planned in 1913, as many as 400 aircraft being ordered, less than half of which had been produced by August 1914, when the twenty-one squadrons available could only produce 132 aircraft of eleven different types to face the German 200, organized in thirty-four *Staffeln*.

The French navy was less favoured than the army in the decade following the Franco-Prussian War. Apart from providing sailors to fight on land, it had been unable to affect the issue, and in 1878 an intense argument about its strategy and composition was initiated by the supporters of the *Jeune Ecole*, led by Admiral Aube, who argued that money spent on the new ironclad battleships was wasted, and that the future fleet should consist of larger numbers of smaller vessels, equipped with the new weapon, the torpedo. These could both provide coastal protection and prosecute a *guerre de course* against the trade of the enemy, who in naval eyes was, by tradition, Britain. Their argument was reinforced by the French lead in development of the submarine, the earliest version of which, driven by compressed air, had been produced in 1863. As with France's first aircraft, her erratic performance discouraged further experiment until the 1880s when Gustave Zédé produced a promising design, based on electric power, which was developed in 1898 by Laubeuf into one steam-driven on the surface, followed soon after by a similar type using a diesel internal combustion engine.

Admiral Aube became Minister of Marine in 1886, but held the post for only a year and a half and, after his departure, the influence of the *Jeune Ecole* faded, as the ideas of Mahan penetrated the newly established *Ecole Supérieure de Marine*, founded in 1896. France could not expect to compete with Britain in the naval battleship race that followed but, in alliance with Russia, could protect her interests in the Mediterranean, increased by her colonial exploits in North Africa, against a combination of the fleets of Austria-Hungary and Italy. 1900 saw a clear departure from Aube's ideas, and in 1901 the first six of the *Patrie* class of battleship of 15,000 tons, with 12-inch guns and a speed of over eighteen knots, were laid down. The programme, to be achieved in 1907, provided for a fleet of twenty-eight battleships, twenty armoured cruisers, fifty-two destroyers and a host of smaller vessels, manned by 3,000 officers and 112,000 *matelots*.

The Anglo-French Entente in 1904 meant that France could switch the emphasis of her naval strategy, not only from the Channel and the eastern Atlantic, but also from guarding French imperial interests all over the world from British encroachment or interference. However, there were no staff talks between the navies, paralleling those between the armies, until the Agadir crisis of 1911. When they did take place, the result was an agreement generally to leave the Mediterranean as the concern of the French navy, a significant proportion of the fleet normally based at Brest being transferred to Toulon, in return for a British guarantee to protect the French coast in the Channel from the German navy, a guarantee which, as we have seen from Chapter 1, was to have an important influence on the British government's decision to give France direct support in August 1914. At the outbreak of war the fleet fell short of the programme proposed. Of its twenty-four battleships, nine dated from before 1900, eleven were of the *Patrie* class, carrying four 12-inch guns, and only four were dreadnoughts, with twelve 12-inch, and a maximum speed of twenty-one knots. There were only fourteen armoured cruisers with 7.6-inch or 6.4-inch guns, none designed later than 1907, of which the most modern could steam at twenty-three knots, and some older cruisers, not all armoured. There were over eighty destroyers, including those building, more than half of pre-1907 design. There was a large number of torpedo-boats, but few of these were ocean-going, and over seventy submarines, only eighteen of which were of post-1907 design.

August 1914 saw the theories of the *Ecole Supérieure de Guerre* put to the test. Those, like de Grandmaison, who believed that *élan* and the *furia francese* (which had lost François I the Battle of Pavia in 1525) would prevail against the invaders, saw their theories rudely shattered as the fields of eastern France were strewn with the corpses of white-gloved officers and soldiers in blue coats and red trousers, garments supposed to boost their morale. In the first fortnight Joffre lost 300,000 men and 4,778 officers, a tenth of France's total officer strength. Fortunately for France, overall command was in the hands of a man who was not addicted to theories, whose experience had been the pragmatic one of organizing the logistics of colonial campaigns, and who was not upset when things did not go according to predictions or plans. Showing more flexibility than his opponent General Helmuth von Moltke, Joffre reoriented his forces and took the bold decision to make a deep withdrawal to the Seine, while planning an envelopment from the west of General Alexander von Kluck's army on Moltke's right. By his decision to move east of Paris, Kluck played into the hand of Joffre, who, influenced and helped by Gallieni, timed his decision to counterattack with finesse. Had the implementation at a lower level, including that by the reluctant C-in-C of the BEF, Sir John French, matched the decisiveness and energy of Joffre and Gallieni, the future course of what came to be known as the Great War might have been very different. But slow reactions led to missed opportunities, and, after the 'race to the sea', the campaign settled down to that 'battle of lines' which Foch had so rightly considered inferior to that of manœuvre. 1915 was characterized by attempts to bring about the breakthrough (*la percée*) which would restore the superior form of warfare, but not even the great losses incurred in the offensive battles by both the British and the French during the year, which brought the total casualties of the former, since the war had started, to 21,747 officers and 490,673 soldiers, and of the latter to 50,355 and 1,911,332, destroyed the faith of those who believed that better organized offensives, with more artillery support, could force a decision, in spite of the ability of the enemy to marshal reserves by rail to reinforce a threatened sector long before there was any danger of its defences, organized in depth, being penetrated.

In any case, what was the alternative? To adopt a purely defensive stance would have left Germany in possession not only of the lost provinces, but of Belgium and a substantial area of north-eastern France, of great importance to its economy, and would have made it possible for her to transfer forces to the Eastern Front, in the face of which Russia might come to terms, leaving France and Britain isolated. Opening up an alternative theatre of war in the Balkans, in the hope of 'knocking away the props' and attacking Germany through her weaker allies, Austria-Hungary and Turkey, had been tried at Gallipoli and Salonika with no success. Failure to exploit the latter was blamed on the unsatisfactory organization of the Higher Command.

The disadvantage of making one man both Chief of the General Staff and C-in-C of the forces in the field in France was becoming evident. In the second capacity he was obviously disinclined to favour demands on resources which were liable to be met at the expense of his own front, nor in the former capacity was he likely to suggest that, because his own efforts were not being crowned with success, an alternative strategy should be applied. Even if General Sarrail, a rival to Joffre, favoured by the anti-Catholic politicians, who commanded the Franco-British force based on Salonika, had been given much greater resources and support than he was, it is highly unlikely that his operations would have been decisive and allowed the Western Front to adopt a purely defensive strategy. Criticism of Joffre and of his method of exercising command grew, and, when General Gallieni, his old sponsor, became Minister of War in October 1915, he began to express severe criticism of Joffre and of the fact that so much power was concentrated in his hands; but he died in March 1916 before he could change anything, other than appointing General Noël de Castelnau, an aristocratic, devout Catholic cavalry officer, fervent believer in the *offensive à outrance*, as Joffre's chief of staff and deputy.

The strategy for 1916, adopted at a meeting of Allied military chiefs at Chantilly in December 1915, was for a series of co-ordinated offensives, British and French on the Western and Russian on the Eastern Front, to be launched 'as soon as possible', but its implementation was pre-empted by Falkenhayn's offensive directed at Verdun, which dominated Allied strategy throughout the year. Whether or not it would have done so to the extent that it did depended on the decision taken by de Castelnau in the first few days of the attack, when it appeared that the defence was crumbling, the great fort of Douaumont having fallen into German hands almost without a struggle. Joffre appears merely to have acquiesced in two important decisions taken by de Castelnau: to hang on to the defences on the east bank of the River Meuse, rather than withdraw behind the river, and to call on Pétain, whose army, having been relieved by the British, was in reserve, to command the defences on both banks with the strict order that Verdun was to be defended at all costs. Left to himself, it is likely that Pétain would have preferred a withdrawal, at least to the left bank, abandoning the citadel of Verdun itself; but more than territory was at stake. There were fears that the morale of the army, after the costly offensive of 1915, was crumbling, and that a withdrawal could turn into rout. The loss of so significant a symbol of the nation's defence as the fortress of Verdun would affect the morale not only of the army but of the whole nation. The cry must be '*Ils ne passeront pas*'.

In adopting it, the French played into the hands of Falkenhayn, whose declared aim was 'to bleed the French armies white'. He succeeded; but in the process bled his own to the same colour. Pétain set out in his methodical, cautious way to stabilize the situation, instil confidence, organize supplies and reliefs, and gradually, in a series of attacks to a limited depth, with strictly limited objectives, supported by intense and highly organized artillery fire, to restore both the situation on the ground, including Nivelle's recapture of Fort Douaumont in June, and the morale of his soldiers, to whose welfare he devoted more attention than had, hitherto, any other French general. Joffre's part in this was to exert the maximum pressure on Haig for the British to launch an offensive which could directly draw off German forces from the Verdun sector, and to keep up the pressure. The result was the Battle of the Somme, which started on 1 July and continued until November, by which time it had cost the British 419,564 and the French 204,253 casualties: estimates of German losses vary from 465,000 to 680,000. The fighting round Verdun continued until the end of the year, by which time it had cost the French army 377,231 casualties and the German 337,000, the former including 162,308 killed or missing and the latter over 100,000.

Another casualty of a different kind was Joffre. For some time Aristide Briand, the Prime Minister, had been trying to persuade him that he should hand over command and become the government's supreme military adviser, covering all fields. He had refused, maintaining that, without the prestige of his appointment in command of the principal front, he would lack authority. Briand succeeded in manœuvring him into resignation at the end of December, compensated with the baton of a *Maréchal de France*. Foch and de Castelnau, both too Catholic and too tainted by association with unsuccessful offensives, were passed over, as also was Pétain, whose undisguised contempt for politicians stuck in Briand's gullet. The choice fell on the glib artillery general Nivelle, whose plan for the recapture of Douaumont, with its creeping artillery barrage, appeared to presage a revolution in the possibility of launching a successful offensive without crippling losses. It was more ready to hand than the British invention of the tank, of which the few that had taken part in the Battle of the Somme had mostly broken down or become bogged, and it was more readily controlled than gas, against which the Germans, who had initiated its use in April 1915, were well prepared.

Nivelle's offensive on the Chemin des Dames in April 1917, for which he had claimed so much, failed dismally. The Germans had also learnt lessons from Verdun. Not only had Ludendorff, Falkenhayn's successor, effected a major withdrawal from the salient astride the Somme and Oise, but the design of the German defences accentuated their preference for manning the front line thinly, placing their principal strength further in the rear, where it escaped the bulk of the enemy's preparatory artillery-fire, including Nivelle's favourite creeping barrage. When, therefore, on 16 April 1917, Mangin's Sixth Army, spurred on by confident assurances from him and from Nivelle, left their trenches with a restored *élan*, they met no opposition for the first 800 metres, and were then mown down by machine-guns untouched by the artillery programme; 120,000 men fell on that first day, Nivelle having forecast 10,000 and his medical services having, as they thought prudently, added another 5,000. This bitter blow to hopes raised so high led swiftly to demoralization and worse. On 3 May mutinies broke out and soon became widespread. The 21st Division, which had distinguished itself at Verdun the previous June, refused to return to the front line. After the ringleaders were shot or sent to prison on Devil's Island, the division went into battle and was decimated. By June fifty-four divisions were affected.

In a panic, Nivelle was dismissed, Pétain installed in his place on 15 May and any idea of continuing the offensive abandoned. The British would have to bear the brunt for the rest of 1917, and did so in what was known as the Third Battle of Ypres, or Passchendaele, at a cost of 244,897 casualties, one more offensive that had gone wrong. However, on 20 November 1917 a ray of hope was lit by the success of the 375 tanks supporting the British Third Army's attack towards Cambrai, the initial success of which, at a low cost in casualties, was dramatic. Unfortunately all that had been gained was lost when the Germans counterattacked at the end of the month.

The mutinies were the culmination of a number of influences which had been undermining the morale of the French army: the contrast between life in the trenches not only with that in areas untouched by the war, where many citizens were enriching themselves, but with the picture painted by politicians and the press of the heroism, glory and high morale of the *poilu* in comparison with that of the despised *Boche*, who unfortunately proved so formidable an opponent in the field; the general unrest in the labour force, which both affected the soldiers themselves and was resented by them; the political and press pressure, some of which was believed to be subsidized from German sources, for an end to the war and for a united move by workers in all countries to sabotage its continuance. All these influences had helped to create a climate in which the abject failure of an operation, the claims for which had been so extravagant, proved to be the straw that broke the camel's back.

Pétain took immediate steps to restore morale. He combined a sharp imposition of strict discipline, although limiting the severest disciplinary measures to the minimum – fifty-five executions only, out of a total of 413 death sentences, were carried out – with a series of reforms designed to improve the conditions of the soldier's life, introducing welfare measures similar to those enjoyed by the British. He made a point of visiting every division of the army, promising that there would be no more offensives *à la Nivelle*, saying, 'We must wait for the Americans and the tanks', for the USA had entered the war in April 1917 and the first American troops had arrived in June. By January 1918, 175,000 had reached France, forming four divisions, of which only one was in the line.

That month saw an important meeting at Compiègne to consider allied strategy for 1918. Russia was now virtually out of the war, although the treaty of Brest-Litovsk was not signed until March 1918, and Germany could switch more forces to the west. Pétain's defensive stance had already incurred criticism, one result of which was the replacement in November of Paul Painlevé, who had succeeded Briand in March, by the aggressive radical Georges Clemenceau, who paid more attention to the offensively minded Foch. The latter not only disagreed with Pétain's generally passive strategy, but also with his imitation of Ludendorff's defensive tactics, being prepared to sacrifice the forward line of defences for the sake of a strengthened defence in depth. He thought that this might only lead to a succession of withdrawals in the face of the superior numbers which the German forces could now deploy, unless the initiative were seized from them by a resumption of the offensive. For this to be effective, it was necessary to establish a unified allied command with full authority over British, French and American forces.

As a step towards this, a Supreme War Council (*Conseil Supérieur de Guerre*) had been established in November 1917. It had recommended the formation of a general reserve for both the Western and the Italian Front, the latter having been consolidated after the retreat from Caporetto in May by the despatch of British and French reinforcements. The employment of this reserve was to be decided by an executive committee (*Conseil Militaire Interallié*) established at Versailles, consisting of Foch for France, Wilson for Britain and Pershing for the USA. Pétain acted in collusion with Haig to frustrate this, by resisting pressure to give up divisions to form the reserve, both being justifiably concerned at the increase in the number of German divisions facing them, and by agreeing with each other's plans for mutual co-operation, which they presented to the committee as a *fait accompli*.

Supported by Clemenceau, and his authority having been enhanced, both by having restored the situation in Italy and by these new arrangements, Foch began to interfere more and more with the decisions of Pétain, whom he was scheming to remove. He had not done so by the time that Ludendorff launched his offensive against the British in March 1918, designed to cut them off from the French and from the Channel ports. The weakness of relying on mutual co-operation between the two national C-in-Cs immediately became apparent, as Haig was primarily concerned with his lines of communication to the ports, more anxious about the danger of a second thrust in Flanders, which might cut him off from them, than of further loss of ground on his right, where Amiens was threatened; while Pétain, convinced that Ludendorff's next major thrust would be on his front, in Champagne, was primarily concerned to cover Paris. The danger, clear to Foch, was that Ludendorff would penetrate between the two and separate them. Haig, his direct approach to Pétain having failed, became an instant convert to giving full authority to Foch, and the situation was saved as the latter, after a Supreme War Council meeting at Doullens on 26 March which charged him with 'co-ordinating the nature of the Allied Armies on the Western Front', switched French divisions to Haig's command. Foch's star was now in the ascendant. Both he and Haig wished to take advantage of the situation in which Ludendorff found himself when his offensive ran out of steam at the end of April, his forces extended and having suffered about 250,000 casualties. They were no longer entrenched in well-prepared defences, the approaches to which had been reduced to a quagmire. The tank, with which all three Allied armies were now equipped in some quantity, could come into its own. Foch and Haig pressed their respective political masters to authorize an immediate return to the offensive, which could end the war in 1918, but met with considerable resistance. Both Lloyd George and Clemenceau feared the political consequences of further long casualty lists, and were inclined to lend a sympathetic ear to Pétain's preference for a policy of 'waiting for the Americans and the tanks', both of which would be available in quantity in 1919. However, Foch's enthusiasm won them both over, and in a series of well-orchestrated offensive thrusts, led by the French and the Americans near Château-Thierry on 18 July, the Germans were driven back until the war came to an end on 11 November. Foch was the hero of the hour, but Pétain remained the hero of Verdun, his glory undimmed by the pessimism which dominated his attitude after the mutinies.

Apart from Nivelle's development of new artillery methods, the French army had not been in the lead in innovations either of equipment or of tactics, as both the Germans and, notably with the tank, the British had. This may have been because they were preoccupied with the major task of providing the quantity of equipment and stores, especially artillery ammunition, which their large armies demanded. However, in one field, aviation, the French army showed admirable enterprise. One of the early examples of its value was the information provided, the British RFC claiming to be the first to do so, of Kluck's inward wheel before the Battle of the Marne. General Hirschauer, who had made a significant contribution to the pre-war development of the 'Aéronautique Militaire', was appointed its head towards the end of 1914. He reopened the flying schools which, in the expectation of a short war, had been closed and, taking a long view, ordered 2,300 aircraft and 3,400 engines. The majority were intended for general observation duties, but some were to specialize as fighters or bombers, one of the earliest strategic bombing sorties taking place on 26 May 1915 against a chemical works at Ludwigshafen. France was first in the field with an important innovation, a machine-gun which was synchronized to fire through the propeller. Unfortunately a Morane Parasol fighter, piloted by Roland Garros, carrying this invention was shot down over the German lines in April 1915, and a more efficient development of the device was produced by Fokker for the Germans.

The beginning of 1916 saw a considerable development in all aspects of aviation, 1,149 aircraft being operational: 826 observation, 135 fighter and 188 bomber. The German attack on Verdun was strongly supported by considerable numbers of their efficient Fokker fighters, in reaction to which the French squadrons began to fly in larger formations. At this time they had a large variety of different aircraft, among which the names of Nieuport, Bréguet and Spad were becoming prominent.

This variety caused headaches to the *Service de Fabrication des Avions* (SFA) responsible for production, which was often at loggerheads with the *Service Technique Aérienne* (STAé), responsible for design, neither of which always saw eye to eye with the *Service Aéronautique* at *Grand Quartier-Général* (GQG). Hirschauer had resigned in September 1915 and responsibility for SFA and STAé had passed to M.Besnard as Under-Secretary for Military Aviation in the War Ministry, but that did not prevent quarrels between Colonel Regnier, head of the *Aéronautique Militaire* at GQG, and Colonel Barès of the STAé. Matters improved in March 1917 when M.Vincent succeeded Besnard. He stopped production of the older types and concentrated on the new Spad C1 fighters and Bréguet 14 bombers, buying some Sopwith bombers from Britain and manufacturing more under licence. At the start of the year, the number of aircraft operational on the Western Front was 2,263, increased by August to 3,556, the fighters being organized into five groups each of four squadrons of Spads. Arguments continued about responsibility for design and production, the former being that of the War Ministry and the latter of the Ministry of Armament. The situation at GQG was gripped firmly by Colonel Duval. He was to be the first commander, in 1918, of the French equivalent of Trenchard's Independent Air Force, the *1ère Division Aérienne*, consisting of two *groupements*, one of 12 fighter and 9 bomber, the other of 12 fighter and 6 bomber squadrons, totalling some 600 aircraft. It was not allocated to any army or corps, as all other groups were, and came into its own when the Germans attacked in March 1918. A sad loss in September 1917 was the death of Lieutenant Guynemer, perhaps the most famous of a long list of aces. It was as a corporal that he had piloted his first Morane and had claimed the first of his eighty-nine successful combats in July 1915.

At the end of the war the French air force had 3,222 aircraft available, organized in 13 fighter and 10 bomber groups, 124 general observation and 24 artillery observation squadrons, manned by 127,630 men, of whom 12,000 were aircrew, and aircraft production was running at 100 a day.

The French navy cannot be said to have had a glorious, nor even a satisfactory, war. It took no part in the principal naval operations of the war, neither in the clash between the battle-fleets, nor in the less glamorous but nevertheless crucial anti-submarine war in the Atlantic. Its activities were almost exclusively confined to the Mediterranean, where the threat from the Austro-Hungarian and Turkish fleets was negligible, but that from German submarines not to be ignored. Failure to appreciate it led to the loss of the battleship *Jean Bart* in December 1914, when Admiral Boué de Lapeyrère took his fleet into the Adriatic to attack the Austrian naval base of Pola. In spite of that warning, the armoured cruiser *Léon Gambetta* was sunk in the same area when steaming at only seven knots, and as late as November 1916 the old battleship *Suffren*, having survived the hazards of Gallipoli, was steaming unescorted at nine knots, when she was sunk with all hands lost.

The principal French naval action of the war was the Gallipoli operation. For the initial purely naval attempt to force the straits, Admiral Guépratte had four old battleships, a flotilla of small destroyers and one of minesweepers. Two of the battleships were holed by Turkish gunfire and one sank when its ammunition blew up after it had hit a mine. Thereafter naval action was confined to gunfire support of the landings, similar support being given to the British forces facing the Turks across the Suez Canal. Apart from these actions, French naval activity consisted of escorts to shipping, their Admiralty being as reluctant as the British to introduce convoys, and attempts by submarines to penetrate Austrian and Turkish coastal defences. Neither their submarine nor their anti-submarine operations were notably successful, enemy submarine losses in the Mediterranean amounting only to eleven, while the French lost seven in that area and five elsewhere. Their losses of surface ships – four battleships, five cruisers and sixteen destroyers – were mostly in older vessels. Total casualties amounted to 11,500, at least a quarter of which were incurred on land. This represented a drop in the ocean compared to French army losses which totalled over six million, of whom 1,357,800 died.

1919–45

The twenty-five years that followed the Great War were to prove an unhappy period for the French army, but at the moment of victory and for a few years thereafter its reputation and popularity were higher than they had been since Napoleon's day, and the two leading figures, Foch and Pétain, were national heroes, the latter retaining an immense prestige throughout the period, with baneful results. The troubles that beset the army during this period derived principally from three factors, political, economic and military, all intertwined. Even before the armistice, a division between the politicians and the military began to appear, and this widened as the nation itself became sharply divided on political issues and the political scene became dominated by instability and intrigue. Economic difficulties contributed heavily to this, to which was added a military factor of major importance: the adoption of a purely defensive strategy based on the rigidly conservative attitude of the old men who dominated the higher posts of the army. The result was correctly described by the C-in-C, General Maxime Weygand, to the Prime Minister, Paul Reynaud, at the height of the crisis on 25 May 1940 in the words: 'This war is sheer madness: we have gone to war with a 1918 army against a German army of 1939. It is sheer madness.' Few people – perhaps only Pétain – had contributed more to that state of affairs than Weygand himself.

The influence of Foch, whose faithful and efficient Chief of Staff Weygand had been, began to wane after the great victory parade in July 1919, in which he and Joffre, whom he had magnanimously insisted must ride by his side, preceded the armies of France led by Pétain, their C-in-C. This was largely due to the fierce quarrel between him and Clemenceau over what he, and most of the generals of the French army, regarded as the essential military measures needed to insure against a repetition of 1870 or 1914. He and his colleagues were fully aware that, even when France had regained Alsace and Lorraine (and, it was hoped, the Saarland), the demographic and economic balance would still be weighted in favour of Germany, dominated by Prussia. Ideally the Rhineland, Catholic and anti-Prussian, key to the routes of invasion of France, must be removed from the nation which Bismarck's politico-military exploits had created. Failing that, France (or failing even that an international force) must occupy the Rhine crossings, as Germany had occupied strategic areas of France after 1871.

Although Clemenceau initially supported this view, he backed further and further away from it as he realized that Britain and the USA would never support it and, if they did not, that France's expectation that 'reparations' from Germany would compensate for the economic losses she had suffered would be gravely compromised. Woodrow Wilson and Lloyd George (in spite of the latter's promise to the British electorate that he would squeeze Germany like a lemon until the pips squeaked) were united in wishing to avoid producing an Alsace-Lorraine situation in reverse, a burning desire for revenge in Germany which would provoke a repetition of the holocaust which had just finished. There was a clear conflict between French demands for security and the declared Allied war aim, deriving from Wilson's Fourteen Points, and strongly backed by the political left in France, of self-determination, a principle which France supported in detaching non-German communities in Eastern Europe from Germany and Austria.

Foch persisted, by direct and indirect means, in attempting to frustrate the completion of a peace treaty which did not meet his requirements, supported by General 'Butcher' Mangin, who, commanding the

French occupation troops in the Rhineland, engaged in direct plotting
with German supporters of an independent Rhineland, among whom
was Konrad Adenauer. But Clemenceau prevailed, relying on an assurance of support from Britain and the USA in the event of a German
breach of the treaty, an assurance soon undermined by the refusal
of the US Senate to ratify it. American suspicions of Foch's designs
had undoubtedly contributed to that. In the light of subsequent events
one must have some sympathy with the prophetic view he expressed
in an interview with the British *Daily Mail* while the peace treaty
negotiations were in progress:

> What was it which saved the Allies at the beginning of the war? Russia.
> Well, on whose side will Russia be in the future? With us or with the
> Germans? ... The Allied armies. Where will the Allied armies be? The
> British army will be in Canada, Australia, in New Zealand. The American
> armies will be in the United States. ... And next time, remember, the
> Germans will make no mistake. They will break through Northern France
> and seize the Channel ports as a base of operations against England.[2]

The Peace Treaty, signed on 28 June 1919, provided among other
things for a special status for the Saarland, to be occupied by an international force under the authority of the League of Nations, while France
operated its coal mines (ended by a plebiscite in 1935), and for the
demilitarization and fifteen years' occupation of the area of Germany
west of the Rhine, while Germany's army was to be restricted to an
internal security force of not more than 100,000 men. As significant
to the French as these military terms were the economic ones. To
compensate for both the actual damage to her resources and for the
general economic cost of the war, France relied on the 'reparations'
exacted by the Treaty, set at £6,600 million for all the Allies, of which
much the largest share was due to France. As the franc fell on the
exchanges against the dollar and the pound, and the country's economic plight became evident, the demand for prompt payment of these
reparations became more insistent, fuelled by the severe unrest in the
labour force, hit by inflation and unemployment and sympathetic to
the workers' revolution in Russia.

These political and economic difficulties had their repercussions
on the armed forces, some of whom were engaged in supporting
counter-revolutionary forces in Russia, where sailors of the French
navy, based at Odessa, mutinied. Germany's political and economic
difficulties at the same time were even greater, and, when she defaulted
on her reparations payment in 1923, French troops occupied the Ruhr.
Their intervention was strongly opposed by German communists, who
were helped by political sympathizers within the ranks of the French
army. With memories of 1917 still fresh in their minds, many French
officers began to see communism, with which they linked socialists

and radicals, as a greater enemy than Germany, an attitude which, much later, in the days of the Popular Front, was to be reflected in the slogan '*plutôt Hitler que Blum*'. Weygand was sent to Poland to support Marshal Josef Pilsudski's campaign against the Red Army; Charles de Gaulle served there as a captain in the Polish cavalry.

Hard hit by inflation and severe economies imposed by the financial situation, with idle soldiers affected by the anti-militarist and revolutionary trends of the 1920s, army officers either lost interest in their profession or sought service overseas, in Indo-China, Syria, Madagascar or North Africa, where anti-colonial movements were, or were suspected of being, associated with Marxism. Morocco was the scene of the most extensive campaign of this kind, where the Riff War, against Abd-el-Krim, lasted from April 1924 to July 1926, Pétain having taken over command of the troops to the chagrin of Marshal Lyautey. The latter's great reputation as a colonial administrator rested on his policy of gaining the support of the inhabitants, spreading French influence from one '*tache d'huile*' to another, rather than by conquest and repression with overwhelming military force. Pétain's victory enhanced his authority, with consequences which were to prove fatal both for the army and for the nation.

His contempt for politicians had been increased by events, as one Minister for War succeeded another – there were eight between 1920 and 1926 in the kaleidoscopic merry-go-round which characterized French politics of the period. When the wartime post of C-in-C came to an end, Pétain was appointed Vice-President of the *Conseil Supérieur de Guerre*. It was an unwieldy and geriatric body, consisting of all six marshals, as life members, certain senior generals retained beyond the age limit, and the serving generals, headed by the Chief of the General Staff. In 1922 he was also made Inspector-General of the Army, the Chief of the General Staff having to submit to him anything concerned with its structure or mobilization, before referring it to the Minister for signature. He was, therefore, in fact the professional head of the army until succeeded by Weygand in 1931.

His thinking, expressed in the famous '*Instruction provisoire sur la conduite des grandes unités*' (Provisional instruction on the employment of formations), issued in 1921, remained the bible of French strategy until 1936, when General Maurice Gamelin issued a very similar *Instruction*, omitting the '*provisoire*'. It stressed the strength that modern firepower had conferred upon the defence, making it impossible to break it without massing an overwhelming strength of artillery, with adequate supplies of ammunition, and the support of tanks to clear the way for infantry and operate closely with them. The primary task of aircraft was to provide observation for the artillery. This military emphasis on defence fitted in well with political factors: the need to avoid any repetition of the costly offensives of 1914–15; the desire to reduce the length of conscript service; the prospective shortage of manpower deriving from the '*années creuses*' (1932–6), when the lower birthrate of the war years would reduce the number of young men available for service; and the pressure from within and without to adopt an international pacific solution to the problems of Europe, rather than a nationally offensive one, in accord with the 1925 Locarno Treaty, which 'guaranteed' the frontiers of Germany with Belgium and France and confirmed the demilitarization of the Rhineland, from which the French withdrew their occupation forces in 1930, the British having removed theirs in 1926.

The prospect of a reduction in the length of conscript service led the Army Council to propose, in 1925, permanent fortifications in the north-east. The first (1926) plan was for a series of fortified regions but, at Pétain's insistence, it was later altered to a continuous line, as far north as Longwy. This was finally approved in 1930, when André Maginot, who had served as a sergeant under Pétain at Verdun, was War Minister, and construction was immediately put in hand. The line was a product, rather than a cause, of France's purely defensive strategy. Not only did its cost – originally estimated at Fr 3,000 million (£24 million at the time) for the first four years, but actually Fr 7,000 million by 1935 – absorb money that should have been devoted to modernizing the army, including the air force, which did not become independent until 1933, but to the garrison troops who manned the fortresses had to be added what were called 'internal' troops to operate between and in support of the forts. In 1939 the army groups manning the line absorbed thirty-five divisions, half the army, while the 250 miles of the Belgian frontier, not covered by the line, had to make do with the other half. It did not, therefore, compensate for the reductions brought about by shortening the length of conscript service. In 1923 that was reduced from two years to eighteen months, bringing the annual intake down to 250,000 out of a total army strength of 360,000, and the number of divisions from fifty-two to thirty-six. The only compensation was that, for two years after the end of their service, conscripts could, with parliament's authority, be called up without involving general mobilization. They were known as *disponibles*. In 1928 this was cut to one year, reducing the number of divisions in the standing army to twenty.

When Weygand succeeded Pétain as Vice-President of the Army Council and Inspector General of the Army in 1931, handing over as Chief of the General Staff to Gamelin, he was keenly aware of the deficiencies of the army, but in the next few years budgetary reductions, resulting from the world economic crisis, weakened it even further. This coincided with Hitler's appointment as Chancellor in Germany and clear indications of German rearmament. Hope that increased recruitment of soldiers from North Africa could compensate for reductions in conscripts were not fulfilled, as recruitment fell there and there were signs of unrest in the ranks. The appointment of Pétain as War Minister in the Gaston Doumergue government that succeeded that of Edouard Daladier after the right-wing-inspired riots of 6 February 1934, did little to improve matters and, when Gamelin replaced Weygand in January 1935, the situation seemed extremely serious. In that month the Saar plebiscite resulted in its return to Germany, which had withdrawn from the Geneva disarmament conference, bringing it to a grinding halt. In May 1935 the Franco-Soviet Pact was signed, but Pierre Laval showed no enthusiasm for giving it any military teeth, nor did the French military press him to do so, although Gamelin had persuaded Weygand not to oppose it openly. Hitler used the signature of the pact as an excuse to introduce conscription for two years' service.

The caution of the French military, apart from right-wing fears of the internal effect of association with the Soviet Union, was linked to two factors: the opposition of France's allies in Eastern Europe, notably Poland and Romania, to any agreement which could envisage the passage of Soviet forces through their countries, or even an increase in Soviet influence, and the difficulty of offering the prospect of any reciprocal French military action, if Germany moved eastward. The latter problem had been raised in the *Haut Comité Militaire*, the French equivalent of the British Chiefs of Staff Committee, when co-operation with Italy was being discussed. Pétain, who retained membership of the committee on a personal basis, had stated that, as the army was organized as a defensive force, it would 'find it difficult' to undertake an offensive; but Gamelin, always one for compromising with political demands, said that he considered that the army was strong enough 'to enter the field in accord with whatever decisions the government would take'.

The army was strengthened in 1935 by a return to two years' conscript service, but no action was taken to change its fundamentally defensive stance. It is true that Weygand, in 1933, had laid plans for the mechanization of five infantry divisions and the conversion of a cavalry division to a *Division Légère Mécanique* (DLM), but these measures were not intended to produce an offensive force similar to that proposed by Lieutenant-Colonel de Gaulle in an article, published in May 1933 and expanded next year into a small book, entitled *Vers l'Armée de Métier* (*Towards a Professional Army*). Having pointed out that demography, industrial resources and geography all tilted the balance in favour of Germany, he proposed that military strength should not be assessed solely in terms of quantity, in which France could never be superior, but by the creation of a highly professional force of high quality, equipped entirely with tracked vehicles. He wrote:

> The moment has come when to our mass of reserves and of recruits (the principal element of national defence, but one which is slow to mobilize and clumsy to set going, and whose gigantic effort can only be used in the last degree of danger) we must add a manœuvring instrument which is capable of acting without delay, that is to say, one which is permanent in its force, coherent, broken to battle. Without a professional army there can be no defence.

He proposed a force of six divisions, 'completely motorized and tracked, and partly armoured',* requiring 100,000 men. His proposal met with no more favour in high military or political circles than had the suggestions of General Estienne in 1922 or of General Doumenc in 1928 for the creation of armoured divisions.

The vulnerability of tanks to modern anti-tank weapons, the unsuitability of the terrain in the north-east, the undesirability of creating a force separate from that of the army as a whole, were all cited in an important debate which took place in parliament on 15 March 1935, when de Gaulle's concept was supported by Reynaud. General Maurin, then War Minister, dismissed it with the words, 'When we have lavished so much effort on building a fortified barrier, who could believe us foolish enough to sally out in front of this barrier, in search of heaven knows what adventure', while Weygand's comment was, 'What a hotbed of communism, this troop of mechanics!' Political objections were added to the military: not only that it would be expensive (which was true) and could only be provided at the expense of the rest of the army, which was badly in need of modernization, while it was also contended that the Maginot Line should be extended to cover the Belgian border; but also that such a professional force could be a threat to democratic liberties and to the constitution. Had such a force been in existence in 1935 the whole history of Europe might have been different.

In October the General Staff warned the government that it was probable that Germany would repudiate the demilitarization of the Rhineland within the next twelve months, and in December that actual steps were being taken. This set off a discussion about what France's reaction should be. Gamelin stated that the army's peacetime strength (370,000 men, of whom 62,000 were regular soldiers and 72,000 colonial troops) was insufficient for him to launch an offensive. He would need to call up the *disponibles*, adding 540,000 men within a week, but equipment even for the peacetime army was seriously deficient, as it was for the newly independent air force, which, although it totalled

---

* Each division would have a tank brigade (two heavy battalions, one medium and one reconnaissance), an infantry brigade (three battalions with a high proportion of anti-tank and automatic weapons) and an artillery brigade (one battalion of howitzers, one of light field guns and an anti-aircraft group).

about 3,500 aircraft world-wide, could only make 1,500 available in support of a move into the Rhineland, the majority of which were purely for observation.

Although Germany could produce more men in uniform than France, she was inferior in equipment in every field. Gamelin's argument, which was accepted by the government, was that, although France could occupy the Saar in retaliation for a German re-militarization of the Rhineland, it would inevitably lead to a general war with Germany, in which the balance would be tilted against France, unless she were supported by allies, including Britain. It would be imprudent to engage in any military operation unless the government was prepared to face a general war, in which case mobilization should be ordered. When, on 7 March, German troops entered the demilitarized zone, the government hesitated and would not even authorize call-up of the *disponibles*. By 12 March the decision was taken to refer the whole matter to the League of Nations, invoking the Locarno Pact. It was a fatal decision or, more accurately, indecision: as Alistair Horne wrote, 'the road to disaster was clearly signposted',[3] and the first result was Belgium's declaration of neutrality.

This crisis point in French defence coincided with the outbreak of the Spanish Civil War, which was also to have significant effects both on her external relations and on her domestic situation, political and military, the gap between right and left in both widening into a yawning chasm. But at least it was clear to both that the nation was in danger and that a programme of rearmament had to be adopted. It was Léon Blum's Popular Front government, with Daladier as War Minister, which set it in train. But, although expenditure on arms was increased from Fr 1,492 million in 1936 to Fr 5,152 million in 1938, production was seriously affected by strikes (in 1936 tank production was running at 120 a month, but by January 1937 had fallen to 19), and a large sum was wasted on the creation of a line of unconnected pill-boxes along the Belgian frontier, deluding the public into thinking that it represented an extension of the Maginot Line. Two armoured divisions and three more DLMs were authorized, and work started on the heavy 'B' tank, both it and the more mobile medium *Somua* mounting the 47-mm gun, which was also produced as an anti-tank gun and was much superior to anything the British and Germans were to field in 1939. No change, however, in France's fundamentally defensive strategy took place, although Gamelin's 1936 *Instruction*, which succeeded Pétain's 1921 *Instruction provisoire*, emphasized that the offensive was the superior form of warfare, but that it could not be undertaken until a sufficient overall superiority had been built up behind an impenetrable, continuous line of defence. Although Daladier described the cover of 'the organization in concrete, this powerful network of automatic weapons whose terrible efficacy against attacking troops has been revealed in war' as 'inviolable' and pretended that it ran 'from Dunkirk to Nice', Gamelin knew well that, at any rate from Longwy north to the sea, that was not true. The strategy which he developed, particularly after the Munich crisis of September 1938, was designed both to increase the forces available to oppose the Germans and to shorten the front on which they had to do so by involving Britain and the Low Countries. Staff talks with Britain were opened in May 1938, but, as the previous chapter explained, produced initially only a non-committal offer of two divisions within sixteen days of mobilization, and Belgium and Holland remained strictly neutral, refusing any discussion of co-operation.

When Hitler invaded Poland in September 1939, and Britain and France found themselves at war with Germany on that account, France's fulfilment of the promise that Gamelin had given in May of that year to the Polish High Command that 'immediately on the outbreak of war the French Army would assume the offensive against Germany and that, at the latest by the fifteenth day after mobilization, it would throw in the full weight of the majority of its forces' was whittled down to a limited move by nine divisions into the Saar on a front of sixteen miles with orders merely to close up to, but not attack, the German Siegfried Line five miles away. When Poland capitulated a month later they were withdrawn to the Maginot Line. At that time France had mobilized sixty-seven divisions and five British divisions had moved across the Channel to join them, while, as long as the Polish campaign lasted, Germany had forty-four divisions, of which twelve were regular, in the west. The only action taken by the French armed forces between then and May 1940, when Germany launched its offensive, was intervention with Britain in Norway, initially intended to help Finland when it was attacked by the Soviet Union. Offensive air action of any kind against targets in Germany was forbidden, for fear of retaliation against French cities.

When the blow fell on France itself, the balance of forces in the west was very different from what it had been nine months earlier. Out of the German army's 157 divisions, 136 were available, of which about one-third could be considered first-class. Among them were 10 Panzer divisions, totalling some 2,500 tanks, but only 627 of them were medium tanks: the rest were light tanks, armed only with a heavy machine-gun. Against these Gamelin could field 94 French divisions and 10 British, and hoped also to involve the 22 Belgian and 10 Dutch divisions. He had 3,160 French tanks including 800 of the new heavy 'B' and medium Somua type, both of which, in armour and armament, were superior to the German medium tanks: their failing lay in having only one-man turrets and no radio sets in four-fifths of the total. Some 1,300 of the tanks were distributed among four armoured and three light mechanized divisions, the remainder in individual battalions scattered about in support of infantry divisions. Gort's BEF had only 100 infantry support tanks and 350 light reconnaissance tanks. In artillery the French army outnumbered the Germany by 11,200 to 7,710 pieces, but was greatly inferior in numbers of anti-tank and anti-aircraft weapons. Only sixteen divisions had received the good 47-mm anti-tank gun; others had to rely on the old horse-drawn 25-mm. The anti-aircraft situation was even worse: 1,500 heavy guns were available, almost all 75-mm of 1918 vintage, of which most were in army reserve, while, for light anti-aircraft weapons, only twenty-two divisions had twelve 20-mm heavy machine-guns each, and thirteen others six new 25-mm, with which to deal with the 1,300 bombers and 380 dive-bombers, escorted by 1,200 fighters, which the *Luftwaffe* could produce.

Although the French air force had a total of 4,802 aircraft world-wide, most of them were short-range reconnaissance machines, and a very large number were obsolete or obsolescent. Sources of authoritative figures of the number of modern aircraft available for operations in France in May 1940 conflict, but it appears that approximately correct figures are 140 bombers, 660 fighters and 500 reconnaissance aircraft. Both fighter and reconnaissance were distributed in support of the army groups, with greater emphasis on the two northern ones, the First, covering the Belgian frontier, and the Second the Maginot Line. The RAF in France had 200 bombers, mostly the slow and vulnerable *Battles*, 130 fighters including two squadrons of *Hurricanes*, and fifty reconnaissance aircraft in 'army co-operation' squadrons.

The poor state of the French air force stemmed from a variety of sources. As the army's *Aéronautique Militaire*, it had been left with a large stock of aircraft in 1918, and was not favoured by the army when it came to distributing the meagre sums of money available for new equipment. The 1921 *Instruction provisoire* emphasized that the task of aircraft, like that of tanks, was to give direct support to the army, particularly to the artillery, with a strong emphasis on observation. This role was confirmed by experience in the Riff War, in which aircraft played a significant part. In 1928 a separate Air Ministry was formed, as much for the development of civil as of military aviation. In the face of strong opposition from the army supported by the navy, the *Armée de l'Air* was formed as a separate service in April 1933, headed by General Denain, but the influence of Pétain, backed by Weygand and Gamelin, prevented implementation of his proposal for a strategic bomber force separate from the forces supporting the army. It was not until 1936, after Pierre Cot had been Air Minister for three years, that all the formations of the air force were grouped into a general air reserve. In 1937 General Jouneaud, Chief of the Air General Staff, tried to obtain the approval of the *Conseil Supérieur de la Défense Nationale* to a draft *Instruction* which suggested that the air force could itself neutralize an enemy army, navy or air force by direct action against them, attack on their bases and communications, and on the production resources supplying them. For this a large fleet of heavy bombers would be needed. The *Instruction* also suggested that offensive action against the enemy air force, on the ground and in the air, was preferable to emphasis on defence of specific areas and that the intervention of the air force in the ground battle was best effected by support of mobile operations, including the use of airborne troops. The *Conseil* flatly rejected this concept and gave priority to expenditure on the navy. In 1938, when Guy la Chambre replaced Cot and General Vuillemin became Chief of Air Staff, he was more successful in obtaining money and some of it was devoted to buying aircraft from America, as production in France, which had not been improved by Cot's nationalization of the air industry, was slow and severely interrupted by strikes. The general structure of the air force was not changed and too high a proportion of the newer aircraft were in training units or in storage. At the outbreak of war the force had no radar, and its airfields were ill-defended against air attack.

Although, therefore, the French forces facing the Germans were not greatly inferior in quantity, they suffered from several inherent weaknesses, the chief of which were their devotion to a linear defensive strategy; their antiquated equipment and methods of command, communication and operation generally; their equally antiquated senior officers; their heavy reliance on large numbers of untrained reservists, and a general *malaise* among all ranks. These were compounded by strategic errors, the first, and principal, of which was Gamelin's obsession with a move forward of his left wing into the Low Countries. The bulk of the most modern part of the army, under his best generais, was devoted to this move: Giraud's Seventh Army, on the extreme left into Holland, and General Blanchard's First Army, on the right of the BEF into southern Belgium. All three DLMs were allotted to this and the infantry divisions were regular (i.e. peace-time formed with conscripts). The three armoured divisions (and a fourth forming under de Gaulle) were in reserve with fifteen infantry divisions. Even if the Germans had done what the General Staff expected, that is repeat their 1914 performance, the chances of this plan being successful were always slender: so much depended on the initial resistance by the Belgians and Dutch. Throughout the campaign the French underestimated the speed with which the Germans could act, basing their forecasts on their own ponderous methods.

For at least two days after General Paul von Kleist's Panzer Group had launched its thrust through the Ardennes towards Sedan, Gamelin and General Joseph Georges remained preoccupied with the situation further north, even after the latter realized, early in the morning of 14 May, that General Heinz Guderian at Sedan and Rommel further north near Dinant had crossed the Meuse. Both he and Gamelin suffered from two obsessions: the need to reconstitute any breach of a continuous front, and the threat to Paris, towards which they assumed that the thrust was directed. The result was the despatch of such reserves as were available piecemeal to different destinations. The inefficiency of the staff system, the dependence on movement by rail, and the vulnerability of the latter to air attack, all contributed to the fact that when, belatedly, reserves were directed to the right place, they arrived too late, almost always forestalled by the foot-slogging German infantry divisions with their horsed transport, following up to protect the vulnerable flanks, particularly the southern one, of the ever-extending narrow thrust made by Guderian's armour. Most of the reserves were used to plug holes in an attempt to maintain or create a continuous defensive line, few of them for counteroffensive action, and then never in sufficient strength or adequately supported by artillery or aircraft. The armoured divisions were sent, one by one, to different sectors and, for a variety of reasons, their action was ineffective. The Germans had no difficulty in holding off the ill-executed counterattack of the 3rd Armoured and 3rd Motorized Infantry divisions south of Sedan on 14 May, potentially the most dangerous threat they faced. The two attacks by de Gaulle's half-formed 4th Armoured Division, at Montcornet on 17 May and Crécy two days later, were made with a force of no more than 150 tanks, the crews of which had had little training, meagrely supported by artillery and infantry and not at all by the air force. Both were held without difficulty by the rear elements of the Panzer divisions, strongly supported by the *Luftwaffe*. The British attack round Arras on 21 May, again with insufficient force and ill-co-ordinated, with which the French on their left were meant to co-operate but did not (although some of General Prioux's Cavalry Corps did to good effect), caused temporary consternation to Rommel. It undoubtedly influenced Kleist on 22 May, to Guderian's dismay, to keep the 10th Panzer division in reserve, instead of directing it straight to Dunkirk, while the 1st made for Boulogne and the 2nd for Calais.

The general mismanagement of the battle was contributed to both by the structure of the French higher command and by the elderly occupants of its posts. At the summit was Gamelin, aged sixty-seven, Generalissimo of the army in all theatres of war, with his headquarters in the gloomy fortress of Vincennes in Paris where, according to de Gaulle, he 'dwelt in an atmosphere very akin to that of a convent, surrounded by only a few officers, working and meditating, completely insulated from current events. ... In his ivory tower at Vincennes, General Gamelin gave me the impression of a savant, testing the chemical reactions of his strategy in a laboratory.' He had no authority over the air force under General Vuillemin, 60 km away to the east at

Coulommiers. Responsible to Gamelin for 'the North-East Front' was the sixty-five-year-old General Georges, who had initially been both *Major Général des Armées*, that is land force commander in France, and Gamelin's deputy; but in January he had handed over the former function, with headquarters at Montry, 30 km east of Paris, to General Doumenc, and set up his own headquarters as C-in-C North-East Front in a large villa at La Ferté-sous-Jouarre, 20 km further north-east. Georges had never fully recovered from being wounded in the assassin's attack on King Alexander of Yugoslavia in October 1934, as he began a state visit to France at Marseilles. His air support was provided by General Têtu, subordinate to Vuillemin, co-ordinating the action of 'zones of air operations', corresponding to the sectors of army groups. Under Georges were three army groups: the 1st, under Billotte, covering the whole front north of the Maginot Line, in which there were four French armies, the BEF, and, after 10 May, the Belgian army. The peacetime General Staff was split up between Gamelin, Georges and Doumenc and the respective responsibilities of all three in regard to operations in Billotte's area were vague and ill-defined. Gamelin made no attempt to impose his will on Georges, nor Georges on Billotte, and the only orders that Gort ever received from either were the one to advance into Belgium to the Dyle, and then to withdraw from it to the Scheldt.

It was typical of the situation that when, on 18 May, Doumenc, disillusioned with Georges, suggested to Gamelin that he should take direction into his own hands, Gamelin's reply was, 'Of course, let me know the right moment'. But by that time Reynaud had taken the fatal step of summoning the seventy-three-year-old Weygand back from Syria and the eight-four-year-old Pétain from his post as ambassador in Madrid, the former to succeed Gamelin and the latter as Deputy Premier. Although Weygand initially showed a more lively and aggressive spirit and was prepared to try and exert control of the battle himself, the delay caused by the change and his unfamiliarity with the situation, leading to a lack of realism in his decisions, only made a hopeless situation worse. The French army and air force, the former demoralized by the general confusion, frequent attacks by the dreaded Stuka dive-bombers, and crowds of refugees, among whom alarmist rumours of every kind were rife, had lost heart. The air force, never concentrating its force as the *Luftwaffe* did, the theoretically vulnerable Stukas being covered by strong fighter escorts, had failed to provide any effective support to the army and had suffered heavy losses, in the air and on the ground, while its communication system, dependent on the telephone between headquarters and airfields, was only functioning intermittently. British air force losses had also been heavy, and on 20 May Air Marshal Barratt's force was moved to airfields in England.

With the failure of his attempt to organize a pincer attack between Arras and the Somme between 21 and 23 May, Weygand realized that the machine he commanded would not respond, and his sole care became how to end the war with honour. Both he and Pétain were primarily concerned to ensure that the army was preserved in being in order to avert internal chaos or subversion and so that there should be an effective French administration, capable of representing France in seeking an armistice and governing it thereafter. Against the pressure exerted both by these prestigious military figures and by his relentless mistress, the Comtesse de Portes, the increasingly exhausted Reynaud could not prevail and resigned on 16 June, handing over to Pétain, who immediately sought an armistice. He and Weygand opposed the proposal, backed by de Gaulle and Georges Mandel and by all the military commanders overseas, that the government and as many troops as possible should be transferred to North Africa and continue the fight from there. Admiral Darlan, Chief of the Naval Staff, at first supported it, but once Pétain had replaced Reynaud and decided to seek an armistice, he changed his attitude. He had however promised the British that he would not allow any part of the fleet to fall into German hands.

The unfortunate General Huntziger, commander of the Second Army which had been responsible for the Sedan sector, was deputed to negotiate the armistice terms on 21 June in the very wagon-lit and on the very spot near Compiègne where Foch, accompanied by Weygand, had received the German emissaries in November 1918. It was signed three days later. French casualties in the six weeks of disastrous fighting had been about 90,000 dead, 200,000 wounded and 1,900,000 taken prisoner or missing, while the Germans lost 27,000 dead, 111,000 wounded and 18,000 missing. British casualties of all categories totalled 68,000, Belgian 23,000 and Dutch 10,000.

The French navy had taken virtually no part in the campaign, and its future was of great concern to the British. This was not surprising, as Darlan had built it up into an efficient, powerful force, after it had been in the doldrums since 1918. It greatly resented the terms of the Washington Naval Treaty of 1922, which restricted the tonnage of its capital ships to a third of that of Britain, half of that of Japan and equality with Italy, which, considering France's widespread colonial commitments, appeared almost humiliating. Further naval conferences, at Geneva in 1927 and in London in 1930, failed to reach agreement, and the French navy greatly resented not even being consulted before Britain reached a naval agreement with Germany in 1935, which abandoned all the restrictions imposed on the German navy by the Versailles Treaty.

Georges Leygues, Minister of Marine from 1925 to 1930 and 1932 to 1933, was impressed by the vigour and intelligence of Darlan, whom he promoted to the rank of rear-admiral at the age of forty-eight in 1929. Thereafter he was the prime mover in modernizing and invigorating the navy, which tended to be even more conservative and sympathetic to the right wing than the army. In the shadow of Admiral Durand-Viel, Chief of Naval Staff from 1932 to 1937, a proper naval staff was created, backed by an *Ecole des Hautes Etudes Navales*, and a building programme embarked upon to provide six capital ships, six new cruisers and several more submarines and destroyers, and more attention was given to the fleet air arm. When Darlan succeeded Durand-Viel, he created the rank of Fleet Admiral for himself, and at the outbreak of war he had two first-class capital ships, *Dunkerque* and *Strasbourg*, each with eight 13-inch guns, two others, *Jean Bart* and *Richelieu* (eight 15-inch), nearing completion, three old battleships, modernized to carry eight or ten 13.4-inch guns, and two older ones with 12-inch guns.* The regular fleet was manned by 91,000 men, joined by 61,000 reservists on mobilization. One cruiser and two destroyers were lost off Norway, and seven in the Channel in May 1940. The navy's principal feat was to get almost all the Atlantic Fleet away from Brest, including the *Richelieu*, just completed, and the unfinished *Jean Bart* from St Nazaire, the former sailing to Dakar and the latter to Casablanca.

Most of the Mediterranean Fleet left Toulon for Mers-el-Kebir, but the armistice terms required it to return 'to ports to be specified and then to be demobilized and disarmed under German and Italian control'. The British government decided that, if that were to happen, they could not rely on Darlan's word, given personally both to Churchill at Briare on 10 June and to the British Chief of Naval Staff, Admiral Sir Dudley Pound, at Bordeaux on 17 June, that he 'would never surrender the fleet'. Even if he tried to keep his word, *force majeure* might prevent him from doing so. French ships in British ports, and at Alexandria, were disarmed, not entirely without incident, and a force from Gibraltar, commanded by Admiral Somerville, sailed to Mers-el-Kebir and presented Admiral Marcel Gensoul with the choice between bringing his ships over to join the Royal Navy and fight the Axis (Italy having joined in the war on 10 June), sending them with reduced crews to a British port, the crews to be repatriated and the ships restored to France after the war, or to sail, with reduced crews, under British supervision, to a French overseas possession, probably Martinique, where the *Béarn* already was, where they would be demilitarized or entrusted to the USA.

---

* These were supported by nineteen cruisers, including six modern ones carrying nine 6-inch with a maximum speed of over thirty-five knots; sixty-six destroyers, mostly modern; seventy to eighty submarines; and only one aircraft-carrier, the *Béarn*, a battleship converted to carry up to forty aircraft.

Gensoul refused formally to enter into negotiations, but received a copy of the terms and transmitted a version, omitting the third choice, to Darlan. Either he was told to refuse, or had not been given an answer before the British Admiral's time limit ran out. An hour and a half before that, Gensoul told Somerville's emissary that Darlan's promise not to let the ships fall into Axis hands would be kept, but that, if he were attacked, he would fight, and, after the emissary left, cleared decks for action and began to steam out of harbour. The old battleship, *Strasbourg*, got away to Toulon and some destroyers to Algiers, but most of the ships were casualties and 1,300 men were lost. These events and the subsequent abortive expedition, in which de Gaulle's Free French forces were involved, to Dakar in September, strengthened the anti-British feeling which had always been latent in the French navy. Darlan, in particular, felt that his word had not been trusted and gladly went along with the line, peddled by Vichy, that all their troubles were due to Britain, who had left them in the lurch and was planning to acquire their colonies. Having become Vice-Premier in February 1941, he had an interview with Hitler, following which the Council of Ministers accepted the policy which he outlined in the following words:

> If we favour English policy, France will be crushed, dislocated, and will cease to be a nation. ... If we try to carry out a policy of alternation between the two adversaries, Germany will cause a thousand difficulties in the exercise of our sovereignty and will stir up trouble. ... The peace will be disastrous. If we collaborate with Germany, without going so far as to place ourselves beside her to make war against England ... if we accord Germany certain facilities, we can save the French nation, reduce to a minimum our territorial losses, in metropolitan France and in the colonies, and play an honourable part in the future Europe.

It was in accordance with that policy that Darlan agreed with the Germans to let them use Tunisia for the supply of their forces in Libya, including selling them some of the stores of the French forces in North Africa; use Dakar as a submarine base; and use the airfields in Syria. The last provoked the British attack on General Dentz's forces there in June, but Darlan, sticking to his policy, refused the offer of German air support for them, which would have been the thin end of the wedge and which the Germans had been trying to insert since the armistice, and would involve France as an ally of Germany in war against Britain.

The terms of the armistice had allowed France to keep an army in the 'free' zone, that is, generally south of the Loire, of the same size and nature as Germany had been allowed by the Versailles Treaty – 100,000 men with no heavy equipment – primarily for the purpose of maintaining internal order. The first reaction of its senior officers, disapproved of by Weygand, was clandestinely to prepare for a resumption of hostilities by hiding stacks of equipment and stores, physically or by transfer to civilian organizations, and by trying to instil a more martial and patriotic spirit into the ranks than they had shown in action. As to the former, the Germans soon discovered what was going on and put a stop to it, although it continued on a limited scale. The senior officers had no great success either in the latter, as the majority of middle-ranking and junior officers had lost faith in their seniors, as the soldiers had in them, and the general public had in the whole military establishment. Some of the more charismatic senior officers, like General Jean de Lattre de Tassigny, managed to keep the idea of a resuscitation of the army's war effort alive, helped by Germany's reabsorption of Alsace and Lorraine; but those who wished to continue in action either escaped to join de Gaulle, or associated themselves with one of the resistance movements. The army had no wish to see either the communists or de Gaulle gain control of resistance movements, and set up its own resistance organization, which at the time of the Anglo-American invasion of French North Africa on 8 November 1942 was preferred by the Allies, especially by the Americans, to the *Partisans* and *Franc-tireurs*. The organization was however taken by surprise and, unrealistically, expected the Allies to follow it up rapidly by landings in France. In expectation of this General Verneau, who in the Army's General Staff had been in charge of secret arrangements since 1940, alerted all senior commanders and moved the staff out of Vichy, while Pétain ordered the forces in North Africa to resist and Laval went off to see Hitler, who offered him a full-scale alliance. Pétain rejected both the suggestion that he should fly to North Africa himself and Hitler's offer, and while Laval was still away the Germans, at dawn on 17 November, invaded the free zone. Pétain protested and announced publicly that the German action was 'incompatible with the armistice conventions'. This was interpreted by Darlan, who was in North Africa, as releasing the forces there from their duty to observe Pétain's previous order to resist the Anglo-American incursion. It appears that he had in fact received a secret message from Pétain expressing his tacit approval of the action he had already taken, with the co-operation of General Juin commanding the army in North Africa, to bring about a cease-fire. Prompted by the Americans he had also sent messages to Admiral Laborde in Toulon 'suggesting' that he might sail to North Africa. The question which then immediately arose for the officers in North Africa was where their loyalty lay. The Americans were trying to impose the authority of General Giraud, who, taken prisoner after succeeding General Corap in command of the Ninth Army in May 1940 and having escaped from the castle in which he had been held, had been allowed to live at home in the free zone where he had been contacted by the Americans. Under the impression that he would be placed in command of all Allied forces in North Africa, which would almost immediately land in southern France, he had left home and been conveyed to Gibraltar by a British submarine. On 13 November an agreement was reached between Eisenhower, Darlan and Giraud that Darlan would be recognized as exercising civil authority over Algeria, Morocco and Senegal, through their Residents General, Chatel in Algeria, Noguès in Morocco and Boisson in Dakar, while Giraud would be C-in-C of all French armed forces in the area. The army remained under command of Juin, who had been instrumental in bringing about this agreement.

Meanwhile, as Vichy passively accepted the German occupation of the unoccupied zone, General Verneau and subordinate commanders returned to their barracks. The army in France now had even less *raison d'être* than before, and a steady trickle of officers left to join former colleagues, most of them to North Africa. De Lattre was arrested, but escaped and joined de Gaulle. On 27 November, after the Germans had demanded the withdrawal of troops from the fortified area of Toulon, Laborde scuttled his fleet.

The arrangement made with Darlan in North Africa not only produced a sharp reaction in Britain and America, but was anathema to de Gaulle, already angry at having been kept in the dark about the whole operation. He had accepted that it was sensible to deal with Darlan over a cease-fire, but regarded as intolerable an agreement which not only kept in office men who had been loyal, even enthusiastic, supporters of Vichy, but preserved all Vichy's objectionable laws and decrees, including anti-Semitic ones. When Darlan was assassinated on 24 December, the Americans recognized Giraud as his successor, and the latter wrote a foolish letter to Pétain, saying that he would never act against the Marshal's policies. De Gaulle was incensed, as he saw the Americans encouraging a process by which Vichy was still recognized as the fount of authority, legitimacy and loyalty, undermining the whole basis of his Free French movement. By a combination of intransigence and adroit manœuvring, he turned the tables on Giraud, first accepting that the Free French forces should be subordinate to Giraud as military C-in-C, then getting Giraud to accept that he was subordinate to the Free French Committee of Liberation. He then got the Committee to agree that de Gaulle alone, and not the two together as had been agreed since the Casablanca conference in January 1943, should preside over it.

What were the Free French forces? They were exiguous and not wholly French, in a strict use of the term. Only a very small proportion of the French army, air force and navy, who found themselves in Britain after the armistice, opted to join de Gaulle rather than be repatriated. In July 1940 the total was 7,000, a high proportion of them officers. The great majority of the army which gave its allegiance to de Gaulle came from French colonies, other than Indo-China, which was occupied by the Japanese, most of them from south of the Sahara, where they were beyond the reach of Vichy and economically dependent on neighbouring British colonies. De Gaulle's claim to lead his own active army rested on three very small forces, all serving with Montgomery's Eighth Army: Brigadier-General Marie-Pierre Koenig's 1st and Colonel de Larminat's 2nd Free French brigades (which de Gaulle insisted on calling divisions) and the force of a few companies which Leclerc* brought across the desert from Chad to join hands with the Eighth Army at Tripoli. Although most of the officers came from Metropolitan France, the soldiers were almost entirely either foreign legionnaires or natives of the colonies. Of the infantry battalions which formed Koenig's brigade, the heroes of Bir Hacheim, the reputation of which did so much to boost de Gaulle's cause, two were Foreign Legion and the third Senegalese. The fourth, a Pacific Ocean Islands battalion, was reserved for duties in the rear, and operational command was exercised, not by Koenig himself, but by Colonel Amilakvari, a Georgian. For obvious reasons the navy did not play a large part: only one admiral, Emile Muselier, joined de Gaulle and, at one stage, he tried to pull his rank (Vice-Admiral) and usurp the upstart *général de brigade*.

Although only 300 men and 6 aircraft strong in July 1940, the air force had become an important part of de Gaulle's forces by January 1942, by which time two squadrons were operating Spitfires from Turnhouse in Scotland. They were too apt to be regarded, and regard themselves, as RAF squadrons for de Gaulle's liking, and, to offset this, a squadron was formed to operate with Soviet forces in Russia. These squadrons, however, were greatly outnumbered by those which had been available to Vichy in North Africa. Many of them had been transferred from France before the armistice had been signed in 1940 and, after Mers-el-Kebir, the Germans had agreed that they should remain active in order to defend the fleet in other African ports. An active air force was maintained in the free zone, but was disbanded by the Germans when they occupied it, many of its personnel making their way to North Africa.

In the forces that were now united under Giraud, acknowledging the authority of de Gaulle's Committee of National Liberation, those who had originally rallied to de Gaulle were very much in a minority. At the Casablanca conference the Americans had agreed to supply equipment for an eventual French force of eleven divisions and twenty squadrons of aircraft. It was to be some time before the men to man them all could be found, those immediately available, apart from the few thousand in Montgomery's army, being the forces inherited from Vichy. In the final battle for Tunis, most of them were grouped in

---

* He was really Jacques, Vicomte de Hautecloque, but used his surname in order to protect his family, who lived in the occupied zone.

General Koeltz's 19th French Corps, which had one Moroccan and two Algerian divisions. The French Expeditionary Force (or First Army as they called it), which Juin took to join General Sir Harold Alexander's forces in Italy, was much the same size, and consisted initially of two divisions, one Algerian and one Moroccan, later increased by the addition of a Moroccan mountain division and a motorized division with European soldiers. Its operations, under command of General Mark W.Clark's Fifth US Army, spearheaded the breakthrough into the Liri valley in May 1944, which led to the fall of Rome.

While Juin's African soldiers were adding laurels to de Gaulle's crown in Italy, de Lattre, in North Africa, was building the force destined to play a major part in the landings in southern France, for which Juin's forces were withdrawn from Italy, to the annoyance of the British, and transferred to de Lattre, giving him seven French divisions for his First French Army which landed in Provence, as part of Operation DRAGOON, on 15 August 1944. Ten days later Leclerc's 2nd French Armoured Division entered Paris.

The participation of French forces, acknowledging the authority of the Committee of National Liberation, was of major political importance to de Gaulle. His concern was not so much fighting the Germans – the British and Americans had ample resources for that – but to fight the tendency of the Americans to treat France as an occupied country in which, in the absence of a legitimate elected government, the writ of the Allied Supreme Commander would run. His threat to remove de Lattre's army from Eisenhower's command, when he was ordered to withdraw from Strasbourg in January 1945, was symptomatic of this. Equally important to him was the need to keep that army involved in the final operations in Germany in order to back his claim for a seat at the table when the future of Germany and of Europe was to be decided, a claim that was not successful, although a compromise was found over the garrison of Berlin and the stationing of French troops in Germany.

While de Gaulle's attention was focused on restoring France's standing as a major power, the armed forces faced a dual problem: their standing vis-à-vis the public, and the internal problem of reconciliation between the different elements: those who had been prisoners of war; those who had joined de Gaulle in the early days; and those who had served in the Vichy forces, which included those in North Africa with Juin, who had moved easily from one allegiance to another; those who had been associated with some form of resistance in the free zone; those who, after the Germans had occupied it, escaped to join the Free French forces; and those who passively remained in France. In addition to these was the resistance proper, the French Forces of the Interior (FFI), 137,000 of whom had been absorbed into de Lattre's army after he had landed in France. The communists, who achieved 26 per cent of the vote at the elections for the Constituent Assembly, demanded that the future army should be based solely on the FFI. There was little political support for spending much money on forces, which had not proved effective in defending the country, at a time when its whole structure and economy needed restoration and there appeared to be no military threat. De Gaulle's resignation in January 1946 had partly been provoked by the attempt of the socialist deputy, André Philip, to reduce the military budget. The measures taken by his successor, Felix Gouin, to limit it shattered the morale of the delicate structure of the forces by a '*dégagement des cadres*' which reduced the army's commissioned and non-commissioned officer strength by 45 and 40 per cent respectively over ten years, linked to an increase in their abysmally low rate of pay, tied to that of *petits fonctionnaires*. In reluctantly accepting this '*dégagement*' senior officers alienated their subordinates, further widening the gap between them and the lower ranks, inherited from the events of 1940. Morale was further lowered by frequent moves from one post to another, in France and overseas, and by unorthodox training introduced by de Lattre.

The lack of purpose, which contributed both to the absence of national interest in the armed forces and to their internal *malaise*, was, to a degree, remedied by the formation of the North Atlantic Treaty Organization (NATO). There was now a potential enemy, not merely the Soviet Union as a military power but communism world-wide: this chimed in with the sentiments of the majority of officers. The armed forces could, *en principe*, now be on the same political side as the government. However an issue soon arose which ranged them opposite each other: the rearmament of Germany. The USA made it clear that their support of the defence of Western Europe depended on finding a way by which the Germans would contribute to it. Objections to the principle of this came from the politicians and those who elected them, particularly on the left, which was a large proportion of the total. The military did not object to the principle, but to the method: that proposed in the Pleven plan for a European army, in which there would be no national units above the level of a battalion. In its attempt to prevent the creation of a German army, it would destroy the French. In the long-drawn-out negotiations and discussions, which ended in 1954 with the rejection by the National Assembly of the plan which it had accepted in principle in 1952, the senior officers of the armed forces, including the army which was most affected, were excluded from the negotiations and hesitant to voice their views publicly, until Marshal Juin, in March 1954, correctly sensing the political mood, came out openly against it and was dismissed for doing so. In return for a guarantee from Britain to station significant forces permanently on the continent, France accepted the admittance of Germany to NATO on a national basis.

While NATO had brought a sense of purpose, it had also produced frustrations for the French. It was dominated by the Americans, whose idea of co-operation was that their allies should adopt the methods of their own armed forces, the provision of military aid, on which France and most of the members of NATO, other than Britain, then depended for their re-equipment, being used as a lever to promote this. Their nuclear monopoly added weight to it. The only significant French post in the NATO military hierarchy was the command of the Central Sector, held by Juin, double-hatted with his post as Chief of Staff of the French Armed Forces, and it was totally overshadowed by the American Supreme Allied Commander Europe, with his headquarters on the outskirts of Paris. All the anti-American sentiments, which de Gaulle had felt so keenly throughout the war, were revived, and not confined to those who had initially been recruited under the Cross of Lorraine.

Germany was no longer seen as France's enemy, while America, of whom Britain was seen as an obsequious ally, was being regarded, culturally, politically, and in a certain military sense, as one. Communism, at home, in Eastern Europe where France's traditional friends Poland and Romania were held captive, and in all her overseas territories, was the real threat. By 1954 France was deeply involved in countering the activities of the anti-colonial, communist-backed movement in Indo-China. De Gaulle, in his ardour to restore France's position as a world power, had rejected the advice of Leclerc, sent there to command the forces, to come to an agreement with Ho Chi Minh, the leader of the movement in Tonkin. The ultra-Catholic High Commissioner, Admiral d'Argenlieu, was surrounded by a group of civilians determined to reassert French authority on the whole area, first by keeping the different provinces separate, and later by holding out the promise of independence to puppet regimes, through whom France could preserve her interests and counter the influence of the Viet Minh, which, although based in Tonkin, had established an organization covering all Vietnam. Successive senior army officers had little faith in this policy, but, particularly after Mao Tse-tung had established his authority over the whole of China in 1949 and the Korean War had broken out in 1950, became converted to the concept that the struggle was not one for the re-establishment of French imperial power, but France's principal contribution to the world-wide fight against communism. That precluded the sort of negotiation with the Viet Minh which Leclerc (who was killed in an air crash in 1947) had favoured. It was a cause which appealed to the romantic soul of de Lattre, who on Juin's recommendation was appointed High Commissioner and C-in-C in November 1950. He brought imagination and vigour to the campaign in the Red River delta, but, when he left a

year later, to die within a few months – his son having been killed
there – the situation had not basically improved. In spite of being
granted additional resources, especially of aircraft, his successor,
General Raoul Salan, who in the early days had been for negotiation,
never succeeded in gaining the initiative from General Vo Nguyen Giap,
a large proportion of his 175,000 men being tied to the defences of
the Red River delta which de Lattre had fortified. It was hoped that
the Vietnam National Army, built up to support Bao Dai's puppet
regime, would free the French forces, most of whom were either North
African or locally recruited, for offensive operations. This was the basic
plan, and when General Henri Navarre succeeded the discredited Salan
in May 1953, he expected to take the offensive with six or seven mobile
divisions when the rainy season finished in April 1954. A stepping-off
point for this was to be a fortified base at Dien Bien Phu on the border
between Laos and Tonkin. By his long period of inactivity and his
misjudgement of Giap's intentions and capability, he brought France's
involvement in Vietnam to a disastrous end. It was typical of the
nature of the army involved that, of the 7,184 casualties and 11,000
men taken prisoner at Dien Bien Phu, only 3,000 came from Metropolitan France. French casualties in Indo-China, in the whole period from
1945 to 1954, had totalled 92,000, of whom 20,000 were from Metropolitan France. Almost all were commissioned and non-commissioned
officers, the number of the former killed averaging one a day for over
seven years, 800 of them, graduates from St Cyr. Conscripts were not
involved.

1945–86

The war had a traumatic effect on the armed forces, particularly on
the army. The ill-starred direction of policy, pursuing one vain hope
after another, was blamed on the in-fighting which characterized the
political stage of the Fourth Republic, and the generals were blamed
for accepting it. The fact that all Frenchmen serving in Indo-China
were from the regular cadre of the army accentuated the feeling of
separation from the nation, engendered by the indifference of politicians and the public, and the hostility of the press, to the crusade
which they were fighting against communism. The enterprising officers
who, breaking away from the conventional methods of defensive and
offensive operations, took a leaf out of the communist book and established a close relationship with local inhabitants, particularly in the
mountains, led the field in proclaiming that the army, and the nation,
should adopt a 'total strategy' of revolutionary warfare that propagated
an anti-communist way of life by every means, cultural, economic,
political and military. That way of life was by no means a straightforward defence of the political democracy and economic capitalism

which prevailed in the Western world, but was tinged with the sort of ideas which had originally lain behind fascism and national socialism, though without the racial superiority and anti-Semitic elements which had characterized Mussolini's and Hitler's movements. The cadres in Indo-China felt more at home in the field with their comrades, French, Foreign Legionaries, African or Indo-Chinese, than they did in France itself. The seeds of discontent, which were to grow in Algeria, were sown in Vietnam. De Gaulle's insistence on regaining *grandeur* for France in Indo-China had produced a bitter *servitude militaire*.

One thought was shared by the army, the politicians and the public: that the humiliation and the waste of immense effort, human and financial, must not be repeated; but the army saw the answer to that in a different light from most of the politicians and the public. To the army it meant that, in other colonial problems – and North Africa was already proving troublesome – there must be no more shilly-shally-ing over policy, no more resting of hopes on broken reeds, and no more old-fashioned military methods unsuited to countering revolutionary warfare.

Less than four months after the armistice in Indo-China had been signed in Geneva, the Algerian *Comité Révolutionnaire d'Unité et d'Action* (CRUA) changed its name to the *Front de la Libération Nationale* (FLN) and launched a series of attacks in the Aurès mountains on 1 November 1954. Of General Cherrière's garrison of 57,000 men, only 3,500 could be made available for operations, and he was reinforced with Colonel Ducournau's 25th Parachute Division, fresh from Indo-China. The government of Pierre Mendès France was negotiating with Tunis and Morocco over proposals for some form of local autonomy and was quick to state categorically that there was no question of the same applying to Algeria, a department of France. But as operations made slow progress and the forces were increased – by February 1956 they had reached a strength of 200,000 – and it had become clear, from the examples of Tunis and Morocco, that any form of local autonomy led rapidly to total independence, the army's attitude hardened. By then almost all the soldiers in Algeria were veterans of Indo-China, the hard core being the regular parachute troops, the hardest of whom were those of the Foreign Legion, and almost half of these were German and German Swiss.

The army had had a love affair with North Africa since Bugeaud's campaign in the 1830s. Lyautey's example fitted exactly the concept of 'revolutionary warfare', of winning the hearts and minds of the people and not allowing them to be intimidated by a handful of extremists who, if they were not overtly communist, as the Viet Minh had been, were in league with them and whose success, if it were permitted, would open the door to communism. At the same time as acting ruthlessly against these extremists, with whom they were at war, the army must win over the Muslim population so that not only did it not demand separation from France but actively sought a closer relationship. In pursuit of this policy, and because many aspects of civil administration, including the police, did not function in areas where operations were in progress, the army increasingly took over the attributes of the civil authority, including the administration of justice, if it could be called that.

The major difficulty in executing this policy was how to reconcile it with support of the interests, security and wishes of the European population, who took a very different view from the army about what should be offered to the Muslims, although both were agreed that any form of autonomy or independence for Algeria must be resisted. It was primarily in this field that the army's policy failed, as the aspirations of the two communities in the towns were diametrically opposed, and urban terrorism, aimed at the European community, was the army's Achilles' heel. In the countryside the army had had a certain degree of success as a result of the introduction of *quadrillage*: dividing the country up into a large number of sectors, in each of which a policy of trying to 'swing over' the population and of intensive operations was instituted. As part of this policy, in order to separate the people from the FLN, they were herded into fortified areas in which it was intended that they would be provided with the amenities of civilization which they had previously lacked. Unfortunately they were more like concentration camps and achieved the opposite of what had been intended. The demands of this policy and the hostility of the Europeans prompted Guy Mollet, the Socialist Prime Minister, in February 1956, to increase the forces to 500,000 by the use of conscripts, whose service was extended to twenty-seven months, and by calling up reservists, a step which was to have significant results.

The *cadres* viewed this with some concern. They feared that conscripts would not be as dedicated as they were to the concept of gaining the goodwill of the 'population' and of driving them away from the clutches of the 'extremists'. A great effort was therefore devoted to a propaganda and education process to convince the conscripts that in Algeria they were fighting the same battle as they had been prepared to in France or Germany: that against international communism. Stress also was laid on the strategic importance of North Africa to NATO and to France itself, a theme which was strongly supported by the air force and the navy. This was in opposition to the widely held view that, unless concessions were made to the nationalists, they would be driven into the hands of international communism.

The arrival in Algeria of the conscripts emphasized the feeling of exclusiveness of the regular formations, particularly of the parachutists. The former were principally employed on *quadrillage* duties of a routine nature, while offensive operations against the formed bodies of the FLN were carried out by the latter, who regarded themselves as a force apart. It was they who provided the Special and Urban Administrative Sections (SAS and UAS), whose task was to set up 'parallel hierarchies' to those of the FLN. 'Handsome on parade, valiant in war, we know how to die' is how they described themselves.

The absence of General Jacques Massu's 10th Parachute Division, taking part in the Suez operation in October and November 1956, caused an interruption to offensive operations, of which the FLN took advantage. The French initiative which led to the Suez affair was directly related to events in Algeria, as they accused Gamal Abdel Nasser, who was supported by the Soviet Union, of fomenting and supporting anti-French movements in North Africa. The diversion of so much military effort to Algeria came as a severe disappointment to the elements in the army who were more concerned with Europe than with Africa. With the war in Indo-China over, they had hoped to be able to give priority to the formation of *Javelot* divisions, light armoured divisions based on concepts developed by General Beaufre, which it was hoped would be able to operate, highly dispersed, on a nuclear battlefield.

In the last few days of 1956 the police lost control of the situation in Algiers, FLN terrorist attacks provoking a violent backlash from the European community, and the Governor-General, Robert Lacoste, ordered the army, of which General Salan had recently taken command from General Henri Lorillot, to take over, Massu moving his division with its band of intensely dedicated and determined colonels into the city. It was a decisive turning-point in the campaign. The ruthless methods they employed, using the argument that swift, exemplary action and rough justice was the most effective way not only of defeating the FLN but of persuading the 'population' to keep clear of them, created an outcry in France, particularly on the political left, on which the government relied for support. There was little backing in France and in the international community for the uncompromising policy which the army and the Europeans in Algeria demanded, the first step away from it being the introduction of the *loi-cadres*, which would grant a carefully qualified degree of autonomy to an Algeria, in which all elements of the population would share power.

The army now faced the possibility that, just as their operations were beginning to show a real prospect of success, they were once more to be betrayed by the politicians. The hard-bitten 'paras' faced not only the prospect of 'the nightmare of a war fought for nothing' but also the possibility that, when it was over, they might be called to account for some of their more ruthless activities. They became more hard-line than ever and, in an attempt to escape from the dilemma of the apparently irreconcilable interests of the European community and the need to offer something to the Muslims to compete with the lure of independence, they plumped for *intégration*.

Algeria must become fully and totally part of France, all citizens of Algeria having the same rights and status as all citizens of France. They were not the only type of officer to be dismayed. The earnest conscientious officers of the SAS, who saw themselves as inheritors of the traditions of Lyautey, or perhaps Père Charles Foucauld, in their *mission civilisatrice*, were equally concerned that the cause into which they poured heart and soul might be undermined: theirs was the typical anguish of the paternal imperialist that all that he had striven for might have been in vain.

Political and international dislike of the way things were going, as the army assumed an ever greater dominance over policy in Algeria, was heightened by the incident of the bombing of Sakhiet, over the Tunisian border, in February 1958. The army, whose operations had considerably weakened the forces of the FLN in Algeria, was frustrated at the sanctuary which Tunisia provided to them, and the bombing took place, without political clearance, after their reconnaissance aircraft had been shot at. As a result of Sakhiet and other issues the government headed by Felix Gaillard resigned on 15 April, and a long period of political haggling ensued. In this political hiatus the army's decisive entry into politics occurred on 9 May. Salan, General Jacques Allard, his deputy Massu, and General Edmond Jouhaud, commanding the air force, sent a telegram to General Paul Ely, Chief of Staff of the Armed Forces, asking him to deliver it to the President, which ran:

> The Army in Algeria is disturbed ... concerning the French population of the interior, which feels deserted, and the Moslem French who, in greater numbers every day, have been once more placing their trust in France, confident in our reiterated promises never to abandon them. The French Army, as one man, would look on the abandonment of this national heritage as an outrage, and it would be impossible to predict how it might react in its despair.

They asked Ely to 'call the attention of the President of the Republic to our anguish, which could be removed only by a government resolutely decided to maintain our flag in Algeria'. Salan showed the telegram to Lacoste and told him that he intended on 13 May to lay a wreath in memory of three French soldiers recently executed by the FLN after their capture, at which he expected a large turn-out of local *anciens combattants*. In fact the whole affair had been set up by a group of colonels, in league with both Pierre Lagaillarde's 'Group of Seven' *colons* and the Gaullists, of whom Léon Delbecque was the local agent, sent over for that purpose by Jacques Chaban Delmas, the Gaullist Minister of Defence in Gaillard's government. De Gaulle had kept himself aloof from plots to bring him to power, and the army had its eyes on Georges Bidault, distrusting de Gaulle's Delphic utterances which gave no clue to his attitude to Algeria, which they suspected might not coincide with theirs.

A series of staged incidents in Algiers led to the establishment of a 'Committee of Public Safety' headed by Massu, now openly defying the government. President René Coty realized that only de Gaulle could exert the authority to prevent a civil war. As the General bided his time and the politicians desperately tried to prevent his advent, parachutists from the army in North Africa occupied Corsica and threatened to intervene in France itself, where they counted on the support of at least four of the military region commanders.

The imminent threat of a civil war, which would divide the army from the nation and divide both internally, allowed de Gaulle to return unconditionally to power at the request of the constitutional authority, and to receive from a somewhat reluctant National Assembly a mandate to rule by decree for six months and propose a new constitution. His utterances, when he visited Algeria within a week of being accepted as President, were almost as Delphic as ever, but appeared to give some support to the cause upheld by the army and the *pieds noirs*.

De Gaulle, however, was not primarily interested in Algeria, but in France itself. His aim was to take up where he had left off on 20 January 1946, when, in his farewell to his ministers as he resigned, he said, 'the exclusive regime of the parties has come back. I disapprove of it.' His aim was to get rid of a political system, which he believed to be fatally flawed, and to restore and strengthen France in the image he held of it. The Algerian problem had to be dealt with in that context. He had no love for the *grands colons*, the *petits pieds noirs* or for that element of the army that was more interested in colonial exploits than in France itself.

His first task was to bring the army under control, but he had to move carefully in dealing with the recalcitrants, loyally helped by Ely, whose abiding concern was to maintain the unity of the army. Salan, having been flattered by being made *Délégué Général* and awarded the *Médaille Militaire*, was shunted off, after a decent interval, into the honorific post of Military Governor of Paris. Other suspect officers were posted away from North Africa; Massu, a wartime Free French officer, not until January 1959, when, in an interview with a German press correspondent he said, 'Myself, and the majority of officers in a position of command, will not execute unconditionally the orders of the Head of State.'

To replace Salan, de Gaulle had appointed a general of noted ability, Maurice Challe. By extensive use of helicopters, airborne operations, air support and freeing more soldiers from static *quadrillage* duties for offensive operations, he achieved a significant improvement in the operational situation against the formed bodies of FLN armed men in the hinterland. De Gaulle had no illusions about the fact that he must, for France's sake, bring the war to an end, and this could only be done by negotiation with the FLN. He looked to Challe to see that, when he entered those negotiations, he had the strongest possible hand to play against the FLN, weakened militarily by Challe's operations and politically by what he had to offer to the majority of the population of Algeria. The fly in the ointment was the European minority. The crisis came as a side-effect of the referendum on the new constitution, when de Gaulle made a public offer to the Algerians of three options: complete secession, which would entail rupture of all relations with France, integration with France, or self-determination, which he described as 'the government of Algeria by the Algerians, backed up by French help and in close relationship with her', making clear that he hoped they would choose the last. The army was aghast, as was the European community, and Challe wrote to de Gaulle's Prime Minister, Michel Debré, 'One does not propose to soldiers to go and get killed for an imprecise final objective. ... One can thus only ask of soldiers of the army of Algeria today that they die in order for Algeria to remain French.' This crisis, heightened by the removal of Massu, led to the affair of the barricades in Algiers, forcing de Gaulle to come to the microphone in his uniform of *général de brigade*, to state categorically that 'the Algerians shall have free choice of their destiny' and that if he were to yield to those 'who dream of being usurpers, France would become but a poor broken toy adrift on the sea of hazard'.

From then on the aim of the army in North Africa, the generals pushed along by the parachute colonels, of whom Antoine Argaud, Jean Gardes and Yves Godard were the leaders, was to get rid of de Gaulle. Their hopes of 'swinging over' the Muslim population were rapidly fading and they became more and more closely identified with the *pieds noirs*. They were also greatly concerned about the feelings of their own men and of those Muslims who had put their trust in the continuation of a French presence, including the *harkis*, a form of locally recruited home guard. The 'nightmare of a war fought for nothing' was now becoming a reality, as de Gaulle made the first overtures for direct negotiations with the FLN, and all this at a time when Challe's operations seemed to be bringing about a decisive victory over the FLN's forces, most of which were now taking refuge in Tunisia.

Challe was removed in April 1960 to succeed General Etienne Valluy in Juin's old post as C-in-C of NATO forces on the Central Front, where he supported the line Valluy had publicly expounded, that, in opposition to de Gaulle's policy, France should become more integrated into NATO and that the future of Algeria should be decided in a NATO context, hoping that somehow other NATO nations could become involved. Meanwhile the extremist leaders of the Europeans, notably Joseph Ortiz, were turning increasingly to violent means and proposing fascist-type political solutions, which alienated the moderate elements in the army, principally the officers who had had links with the Free French or had radical or socialist leanings, and the conscripts. The army was being split into two political factions, to the dismay of Ely.

This was accentuated by the campaign, initiated on the political and intellectual left, to try to suborn the conscripts and encourage them to desert or disobey their orders. The demonstration which accompanied de Gaulle's visit to Algeria in December 1960, in which he evaded attempts both to kidnap and to assassinate him, had provided clear evidence both of the strong and well-organized Muslim support for independence and of the lengths of violence to which the *pied noir* FAF (*Front de l'Algérie Française*) was prepared to go. The former was confirmed by the referendum of January 1961, which returned a 75 per cent 'Oui' to the question 'Do you approve the Bill ... concerning the self-determination of the Algerian population?' The truce, imposed while negotiations with the FLN began, was the last straw which drove the plotters to put in hand the conspiracy which resulted in the attempted coup of 22 April, involving Challe and Salan. De Gaulle defeated it by once more appearing in uniform on television, claiming the loyalty of the nation: 'The nation defied, our strength shaken, our international prestige debased and our role in Africa compromised. And by whom? Alas! Alas! By men whose duty, honour and *raison d'être* it was to serve and obey.' With great force and vigour he forbade all Frenchmen, and especially soldiers, to give them aid and appealed to all Frenchmen to help him.

His words had a decisive effect on the waverers who, especially among the senior officers, were numerous. In their hearts they sympathized with those who fought to keep Algeria French, and most of them had no great love for de Gaulle, having been either in the Armistice army under Vichy or prisoners of war; but they were highly sceptical of Challe's chances of success, and what then? As they prevaricated, the conscripts and many of the junior officers, who had heard de Gaulle on their transistor radios, made it clear that they would not obey any orders to support the conspiracy. Challe threw up the sponge and flew to Paris to surrender, while Salan, arriving at the last moment from Spain with General Marie-André Zeller, a former Chief of Staff of the Armed Forces, the Algerian-born General Jouhaud and Colonel Dufour, joined the OAS (*Organisation Armée Secrète*), the FAF's underground army. It was a victory for de Gaulle, but his hand was fatally weakened in negotiations with the FLN, to whose unflinching demands he had to accede before the agreement was reached in March 1962 which gave Algeria its independence in July. The army had defeated itself.

Having rid himself, France and the army of the burden of Algeria, de Gaulle was able to turn to his main task, the restoration of France in the image in which he conceived it, the armed forces occupying a central place. His outlook had not changed since 1946, and fundamental to it was the assertion of France's independence and status as an equal of the USA and Britain. France's first atomic weapon, work on which had been authorized by the Socialist Mollet in the aftermath of the Suez affair, had been exploded in the Sahara in 1960, providing de Gaulle with a powerful lever to support his policy. He was determined to extract the French forces from the Anglo-Saxon-dominated military machine of NATO, although he wished to remain within the alliance 'as an insurance'. His public justification for this was the refusal of the USA and Britain to agree to his proposed three-power directorate, which would control the use of nuclear weapons, aggravated by the American decision to supply the Polaris missile to Britain. He appeared little concerned with the military effect on NATO's capability to meet the threat of Soviet aggression. He thought it unlikely in any case, and was prepared to leave that task to the Americans, British and Germans. If they failed, or looked like failing, France would make its own decision. The function of the French armed forces should be to enhance the strength and independence of France and its ability to pursue a policy distinct from that of the Anglo-Saxons. In that light, the militarily absurd nuclear strategy of facing 'tous azimuts' was logical.

Ever since then the organization and equipment of the French armed forces has not been primarily designed to fight those of the Soviet Union. As far as the equipment is concerned, it has resulted in a policy which has placed highest priority on its potential for world-wide arms sales, often to the annoyance of the French armed forces, who realized its shortcomings for use against Soviet forces in Europe – their tanks and aircraft are good examples. This policy has also been reflected in organization and in attempts to develop a coherent military strategy. For some time considerable emphasis was given to the Force d'Intervention, a regular force which could intervene, and has on several occasions, to support French interests or policy in former colonial territories. It has proved more difficult to devise a strategy for the French Army Corps, stationed across the Rhine in Germany, and the rest of the First French Army which backs it up, which was both rational and acceptable to the Federal Republic, particularly when the French army was provided with its own nuclear weapons. Since de Gaulle ceased to be President in 1969 and even more so since his death the following year, France's attitude to the co-operation of its forces with those of other members of the Atlantic Alliance has softened from the hard line taken in 1966. For some time a close liaison has been maintained both with the American Supreme Allied

Commander Europe and the German C-in-C Centre, on the assumption that French forces will be available as a reserve in the event of a Soviet attack. The navy had all along maintained direct links with the headquarters of the British C-in-C Fleet, who is also both the NATO C-in-C Channel and a subordinate, as Commander Eastern Atlantic, of the American NATO Supreme Allied Commander Atlantic; and the air force had established direct links with the NATO air defence networks.

The impetus for closer co-operation came from two sources: the professional service officers who saw the need for it, if their forces were to make operational sense, and the fear that France was in danger of losing opportunities in the arms supply market within NATO, which was dominated by the Americans. While adhering rigidly to the principle that France would not participate in any part of the military organization of the Alliance, the French government has constantly urged the need for greater European procurement of arms and made several attempts to resuscitate the almost defunct Western European Union, in order to develop a specifically European view of defence, although she consistently refused to join the Eurogroup, formed within NATO for just that purpose. But when joint procurement projects were discussed, the fact that the French appeared to have their eyes more on arms sales outside the Atlantic Alliance than on developing equipment to match that of the Soviet forces, often proved a stumbling-block.

The urgent task facing de Gaulle in 1962 was to restore unity, discipline and morale to the armed forces, primarily in the army. A number of officers were court-martialled, others dismissed and many more posted elsewhere, while some of the units which had taken a leading part in the rebellion, notably Dufour's 1st Foreign Legion Parachute Battalion, were disbanded. But the emphasis was on healing the wounds rather than on retribution, the final act being the amnesty in 1968 after de Gaulle's extraordinary flight to see Massu, then commanding the corps in Germany, to seek assurances of the army's support when the student riots made him fear another *commune*. An important act in this respect was to try to enhance the importance of each service by giving it a nuclear role, then thought to be the status symbol of a modern force. Although popular initially, this has become less so with the army, which, although it has been given its own missile, has seen high priority and a significant slice of defence expenditure allotted to the navy's ballistic missile submarines and to the air force's aircraft and missile systems, the latter, based on the Plateau d'Albion proving to be a veritable white elephant in the days of the Multiple Independently-targeted Re-entry Vehicle (MIRV). In 1985 these strategic nuclear forces absorbed 20 per cent of the defence budget and 5.5 per cent of the manpower of the armed forces, 16,700 regulars and 10,900 conscripts.

In 1973 conscript service was reduced from two years to one, the number called up annually being about 250,000. Until 1977 the field army was divided between the *Force de Manœuvre*, generally designed for European warfare, and the *Force d'Intervention* for use elsewhere, the latter being air portable. The reorganization put in hand in that year abolished the distinction between the two to provide fifteen small divisions, in fact of the size of a brigade, of which three, all armoured, were stationed in Germany. A more recent reorganization has divided the army into two major formations, 1st Army and the Rapid Action Force (FAR). The former has three corps, one in Germany, all but two of its ten so-called divisions being classified as armoured. The FAR has five so-called divisions, one light armoured, one marine, one parachute, one alpine and one air-mobile. It is intended for use either in Germany or overseas, where some 17,000 men are stationed in French overseas territories and former colonies. It is well-equipped with helicopters. The army has some 342,000 men, of whom 114,000 are regulars. The army's air support is provided by a tactical air force of 270 combat aircraft, equipped with Mirage IIIs and Jaguars. The strategic air force mans 21 supersonic Mirage IVs and the 18 S2 ballistic missiles, while air defence of France is provided by 135 Mirage FIs, 15 FIIIs and 15 Mirage 2000s, which will eventually replace the others. The air force is manned by 96,500 men, two-thirds regulars.

The navy has built up an impressive force, in spite of much of its funds having been devoted to its six ballistic missile nuclear submarines (to be increased to seven). Not only does it operate three aircraft-carriers, two with fixed-wing aircraft and one with helicopters, but the construction programme includes two nuclear-propelled ones. It has sixteen attack submarines, of which two are nuclear-powered, the construction programme providing for one new nuclear-powered submarine a year, forty-two other surface combat vessels, fourteen patrol craft, and twenty-nine minehunters and sweepers, manned by 32,804 sailors, three-quarters of them regulars.

De Gaulle's military policy of independence, conventional and nuclear, has been followed by his successors, for it has the supreme political value of uniting left and right: both wings are fundamentally anti-American. This has obvious advantages for the military, their existence, needs and policies no longer being the centre of a political argument, threatening the unity of the forces themselves; but it has disadvantages also. Not only do they feel isolated internationally, unable to take their share either of prestigious Allied commands or in debates on Allied military policy, but they are faced with difficulties in developing strategies which make sense to their intelligent subordinates. Knowing that France cannot defend herself by herself, they are forced to fall back on the unsatisfactory argument of the importance of ensuring that the President retains the freedom to make his own decision for France, affecting both the nuclear and the conventional forces, in the light of circumstances at the time. The more they stress the effectiveness of the *Force de Dissuasion* in ensuring that France will not be attacked, the more difficult it is to justify any French forces except those intended for use outside Europe. The general result is to think up organizations and methods of employing and equipping them which they maintain have a relevance for European warfare, but in fact derive their origins from the development of equipment for use elsewhere. In the 1980s, although there has been no deviation from the principle of national independence in defence, both political and military leaders have shown a greater willingness to co-operate with the USA and NATO and a concern to avoid offending German susceptibilities in defence matters.

3

GERMANY

1900–18

In 1900 Germany had every reason to be confident in her armed forces and her general military situation, both fundamentally based on the military machine which Frederick the Great of Prussia had created in the eighteenth century, and which had been transformed, after Prussia's defeat by Napoleon at Jena in 1806, from the private army of the monarch into a national army, in imitation of the French. Gerhard von Scharnhorst and Graf Augustus von Gneisenau had been the architects and builders, while Clausewitz provided the theoretical basis for its employment. But after Waterloo, to which Field Marshal Gebhard von Blücher's contribution was decisive, the Prussian army saw no action until the triad of Kaiser Wilhelm I, Otto von Bismarck as his Chancellor, and Helmuth von Moltke as his Chief of the General Staff, created modern Germany by their victories over Denmark, Austria-Hungary and France. Although the German army, from 1900 to 1914, was in fact four armies, those of Bavaria, Württemberg, Saxony and Prussia, it was the last which dominated the whole, not only by its size and by the political dominance of Prussia, whose King was the German Emperor, but, very significantly, by the existence and prestige of the Prussian General Staff. The army was seen as the embodiment of the Reich. Ambitious officers of the other armies were only too anxious to gain admittance to the General Staff, and the Prussian leaders of it appreciated the value of placing Prussian members of that military élite in influential positions within the other armies.

That formidable organization was the brainchild of Scharnhorst, determined to emulate Napoleon's staff organization and improve on it by a typically German attention to methodical detail. Others, including Gneisenau, Blücher's Chief of Staff, Kleist and Bernhard von Bülow, contributed to its development, few more significantly than Clausewitz by his posthumous provision of an extremely detailed theoretical foundation. But the true credit must go to the elder Moltke, its head from 1857 until 1888. He was a cultivated intelligent man, who appreciated the subtlety of Clausewitz's analysis of the essence of strategy and realized the revolutionary effect of modern technical develop-ments, which had followed the publication of *Vom Kriege* in 1832, notably railways and the electric telegraph. The defeat of the Austrian army at Königgrätz (Sadowa) in 1866, which resulted from his judicious manœuvring by rail, organized from Berlin by telegraph, established the dominating influence of the Prussian General Staff and the position of its Chief. Both were reinforced by the defeat of France at Sedan five years later, although the events which followed that encouraged an arrogant attitude on the part of the General Staff to political direc-tion, which was to have a fatal influence in later years.

Bismarck has often been thought of as a brutish, autocratic Junker, trampling Prussia's neighbours below the jackboot's heel. Neverthe-less, he was above all a practitioner of *realpolitik*. He realized that the creation of a German Reich, endowed with a combination of sterling human qualities and important mineral, and, to a lesser extent, agrar-ian resources, would arouse strong enmities in its neighbours, about whose strengths and weaknesses he had no illusions. He wished to remain on friendly terms with Russia, to ensure that Austria-Hungary did not drag Germany into involvement in Balkan adventures and that Britain should not side with France, whose desire to regain Alsace and Lorraine inevitably made her a potential enemy. This prudent policy was undermined in his last years as Chancellor, and then reversed after his dismissal in 1890, two years after Kaiser Wilhelm II came to the throne. Economic factors played some part, Russia being antagonized by tariffs imposed on her agricultural products to protect German agrarian interests, while Britain was antagonized both by com-petition from Germany's powerful industrial cartels, which had been built up behind protective barriers, and by her naval and colonial ambi-tions. But the prime influence came from within the German armed forces, encouraged by the ideas propagated by generals like Bernhardi and von der Goltz. The worst offender was General Alfred, Graf von Waldersee, who, before he became Chief of the General Staff in 1888, had been its most influential member in the elder Moltke's declining years. He was typical of the reactionary Junker outlook, which, in con-trast to the enlightened views of Scharnhorst and the elder Moltke, wished at one and the same time to assert the claim of German-speak-ing people to a more powerful position in Europe and in the world generally, while deeply mistrusting not only all Germany's neighbours but also those elements in her own population which provided her economic strength, the industrialists and their workers.

Unable to preserve the officer corps as an exclusive aristocratic body, the army opened its ranks to that element of the bourgeoisie which, especially after the victories of 1864–71, looked to a commission in the army, regular or reserve, as a sign of social status. This had accen-

tuated the division between the army and liberal elements in political
and intellectual life. In Bismarck's day there had been a running battle
between the army and both the Prussian parliament and the Reichstag
over parliamentary control of military expenditure. Wilhelm II encouraged his generals, and was himself encouraged by them in the attitude
expressed in its extreme form by General Hahnke, Chief of the Military
Cabinet: 'The army must remain an insulated body into which no
one dare peer with critical eyes.' In their eyes the army's task was
to uphold the monarchy and the traditional values of the Prussian
landowners associated with it, in its control of as many of the German-
speaking people as possible, and in extending it to assert Germany's
international position as an equal of the older European powers,
Britain, France and Russia. Wilhelm II was a willing accomplice,
encouraging Waldersee in his intrigues, not only to by-pass the Foreign
Ministry through direct contact with military attachés, but also to
evade or diminish any form of civilian influence on the conduct of
policies affecting the army. The ghost of Waldersee's legacy of intrigue,
directed against politicians and civilian influence of any kind, was
to stalk the corridors of the German General Staff for years after his
death in 1904. One of the factors encouraging intrigue of this nature
was the division of responsibility for the affairs of the army between
the Chancellor's office, the Prussian War Minister, the Chief of the
General Staff, the Chief of the Military Cabinet and Wilhelm II's creation of a Royal Headquarters, consisting of generals and aides-de-camp,
who accompanied him wherever he went and acted as a channel to
and from him on military matters.

After Waldersee's dismissal in 1891, Leo von Caprivi, himself a
general, who succeeded Bismarck as Chancellor in 1890, managed with
difficulty to exert a sobering influence, opposed by those who saw
any concession to the political parties as a form of treason. His reduction of conscript service from three years to two in return for parliamentary approval of an increase in the peacetime strength of the army
of 65,000 (bringing it to 589,000) was an example of this, leading to
his resignation in 1894.

Waldersee's successor, Schlieffen, was cast in a very different mould:
studious, hard-working, a master of detail with a penetrating insight
and interest in military theory and the application to it of modern
technical developments. Far from wishing to indulge in political intrigue, his fault was to ignore political factors, especially the international ones. He underestimated the importance of the international
political implications of his plans and failed to suggest, as a more
politically minded general might have done, that there were diplomatic or political solutions which might more easily, and at less cost,
remove the threats which, in spite of Germany's military strength,
caused him great concern. They were the product of the policies which

his predecessor had pursued. One was external: that of a two-front war against Russia and France, the latter perhaps also involving Britain. The other was internal: that of social democracy, the threat from which was regarded by the army at that time in the same light as the capitalist world regards communism today, any combined activity by 'workers' being suspected of being an incipient revolution.

This dual threat meant that a war must be short. Not only was it thought that modern economies could not support long wars, but a long-drawn-out affair could not be decisive. The potential opponents, particularly if combined, could mobilize greater manpower and resources generally, and a long struggle, perhaps involving hardship for civilians, could breed revolution among the workers.

Moltke's plan had been to stand on the defensive initially in the west, while carrying out a limited offensive against Russia, inflicting sufficient damage on her to prevent her from attacking Austria-Hungary and thus forcing Germany to commit large forces to the latter's support. That having been done, the efficient railway network, designed primarily for military purposes, could switch the bulk of the army westward to deal a knock-out blow to France. Schlieffen reversed the order of events. He feared that a campaign in Russia, even if intended as limited, could become long-drawn-out and, while it was in progress, the French might regain Alsace and Lorraine. The essence of Moltke's strategy had been to bring about, by railway movement directed by telegraphic orders, a last-minute concentration of superior forces which would surprise the enemy, the different army commanders being left free to use their initiative to exploit the situation created by this concentration. That had brought success at Königgrätz, although it had also risked some confusion and lack of co-ordination. Schlieffen's reflective and methodical mind, together with the more complicated organization of the army thirty years later, led him to try to impose a firmer control on a campaign which could not rely so much on the chance of battle. Conscious of the speed with which France had recovered from her defeat in 1871 and obsessed by the need for a short, decisive war, he sought to emulate Hannibal's victory at Cannae and encircle the whole French army in its defences, by a wide out-flanking movement hinged on Metz, the right flank being protected by resting on the Channel coast. This ambitious plan, as completed in 1905, would employ eight armies of seventy-two infantry and eleven cavalry divisions and twenty-and-a-half brigades of the Landwehr, the reserve army into which conscripts passed after their regular service, to which eight Ersatz corps, consisting entirely of reservists, would be added. Of these only nine infantry and three cavalry divisions and four-and-a-half Landwehr brigades would be allotted to the sector south of Metz. By the thirty-first day after mobilization, the army's right wing was expected to reach the River Somme, after which it would cross the Seine below Paris and then turn east to advance from the rear against the French forces manning their defences between Verdun and Belfort.

To implement this ambitious plan, while guarding the eastern frontier in alliance with Austria-Hungary, Schlieffen constantly pressed for increases in the army's strength. In 1899, peacetime numbers had been raised to 612,000, but even then little more than half of the young men eligible for conscript service were being called up, compared to three-quarters of those eligible in France. The reasons for this were not only the financial and political objections to having a larger army, but the competition for funds from Tirpitz's naval programme and objections from the right of the political spectrum, averse to having too high a proportion of recruits from urban areas, and of officers who might be affected by sympathy for left-wing movements.

By this time the naval programme had become a serious competitor for funds. It had hardly existed before the war against Denmark in 1804, as a result of which Prussia acquired Kiel. From its earliest years the navy was an organ of the North German Confederation, not of Prussia and, lacking admirals, was controlled by generals, Caprivi heading the Imperial Admiralty from 1883 to 1888. The accession of Wilhelm II to the throne in that year was a decisive event in the development of the navy. In visits to England in his youth he had been greatly impressed by the connection between the strength of the Royal Navy and Britain's world-wide power and influence. When his grandmother, Queen Victoria, conferred on him in 1889 the honorary rank of Admiral of the Fleet in the Royal Navy, it encouraged a naval ambition already latent. He decided to do for the navy what his grandfather, Wilhelm I, had done for the army.

The first step was to establish his own personal control, which he did by abolishing the Admiralty and dividing the higher control between the Navy Office, corresponding to the Prussian Ministry of War, and the High Command, equivalent of the General Staff, to the head of which the torpedo expert Admiral Alfred von Tirpitz was appointed, at the same time establishing a Naval Cabinet. After Tirpitz became head of the Navy Office as Secretary of State in 1897, a post he held until 1916, an even greater division of responsibility was introduced. By 1899 the following were all directly subordinate to the Kaiser: the Secretary of State of the Imperial Naval Office (Tirpitz), the Chief of the Naval Cabinet, the Chief of the Admiralty Staff, and the Chiefs of all naval commands. It is not surprising that controversies raged about every naval issue: the size and shape of the fleet, its purpose, strategy, deployment and construction programme, and training. Tirpitz, a faithful pupil of Mahan, consistently pressed for the development of a fleet of battleships which could challenge Britain's dominant position. He believed that, if Germany could deploy in the North Sea a battle fleet two-thirds the strength of Britain's total, the latter would be forced to accept Germany as an equal in imperialist terms. Not only would it remove the threat, which obsessed many Germans of the time, that Britain was intent on depriving Germany of her overseas trade and possessions and might even attack her directly to prevent the creation of such a fleet, but that the possession of a fleet would persuade either Russia or Britain itself to ally themselves with Germany rather than France.

Tirpitz had to argue against followers of the *jeune école*, which included the Kaiser himself, who wished greater emphasis placed on a cruiser programme to 'fly the flag', give both a more direct support to overseas trade and possessions, and threaten Britain's world-wide naval supremacy, forcing her to disperse her naval strength, as well as act as a counter to the Russian threat in the Baltic and the Mediterranean. The acquisition of Heligoland from Britain in 1890 (in exchange for Germany's renunciation of her claim to Zanzibar and the coastal strip of Kenya) and the construction of the Kiel canal in 1895 both initiated further controversy: about the degree to which effort should be devoted to coastal defences, how the fleet should be deployed – between Kiel and Wilhelmshaven – where it should deploy, concentrate and seek battle; the proportion of lighter craft – torpedo-boats, destroyers and submarines – to cruisers and battleships; and, in the case of the last two, the competition between numbers and size of ships and their armament. Tirpitz was consistent in pressing for the development of the largest possible number of battleships, based on a construction programme of three a year, which would create a fleet of sixty battleships by the 1920s. Until then he saw that Germany could not challenge the Royal Navy directly in battle in the open sea with real prospects of success. At first he believed that Britain could not afford to withdraw battleships from elsewhere to concentrate in the North Sea, without risking loss of her colonies (which the Germans always assumed to be on the verge of revolt, like the Boers) and thereby of her overseas trade. When Fisher's concentration proved that wrong, he deluded himself into thinking that Britain would not be able to provide the manpower and money to maintain its ratio of 2:1.

Early naval plans for action against Britain alone (even including an invasion), or France alone, which assumed that Britain would do nothing while the German fleet sailed down the Channel to bombard French ports and naval bases, gave way to more realistic assessments after the Agadir crisis and the establishment of the Entente. In a war against Britain alone, consideration was given not only to the occupation of Denmark but possibly of Holland and Belgium also. Both were soon discarded in favour of a policy which ensured their neutrality, which led to a difference of opinion with Schlieffen. By 1905 these fantasies had been discarded in favour of a defensive strategy, based

on the assumption that, on the outbreak of war, Britain would attempt
to attack German naval bases and the Kiel canal, establish a close
blockade of both the North Sea and Baltic coasts or even attempt
to land troops on them: not such a far-fetched concept, given Fisher's
wild ideas. Arguments continued about the type of navy needed for
this – torpedo-boats, destroyers and submarines were clearly suitable
– the balance to be struck between purely defensive measures, such
as mining and fortification of bases and of Heligoland, and the distribution of forces between the Baltic and the North Sea. Tirpitz stuck
firmly to his fundamental aim of building up a battle fleet of capital
ships. Support for his naval construction programme came from the
industry which stood to benefit from it and by the establishment of
a Navy League (*Flottenverein*) to encourage popular support.

Schlieffen inherited from Waldersee the General Staff's belief that
a war with France and Russia was inevitable, and that Germany should
choose the moment that suited her best to face it. Russia's involvement
in war with Japan in 1904–5 seemed the ideal moment, the Anglo-
French Entente not having by then developed into anything concrete.
The Moroccan crisis of 1905 seemed to provide the *casus belli*, but
Wilhelm and his Chancellor Bülow got cold feet, and both Friedrich
von Holstein, Under-Secretary in the Foreign Ministry, who had engineered it, and Schlieffen, who had supported him, were forced into
resignation.

French and British fears were now thoroughly aroused, and as they
reorganized and strengthened their forces, Moltke's nephew and namesake, who succeeded Schlieffen, demanded and received increases in
the army's strength. He asked for a huge one – 300,000 – but got only
10,000 in 1911, 29,000 in 1912 and 134,850, including 4,000 officers,
in 1913, bringing its peacetime strength up to 800,000. While expanding, the army was also modernizing its organization and equipment.
Schlieffen, both in office and in retirement until his death in 1913, and
Moltke had to fight the conservatives in order to introduce more artillery, notably mobile heavy howitzers, machine-guns and airships and
aircraft for observation. They were successful, so that the German army
that invaded Belgium and France in 1914 was supported by thirty-eight
airships and 800 aircraft, all unarmed observation machines, a flight
of six aircraft being assigned to each army and corps headquarters.
Schlieffen's appreciation of the strength which modern weapons had
conferred on the defence was reflected in his plan, both in his desire
to avoid frontal attacks and in his confidence that he could allot only
a limited force to the defensive role of the left wing, south of Metz.
Moltke, altogether a less confident character, was not prepared to take
such risks, and, as France increased the size of its army, strengthening
its right wing, dedicated to an offensive into Lorraine, he modified
Schlieffen's plan. He attempted to emulate his uncle in not tying his

army commanders down to a rigid plan, once they had been set in motion. He had justified doubts about whether Kluck's army on the right wing could maintain the rate of advance Schlieffen's plan demanded. Even if it did, the decisive battle could not take place until well into the second month after mobilization. An earlier decision could be reached by a shorter envelopment to turn the flank of the French force which advanced into Lorraine. It was there that he thought the decisive battle could be fought. With these factors in mind, he allocated almost all the newly formed divisions to the left wing, so that the ratio between the right and the left wings changed from Schlieffen's 7 :1 to 3 :1.

Another most significant change effected by Moltke lay in his attitude to the Austrians. Schlieffen's relations with Field Marshal Beck, his Austrian counterpart, had been cool ever since he had abandoned the elder Moltke's plan for giving priority to the Eastern Front. When the younger Moltke succeeded him, Germany was becoming dangerously isolated, the Triple Entente of Britain, France and Russia having been established, and Italy showing signs of moving in that direction also. Germany could not afford to see Austria-Hungary drift away. Moltke established good relations with Beck's bellicose successor, Conrad von Hötzendorf. In 1909 the latter wrote to Moltke, asking him to clarify Germany's position in the event of an Austro-Serbian war, which he envisaged, leading to Russian intervention. He assumed that 'in conformity with the treaty of 1879', Germany would support Austria against Russia, and France might then also become involved. In that case, where would Germany make its main effort?

Such a question would have earned a sharp rebuke from Bismarck, who had always insisted that the treaty was purely defensive and had reminded Vienna, at the height of the Bulgarian crisis in 1887, that 'For us, Balkan questions can in no case constitute a motive for war.' But Moltke, far from discouraging Conrad, accepted the idea of a war against Serbia, and said that if Russia intervened it would 'constitute the *casus foederis* for Germany' who would mobilize her whole army as soon as the Russians began to do so. He assumed that German mobilization would inevitably involve war against France, against whom the bulk of German forces would have to be deployed. His reply was shown to the Kaiser and to the Chancellor, Bülow, before it was despatched, neither of whom objected to this encouragement to the Austrians to embark on the fatal course which five years later initiated the First World War. Europe was now in the grip of the rigid plans of the German General Staff, the wider implications of which escaped their minds, so immersed were they in the fascinating detail of how to make them work.

In his letter Conrad had proposed that if Russia intervened an immediate offensive should be launched, by the Austrians from Galicia in the south and by the Germans from East Prussia in the north, cutting off Russian forces in the salient formed by Russia and Poland between the two. Moltke weakly assented to this, but made no plans and provided no forces for a German offensive which he did not intend to launch until after he had finished with France. He estimated that six weeks would suffice for that, after which ample forces could be transferred by rail to the east. He expected Russian mobilization to be so slow that there would be no real danger to East Prussia before then, particularly if the Austrian offensive disrupted Russian plans. The force deployed in East Prussia in August 1914 consisted of one cavalry and nine infantry divisions, supplemented by *Landwehr* brigades and garrison troops, organized in four corps under General von Prittwitz's Eighth Army.

Although the army's eyes were constantly turned towards France and Russia, the navy had come to realize that, whatever the origin of a war and whoever the initial opponents might be, their eyes must be kept firmly on Britain. Even if the latter were not initially involved, she would be tempted to take advantage of a European war to eliminate German rivalry. Germany could not afford, either by losses in action or by deployment to face other threats, to weaken her naval position *vis-à-vis* Britain. The irony of the situation was that Germany's exaggerated fears of British hostility, which had led (or been exploited) to the development of Germany's fleet, had brought about a situation in which Britain was inevitably hostile to Germany and had strengthened her alliance with Germany's natural enemies, France and Russia.

Naval exercises by both navies in 1909 made it clear that the threat from torpedo-boats, mines and submarines would severely limit the freedom of the battle fleets to operate near enemy coasts, and that the Royal Navy planned to establish a more distant blockade, based on the eastern end of the Channel and the Shetlands–Norway exit from the North Sea. This need to change the assumption on which German naval strategy was based coincided with another factor to force a change. In order to compete with the British *Dreadnought* class of battleship, Tirpitz was reluctantly forced to agree to the construction of battleships and battle-cruisers with heavier armament and greater displacement, which could not navigate the Kiel canal until it had been deepened. Admiral Friedrich Graf von Baudissin, the new Chief of the Admiralty Staff, adopted an offensive strategy by which he hoped to provoke the British battle fleet to venture out into the North Sea by attacks on the east coast of England and on the blockading forces. There were still arguments about whether the German fleet should be concentrated beforehand in the North Sea, in German waters, or effect a wartime concentration via the Skagerrak. The 1911 Moroccan crisis and the clear evidence that Britain was not limited, either in her construction programme or in her preparedness to concentrate her naval strength in the North Sea, by the factors which Tirpitz had assumed would operate, led not only to a reversion to a defensive strategy, but persuaded Tirpitz to accept reductions of his construction programme in order not to provoke Britain. Not only were there fears that Britain might take pre-emptive action to disrupt Tirpitz's plans, but there were hopes that it might be possible to wean her away from the Entente, if not to be an ally, at least to be neutral. To this end Tirpitz's capital ship programme was cut to two a year, but after the Haldane mission in 1912 had refused to agree that Britain would remain neutral in a war between Germany and France, the tap was turned on again and the deepening of the Kiel canal anxiously awaited. It was reopened in April 1914. By then Admiral Friedrich Ingenohl's High Seas Fleet had thirteen Dreadnoughts and three battle-cruisers to face Jellicoe's twenty-one and four. On mobilization a further squadron of eight older battleships and twelve cruisers could join it. In July 1914 the fleet was accompanying the Kaiser on a cruise off the Norwegian coast, and was not ordered back to its bases until 26 July.

Gavrilo Princip's bullet, which killed the Austrian Archduke Franz Ferdinand at Sarajevo on 28 June 1914, set in motion a train of events dictated by the rigid plans of the German General Staff, based on their long-held belief, shared by the Kaiser, of not merely the inevitability, but the desirability, of a pre-emptive war. The result of which would be the establishment of Germany as the dominant European power, endowed with enough *lebensraum*, in Europe and in the world at large, to satisfy the economic, social, political and cultural demands of the German-speaking people, whose industry had developed a dominant position on the continent. The nonchalance with which the leading figures, royal, military and political, continued their holidays in July concealed the pressure which all the leading German officials were bringing to bear on the Austrians towards war, the blame for which they were intent should rest on Russia. If that could be achieved they hoped that Britain would remain neutral. Austria's ultimatum to Serbia was delivered on 28 July, but Germany's hopes of British neutrality were dashed by the British Foreign Secretary's reply to an approach made on 29 July. He made it clear that, if France were drawn into the war, Britain would not be able to stand aside. As early as 26 July, Moltke had drafted a message to be sent to the German Minister in Brussels, instructing him to demand from the Belgian government free passage of German troops through their country. It was sent on 29 July and presented to the Belgian government on 2 August.

Everything was linked to Russian mobilization and, like Schlieffen's predecessor, Waldersee, Moltke preferred to listen to the military attachés, particularly the one in Petersburg. He was obsessed by the fear that, unless Austria and Germany mobilized without delay, Russia could crush the former and thus cause the collapse of his strategic plan, which was based on knocking out France before he and the Austrians had to face the fully mobilized Russian forces. In response to suggestions that changes should be made which would not immediately involve Britain, Belgium and France, Moltke insisted that it was impossible to change the plans, based on intricate rail movements.

The crucial day was 31 July, the Russian order for mobilization in response to Austria's ultimatum to Serbia. This was followed within a few hours by the Kaiser's order putting Germany on a war footing, mobilizing a total of 84,000 officers and 2,314,000 men. In the first three weeks of war the railways moved 2,000,000 men, 400,000 tons of war material and 118,000 horses. On 2 August Luxemburg was occupied and two days later the first German forces crossed the Belgian frontier. The resistance of the forts of Liège and Namur, and the threat of the Belgian army, which withdrew towards Antwerp on his right flank, provoked the first of Moltke's hesitations which altered the operation even further away from Schlieffen's plan than had his previous changes to it. It was not until 20 August that Brussels was reached and the French frontier crossed. A further delay was imposed on the right wing by the resistance of the British at Mons and Le Cateau and the French at Maubeuge. Mistakes were being made on the left also. Instead of letting the French offensive in Lorraine extend itself into a vulnerable salient, he allowed both the German Crown Prince Wilhelm, commanding the Fifth Army between Metz and Thionville, and Crown Prince Rupprecht of Bavaria, commanding the Sixth on his left, to attack. Although they inflicted heavy casualties on the French, the general result was to cause the latter to withdraw to their own defences and allow Joffre to transfer divisions to strengthen his left, retreating to the Marne. Moltke's emulation of his uncle's method of letting army commanders have their head did not have the happy consequences it had had at Königgrätz. The centre of gravity was shifting to the centre and, as General Karl von Bülow's Second Army pulled to its left, he became concerned at the gap appearing between his army and Kluck's First, which was supposed to move south-west and swing round behind Paris. He appealed to Kluck to close in to the left. The latter was only too happy to oblige, having become increasingly anxious about being stuck out on the right flank all on his own, his men tiring from their long marches in the hot August sun, and his logistic support failing to keep up. But, having turned south, he became concerned about his open right flank. Uncertain of the whereabouts of the British and assuming that there were French forces covering Paris, he detached an increasing proportion of his strength to guard his right flank, with the result that the gap between his left and Bülow's right was not filled. It could have been filled by the six *Ersatz* divisions which had become available since 4 August, but which Moltke had directed to Lorraine, and the four which he had sent to East Prussia, alarmed at Prittwitz's announcement that he proposed to withdraw to the Vistula, a decision the latter reversed within twenty-four hours, on the recommendation of a comparatively junior staff officer, Lieutenant-Colonel Hoffmann, but omitted to inform Moltke that he had done so.

On 6 September Joffre launched his counterattack across the Marne, which Kluck had already crossed, in danger of being trapped between the British and the French Sixth Army, taxied out from Paris. Only a failure by the British and French to exploit vigorously the opportunity provided by the gap between the First and Second Armies saved Kluck from destruction. Alarmed at the situation, Bülow, having consulted a liaison officer from Moltke's headquarters, Colonel Hentsch, but without consulting Moltke himself, ordered a withdrawal to the Aisne, forcing Kluck to withdraw also and setting the seal on the failure of Moltke's plan. On 14 September Moltke was replaced by Falkenhayn, who retained his post as Minister of War, which he had held since 1913. His attempts to envelop Joffre's left flank failed, as it was extended north to reach the Channel and to join hands with the Belgians, whose country was completely occupied by the Germans as they extended their line northwards. Two attempts to break through to the French Channel ports failed with bloody losses, by which time events in the east were demanding his attention.

Germany's basic strategy had failed. Falkenhayn found that he could not deal first with one front and then the other. He was committed on both and could never run the risk of removing sufficient forces from one in order to deal a decisive blow on the other. Moltke's despatch of Ludendorff and four divisions to East Prussia had helped to achieve the victory of Tannenberg (to which General Paul von Hindenburg, who replaced Prittwitz, contributed nothing). The real credit should go to Hoffmann and General Hermann von François, commander of the 1st Corps. But victory at Tannenberg was offset by the failure of the Austrian offensive in Galicia. Conrad attempted more than his armies were capable of achieving against superior Russian numbers. Instead of allowing his forces to be cut off in Poland, the Russian Grand Duke Nicholas withdrew them east of the Vistula, which allowed him to concentrate four armies against Conrad. The latter persisted too long in his offensive round Lemberg, and was then forced to retreat behind the River Dunajet, a withdrawal of 150 miles, having lost 350,000 men of his original 900,000. There were rumours of further withdrawal, even of Austria seeking a separate peace with Russia. There was clearly a danger that she might collapse unless supported, and the Hindenburg–Ludendorff partnership of *Oberöst* pressed for

The Eastern Front 1914-18

BALTIC SEA

Danzig

Königsberg

Pregel

Gumbinnen

EAST PRUSSIA

Masurian Lakes

X Tannenberg

Vistula

Narew

Bialystok

Nieman

• Vilna

BYELORUSSIA

Grodno •

Bug

WARSAW

Brest-Litovsk

Pripet Marshes

P O L A N D

Lublin •

• Kovel

UKRAINE

Lutsk •

San

Cracow •

G A L I C I A

Przemysl

Lwow (Lemberg)

CARPATHIAN

MOUNTAINS

0   50   100 miles

50   150 km

reinforcements so that a counteroffensive could be launched before the winter set in, on the lines that Conrad had originally proposed.

Falkenhayn had no love for the Austrians and did not believe that a decisive victory could be won on the Eastern Front. Only the defeat of France and Britain could achieve that. Reluctantly he agreed to an offensive west of the Vistula south-east towards Lodz in November, to be followed in January 1915 by a renewed attack by Conrad, reinforced by German divisions, north-east from the Carpathians. For this he transferred four corps to Hindenburg at a time when their use round Ypres might have achieved a breakthrough. Ludendorff complained bitterly that the reinforcement was too little and too late. Conrad's offensive, combined with a renewed German thrust south from East Prussia, failed with the loss of 600,000 men, of whom 100,000 were killed. The Russians lost more, but held their general line and maintained their pressure against the Austrians, forcing Falkenhayn, against all his inclinations, to agree to a further offensive.

The Hindenburg–Ludendorff partnership exploited the laurels gained at Tannenberg, which had made them popular, to press the Kaiser to force Falkenhayn to devote more effort to the Eastern Front than he wished. However, he was now persuaded that having deprived her of Poland there was a chance of forcing Russia to make a separate peace, which would allow him to concentrate effort where it could be decisive. Hindenburg and Ludendorff had more grandiose ideas. They wished to launch a pincer movement in great strength into the Ukraine, from East Prussia in the north and with General August von Mackensen, commanding a Germano-Austrian army group, from the Carpathians. Falkenhayn rejected this, not merely because he was not prepared to make the forces available at the expense of the Western Front, but because he believed it to be impracticable. Movement would depend on horse and foot in country ill-equipped with roads running in the right direction, and the German railway system, the key to her army's mobility, could not contribute.

The series of offensives in the east in 1915, following Falkenhayn's plan, were successful, and he had little difficulty in holding off the attacks by British and French in the west, largely designed to relieve pressure on their ally. But he was right in thinking that however many casualties were inflicted on the Russians – and a great many were – it did not bring decisive victory. Even with the whole of Poland in German hands in August, the Tsar rejected overtures for a separate peace.

By then it was clear that another element of German strategy had been undermined, that of a short war. Falkenhayn was resigned to the prospect of a war of attrition, and the German economy had to be directed towards both the prospect of a long war, in which the enemy's blockade might not be broken, and the unparalleled demand made by operations, whether offensive or defensive, on manpower and material. The nation responded in a remarkable fashion, but it had social consequences which were, finally, to be very significant.

A new factor had been introduced in 1915 on which the Germans placed high hopes: that of subversion of the enemy. The principal value of the adherence of Turkey to the Central Powers was seen as opening up the possibility of subverting the Muslim population of both the British and the Russian empires, on which the Germans placed exaggerated hopes, although the British authorities in India took the threat seriously. The other possibility of subversion lay in exploiting the general discontent within Russia itself, notably in the Ukraine. That was to prove a more fertile seed-bed for revolution, with long-term consequences that were to be fatal to Germany. German help in Lenin's rail journey from Switzerland to Leningrad in 1917 was to prove one of the greatest of the errors to which all wartime undercover operations, taking a short-term view influenced by military advantage, are so prone. A considerable debate took place as to what Germany's policy towards Russia should be. Bethmann-Hollweg's hope of persuading the Tsar to seek a separate peace was not entirely abandoned, but the harsher alternative, favoured by Hindenburg and Ludendorff, of pushing Russia back behind the Bug, gained favour. It was linked to the Chancellor's concept of a *Mitteleuropa* federation, in which Austria-Hungary would absorb a Greater Poland, extending into Byelorussia and the Ukraine, while its German-speaking peoples would be more closely associated with Germany itself, into which Belgium would be incorporated. This concept was not popular in Vienna. The result of the argument was a compromise. Against Falkenhayn's wishes, Ludendorff renewed the offensive in the north to capture Vilna, while the southern force, commanded by Mackensen over the protests of Conrad, was diverted to finish off Serbia. Conrad, to Falkenhayn's anger, insisted on continuing independently to conquer Montenegro. Relations between the two could not have been worse.

From June onwards, Falkenhayn's principal concern had been the Western Front, to which he proceeded to transfer as many German formations as he could. By December he had formulated his plan. He realized that his chances of achieving a breakthrough there were slim. If he had concentrated on such an attempt in 1915, instead of switching priority to the east, he might well have succeeded; but by 1916, in spite of their losses in the 1915 offensives, the Franco-British armies were strong enough to prevent it. He saw the British as the principal enemy, but the most difficult to get at by land warfare. The navy should assume the main responsibility for action against her, unrestricted submarine warfare being the prime weapon, even at the risk of involving America. France was the weak link, having already suffered heavy casualties. In a memorandum to the Kaiser he wrote:

The uncertain method of a mass breakthrough, in any case beyond our means, is unnecessary. We can probably do enough for our purposes with limited resources. Within our reach behind the French sector of the Western Front there are objectives for the retention of which the French General Staff would be compelled to throw in every man they have. If they do so the forces of France will bleed to death – as there can be no question of a voluntary withdrawal – whether we reach our goal or not.

Verdun was the area chosen, and the Fifth Army, commanded by the Crown Prince, was entrusted with the task. But, in limiting the resources that he allotted to him (only nine divisions), in restricting the frontage of attack to the right bank of the Meuse and in withholding reserves in the early stages, he went a long way towards undermining his own assumption that he could bleed the French to death with limited resources. The latter were given the chance to build up their own strength, so that, although Falkenhayn was correct in judging that Verdun was an objective 'for the retention of which the French General Staff would be compelled to throw in every man they have', and that it did very nearly bleed France to death, the cost to his own forces was almost as great. The allotment of an unprecedented concentration of artillery – 1,220 guns, most of them of large calibre, deployed on the initial frontage of only eight miles – did not save the Fifth Army from casualties which mounted to 120,000. One reason for this was that the Crown Prince believed that his task was to capture Verdun, while Falkenhayn's true aim was to force the French to fight for it.

From the start of the war the German navy felt frustrated. It would have been folly for the High Seas Fleet to venture out in the hope of interfering with the passage of British troops to France. Jellicoe would have easily cut them off from their bases. The first naval surface action, off Heligoland on 28 August, when three of his light cruisers were blown to pieces by Beatty's battle-cruisers, persuaded Ingenohl, backed by the Kaiser, not to risk his fleet in the open sea but to limit action to surprise attack by light units. In December, however, he became more adventurous, and planned a major raid commanded by Admiral Franz von Hipper to bombard the seaside resort of Scarborough on England's north-east coast. Although by then the British had broken the German naval code and were intercepting Ingenohl's orders, Beatty's force, then based in the Cromarty Firth, failed to make contact. As a result it was moved further south to Rosyth. From there it carried out a series of raids into the German Bight, in retaliation for which Ingenohl was allowed to send Hipper four battle-cruisers, escorted by four light cruisers and eighteen destroyers, to try to surprise British light forces off the Dogger Bank in January 1915. The result was a rather inconclusive engagement in which Jellicoe's main fleet took no part. Hipper lost one battle-cruiser and had one badly damaged. Beatty's flagship was also damaged. Ingenohl was replaced by Admiral Hugo von Pohl. In more distant waters the cruisers *Goeben* and *Breslau* succeeded in evading the British and French in the Mediterranean and sought refuge at Constantinople, the *Königsberg* doing the same in Dar es Salaam. The Far Eastern squadron, when Japan declared war on Germany, after a series of clashes in the Pacific met its fate off the Falkland Islands in December, ignorant of the fact that two British battle-cruisers had been detached from Jellicoe's fleet to the South Atlantic after Admiral Maximilian von Spee's successful action at Coronel in November.

The disappointing results of all the efforts which had been devoted to the Imperial Navy provoked a fierce argument between the advocates of the *guerre de course* encouraged by the feats of the few submarines then available, and those who called for both greater activity by and an enlargement of the High Seas Fleet. Tirpitz was finally won over to the admission that a successful action against Jellicoe's fleet was unlikely and that priority should be given to cruiser warfare in the Atlantic and Zeppelin attacks on London, postponing a submarine campaign against merchant shipping until the summer of 1915 when the twenty-nine boats available in January should have been significantly increased. Against his advice it was launched in February and led to protests from neutral countries, notably the USA after the sinking of the *Lusitania* in May. As a result of this restrictions were imposed, orders being given that passengers should be allowed to take to boats before a ship was sunk. Tirpitz protested vigorously, his submarine fleet, increased to fifty-four by the end of the year in spite of twenty-seven losses, having sunk 748,000 tons of British shipping. Tirpitz's anger at the Kaiser's rejection of his advice led to his resignation in March 1916. A month later the sinking off Boulogne of a French liner carrying many American passengers led to a further restriction, by which time 131 ships of 442,000 tons had been sunk in 1916, against the loss of seven submarines, while thirty-four had been added to the fleet.

The replacement of Pohl by Admiral Reinhardt Scheer in January 1916 presaged a more active role for the High Seas Fleet, sorties into the southern North Sea in March and April resulting in no more than the bombardment of Lowestoft and Yarmouth in East Anglia. A southern sweep by Jellicoe off the Skagerrak in May failed to bring about a contact, Scheer arriving after Jellicoe had left. The High Seas Fleet, increased to eighteen dreadnoughts and four battle-cruisers, was coming under increasing criticism for its inactivity, and Scheer decided on a further raid at the end of May, hoping that it would entice Beatty's force of four dreadnoughts and six battle-cruisers into action before Jellicoe's twenty-four dreadnoughts and three battle-cruisers could intervene. Scheer's force of sixteen dreadnoughts and five battle-cruisers put to sea at 2 a.m. on 31 May, three hours after Jellicoe, warned by British Admiralty intelligence, sailed from Scapa Flow. The result-ing battle, known by the British as of Jutland and the Germans as of the Skagerrak, was a tactical victory for Scheer, whose losses were less than Jellicoe's (one battleship, one battle-cruiser, four light cruisers and five torpedo-boats, totalling 62,000 tons and 3,058 men), but produced no strategic gain, since the stranglehold of the blockade was maintained. The High Seas Fleet only ventured out again on three occasions, in August and October 1916 and April 1918, all without incident. The Imperial Navy switched its effort to the *guerre de course* by submarines in which it was to have greater success.

By June 1916 Falkenhayn's strategy was failing. The French had not collapsed and threats appeared elsewhere which demanded the diver-sion of forces to meet them. On 4 June the Russian General Alexei Brusilov launched a forty-division offensive against the Austrians in Galicia, weakened by Conrad's diversion of forces for an offensive against Italy, which had joined the Allies in May 1915. Falkenhayn was forced to transfer divisions to the east to prevent a total collapse, and, just as he did so, Haig launched his offensive on the Somme. Thereafter the effort against Verdun had to be reduced, each side hav-ing lost of the order of a quarter of a million men in a struggle which, although it ended with no victory for either, had a profound influence on the course of the war.

The enmity between Ludendorff, Falkenhayn and Conrad had contri-buted to the crisis on the Eastern Front. When Brusilov launched his offensive south of the Pripet marshes, Ludendorff, faced with a superior Russian force in the north, refused to come to Conrad's help by sending any of his forty-four divisions to support him, although urged by Fal-kenhayn to do so. He was motivated by his ambition that Hindenburg–Ludendorff should be given authority over the whole front and, through that perhaps, replace Falkenhayn in supreme command. Fal-kenhayn, who was hardly on speaking terms with Conrad, was deter-mined that the latter should sort out his own affairs, by abandoning the offensive in Italy to which Falkenhayn had been opposed, before he received any substantial help from German forces. The result was that reinforcing divisions were fed in piecemeal merely to block the thrusts of Brusilov's armies. This suited the latter's strategy, which was to carry out a series of attacks to a limited depth on a broad front, switching the point of main effort as soon as one attack began to flag, keeping the enemy off balance and limiting his own casualties. It proved to be an effective method, and it was unfortunate for the Russians that it was abandoned at the height of Brusilov's success in mid July in favour of a concentrated attack on a narrow front in an attempt to break through the northern sector of the Austrian Front, by then reinforced by Germans, to reach the important rail junction of Kovel.

At this stage the first step in Ludendorff's ambitions was achieved, Hindenburg and Ludendorff assumed responsibility for the whole front, although Hindenburg was officially subordinate to Conrad. August saw the end of the Russian successes, which had prompted Romania, very imprudently, to join the Allies, partly for fear that Russia would acquire the Transylvanian area of Hungary which she herself coveted. Retribution was swift. While Austrian and German forces, the latter commanded by Falkenhayn (who was replaced in supreme command by Hindenburg–Ludendorff at the end of August under the overall command of the Kaiser himself, exercised through Prince Leopold of Bavaria), halted and then drove back the Romanian advance into Hungary, Mackensen directed a force of German, Bulgarian and Turkish troops northwards from Bulgaria into the Dobrudja, the Danubian plain in which Bucharest lies. The Russians did little to help the Romanians, whom they despised, being solely concerned to guard their own southern flank in Bessarabia. Winter weather and quarrels between the invading powers brought active fighting to an end after Bucharest had fallen to Mackensen in December.

Political as well as military factors had led to the fall of Falkenhayn. The failure of the high hopes with which the largely Prussian-based political right had embarked upon war, the discontent caused to both middle and labouring classes by inflation, to which the naval blockade contributed, and the mounting toll of casualties had bred an intense feeling of frustration. Much of this was directed against those who like Falkenhayn thought that the war could be fought in some limited fashion, or who favoured attempts to negotiate an end to the fighting. From 1914 onwards Bethmann-Hollweg had sought diplomatic solutions, but the terms he attempted to establish were so harsh as to be clearly unacceptable. Attempts to reach a separate peace with Russia have been mentioned. A long-drawn-out attempt to get the Belgian King Albert to agree that Germany should have rights in his country, which would have amounted to military occupation, met with no success; but that – or outright annexation, absorbing the Flemish population and expelling the Walloons into France – remained a condition from which the army and navy would not retract. Discussions, initiated in 1916, with French socialists also broke down on the clearly unacceptable condition that the iron-ore producing area of Longwy-Briey, between Verdun and Metz, and Belfort should, in addition to Alsace-Lorraine, be transferred to Germany. By the end of 1916 the Chancellor's hope of detaching France from the Entente, either by force, as Falkenhayn had tried, or through a desire to avoid further losses, had clearly failed.

The defeat of Romania, the exhaustion of the Russian offensive, and signs of discontent and a general breakdown of administration within Russia still held out hopes of a non-military solution in the east. The implacable element appeared to be Britain. Unable to attack her directly, the German hard-liners, backed by the army and navy, demanded a resumption of unrestricted submarine warfare, combined with air attack by Zeppelin dirigibles on her cities. Tentative peace feelers were put out towards Lloyd George and Grey in February 1916, after the US President Wilson's representative, Colonel House, had reported that they might be favourably received. But when it became clear that the minimum terms acceptable to them would be the German evacuation of Belgium, France and Poland, they were discontinued. The Chancellor switched his attention to how to reintroduce unrestricted submarine warfare without antagonizing the USA, under the influence of the army and navy, who believed that it would bring Britain to her knees in a matter of months. His first attempt was to try to obtain a deal with the USA by which Germany would continue to observe the restriction on attacks on neutral shipping if the Americans would force Britain to lift her blockade. When that proved fruitless, he embarked on a complicated process by which, under cover of supporting peace proposals initiated by Austria-Hungary, the blame for rejection of them would fall on the Entente, which would thus incur the displeasure of the US sufficiently for it to seem reasonable for Germany to resume its unrestricted submarine campaign. His problem was that he could never disclose the terms on which Germany would settle, as they would be as clearly unacceptable to American opinion as they would be not only to Germany's enemies, but even to some of her allies. But any less demanding terms would never be accepted by the political right, including much of industry, the Kaiser and the military, now led by the very hard-line Hindenburg–Ludendorff partnership, which had always favoured annexation of occupied territory, especially in the east, and insisted that only a peace dictated on Germany's own terms could justify the sacrifices made.

The proposals put forward by the Central Powers on 12 December 1916 merely stated that they were ready to enter into immediate peace negotiations with the enemy 'under conditions calculated to assure the existence, honour and freedom of their peoples'. Four days later Woodrow Wilson, having been re-elected as President, offered to mediate if both sides would state their terms. The Germans, knowing that Wilson would not look favourably on theirs, although formally accepting his offer were determined that he should not become involved, and merely referred him to the statement of 12 December. It was seen, as it was meant to be, as a rejection. On 30 December, the Entente turned down their so-called proposal, and the decision was taken to resume unrestricted submarine warfare on 1 February 1917. By then the Entente's reply to Wilson had made it clear that their terms were totally unacceptable to the Central Powers, not only in demanding the restoration of all Germany's territorial gains, including Alsace-Lorraine, but in implying also the break-up of the Austro-Hungarian and Ottoman Empires.

The nonchalance with which the involvement of the United States in the war was risked, increased by the revelation (through British wireless intercept) of German attempts to enlist the help of Mexico and Japan against the USA in its event, was based on the wholly unrealistic overestimate by the German military of the effect of unrestricted submarine warfare and Zeppelin attack, and an even more unrealistic underestimate of America's military potential and the speed with which she could convert potential to actuality. They argued that they could finish off Britain before America could intervene; that even if they had not, they could sink all her troopships attempting to cross the Atlantic; and that in any case the US army was so small, inexperienced and out of date as to be insignificant. For the unrealistic arrogance of her leaders, political, military and industrial, at this crucial turning-point of the war, Germany, Europe and the world as a whole were to pay a heavy price.

Nevertheless the results of the resumption of unrestricted submarine warfare in the spring of 1917 seemed to confirm the claims made for it. In February 540,000 tons were sunk; in March 600,000 and in April the 107 U-boats operating undertook 133 sorties, destroying 458 ships totalling 841,118 tons. But these startling successes at last prompted the British Admiralty to adopt convoys and devote greater effort to anti-submarine devices. Sinkings in May dropped to 600,000, rose to 700,000 in June, and settled down to an average of 350,000 a month in the second half of the year, in which forty-three submarines were lost, compared to twenty in the first half. As convoys and countermeasures reduced the effectiveness of the submarine campaign several commanders suggested operation in teams, instead of as single boats; but this method, which Karl Doenitz was to develop to great effect in the Second World War, was rejected.

While the submariners had been active and gaining glory, the sailors of the High Seas Fleet, affected by the news of the Russian revolution, became discontented, frustrated at inaction and disgusted with their food and general living conditions. This culminated in a series of strikes in August, for which the leaders were sentenced to death, two being executed on the orders of Scheer. His reaction was to improve conditions of service and stimulate more activity at sea, where the fleet in the Baltic supported the activities of the army in extending its hold over the territories previously under Russian control.

Reliance on the submarine campaign lay behind Ludendorff's decision to remain on the defensive in the west in 1917. He made his position secure against the offensives launched against it, both by shortening his line in the centre and by the introduction of new defensive tactics which limited the number of men exposed to the enemy's artillery bombardment. The length and weight of the latter always gave a clear signal as to where and when the blow was about to fall, allowing the German commanders to move up reinforcements to strengthen their reserve positions, just as the exhausted French and British infantry had reached the limits both of their physical capacity and of their artillery support. These tactics of defence were generally successful until the British Third Army launched its attack towards Cambrai in November, supported by 375 tanks, a machine to which the Germans had paid little attention when it had first been used the year before. Nevertheless, successful as his strategy and tactics were in preventing the Allies from making any serious inroads into his positions, the toll of casualties continued to rise and the strain on the soldiers' morale of never themselves moving forward was considerable.

While holding in the west, the Hindenburg–Ludendorff partnership's principal concern was to settle affairs in the east, where the internal difficulties of Russia appeared to offer opportunities, although every one also brought with it the basis of disagreement with Austria-Hungary as to how it should be exploited, the issue of Poland being the most contentious. When the revolution of March 1917 led to the abdication of the Tsar, Germany had no wish to see his regime followed by one which might be able to organize Russia's chaotic economy more efficiently in order to strengthen its war effort. The choice for Germany lay between trying to come to an agreement with the Liberal-Democrat Lvov-Milyukov government for a separate peace, or to attempt to undermine it in order further to weaken Russia's ability or willingness to continue fighting. The first would prejudice the achievement of Hindenburg–Ludendorff's extravagant ambitions for annexation of territory in the Baltic states, Poland, parts of Russia itself and Romania; but, if there were doubts about the ability of the Central Powers to maintain their war effort in the east, it could be the preferred choice. Hindenburg and Ludendorff were confident that, in spite of the sorry state of their allies, Austria-Hungary, Turkey and Bulgaria, their military strength was sufficient to support the alternative policy. The fateful step was therefore taken to assist Lenin to return to Russia to help the Bolsheviks to undermine the government and its support for continuation of the war. Their policy bore fruit in the October revolution, followed by the armistice in November. The German terms, insisted upon by Hindenburg–Ludendorff against the advice of the Foreign Ministry, were so harsh that Lev Trotsky refused to accept them and, when the armistice expired in February 1918, the Germans resumed the offensive with fifty-two divisions, quickly capturing Kiev and forcing the Bolsheviks to accept their terms, which included the evacuation of the Baltic states and of parts of Byelorussia and Poland, recognition of the Ukraine as a separate state and the transfer of areas south of the Caucasus to Turkey. While negotiations were in progress, German troops occupied the Crimea, Rostov, the Donetz basin and Finland.

But the Russian revolution had its repercussions within Germany, strengthening the socialist movement, which exploited the mounting war weariness of the population and resistance to the increasingly heavy burdens imposed by the insatiable demands of the large army, by this time some 250 divisions. America's entry into the war in April 1917 and the failure of the submarine campaign to deliver the goods both produced a mood of depression among the public and drove Hindenburg–Ludendorff into striking an increasingly dictatorial attitude towards the politicians, culminating in their unsuccessful demand for the dismissal of Theodore von Bethmann-Hollweg in July 1917 in favour of the compliant Georg Michaelis. They repeated their threat of resigning themselves in order to force the dismissal of the head of the Kaiser's civil Cabinet, whom they accused of pandering to the political left and, in July 1918, of the Foreign Secretary, Richard von Kuhlmann. By that time Hindenburg–Ludendorff was in all but name a military dictatorship, of which the Kaiser had become a prisoner, dominated by Ludendorff who was attempting to put into practice his concept of Total War: that all the resources of the nation, human, material, cultural and psychological, should be devoted without stint to the prosecution of the war to secure the nation's interests, the nation to him being all the German-speaking people of Central Europe. Opposition to his demands was treated as rebellion, and repression of political dissent was intensified. In his increasingly arrogant attitude, tempered by Hindenburg's more stolid caution and common sense – the only contribution that his titular superior made – Ludendorff sowed the seeds of the political turmoil which finally undermined the overweening ambition of his military strategy. The balance between King, Chancellor and Chief of Staff, which had generally been maintained since the time of Wilhelm I, Bismarck and the elder Moltke, had finally collapsed.

The Russian revolution having apparently solved Germany's basic strategic problem of two fronts, and the submarine campaign having failed to come up to expectations, the only strategy left, if Germany was not to abandon all its war aims, was to split the Alliance in the west before the Americans could deploy their full potential. This was the aim of Ludendorff's offensive which began so promisingly in March 1918. In two successive thrusts he hoped to isolate the British from the French and cut off the former from the Channel ports on which their supply depended. As in the case of his 1917 defensive policy, his strategy was supported by tactical innovation. Instead of the concentrated, rigidly controlled infantry attack, preceded by a long artillery bombardment, he had trained his soldiers to infiltrate the enemy defences under cover of a short bombardment, which included gas shells, first introduced by the Germans in 1915 and used with varying success since then. Penetrations of the enemy defences were to be exploited in a process which Liddell Hart was later to describe as an 'expanding torrent', based on its similarity to water finding its way round obstacles in the bed of a stream. He was helped by early morning mist; but, as on so many other occasions, he overestimated the capability of his own forces, even though they were 200 divisions strong, and underestimated the underlying strength, in spite of all its previous losses, of the Anglo-French-American alliance. All hope of victory had in fact disappeared when the second arm of his thrust in the Ypres sector failed in April.

It was hopeless to think of launching another offensive and, had he been wise, he would have initiated peace negotiations then. When the Allied counteroffensive began in July and the Kaiser, backed by an influential group within Supreme Headquarters, wished to initiate peace talks, Ludendorff argued that a defensive strategy could wear down the Entente until they accepted peace on terms which would let Germany retain the territories that Hindenburg–Ludendorff had consistently demanded should be annexed. But he himself began to lose his grip and when, by the end of September, as the Allies pushed the German armies back, the political situation in the country became volatile and Germany's allies began collapsing, he bowed to the inevitable, characteristically laying all the blame on the socialists for undermining the morale of the soldiers, the Foreign Ministry for bungling relations with other countries and the War Ministry for failing to organize labour sufficiently vigorously to support the war effort. His narrow, reactionary attitude, typical of the mood with which so many Germans had entered the war, proved the truth of Clausewitz's cautionary remark about the danger of divorcing war from political life. 'When that happens in our thinking about war,' he wrote, 'the many links that connect the two elements are destroyed and we are left with something pointless and devoid of sense.'

The first signs of disintegration had appeared in the navy, of which Scheer had become the single head as a result of a reorganization in August, Hipper succeeding him in command of the High Seas Fleet. In a desperate bid to achieve something, he ordered the fleet to sea at the end of October, but the sailors refused. Hipper dispersed the squadrons of the fleet to separate bases, where they joined hands with workers and resisted all attempts to bring them under control. On 9 November Scheer told the Kaiser that his navy 'could no longer be relied upon', to which Wilhelm replied, 'I no longer have a navy.'

As the armistice negotiations were embarked upon, the victors quarrelled about what should happen to the fleet. Initially it was decided that it should be split up into squadrons and interned in neutral ports, but no neutral nation welcomed the prospect of harbouring rebellious sailors. It was agreed that the fleet should surrender at Scapa Flow, where it arrived between 25 and 27 November and was disarmed; its crews were reduced to 220 men, who remained on board under conditions of internment, becoming ever more sullen as the months passed. When the admiral in command, Ludwig von Reuter, learned from the British newspapers that the armistice was due to expire at noon on 21 June 1919, he gave the order to scuttle the ships at 11.20 a.m. Flying the Imperial Flag, sixteen battleships, five battle-cruisers, five light cruisers and forty-six torpedo-boats settled onto the sea-bed or beached themselves. Of their crews, nine were killed and twenty-one wounded by their British guards and 1,860 taken off to prisoner-of-war camps, from which they returned to Germany in June 1920 and were greeted as heroes.

In spite of the lead which Germany had established in several technical fields, there were few in which her forces introduced technical innovations which had any significant effect on the battlefield. The use of gas was an important one, but it was always an unreliable weapon, the value of which had to be balanced against the inconvenience caused to one's own troops by having to take precautions against being affected by it. The chemical industry made a major contribution in improving the efficiency of explosives. In mechanization of transport, including the use of tanks, the Germans were a long way behind their opponents on the Western Front. They had less need, being generally on the defensive for most of the time and supplied by a highly efficient rail system. On the Eastern Front roads were few and incapable of carrying much heavy traffic.

In the important field of aircraft, the Germans kept pace with the Allies, except in the case of numbers, in which they began to fall heavily behind from 1917 onwards. At the start of the war the army in the west was supported by thirty-eight airships and 800 aircraft, all designed for observation duties, many with a crew of two, none of them armed. Machine-guns were soon fitted and a significant advantage gained with the introduction of the Fokker single-seater monoplane in 1915 with its forward-firing machine-gun, developed from the captured French Morane. The concentration of these aircraft into fighter squadrons gave the German Air Force, the *Luftstreitkräfte*, set up as a separate arm under General Höppner, a decided advantage over the British and French. But the latter adopted the same technique, and by the summer of 1916 had reversed the situation. The German reply to this was to form 'circuses', formed from squadrons using pilots personally chosen by the fighter 'ace' who was the circus leader, Oswald

Bölcke and Manfred von Richthofen being the best known. This coincided with the introduction of improved aircraft such as the Albatross D.I with its 160 hp Mercedes engine. Naval interest in aviation dated from 1912, but became side-tracked into the popular and dramatic form of the Zeppelin dirigible. Raids on British cities in 1915 and 1916 forced Britain to direct effort to counter them, which by the end of that year they had succeeded in doing by a combination of fighter aircraft and incendiary ammunition. Naval aviation effort was then switched to the highly successful Heinkel seaplane, while the strategic bombing role was taken over by the air force.

1917 was a significant year. In France the Germans, led by Richthofen, established both a tactical and a technical superiority over the British and French, which was on the point of being mastered when the very manœuvrable Fokker Triplane was introduced, in which Werner Voss made a name for himself before being shot down by superior numbers of British SE5as in September. By the end of the year the superior numbers of Allied machines, by then technically the equal of the Fokker, which suffered from a serious fault, were beginning to tell. The German Air Force's offensive against London had also by then tapered off, after a success which caused grave concern in Britain, although the physical damage inflicted was trivial. The principal instrument was the twin-engined biplane Gotha bomber, which carried 900 lb of bombs and several defensive machine-guns and flew at a height which protected it against both anti-aircraft guns and the fighters based in Britain at the time. After a series of raids in the Home Counties, Captain Brandenburg's fourteen Gothas dropped their bombs on the City of London, untouched by the ninety-two fighters sent up to meet them; but when he broke his leg in a bad landing his successor, Captain Kleine, rashly led the bombers in raids in unsuitable weather, and the losses he suffered persuaded the Supreme Command to turn to moonlight bombing raids. Twenty-one of these took place between September 1917 and May 1918, when forty-three Gothas and Zeppelin-Giants took off to bomb London. Less than half reached the target area and, of them, six were destroyed and two damaged by the improved defences, which included balloons suspending mines. The fifty-two German air raids against Britain, on which such exaggerated hopes had been placed in 1916, dropped 73 tons of bombs, killed 857 people and injured 2,058, a trivial achievement which nevertheless caused a totally disproportionate moral and political effect in Britain, leading to the exaggerated claims in post-war years for such strategic bombing campaigns.

Ludendorff's offensive in March 1918 was accompanied by a major effort on the part of the air force, in which Richthofen's *Geschwader* of Fokker triplanes played the leading part. Among his pilots were Ernst Udet and Hermann Goering, who in July succeeded to command of the group when Richthofen's successor, Reinhard, was killed in an accident. Richthofen himself was shot down and killed by a British pilot, Captain A.P.Brown, on 21 April over the Somme, as he clocked up his eightieth kill. To support the offensive the air force had forty-eight reconnaissance, thirty ground support, seven heavy bomber and seventy-seven fighter units. By July most of the latter had been equipped with the new Fokker D.VII biplane, perhaps the best fighter aircraft produced by any nation in the First World War. Armed with twin Spandau machine-guns firing through the propeller, powered originally by a Mercedes 160 hp and later a BMW 185 hp engine, its performance (110 mph at 10,000 feet) was not remarkable for the time, but its excellent handling characteristics at maximum height gave it a distinct advantage over its opponents. By that time, however, technical quality could not compensate for a growing inferiority in numbers, accentuated by a falling off in aircraft production due to the blockade. At the end of the war the German air force had a total strength of 2,709 aircraft and claimed to have shot down a total of 7,425 enemy planes in the four years it had lasted.

1919–45

Hindenburg–Ludendorff's wish that the war should be brought to an end by a dictated, not a negotiated, peace was fulfilled, but it was their enemies who did the dictating. Germany's internal political situation in November 1918 undermined her military one, and she was left with no cards to play. There was a general breakdown of order, and the politicians of the newly proclaimed republic looked to Wilhelm Gröner at Supreme Headquarters, established first at Kassel and later at Kolberg in Pomerania, to help restore it. This was done by the formation of the *Freikorps*, formed from officers and NCOs who volunteered for the purpose both of maintaining order and resisting Polish encroachment on the eastern border. Its members, on the extreme right of the political spectrum, were equally distrustful of the senior army officers and the provisional government. By the summer of 1919 the corps numbered 400,000 and, although anti-republican, by its defeat of the revolutionary guards in a series of bloody street battles, created the conditions in which the Weimar Republic could be established. Under the latter's constitution, a 'Provisional *Reichswehr*' was to be raised on the pattern of the *Freikorps*: a voluntary force, recruited for a six- or nine-year engagement, some concessions being made to left-wing attitudes by provision for soldiers' councils and, a novelty, the right of the soldier to vote.

No sooner had this been agreed than, on 7 May 1919, the terms of the Versailles Peace Treaty were conveyed to the precariously established Republic. The future German armed forces were to be restricted to 100,000 long-term volunteers, of whom 4,000 could be officers. It was intended to be for internal security duties only and was to have no heavy weapons, such as artillery, tanks and aircraft. The High Seas Fleet had already got rid of itself; submarines were forbidden and the navy reduced to coastal forces. The General Staff was to be dissolved, as were all military schools save one for each of the main arms – infantry, cavalry, artillery and engineers – and all preparations for mobilization forbidden. Almost worse than all this was the demand that Germany admit responsibility for causing the war, and that her leaders should be tried before an Allied tribunal. The reaction of the army was to refuse acceptance and, when the government did so, plans were made for a coup led by the commander of the garrison of Berlin, General W.Freiharr von Lüttwitz. The order for the first reduction of 200,000 men in March 1920 sparked it off, but the situation was saved by General Seeckt, who had been entrusted with the task of forming the new army. He saved the Weimar Republic from collapse, and thereafter established a relationship with its political leaders under which, in return for his support, they would connive at the covert infringement of the terms of the Versailles Treaty, which all three elements of the armed forces were thenceforth to practise. This went hand in hand with a policy of preferring quality to quantity, forced on him, in any case, by circumstances. Perhaps the greatest revolution was in the attitude as to who was suitable to be commissioned as an officer. Seeckt realized that the base must be broadened, and the first principle of training was that a leader, whether commissioned or not at the time, must be trained so that he could assume responsibilities several levels above his current one, when war came.

Even before the secret agreement with Russia, arrived at under cover of the Rapallo Conference in 1922, arrangements had been made for the preservation of armament design and production techniques under the guise of foreign firms. Goering became a salesman for Fokker in Sweden, where Krupp acquired a strong influence in the firm of Bofors. Dornier set up facilities in Italy and Switzerland, and submarine construction in Finland and Spain was organized by a convention of German shipping firms, camouflaged as a Dutch firm. The navy's efficient arrangements for evading the terms of the Versailles Treaty, under the direction of Captain Lohmann, did not come to light until 1927.

The Rapallo arrangement with the Soviet Union granted Seeckt facilities for the development of equipment and training of personnel for the forbidden arms of artillery, tanks and aircraft in return for instruction given to the Russian forces. From 1924 to 1931 it was under the direction of Colonel Niedemayer, and many of the future senior officers of the *Luftwaffe* were trained at the airfield of Lipetsk, 200 miles southeast of Moscow. An exception was Goering who, returning to Germany in 1921, fell under the spell of Hitler, joined his National-Socialist

German Workers' Party and was placed in command of the *Sturm Abtei-lung* (SA) in order to impose some discipline on its brown-shirted members. Other future organizers and commanders of the air force were tucked away in civil aviation, which was permitted from 1922 onwards, a number of different firms concentrating to form the Deutsche Lufthansa in 1926. By then the wartime aircraft firms had re-established themselves, Junkers, Heinkel and Dornier being the leading ones. Anthony Fokker had returned to his native Holland, his place as a fighter aircraft designer/producer being taken by Wilhelm Messerschmitt and Heinrich Focke-Wulf who entered the field with light aircraft, ostensibly for sporting purposes. German interest in dirigibles was maintained, Hugo Eckener's *Graf Zeppelin*, completed in 1928, making a number of spectacular long-distance flights. In the early years of the Weimar Republic the principal external threat came from the newly independent state of Poland, actively supported by France. Both the army and the navy were much concerned at the threat it seemed to pose to the German-speaking populations of East Prussia, the Baltic States and, especially, to Danzig, which had been placed under the administration of the League of Nations. A Franco-Polish agreement provided for French naval support in the Baltic. It was primarily to counter that threat that the navy embarked on the design of what was known as a 'pocket battleship', the first of which was launched with great ceremony at Kiel in 1931.

The army, although anxious about Poland, was primarily concerned with creating the basis for future expansion, once the restrictions of the Versailles Treaty could be lifted; with ceaselessly working for their removal, and with the internal situation which, in a state of deepening economic crisis, remained extremely volatile. Seeckt tried to preserve the army from political entanglement, but in emphasizing that it served 'the state and not the parties' and its subordination to that permanent titular figurehead, President Hindenburg, he created no sense of loyalty to the government of the Republic, for which most of its personnel retained an arrogant disdain. This lack of loyalty to the elected government, combined with the deliberate deception fostered by evasion of the terms of the Versailles Treaty, encouraged an atmosphere of intrigue, never absent in German military circles, which was to prove fatal in the years to come. Seeckt resigned in 1926, having imprudently allowed the Crown Prince's eldest son to participate in manœuvres as a temporary officer, and Dr Otto Gessler, the Minister of War, with whom he had worked in harmony, was forced out over the Lohmann affair in 1928. Seeckt was succeeded by Heye, who was more prepared to co-operate with the politicians, but his influence did not last long.

Working behind the scenes was an arch-intriguer, Colonel Kurt von Schleicher, a brother officer, in Hindenburg's old regiment, the 3rd

Footguards, of the President's son Oskar. His influence was largely responsible for the appointment of General Gröner to succeed Gessler at the War Ministry. A special post was created for Schleicher himself, who set about transferring real power to the Ministry from Wilhelm Heye's army headquarters, the *Heeresleitung*. It was not long before Schleicher's nominees were filling the important posts in the latter, Heye himself being replaced by General K.Freiherr von Hammerstein-Equord, another old friend from the 3rd Footguards. Schleicher did not confine his intrigues to military appointments, but set about influencing the political scene also. He was as much opposed to the extreme right, led by Alfred Hugenberg, as he was to the socialist and radical left, and pinned his hopes on Heinrich Brüning, leader of the right wing of the Catholic Centre Party, whom he persuaded Hindenburg in 1930 to appoint as Chancellor, when the coalition government under the Social Democrat Hermann Müller collapsed in the face of the economic crisis. He achieved that by using the threat of Gröner's resignation if Hindenburg granted Müller emergency powers, as he proposed to do, on the grounds that it would be unacceptable to the army. Schleicher's influence on Hindenburg had eclipsed that of Gröner, the latter having offended the President by marrying a lady, already pregnant, shortly after the death of his wife. At that stage Schleicher and the army hierarchy strongly disapproved of Hitler's NSDAP, and particularly of the disorderly activities of the SA and of its homosexual chief of staff, Ernst Röhm, although it had many sympathizers within the army's ranks, even at high level, General Beck making no bones about it. The court-martial of three officers at Ulm in 1930 for associating themselves with the party, resulting in an eighteen-month prison sentence, split the army on the issue, many officers considering that it was an unwarranted political interference which senior officers should have resisted. From then on Schleicher pursued two aims: to abolish the state government of Prussia, which was in the hands of the Social Democrats, and to avoid driving the NSDAP into opposition by encouraging it to take a less radical line and, as its popular strength grew, to participate in government. The crisis came in 1932 over the SA. Its objectionable activities led to a demand to ban it. Gröner, who had added the Interior Ministry to that of War, wished to ban all paramilitary organizations, which Schleicher opposed, hoping to make use of them, including the SA, as a covert military reserve. Gröner, pressed by the interior ministers of the states, went ahead, opposed by Hammerstein and Schleicher, who threatened the resignation of the whole army leadership. Brüning, Chancellor since 1930, offered his post to Schleicher himself, but the latter was already intriguing with Hitler to oust Brüning, in whom he had lost confidence. The result was a rapid succession of governments, led in turn by Franz von Papen, around whom Hitler ran rings, and then Schleicher, who found himself with no political support and with the army hierarchy deeply divided. General Werner von Blomberg, commander of the East Prussian division, whose Chief of Staff, Walther von Reichenau, was a fervent National Socialist, went directly to Hindenburg and told him that the army had lost confidence in Schleicher and that only a government headed by Hitler, whose party held 230 seats in the Reichstag, could save the country from a civil war and the army from a serious political division. Hammerstein and Schleicher, who had failed to entice one of Hitler's lieutenants, Gregor Strasser, to switch the left wing of the NSDAP to their support, reluctantly accepted the same view, and on 30 January 1933 the old President, who had sworn that he would not appoint 'the Austrian Corporal' to be his Chancellor, did so, Blomberg becoming War Minister.

The navy had watched these events with great anxiety, hoping to remain detached from the political divisions, and putting implicit trust in Hindenburg, that granite idol with feet of clay. Erich Raeder, head of the navy, had no particular reason to be concerned at Hitler's advent to power, and entertained the hope that he might succeed in removing altogether the shackles of the Versailles Treaty, which Papen's government had already allowed him to disregard by approving a future construction programme of six pocket battleships, six cruisers, six half-flotillas of destroyers and torpedo-boats, and three half-flotillas each of minesweepers, E-boats and submarines, the last only to be embarked upon when the international political situation, arising out of the Geneva disarmament conference, permitted it. In his first official interview with Hitler, the guidance Raeder received was: 'I never want to be involved in war with Britain, Italy and Japan. The German navy must therefore be built up for tasks within the framework of a European continental policy,' which, as the Russian navy was not then of any significance, Raeder interpreted to mean that the French fleet was the target that he had to aim at. One of the important milestones in German naval development was the Anglo-German naval agreement, which followed the German government's statement, in March 1935, that it no longer accepted the disarmament provisions of the Versailles Treaty. In June the naval agreement, to which Raeder attached great importance, was signed, stipulating that the German fleet would not, in total or in any individual class of ship, exceed 35 per cent of the combined fleets of the British Empire, except in submarines in which it could build to parity. The immediate result of this was the approval of a construction programme of two pocket battleships (26,000 tons: 11-inch guns), two cruisers (10,000 tons: 8-inch guns), sixteen destroyers and twenty-eight submarines, of which eight were to be ocean-going. The submarine arm was headed by Captain Doenitz, who was already proposing the operation of ocean-going boats in groups, later known as wolf-packs. Raeder's proposal for a fleet air arm of twenty-five squadrons was turned down, Goering winning the battle between the *Luftwaffe* and the navy on this issue. The best the navy could achieve was that nine *Luftwaffe* squadrons were specifically allocated to maritime reconnaissance.

The good relations with the British on naval matters which Raeder hoped would follow from the agreement were affected by German naval action in the Spanish Civil War. It had started, as with other navies, in an attempt to ensure the safety of German nationals in places like Barcelona where fighting took place, but soon developed into covert support of Franco, intensified when German ships were attacked by Republican aircraft.

Raeder continued to assume that there was no question of his navy being employed against Britain's, and to be reassured by Hitler on that point. The change came in 1938, by which time it was becoming clear that Hitler's policies were causing serious concern in France and Britain, and that a war against them both had to be considered a possibility. In response to this, Raeder produced his 'Z Plan', including six of a new type of battleship (50,000 tons: diesel propulsion: 16-inch guns) and three battle-cruisers, primarily designed for commerce-raiding (30,000 tons: mixed diesel and high-pressure steam turbine propulsion: thirty-four knots: 15-inch guns), new light cruisers and other small craft and an acceleration of the submarine programme to reach a total of 249. This programme could not be complete until 1948 and would involve, at a suitable moment, the abrogation of the Anglo-German agreement. The strategy on which it was based was to force a world-wide dispersion of the British fleet by attacking convoys at considerable distances from British ports, with powerful ships of long endurance, as well as by submarines. The keel of the first of the new battleships was laid down in June 1939, two months after Hitler, at the launching of the battleship *Tirpitz*, fruit of the 1932 programme, had publicly threatened to abrogate the Anglo-German agreement, although he was still assuring Raeder that he had no intention of going to war with Britain, pinning his hopes on Chamberlain being no more inclined to go to war over a proposed 'frontier revision' with Poland than he had been over Czechoslovakia the year before. After the annual Kiel regatta, in June, at which the British navy's team won the Hindenburg trophy, the new battle-cruiser *Gneisenau* was sent on a shakedown cruise in the Atlantic, and Doenitz and the commander of all the minesweepers set off on their summer holidays.

Although Raeder had always realized that forcible changes in the territorial provisions of the Versailles Treaty involved the risk of a general European war, he had nurtured the hope that somehow or other Britain could be kept out of it. The latter's declaration of war on 3 September 1939 came as a real shock. If he had thought that war with the Royal Navy was a possibility in 1939, he would never have embarked on the long-term Z plan, but put priority on submarines, as Doenitz had recommended after naval manœuvres in the previous winter. To carry out an effective campaign against British shipping, he had estimated that he needed ninety boats on patrol, which required a total fleet of 300. On 3 September 1939 he had fifty-seven, of which only twenty-six were suitable for ocean-going operations. The maximum number on patrol at any one time would be nine. The balance in surface ships between the German navy and the combined British and French fleets was: battleships 5:22, heavy cruisers 2:22, light cruisers 6:61, destroyers and torpedo-boats 34:255 and aircraft-carriers 0:7. The only hope of doing anything effective was to take immediate action with all the forces available against shipping at sea, before the British and French could establish countermeasures; but Hitler turned that down in the hope of being able to come to some form of accommodation with Britain and France, after he had finished with Poland.

Blomberg's appointment as Minister of War immediately raised issues about both the organization and the personalities of the high command. His first clash was with Hammerstein, C-in-C of the Army (*Chef der Heeresleitung*), who was out of favour with Hindenburg and whom Reichenau, Blomberg's pro-Nazi Chief of Staff, consistently by-passed. He was on no better terms with General W. Adam, Chief of the General Staff (*Truppenamt*), who was an outspoken critic of the proposal to establish an independent air force. Both of them were due to retire in February 1934 and Blomberg clashed with Hindenburg over Hammerstein's replacement. General Ludwig Beck, a well-known pro-Nazi and an able and intelligent man, replaced Adam, and Blomberg wished Reichenau to become C-in-C, but Hindenburg, annoyed by Hitler's lobbying in his favour, objected on the grounds of his lack of command experience and insisted on the appointment of General Werner von Fritsch, a dour bachelor, respected for his ability and, in contrast to the conservative Beck, as a forward-looking innovator. These two were to wield great influence over the expansion and development of the German army which took place between 1934 and 1938. Both, however, were in agreement that German foreign policy had to be tailored to the capability of her armed forces, primarily of the army; that that depended on the state of the economy and of the general morale of the people; and that military aggression in Europe would inevitably develop into a general European or world war, in which the greater resources of Germany's potential enemies would weight the balance against her. Germany's ambitions should therefore be limited to Europe and to the capability of her army, which should control the activities of the other services, the army General Staff being solely responsible to a War Minister who was also C-in-C of the armed forces (*Wehrmacht*), as Blomberg soon became, in which case Fritsch's post should be abolished, or to a C-in-C of the army (*Heer*), who would also be C-in-C of the *Wehrmacht*.

A dispute of a different nature had a much more serious effect on the army: the argument over relations with the SA. Its chief of staff, Röhm, saw it as a people's army which would replace the regular army, the thin end of the wedge being assumption of responsibility for internal order and border control. In the hope of damping down clashes between the army and the SA, Blomberg and Reichenau gave way to several of Röhm's demands, but Röhm persisted in his far-reaching proposals, which were not supported by Hitler. The crisis deepened from February to the end of June 1934, and was brutally settled by Hitler in the night of the long knives on 30 June, when Röhm and about a hundred of his SA associates were murdered, along with others against whom Hitler held a grudge, including General Schleicher and his wife, General Bredow, Reichenau's predecessor, and Gregor Strasser. Hitler justified this brutal act by claiming that he was 'protecting the integrity of the German army', involving Hindenburg by saying that he had promised the President 'to preserve the army as a non-political instrument of the nation'. Hindenburg replied with congratulations and an order forbidding any member of the armed forces to attend Schleicher's funeral. Although the army was not directly implicated, Hitler had managed to entangle it by inference, to give it reason to be grateful to him, and at the same time to give the generals a clear warning as to what might be their fate if they attempted to oppose him. In its silent acceptance of a blatantly illegal and inhuman act, the army gravely weakened its position, both politically and morally.

Its position was further weakened after the death of Hindenburg on 2 August 1934, when Hitler combined the posts of President and Chancellor into that of Führer, held by himself, assuming also that of C-in-C of the *Wehrmacht*. Under the Weimar Republic the soldier's oath of loyalty had been 'to the Reich constitution, the Reich and its lawful institutions' coupled with obedience to the President and the soldier's superior officers. After Hitler's advent to power that had been changed to 'the people and Fatherland', with no mention of institutions or superior officers. Now all officers and soldiers were ordered to swear unconditional obedience to the person of Hitler 'the Führer of the German Reich and Commander-in-Chief of the *Wehrmacht*'. While Blomberg grovelled, suggesting that the army adopt the fascist salute, Fritsch and Beck became increasingly concerned at the development of the SS (*Schutzstaffel*) as a rival armed force, in spite of Hitler's promise that the *Wehrmacht* would be 'the sole bearer of arms'. Hitler's position was confirmed by a plebiscite in which only five out of forty-three million voters opposed it, and the associated measures were retroactively legalized.

Whatever qualms army officers, senior and junior, might have about
these developments, they welcomed Hitler's plans for a major expansion of the army and the prospect of the removal of the restraints
of the Versailles Treaty. At the meeting in February 1934 at which
Hitler had rejected Röhm's plans, he had told Blomberg that the army
had to be ready to defend Germany in five years' time and to take
the offensive by 1942. The first step was to increase the number of
infantry divisions from seven to twenty-one over two years, increasing
the strength from 100,000 to over 300,000 in clear breach of the Versailles Treaty. The adoption of conscription, which would be an
obvious breach, was to be deferred until 1935, recruits until then being
volunteers, many from the SA and from the police. Even before conscription was introduced on 16 March 1935 the army's strength had
risen above that planned. The announcement was accompanied by
a further one that a new target of thirty-six divisions had been set
(without consultation with Fritsch or Beck) and a few months later
the traditional titles were restored, the *Reichswehr* becoming *Wehrmacht*, headed by its *Ober Kommando* (OKW), the *Heeresleitung* being
redesignated *Ober Kommando der Heeres* (OKH) and the *Truppenamt*
the *Generalstab*, while the existence of a *Kriegsakademie* and of the
Luftwaffe was admitted, all in defiance of the Versailles Treaty.

The treaty had also been defied in the development of army weapons,
the expansion of the artillery being based on conversion of horsed
transport units, and the development of Panzer troops by conversion
of motor transport battalions of the *Kraftfahrtruppe*, of which Major
General O.Lutz was the Inspector, his chief of staff being Colonel
Guderian. They had to fight for the concept of the use of tanks and
motorized troops as independent formations against the views of the
General Staff, who, like their French counterparts, wished to restrict
tanks to the support of infantry, cavalry still being considered as having
a part to play on the battlefield, particularly in the east, where the
road network was ill-developed. But from 1933 onwards Lutz and
Guderian received support from Blomberg, Reichenau and Hitler himself, as well as from Fritsch, although Beck opposed them. The hectic
expansion which took place from 1934 to 1938 meant that the army
was incapable of carrying out serious operations in that period. The
many problems it faced caused considerable dissension among the
senior officers, which was not surprising given the personalities concerned and the division of responsibilities between them. Neither
Goering nor Raeder acknowledged that Blomberg had any authority
over the air force or the navy, so that in fact he was left to dispute
control over the army, and over who would be C-in-C in the field in
war, with Fritsch, whom he disliked. After Reichenau had been replaced in October 1935 by the obsequious General Wilhelm Keitel, Blomberg became increasingly isolated, while Beck continually tried to push

Fritsch aside, and to by-pass Keitel by direct dealings with other ministers. From 1935 onwards, in spite of the state of the army, and in
the face of protests from Beck, Blomberg began planning possible operations, including surprise attacks on Czechoslovakia, one against Austria and various defensive plans directed against France, which might
or might not be connected with them.

On 12 February 1936 Hitler, having decided to abrogate the Locarno
Pact, told Blomberg to prepare to send troops into the Rhineland.
Blomberg, Fritsch and Beck were united in opposing it, but Hitler
prevailed, three battalions marching in on 12 March. The international
reaction, weak as it was, caused Blomberg to advise their withdrawal,
finally undermining confidence in him on the part both of Hitler and
of Beck. At Blomberg's request Hitler held an important conference
on 5 November 1937, attended by Fritsch, Goering, Raeder and the
Foreign Minister, Konstantin Freiherr von Neurath. Hitler explained
that rearmament would be complete by 1943, after which the equipment of the forces would progressively become out of date. The real
purpose of the *Wehrmacht* was to exploit opportunities that might
arise in Europe, perhaps from the internal troubles of France, a Franco-
Italian war, or the Spanish Civil War. Before exploiting them Germany
must secure her south-eastern flank in Czechoslovakia and Austria,
neutralizing Russia through the co-operation of Japan. Blomberg and
Fritsch warned Hitler against becoming involved in a war against
France and Britain. Fritsch, Beck and Neurath were so concerned at
Hitler's preparedness to risk such a war that they met together, follow-
ing which Fritsch and Neurath separately asked for interviews with
Hitler. Hitler refused to see Fritsch and delayed the interview with
Neurath until he felt ready to dismiss him, which he did. Blomberg
and Fritsch were soon to follow.

Blomberg, a widower, played into Hitler's hands by not only marrying a lady of doubtful virtue, but by involving Hitler and Goering as
witnesses. Goering wanted Blomberg's job for himself, and was determined that Fritsch, whom Hitler also wanted to be rid of, should not
get it. A trumped-up charge of homosexuality was therefore brought
against him, but Hitler did not want Goering in such a potentially
powerful position and assumed it himself, with Keitel as Minister
and head of OKW. On the latter's recommendation, Fritsch was
replaced by General Walter von Brauchitsch, who was persuaded to
accept the post, it was alleged (he later denied it), by the gift of a
sum of money to enable him to divorce his wife, from whom he was
separated. A major purge of senior posts followed, in which Keitel's
younger brother became head of the Army Personnel Department, and
Guderian succeeded Lutz. Beck, due to be replaced by General Franz
Halder, was left in his post for appearance's sake.

Fritsch's court-martial was delayed by Hitler's invasion of Austria in March 1938, ordered only forty-eight hours beforehand, and against which Beck protested to Keitel. Fortunately for the army there was no opposition, as the route to Vienna was littered with broken-down vehicles. A month later Hitler ordered Keitel to bring up to date the plan for invasion of Czechoslovakia (*Aufmarsch Grün*). Once again Beck protested, pointing out to Brauchitsch the risk of intervention by the Soviet Union and France, with whom Britain would ally herself. He maintained that Germany's forces were not ready for war and that her 'lack of living space' would make it impossible for her to sustain a long war. He continued to subject Brauchitsch to a succession of memoranda on the wider dangers of the course on which Hitler was set. His views led to his isolation and, after a meeting between Hitler and senior army commanders in August at which the Führer reiterated his decision to 'solve the Czech question by force', Beck tendered his resignation which was accepted. He was succeeded by Franz Halder, who had discussed the possibility of a coup to arrest and try Hitler, if he ordered the army to attack Czechoslovakia, with a number of senior commanders and civilians, most of whom were eventually to be involved in the 20 July 1944 plot.

Brauchitsch now found himself ground between the doubters in the army, who included Halder and General Gerd von Rundstedt, and Hitler's admirers and lackeys, led by the Keitel brothers and the elder's Chief of Staff, Alfred Jodl, among whom the Führer's aide-de-camp, Colonel Rudolf Schmundt, exerted a sinister influence. The Munich agreement by which the German army occupied the Sudetenland appeared to prove the doubters wrong. Rundstedt resigned and Halder stifled his doubts, while Brauchitsch became totally subservient to Hitler, who ordered the occupation of the rest of Czechoslovakia, which took place on 15 March 1939. A few days later he told Brauchitsch that he intended to use force against Poland in order to 'gain control of Danzig and a German land-bridge to East Prussia'. His decision was not affected by Chamberlain's announcement on 31 March that Britain 'guaranteed' the defence of Poland, a week later adding Greece and Romania to that empty gesture. The army's plan for Poland was presented to Hitler immediately after he had signed the Pact of Steel with Italy in May. From his retirement Beck tried but failed to persuade Halder to take steps to prevent its implementation.

By that time the order of battle of the field army comprised 103 divisions, of which five were Austrian and six Panzer. Of the eighty-six infantry divisions, thirty-five were regular and had to be ready for war within forty-eight hours of mobilization. The remainder fell into three different categories, almost wholly dependent on reservists, largely from the *Landwehr*, which relied heavily on veterans of the First World War. To man this organization the war establishment required 2,750,000 men in the field army and a million in the *Ersatzheer* stationed within Germany. To meet this the active army had 730,000 men and trained reserves of 1,100,000. The shortfall would have to be made good from the *Landwehr*. The size of the army outran not only the trained manpower available, but also the material resources to support it in the field, including weapons and mechanical transport. The occupation of Czechoslovakia had helped to remedy some of the deficiencies in the latter, both by capture from the Czechoslovak army and from its thriving armaments industry.

Although the army had a formidable number of tanks, 3,195, the vast majority of them (2,886) were light tanks, thinly armoured and mounting only a heavy machine-gun. The medium tanks, on which the Panzer arm was to rely for its offensive action, were the Panzer Kraftwagen (PzKw) III and IV.* In terms of the quality of its tanks, the German army's Panzer force in 1939 was far from being the formidable instrument it was made out to be then and later. Hitler's obsession with quantity prevailed over the army's desire to maintain quality. In no field was this more apparent than in the provision of officers. The vast expansion in the officer corps involved a dilution which resulted in a large proportion of officers, particularly the junior ones, and an even higher proportion of non-commissioned officers, being firm supporters of the Nazi party.

On 22 August, the day before the signature of the Russo-German non-aggression pact, a secret protocol which agreed to a partition of Poland, Hitler summoned the senior officers of the armed forces and told them that he intended to crush Poland, even if it meant war with France and Britain. He thought that, now that he had removed the potential threat posed by Russia, they would not intervene, and in any case could give no direct help to Poland. They could not attack Germany without infringing the neutrality of Belgium and Holland. With its world-wide commitments and falling birthrate Britain could not afford a war. If she tried an economic blockade, he would frustrate it by expansion into Eastern Europe. A long peace was not good for a nation. Wars were won by men, not machines, and Germany had better men. The nation was united behind him, unlike France and Britain which were disunited. Germany had little to lose by war, and it was necessary to give the *Wehrmacht* a trial run in a limited conflict before having to face the major powers. A victory would be good for the morale both of the forces and of the public. Some of his listeners thought a war inevitable in any case, but Rundstedt is said to have remarked to his chief of staff, 'That crackpot wants war.'

---

* The former was equipped with the 37-mm gun which, because it had been chosen as the infantry's anti-tank gun, Guderian had been forced to accept in place of the 50-mm which he wanted and which was soon to replace it. The PzKw IV, of which there were 211, carried a low-velocity 75-mm gun, designed to fire high-explosive or smoke in support of the 98 PzKw IIIs.

A partial mobilization had been in train since April. The force allotted to the Polish campaign consisted of sixty-two divisions, of which only sixteen were mechanized, six of them Panzer divisions – a million and a half men. Two-thirds of the infantry were regular. Forty-four divisions, of which only twelve were regular and none mechanized, totalling 750,000 men, were deployed to guard Germany's western frontier while Poland was attacked. OKH's plan was for Rundstedt's Army Group South – three armies totalling thirty-six divisions – to thrust north-east from Silesia and Slovakia towards Warsaw. In General Fedor von Bock's twenty-one-division-strong Army Group North, General Gunther von Kluge's Fourth Army would sweep across the Danzig corridor and drive up the Vistula towards Warsaw. General G. Küchler's Third Army, spearheaded by Guderian's 19th Panzer Corps, would thrust south-east from East Prussia and cut Warsaw off from the east – the last was a contribution of Hitler's. OKH had originally planned that Küchler should attack the Danzig corridor from the east before embarking on such a move.

The overwhelming superiority of the Germans in the air and in modern equipment, including tanks, combined with the geographical advantage of being able to invade from several different directions, made the campaign a foregone conclusion. Overall command was exercised by Brauchitsch from OKH at Zossen, south of Berlin, while Hitler, who did not interfere with the conduct of operations, took the field with Keitel and Jodl and a small element of OKW in a train. While there, he told Jodl that he had decided to attack in the west in 'a fateful struggle of the German people', and he gave 15 October as the starting date. Delay would allow Britain further to reinforce France and allow both to gain in strength. Brauchitsch's opposition to this, on the grounds that the army needed time to reorganize and train, made him even more unpopular with the Führer than he already was, aggravated by his issue of an order of the day honouring Fritsch when he was killed in Poland with the artillery regiment of which he was honorary colonel.

On 27 September, the day Warsaw fell, Hitler summoned the senior officers of the *Wehrmacht* to tell them of his decision. The superiority of the *Luftwaffe* over the French and British was given as an important factor to be added to the strength of German morale in contrast to the doubts and confusion in France. The general plan would be to strike west-north-west to the Channel ports, accepting the violation of the neutrality of Belgium, which was already, he argued, in collusion with Britain and France. On 10 October Brauchitsch and Halder were summoned to be told that the offensive was to be launched as soon as a period of fine weather for the *Luftwaffe* was forecast. Every formation capable of offensive action was to be used, the general aim of the campaign being to defeat as large an element of the Franco-British forces as possible and clear the coast of northern France, Belgium and Holland, while protecting the Ruhr, as a preliminary to operations against Britain. No attack was to be launched against the Maginot Line.

The two generals raised no objection at the time, but Halder told Jodl of his and Brauchitsch's reservations about embarking on an offensive that year. A week later Hitler gave 12 November as the date for the operation to start, in spite of protests from Bock, Kluge and Reichenau, subject to review on 5 November. When, on that date, Brauchitsch said the army was not ready and compared its general state unfavourably with that of the Imperial army in 1914, the Führer lost his temper. On 23 November he summoned all commanders down to corps level to read them a lecture in which he threatened that he would 'ruthlessly destroy any defeatist and anyone who opposed him and his plans for the offensive'. But he could not control the weather and one postponement followed another.

Up to the meeting on 5 November, Halder had kept in touch with a group of service commanders who plotted to kill Hitler if he persisted in his plans to attack France, and secret plans were held at Zossen, near which Halder ensured that reliable divisions were stationed, to put it into effect. But the demoralization of Brauchitsch and a remark by Hitler, at that meeting, that 'he knew all about the spirit at Zossen', led Halder to refuse to have any more to do with it. From this time onwards the distrust between Hitler and OKW on the one hand, and OKH and the senior generals of the army as a group on the other, deepened. As the breach between them widened, Hitler and OKW took over more and more of the functions of OKH with grave consequences for the army and for the conduct of operations.

While the army had been at almost constant loggerheads with the Führer, the air force had enjoyed a privileged position. There had been no doubt in Hitler's mind that it should receive priority in his rearmament programme. Even if Goering had not been his right-hand man that would probably have been the case. Even those who did not accept the theories of Giulio Douhet *in toto* believed that air forces would have a profound effect on war in the future, and the restrictions imposed by the Versailles Treaty in respect of military aircraft were particularly strict. One of Hitler's first acts was to establish a new department of government, the *Reichsluftfahrtministerium* (RLM), or State Air Transport Ministry, with Goering as Minister and General Erhard Milch as State Secretary and his deputy. In May 1933 the Air Defence Office (*Luftschutzamt*) of the War Ministry was transferred to it, manned by 200 officers, including General Walther Wever, head of the Command Office, General Hans-Jurgen Stumpff, head of the Personnel office and General Albert Kesselring, head of the Administration office. Among others who were to become well known were Generals Hugo Sperrle, Hans Jeschonnek, Freiherr Wolfram von Richthofen and

Kurt Student. Few of them had any previous experience of aviation. Milch, an artilleryman, had served as an observer and aerial photographer and had commanded a reconnaissance group on the Western Front. After service in the Prussian police on the Polish border, he had been one of the three directors of Lufthansa, to the development of which he had made a major contribution. In theory he was supposed to deal with civil aviation and Wever with military, but being a highly efficient and domineering man, he insisted on having a finger in every pie, asserting his position as deputy to Goering, who was idle and did not concern himself with detail. Wever also was highly intelligent and efficient, but managed to maintain an equable relationship with Milch.

The first plan, which would have to be done clandestinely, was to produce an air force of 600 front-line aircraft, of which two-thirds would be bombers, by 1935. This was quickly superseded by the Rhineland programme, the first target of which, 3,715 aircraft by the end of 1935, was increased to 4,021 by September of that year. By January 1935, 3,183 had been produced and the target increased again to 9,853 by October 1936. After Hitler had announced in March 1935 that he was no longer going to observe the disarmament restrictions of the Versailles Treaty, the existence of a separate air force, the *Luftwaffe*, was recognized and the target was increased again, that for April 1936 being set at 12,309, expected to increase over the following years to 18,000.

From the start the emphasis was on bombers, which would form what was called a Risk Fleet (*Risiko Flotte*). The rapid development of an offensive force would, it was held, act as a political and military weapon, threatening any nation which might show signs of attempting to prevent Germany's rearmament. Both Hitler himself and those whom he wished to threaten were primarily impressed by numbers, and this factor had a significant effect on the subsequent organization of the *Luftwaffe* and on the type of aircraft selected. It resulted in emphasis on twin-engined bombers of considerable speed but limited payload, of which the Ju 88 became the mainstay, rather than on long-range four-engined bombers. The ambitious programme ran into a host of difficulties, due partly to the confusion reigning within the RLM and its procurement organization, partly to the attempt to produce too many different types too quickly, and partly to economic and industrial bottlenecks. There was also confusion of responsibility and personal quarrels between the generals in the RLM, beginning with the death of Wever in 1936.* Kesselring, who succeeded him, would

---

* He failed to unlock the controls of his aircraft before take-off, having qualified as a pilot only the previous year.

not yet accept Milch's interference and left, succeeded by the lack-lustre Stumpff, who bickered and established a greater independence for his post, by then re-designated as Chief of the Air General Staff. The quarrels continued, intensified by the appointment of the war ace and stunt flier Ernst Udet to oversee procurement, for which he was temperamentally unsuited. When, in 1939, the brilliant thirty-nine-year-old Jeschonnek, a former favourite of Milch's, succeeded Stumpff, the situation became worse, as he quarrelled endlessly with Milch and Udet.

Given all these difficulties, it is astonishing that the *Luftwaffe* succeeded in developing so rapidly into such an effective force. Its participation in the Spanish Civil War helped to clarify some of the contentious issues. It began with Hitler's agreement to Franco's request for help with Ju 52 transport aircraft to ferry his troops from Morocco to Spain between July and October 1936. This was followed by Sperrle's Condor Legion which, by the time he handed it over to General Volkmann in November 1937, numbered 105 aircraft, most of which were Heinkel 111 bombers and Bayerischeflugwerke (later renamed Messerschmitt) 109s. The success of the Ju 87 dive-bomber (Stuka) had also been proved, although the demand that light bombers, such as the Ju 88, and long-range fighters, like the Me 210, should be capable of dive-bombing had an adverse effect on their performance. One of the principal gains from experience in Spain was the development of procedures and communications for close co-operation with the army. The idea that the *Luftwaffe* was pre-eminently designed and trained for army support, and that it suffered in air to air operations as a result, is misconceived. It saw its primary task as attacking a large number of different types of target of military significance, including industrial plants producing military equipment, well behind the enemy lines, airfields being of high priority. For defence, it relied heavily on anti-aircraft guns, which it manned itself, although in the Bf (later Me) 109 it had one of the best fighters of the time, but methods of detection of enemy aircraft and control of anti-aircraft defences were primitive. Its doctrine did not include attack on cities as such, and it assumed that its targets would be in neighbouring countries. Before 1936 at least, Britain was not thought of as a target. Early in that year Udet said to Ernst Heinkel:

A war against England is completely out of the question. ... It will suffice for any potential conflict if we have a medium bomber with relatively limited range and relatively low bomb-carrying capacity, but with a high degree of diving accuracy, in short, the new Ju 88 ... we can build as many of them as the Führer wants. At the same time it will impress England and France sufficiently, so that they will leave us alone in any case.

He was correct. The threat of *Luftwaffe* action gave Hitler a walk-over in Austria in March, and Udet's prognostication was exactly fulfilled at Munich in October, when the same treatment had been applied to Czechoslovakia. The actual strength of the *Luftwaffe* was then formidable, although not as great as the Germans made it appear and as their potential opponents assumed. Nevertheless, in terms of numbers, it was a match for the combined British, French and Czech air forces: 2,928 front-line combat aircraft, of which 1,284 were bombers, 207 dive-bombers, 173 ground-attack and 643 fighters. The British had 640 bombers and 566 fighters, the French a total of 859 and the Czechs 566. But many of the German bombers were already obsolescent and serviceability was low. The British air staff's estimate that the *Luftwaffe* could drop a daily average of 600 tons of bombs on Britain in the first few weeks of war, causing 20,000 casualties on each raid, was a gross overestimate both of the bomb-load and of its effect.

When war came a year later the only aspect in which the position had improved for Germany's enemies was in the air defences of Britain, fighters, anti-aircraft guns and radar. In numbers and in quality, the *Luftwaffe* had increased its advantage. It had 4,093 front-line combat aircraft, of which 75 per cent were serviceable, more than double the previous year's total. Of these 1,176 were bombers, mostly He 111s, 406 dive-bombers, almost all Ju 87s, and 1,179 fighters, almost all Me 109s. Against this Britain could muster only 1,660, of which 536 were bombers, 608 fighters, 96 reconnaissance and 520 maritime aircraft (304 in the Fleet Air Arm); and France 1,735, of which 463 were bombers, 634 fighters, 444 reconnaissance and 194 maritime. The Czech air force was no longer in the balance.

Favourable as the *Luftwaffe*'s situation was in comparison with the air forces of its enemies in September 1939, it was taken aback by having to engage in war three years before it had planned to do so. It was in the middle of a reorganization with a view to future expansion, had few reserve crews, a very small stock of bombs and an inadequate repair and logistic organization. Unlike the British, its commands were not based on function but were geographic, corresponding to those of the army, army groups being supported by air fleets (*Luftflotte*) in which there were groups (*Geschwader*) of bombers, fighters, reconnaissance and close-support aircraft. On the outbreak of war it was decided that production should be concentrated on four types: the Me 109 fighter and its intended successor, the Me 210, the Ju 88 bomber and its intended successor the He 177. Many other aircraft were under development, including an experimental jet, the Me 262, the engine for which Junkers was developing.

Caught on the hop, *Luftwaffe* strategy demanded a short war. To ensure that, all its resources were to be thrown in from the start with no holding back to meet possible future commitments. It paid off initially, but with consequences which had a serious effect later. Victory in Poland was largely due to the paralysis produced by the rapid destruction of the Polish air force, which lost 330 of its 463 front-line combat aircraft before the rest took refuge in Romania on 17 September. The *Luftwaffe*'s 1,939 aircraft were divided between two *Luftflotte*, one commanded by Kesselring, the other by General A.Löhr, ex-head of the Austrian air force. The campaign was a virtual walk-over, and no real test, although it proved the soundness of *Luftwaffe* training and tactics and procedures for co-operation with the army. The next victory, in Denmark and Norway in April 1940, was a more severe test, involving as it did a much more detailed plan incorporating the use of airborne troops (also part of the *Luftwaffe*) and the need to acquire and bring into use airfields at a long distance from Germany. But again opposition was minimal and the effect of concentrating air effort at the crucial point well demonstrated.

After much argument and several changes, command was exercised by OKW through an army general, Nikolaus von Falkenhorst, who was not allowed to exercise command over General Hans Geisler's *Fliegerkorps X*, but had to request support to the *Luftwaffe* General Staff under Jeschonnek. The army high command, OKH, took no part, being primarily concerned with the impending offensive in France. On the day the operation started, 15 April, this was changed: *Luftflotte 5*, commanded by Milch himself, was established at Hamburg and moved to Oslo on 24 April. He proved a decisive commander. It was the largest air transport operation hitherto undertaken: 3,018 transport sorties by the solid old Ju 52s flying 29,280 men, 2,376 tons of supplies and 259,300 gallons of petrol to Norway.

The navy's contribution to the Polish campaign had been to ensure the security of the sea route to East Prussia, which Poland could not seriously challenge. Its first major test came in the Norwegian campaign in April 1940. Before that the loss of the *Graf Spee* in South America, after her successful commerce-raiding exploits in the Pacific, had been offset by some dramatic submarine exploits, notably the sinking of the British aircraft-carrier *Courageous* in the Bristol Channel in the first few weeks of the war and the battleship *Royal Oak* inside the defences of Scapa Flow a month later. Rumours that the British intended to take action to interfere with the supply of Swedish iron ore to Germany from the Norwegian port of Narvik forced the navy to initiate a study of the implications. Raeder concluded that the ideal was that Norway should remain neutral and defend its neutrality, including the passage of ships through its territorial waters, against all comers; but that at all costs the British must be prevented from establishing bases in Scandinavia from which they could not only prevent the precious iron ore from being shipped from Narvik (and perhaps succeed in exerting pressure on Sweden to prevent it being shipped in the summer from Luleå in the Baltic also), but block the passages from the Baltic to the North Sea and from the latter to the Atlantic to both surface ships and even submarines, as well as make direct air attacks on targets in north Germany, including her naval bases.

Allied plans to send help to Finland in her war with Russia and the British interception of the *Altmark* (the *Graf Spee*'s supply ship, carrying 300 British prisoners of war taken from her) in Norwegian waters stimulated the reluctant army and air force (whose eyes were on the planned attack in the west) to complete the planning for landing a force at Narvik to pre-empt Allied action. Raeder had hoped to avoid challenging the neutrality of Denmark, but the *Luftwaffe* insisted that they needed to use airfields in Jutland. The timing of the operation was linked to the need to pre-empt Allied action, for a moonless period to give the ships long dark nights for approach and withdrawal, and to get the operation over before the offensive in the west was launched. The naval risks were high, and all the ten destroyers which ferried the force to Narvik were sunk, after the oil tankers with fuel for their return journey had suffered the same fate, and the British eventually captured Narvik, evacuating it almost as soon as they had done so. The real significance of the German victory, obtained more by the army and the air force than the navy, was the contribution it made to the naval situation when Germany attacked the Soviet Union, which thus joined the Allies, an eventuality that Raeder had never contemplated.

Between November 1939 and February 1940 the plan for the attack in the West (Operation YELLOW – *Gelb*) underwent a series of changes. In accordance with Hitler's original directive OKH planned to concentrate the bulk of the forces, including most of the Panzer divisions, in Bock's Army Group B which would attack through Belgium, while Rundstedt's Army Group A protected its southern flank through the Ardennes and across the Meuse, heading towards Amiens. Based on the assumption that the British and the best of the French forces would advance into Belgium, this would satisfy Hitler's directive that as large an element as possible of their forces should be defeated and the Channel coast secured. From the beginning Rundstedt and his chief of staff, Erich von Manstein, argued against it as being unlikely to be decisive. Whether the Franco-British force stayed where they were or advanced into Belgium, it would push them back onto their lines of communication. They argued for a transfer of the main effort, including the bulk of the mechanized divisions, to the left, while the right pinned the enemy down where they were. A classic envelopment could thus be executed and the Führer's aims more effectively be achieved. The postponement of the operation through the winter, with the increase in the forces available and in the resources for their logistic support, and the time it granted for major changes and transfers, meant that this alternative became more attractive, while the disadvantages of the original plan became more evident. Both OKH and Hitler himself, although not Jodl, were being brought round to favour it, the capture of an officer carrying the original plans when he made a forced landing in Belgium being, perhaps, one of the factors influencing both to accept the views forcibly expressed by Manstein. Rundstedt was given four armies, totalling forty-five divisions, which included three-quarters of the mechanized ones, including seven Panzer, while Bock was left with two armies, totalling twenty-nine, including three Panzer and two motorized infantry. The tank strength of the two army groups was 2,574, of which 349 were PzKw III and 278 PzKw IV: the rest, three-quarters of the total, were light or command tanks, 334 of which were Czech. The area south of the Ardennes, facing the Maginot Line, was the responsibility of General Wilhelm Ritter von Leeb's Army Group C, which had only nineteen infantry divisions, none of them regular, and four of them fortification divisions. Forty-two infantry divisions were in OKH reserve. Ten low-grade infantry divisions were all that was left in the east, while Norway absorbed seven and Denmark one.

Although Rundstedt and Manstein had won their argument, the plan did not go as far as Guderian wished, in respect of the organization and employment of the tanks. Only half of the ten Panzer divisions available were in the special Panzer group, commanded by Kleist and not Guderian himself, as he hoped and might have been expected. Within Kleist's group, Guderian commanded the 19th Panzer Corps of three Panzer divisions, totalling 828 tanks. He had had to fight hard against his superiors to preserve the key element of his plan, which was that his corps should head the advance through the Ardennes and itself force a crossing of the Meuse near Sedan, not waiting for infantry divisions to break the French defences first. Few of his seniors believed that his Panzer divisions, unaided by anything but air attack, could effect a crossing of a major river, the far bank of which was defended.

But this was what Guderian achieved on 13 May, only three days after the frontier of Belgium and Luxemburg had been crossed. While operations elsewhere were successful, it was the rapid advance of Guderian's corps which, more than any other aspect of the operation, brought about the collapse of the Franco-British defence, to which the Belgians and Dutch had perforce been added at the last moment by the invasion of their countries. The risks Guderian ran on his southern flank, as he pushed rapidly westward towards the Somme, were, in theory, almost reckless, and caused grave concern to his superior Kleist, who on 17 May ordered him to stop, causing Guderian to threaten resignation. Kleist was acting on orders from Rundstedt, endorsed by Brauchitsch, backed by OKW and Hitler himself. Guderian evaded the restraint imposed on him by continuing what he described as a 'reconnaissance in force'.

However, it was not long before OKW became concerned that the enemy was building up a new defensive line to block further progress, and on 20 May, by which time Guderian's forward troops had reached Amiens, he was given the go-ahead. By the end of that day 2nd Panzer Division had reached the coast beyond Abbeville. The British counter-attack at Arras next day had a psychological effect all the way up the command channel to OKW totally out of proportion to the size of force involved and its almost insignificant actual military effect. On 22 May Kleist withdrew from Guderian's command the 10th Panzer Division, which he had planned to direct towards Dunkirk, and Rundstedt gave orders that no advance towards Calais and Boulogne should start until the situation round Arras had been cleared up. On 23 May both these orders were rescinded, but all the doubts of the previous week were revived on the 24th. The Panzer divisions were down to about half their original strength, and they and the mechanized divisions were widely scattered. Kleist, his superior at Twelfth Army, Kluge, and Rundstedt favoured a halt so that the infantry divisions, moving at the pace of the marching man and the horse, could relieve the mechanized forces of flank protection and the latter could close up to concentrate for a further thrust.

Into this atmosphere of caution Goering launched his proposal that the task of finishing off the British and those French cut off with them should be left to his *Luftwaffe*. When Hitler visited Rundstedt's headquarters on 24 May, the two influences which chimed in with his own intuition coincided to lead him to agree with Rundstedt that Bock's Army Group B should push the Anglo-French forces back towards the coast, while the *Luftwaffe* pounded them and Rundstedt's Army Group A with its mechanized divisions 'caught' any trying to escape southwards, and the Panzer divisions reorganized themselves for 'future operations'. Guderian, Brauchitsch and Halder argued against it, but Rundstedt, backed by Hitler and OKW, prevailed, with the result that the British troops escaped the net, leaving all their equipment behind.

OKH had been considering since 21 May what 'future operations' should be. They proposed a wide outflanking movement by Army Group A to the west of Paris, while Army Group B attacked southwards to the east of it, followed by an attack by Army Group C against the Maginot Line. Hitler changed this to give Bock the task of advancing to the Seine on both sides of Paris, while Rundstedt, with the bulk of the mechanized forces, struck south through Rheims, aiming for the Plateau de Langres behind the Maginot Line. Given the parlous state of the French army at the time it would not have mattered much what the plan was. The spirits of the German troops were so high that astonishing feats of speed and endurance were achieved: infantry, who had already marched 300 miles, threatening to catch up the tanks, the speed of whose advance was reduced by constant changes of orders from the Führer's headquarters. The campaign, which ended on 25 June, cost the Germans 27,074 dead, 18,384 missing and 111,034 wounded. Nine army generals were promoted to field marshal, nineteen to colonel-general and seven to full general, an inflation of the rank structure which many thought degraded it. At the same time Hitler made sure that the credit for victory went to none of them, but to himself. Henceforth he regarded himself, and insisted that others did also, as the supreme operational commander.

The navy had played no part in the campaign, but was now faced with a major task in preparation for Operation SEALION, an invasion of Britain which only the army regarded with any enthusiasm. Hitler throughout had reservations, and OKW believed that there were better and less risky ways of putting Britain out of the war: a combination of air attack with a submarine campaign against its supplies, the seizure of Gibraltar and, with help from Italy, of Egypt also.

Not until November 1939 had Raeder even contemplated the idea of an invasion of England. The naval staff's initial study confirmed Raeder in his firm view that Britain's Achilles' heel was its dependence on overseas supply, and that a concentration of effort against that target was more likely to pay a dividend than any direct assault, the naval preparations for which could only be at the expense of the former. When the campaign in France had been crowned with such spectacular success, making all the ports of the Low Countries and northern France available, he stuck to that view, emphasizing that a precondition of a successful invasion was absolute air supremacy over the sea approaches to Britain. Nevertheless he was forced to assemble an invasion fleet of 155 ships, 1,200 barges and lighters towed by 500 tugs, and over 1,100 motor boats, in ports between Antwerp and Le Havre, pointing out as he did so that none of them were suitable for landing troops on the open shore; and he was at constant loggerheads with the army over their desire to land on a very broad front, which he could not provide the resources to support. Even when their first demand to land forty divisions was whittled down to thirteen, which they wanted to land in three different areas on a frontage of sixty-two miles, he maintained his objections. As it became clearer that the *Luftwaffe* was not going to be able to gain the requisite air supremacy, Raeder contented himself with pressing for postponement until the advent of winter ruled Operation SEALION out of court, much to his relief.

He now hoped that priority could be switched to the maritime campaign against Britain, which he urged Hitler should be extended to the Mediterranean with the co-operation of the French, whose navy had been antagonized by the British action designed to neutralize it. Hitler's decision, announced to his senior commanders in December 1940, to 'dispose of' the Soviet Union before finishing off Britain, filled him with dismay. On several occasions Raeder had warned him against becoming involved in a war on two fronts and stressed his conviction that Britain was the principal enemy, and that maritime action against her, involving aircraft as well as surface ships and submarines, was the method by which she could be brought to her knees. A war against Russia, apart from the diversion of naval effort to the Baltic, would involve great demands by the army and the air forces to support them, and on industrial resources which could only be at the expense of the navy.

The help, direct and indirect, which Britain was already receiving from the USA convinced Raeder that the time would come when he might have to face direct American naval participation in the war. The strategy he favoured was to take all the offensive action he could before that occurred by the maximum deployment of forces against Atlantic convoys. Once the US Navy was directly involved, the chances of German battleships and cruisers being able to operate freely would significantly diminish.

To this end the pocket battleship *Admiral Scheer* got out through the Denmark Strait in October 1940, followed by the heavy cruiser *Admiral Hipper* in November, both of them inflicting serious damage to convoys and forcing the British to escort them with battleships. In January 1941 the battle-cruisers *Scharnhorst* and *Gneisenau*, which had been damaged by torpedo attacks in the Norwegian campaign, finished repairs in Kiel and also slipped through the net, sinking or capturing twenty-two ships, amounting to 115,000 tons, before entering Brest in March, where they were subjected to heavy air attack. The new battleship *Bismarck* was ready for action in April 1941. Her sister ship, the *Tirpitz*, would be ready in six months. The problem was whether to wait to sail the *Bismarck* until she was ready also. In view of the increasing risk from air reconnaissance and attack, and of the possibility of American intervention, Raeder decided not to wait and planned to send the *Bismarck* and the new heavy cruiser *Prinz Eugen* out to join the *Scharnhorst* and *Gneisenau* in a major operation in the Atlantic, supported by submarines and supply vessels; but engine trouble kept the *Scharnhorst* in dock and air attacks crippled the *Gneisenau*. Against the advice of Admiral Lütjens, Raeder decided to go ahead with the *Bismarck* and *Prinz Eugen* alone. Although all the ships which the Royal Navy could muster were deployed to intercept them, they cleared the Denmark Strait on 24 May, sinking the British battle-cruiser *Hood* as they did so. But the *Bismarck* had herself suffered damage and was leaking oil. Lütjens decided that he had to make for St Nazaire, which, sending *Prinz Eugen* off to Brest at high speed, he proceeded to do, but was located, crippled by torpedoes and finally sunk by a combination of battleship gunfire and torpedo attack. Only 110 of her crew of 2,000 were saved, Lütjens going down with his ship. From then on Hitler ceased to trust Raeder. He became obsessed with the fear of a British attack on Norway and insisted that the *Tirpitz* and cruisers should be based there, where they were also available to attack British convoys taking supplies to the Russians at Murmansk, which they did with considerable effect. It was a disagreement between Hitler and Raeder over the action of the *Admiral Hipper* in one of these engagements in December 1942 which led to Raeder's resignation the following month. Hitler complained that the navy's heavy ships were more of a liability, in terms of the air defence and escorts they needed, than an asset, and demanded that they be taken out of commission and their guns used for other purposes. Doenitz, Rader's successor, persuaded him to reverse that decision.

The start of the campaign in France and Belgium, before that in Norway had finished, in May 1940 imposed a conflict of priorities for the *Luftwaffe*. This was resolved by the position of Milch, combining his post of Secretary of State with command of *Luftflotte 5* in Norway. The assault in the west, which began on 10 May, was supported by 4,000 aircraft divided between Kesselring's *Luftflotte 2*, in the north, supporting Army Group B and the navy, and Sperrle's *Luftflotte 3* to the south, supporting both Army Groups A and C. The French, British, Dutch and Belgians deployed 1,151 fighters and 1,045 bombers and ground attack aircraft, almost all of inferior performance to the German. In Britain and the surrounding seas were a further 1,200. These air forces were the main target of the *Luftwaffe*, which quickly reduced them to impotence, making it possible to switch effort to the direct support of the Army Groups and the general paralysis of the French and British command and movement organization, in fulfilment of Hitler's directive of October 1939, which had listed the two priority tasks as 'to destroy or put out of action enemy air forces, but also primarily to hinder or prevent the enemy High Command from putting its decisions into effect'.

Kesselring's operations included the use of *Luftwaffe* paratroops and an army air-landed division in Holland and Belgium. Their operations were successful, but at the cost of 109 Ju 52s destroyed or beyond repair and 53 damaged out of the 430 employed. There is no doubt that the *Luftwaffe*'s contribution to Germany's victory was very significant, but Goering was responsible for a major error in persuading Hitler on 23 May that the army should be held back and allowed to secure its flanks, while the *Luftwaffe* alone finished off the British, hemmed in round Dunkirk. If the weather had remained perfect, it might have been successful, but a combination of poor weather conditions and the intervention of fighters based in Britain robbed him of the opportunity to claim the final act of victory for the arm of which he was the head.

Both Milch and Kesselring had been sceptical of their chances of success. The short campaign cost the *Luftwaffe* 1,389 aircraft (521 bombers, 122 Ju 87 Stukas, 367 fighters, 213 transports and 166 reconnaissance), for the loss of which they claimed 4,233 aircraft destroyed, 1,850 of them on the ground. Goering, to his delight, was raised to a special new rank of *Reichsmarschall* as well as a new grade of the Iron Cross, and Milch, Kesselring and Sperrle were made field marshals.

The next task was to deal with Britain. When this had first been studied in 1938 the conclusion was reached that air action alone could not be decisive: it must form part of either a naval campaign or an actual invasion. Initially, after Dunkirk, the former view prevailed. The *Luftwaffe*'s task was to attack ports and shipping and help, as far as its range of action would permit, in the navy's campaign against Britain's Achilles' heel; but on 26 May, after British air raids on the Ruhr, priority was changed to her aircraft industry, although attacks on maritime targets continued. This lasted through June and July, during which each side took the measure of the other, and it became clear to both that the Me 109s had an advantage, where its limited range allowed it to operate, over the Spitfires and Hurricanes. While the *Luftwaffe*'s strategy was to entice the RAF's fighters into combat, the latter, under Dowding, learnt to avoid it. Priorities changed again when Hitler decided on Operation SEALION. It was clear that it could not be launched unless the *Luftwaffe* could achieve air supremacy over the landing areas and the sea approaches to them. The essential preliminary was Operation ADLERANGRIFF (Eagle Attack), launched in August with the aim of destroying RAF Fighter Command in south-eastern England.

For this task *Luftflotte 2* and *3*, assisted by Stumpff's *Luftflotte 5*, based in Norway, had 3,196 aircraft, of which 2,485 were serviceable on 10 August, to oppose which Dowding's Fighter Command could put 350 out of its 503 Hurricanes and Spitfires into the air, assisted by 1,200 heavy and 650 light anti-aircraft guns. The British fighters were excellent machines, but were outmatched by the Me 109 which could operate at higher altitudes (up to 34,000 ft), was faster than the Spitfire above 20,000 ft and than the Hurricane at all heights, and had both a better diving performance and a heavier armament than both. Its weak points were its inferiority in manœuvrability and its limited range. The latter restricted its operations to twenty minutes over south-eastern England, and that only if flying direct. If the Me 109 had to slow down or zig-zag to escort bombers, it became vulnerable and could spend less time over the target area. Dowding exploited this weakness by refusing to challenge fighters on their own, concentrating on attacking the bombers.

The *Luftwaffe* began its assault by failing to appreciate the importance of Fighter Command's radar-based control system and of the airfields from which it operated, wasting their effort on ones of less importance. Sperrle and Kesselring had argued that the operation should start with night attacks on airfields and aircraft factories, mass daylight attacks only being introduced when British fighter strength had been significantly reduced by this means. But Goering brushed their caution aside, confident in the *Luftwaffe*'s superiority and pressed for time. In the second phase of the battle, which started on 24 August and lasted until 6 September, the *Luftwaffe* concentrated on the vital airfields and came near to victory. The RAF lost 273 fighters in the air, 49 more being damaged; the *Luftwaffe* suffered 308 aircraft shot down and 66 damaged, of which only 146 and 27 were Me 109s. Fighter Command was running short of pilots, and ground installations at the key airfields had been very seriously damaged. If the rate of loss and damage suffered in the first week of September had been continued, the *Luftwaffe* might well have gained, before the end of the month, the air supremacy required to launch Operation SEALION, originally planned for the 15th, but postponed on the 3rd until the 21st. But their attempt to lure Dowding's fighters into the air by concentrating on a target apparently even more vital than his sector control airfields – the docks of London, supported by Hitler as a retaliation for a British raid on Berlin on 25 August (itself a retaliation for a German raid on East London caused by a navigational error) – was to rob Goering of a possible victory. On 7 September the first mass daylight raid by 650 bombers, escorted by 1,000 fighter sorties, was aimed on London docks, followed by nightly attacks and major daylight raids on the 9th and 15th, the latter intended to be the *coup de grâce*. A total of 1,300 sorties, 300 by bombers, were made without any diversionary operations to distract the 170 Spitfires and Hurricanes concentrated to meet it, which shot down 58 aircraft (35 bombers) and damaged 25 (22 bombers) against the loss of 26 fighters of their own, and 8 damaged. This was a severe blow to the morale of the *Luftwaffe*, adversely affecting the relations between bombers and fighters, each accusing the other of being the cause of failure. It was accepted by all that the necessary conditions to launch the invasion were not going to be achieved and, after two days of bad weather, Hitler postponed it indefinitely. Excellent as the British fighters and their control system were, and skilful and courageous as were their pilots and commanders, it was the faults of the *Luftwaffe* itself and its commanders, Goering in particular, that robbed it of a victory which the quality of its aircraft and their crews and the superiority of its numbers placed within its grasp, with decisive effect on subsequent events.

Hitler was not prepared to see the army stand idle, waiting for the opportunity to launch Operation SEALION, but his plan to seize Gibraltar, the Canary Islands and perhaps French North Africa was frustrated by the Spanish dictator Franco and by the need to send help to the Italians in North Africa after their defeat by Britain there between December 1940 and February 1941. The Italians had also complicated the Balkan situation by attacking Greece. British help posed a potential threat to the Romanian oilfields, and in November 1940 Hitler ordered OKH to plan Operation MARITA, the despatch of a force of twelve divisions to move into Bulgaria and thence into Greece to forestall British deployment of forces there. This was to be launched in the spring by General Wilhelm List's Twelfth Army of four Panzer, two motorized, two maritime and ten infantry divisions, supported by Richthofen's *Fliegerkorps VIII*.

Early in January Hitler made clear to his service commanders that the aim of these operations was subsidiary to Operation BARBAROSSA, the major campaign he planned for 1941 against the Soviet Union, for which Brauchitsch had been told on 31 July 1940 to expand the army to 180 divisions, double the number to which only six weeks before he had been told to reduce it. The number of Panzer divisions was to be doubled, from ten to twenty, achieved by reducing the existing ones from two Panzer regiments to one each and grafting the regiments thus released onto the motorized divisions, of which eight new ones were formed, making a total of ten. The army therefore underwent a major reorganization during the winter, as existing divisions split to form eighty-four new ones, bringing the total by June 1941 to the staggering figure of 205 divisions, the great bulk of which were infantry, dependent, once they left the railway, on their feet for movement and horse-drawn transport to haul their supplies and weapons.

From the time that planning for the campaign in Russia started in July 1940 until the operational plan was finalized in the second half of March 1941, there were many changes in the concept. Throughout there was a fundamental difference between the importance OKH and most of the army generals attached to the capture of Moscow, and the priority which Hitler gave to envelopment of the Russian armies, so that they could not escape to fight another day, and the occupation of Leningrad, the industrial area of the Donetz basin and the oilfields north of the Caucasus. The generals believed that the Soviet Union would be forced to commit its army to the defence of Moscow, which it could not afford to lose for both political and economic reasons. Once the enemy's army had been defeated there, the economically important areas could be secured without difficulty. Moscow was a key rail centre, and its capture would seriously affect the ability of the Soviet army to transfer troops and supplies from one front to another. The final plan divided the front between three

Army Groups, North under Leeb, Centre under Bock and South under Rundstedt. Leeb had two armies, General E.Busch's 16th of eight divisions and Küchler's 18th of seven, and General Erich Hoeppner's 4th Panzer Group of three Panzer, three motorized and two infantry divisions. His task was to 'attack from East Prussia in the general direction of Leningrad'. Bock had two armies, Kluge's 4th of fourteen divisions and General A.Strauss's 9th of ten, and two Panzer Groups, Guderian's 2nd of five Panzer, three motorized, one cavalry and six infantry, and General H.Hoth's 3rd of four Panzer, three motorized and four infantry. His task was to advance from northern Poland and 'force a breakthrough towards Smolensk'. These two army groups would together 'destroy the enemy formations in the Baltic area'. South of the Pripet marshes Rundstedt had three armies, General Karl-Heinrich Stülpnagel's 17th of thirteen divisions, Reichenau's 6th of six and General E.Ritter von Schobert's 11th of seven, and Kleist's 1st Panzer Group of five Panzer, three motorized and six infantry. He was to advance from southern Poland through Kiev to the bend in the lower Dnieper.

Only after the two northern army groups had completed their joint task would 'freedom of movement for other tasks, perhaps, in co-operation with Army Group South' be granted, which *might* include an advance to Moscow. Committed to the invasion were 134 German, 21 Finnish and 14 Romanian divisions, the German force totalling three million men, supported by 7,184 pieces of artillery and 3,332 tanks. The arguments about aims and methods that had raged during the planning stage surfaced again soon after the invasion was launched on 22 June 1941. Guderian wished to repeat his performance in France by thrusting forward as fast and as far as possible before attempting any envelopment. Hitler soon began to interfere and continually called for shorter pincer moves to surround the enemy. When Bock disagreed, Hitler insisted that both his Panzer Groups should be placed under Kluge in order to ensure that their action was more closely co-ordinated with the infantry. Against their wishes, Kluge forced Guderian and Hoth to converge to surround the enemy at Minsk. Nevertheless Smolensk, 400 miles into Russia, was reached on 16 July. Fighting to eliminate the forces trapped there was fierce, and it was not until 8 August that it was reduced, by which time Army Group Centre had captured 600,000 prisoners and knocked out or taken 5,000 tanks. Army Group North was only sixty miles from Leningrad and Rundstedt fifty miles from the Dnieper, although there were still large Russian forces south of the Pripet marshes.

It was then that the argument about future strategy reached its climax. All the generals, even Jodl, wanted the major thrust to be made to Moscow, while Hitler resisted it. He wished to concentrate first on securing Leningrad and clearing up the situation in the south as a preliminary to securing the Donetz basin and the Crimea as a stepping-stone to the oilfields. In the first few weeks, optimistically believing that the Russian army was already defeated, OKH, OKW and Hitler had for once been in agreement. But after the severe fighting round Smolensk, OKH became concerned that the Soviet army, given an interval in which to do so, could build up an effective defence covering Moscow, while the logistic difficulties which were beginning to beset the German army would prevent it from being able to overcome it before winter set in and for which the Germans were not prepared.

Halder was determined to resist Hitler's order and tried to persuade Brauchitsch that both of them should proffer their resignation, but, dispirited and weary of Hitler's accusations, the C-in-C refused, and it was left to Bock and Guderian to face the Führer on 23 August. The latter dismissed their argument with the words: 'My generals know nothing about the economic aspects of war.' Orders were immediately given that, while Hoth helped Leeb to reach Leningrad, Guderian and two of Bock's armies would strike south to link up with 17th Army and 1st Panzer Group from Army Group South to encircle the Soviet forces round Kiev. Throughout the operation argument continued between Guderian, who wished to exploit opportunities eastward, and demanded for that purpose the return of some of his forces which had been held back, and Halder at OKH who insisted that he concentrate on the more limited double envelopment which had been planned. Guderian's direct appeals to OKW were fruitless and merely angered Bock and Halder. When the operation finished successfully on 26 September, 665,000 more prisoners had been taken, with 884 tanks and 3,178 guns.

Three weeks before that Hitler changed his mind. Banking on the success of both the flanking moves, he stated that favourable conditions had been created

> for a decisive operation against the Timoshenko army group which is attacking on the central front: [it] must be defeated and annihilated in the limited time which remains before the onset of winter weather. For this purpose it is necessary to concentrate all the forces of the Army and Air Force which can be spared on the flanks and which can be brought up in time.

The trouble was that September and October are wet months in that region. Nevertheless the planned double encirclement started well, and by 7 October the initial enveloping operation at Vyazma and Bryansk had yielded another 673,000 prisoners, 1,242 tanks and 5,412 guns, raising high hopes of encircling Moscow itself before the snow came in December. But then the weather broke, so that even Guderian's tanks were only advancing five miles a day, and wheeled and horsed transport alike were stuck in the mud, as were the unfortunate infantrymen, thus deprived of their supplies.

After a two-week pause at the beginning of November, the operation was resumed as the ground hardened under frost, but the bitter cold and early snow and the increasing strength of the Soviet resistance brought it to an end on 5 December when Guderian's leading tanks were only nineteen miles from Moscow. Leeb's troops were on the outskirts of Leningrad, but could not encircle it, and those of Army Group South on the line of the Donetz and in the Crimea. They had reached Rostov on 20 November but, threatened by a Soviet counter-offensive, Rundstedt ordered a withdrawal to the River Mius on 28 November, for which he was replaced by Reichenau, who was allowed to withdraw a few days later, only to die of a stroke after a month. Leeb was also refused permission to withdraw from exposed positions round Leningrad and, early in January, resigned in protest.

The principal problem lay in the centre. The army was almost at the end of its tether. It had suffered severe casualties in men – 750,000, of whom 230,000 died – and in equipment, and neither the men nor the equipment it had left were suitably prepared to face the rigours of the winter. Meanwhile Soviet strength was growing, reinforced by formations brought from Siberia. Bock and almost all his subordinates pressed for a withdrawal to a shorter line, easier to supply, where shelter of some sort might be provided, and to attack which would face the Russians with the problem of advancing through the snow. On 8 December Hitler appeared to accept this, but five days later countermanded the order Brauchitsch had given for a ninety-mile withdrawal and thereafter stubbornly refused to consider it, Guderian and Hoeppner being dismissed for having made limited ones. On 18 December Bock resigned on grounds of ill health and was replaced by the compliant Kluge, the day before Hitler announced that Brauchitsch had retired and that he himself had assumed the post of C-in-C of the army. Brauchitsch, dogged by ill health and dispirited by the insulting way in which he was treated, had tendered his resignation nearly two weeks before, but received no reply until 17 December. Halder remained Chief of the General Staff and Keitel took over all of the C-in-C of the army's non-operational duties. Jodl headed the staff of OKW, which replaced OKH as the authority responsible for army operations in all other theatres. The Führer was now officially the Supreme War Lord, which he had in fact been for some time. Keitel was his man, and henceforth the senior officers of the army had no one to defend them against the whims of the dictator. One of the most important posts, head of the personnel staff (*Personelamt*), held by Keitel's younger brother Bodewin, was transferred to OKW. He was replaced in 1942 by an even more sinister figure, Schmundt, Hitler's personal adjutant.

The gloom and confusion reigning on the Eastern Front at the turn of the year was not relieved by any good news from elsewhere.

Rommel's exploitation of Wavell's diversion of troops to Greece, from which they had been rapidly expelled by Operation MARITA, had brought him to the frontier of Egypt in April; but this success had been reversed by the British offensive in November, driving him back to where he had been a year before. Although Japan's attack on the US fleet and on British possessions in the Far East in December would mean some diversion of their resources from their use against Germany, it decisively shifted the strategic balance against her, recognized by Hitler's immediate declaration of war against America.

In the first three months of 1942 on the Eastern Front, Army Group Centre was forced back some 200 miles, and the whole army in the east suffered 376,000 casualties, of whom 108,000 were dead or missing, and over 500,000 had been evacuated sick. Since the start of the war half the number of horses with which the army had begun the campaign had died, and 2,300 armoured vehicles, including 1,600 PzKw III and IV, had been written off. Hitler paid a high price for his refusal to authorize the withdrawal, which in the event was forced on his armies by the Russians. As early as January he ordered planning for a renewed offensive in the spring. The difficulties of finding the men and equipment to bring divisions up to strength led to its postponement until May. Army Groups North and Centre were to remain on the defensive, while Army Group South, commanded by Bock, recovered from his stomach ulcer, was to eliminate the Soviet salient near Kharkov and complete the occupation of the Crimea. Thereafter General Maximilian Freiherr von Weichs's 2nd Army and Hoth's 4th Panzer were to drive east to Voronezh on the Don, and then follow it down to the Volga at Stalingrad, joining up there with General Friedrich Paulus's 6th Army. General R.Ruoff's 17th Army and Kleist's 1st Panzer would recapture Rostov and then thrust to the Caucasus, the southern half of the Army Group then splitting off into a new Army Group A under List. For this ambitious operation Army Group South commanded sixty-eight German divisions and fifty from its allies, in two Romanian armies, one Italian and one Spanish. The nine Panzer divisions each had an average of 140 tanks, almost all PzKw III and IV, the latter including some with the new high-velocity 75-mm gun.

The first stage of Bock's offensive went well, but when his forces were near Voronezh, but had not yet captured it, he became concerned at the threat to his left flank, and wished to ensure that it was secure before he launched Hoth's 4th Panzer Army south-eastwards down the Don to encircle the Russian armies west of it. The result was that they started to escape eastwards and furious arguments went on between him, Keitel, Halder and Hitler, leading to counter-orders and confusion, Hitler alternating between shorter encircling movements and wider ones. Bock was replaced by Weichs on 14 July, never to be employed again, and the fatal decision taken to turn Hoth southwards towards Rostov instead of heading for Stalingrad, which at that time he could certainly have captured.

While confusion reigned in Army Group Centre and between it and Army Group South, the latter had made good progress and List's Army Group A had split off towards the Caucasus. Hitler now ordered List not merely to reach the mountains, but cross them, secure the oilfields at Batum on the Black Sea coast up to the Turkish frontier, and then thrust east through Georgia and Azerbaijan to the Baku oilfield on the Caspian. While he was doing this, Kleist's Army Group South would redirect Hoth to Stalingrad and beyond that to the mouth of the Volga on the Caspian at Astrakhan. Logistic realities flew out of the window. List refused to split his forces on either side of the Caucasus. Surprisingly he was supported by Jodl, and was dismissed on 9 September, the Führer assuming direct command of Army Group A himself and refusing to shake hands or dine together with Jodl and Keitel from then on. The cynical and pessimistic realist, Halder, was retired and replaced a few weeks later by the rapidly-promoted Führer-favourite Kurt Zeitzler, a former subordinate of Jodl, who hoped in vain that OKW would be given a say in the affairs of the Eastern Front.

Hitler now became fatally obsessed with Stalingrad, the capture of which was entrusted to Paulus's Sixth Army, whose principal task had been to protect the northern flank of List's thrust to the Caucasus, the foothills of which Kleist's 1st Panzer Army had reached in August. While Paulus concentrated his army, assisted by Hoth's 4th Panzer Army from the south, to try to seize Stalingrad, to the defence of which the Russians devoted an ever-increasing effort, the defence of the 400-mile-line of the Don was left to the Romanian, Hungarian and Italian armies. When the Soviet Army launched a counteroffensive on 19 November north and south of Stalingrad, they swiftly broke through, and within two days had encircled the whole of 6th Army and half of 4th Panzer, some 280,000 men, concentrated in an area thirty miles by twenty-five. Hitler refused to consider withdrawal but pinned his hopes on air supply and the creation of a new command under Manstein, called Army Group Don, which would command both the besieged forces and those trying to break through to relieve them. Both failed, insufficient strength being devoted to them as a result of the realities of logistics. Hitler's stubborn refusal to allow Paulus to attempt a break-out, in spite of the advice of both Weichs and Manstein that air supply could not meet his requirements, sealed the fate of his army, which starved to death. More than 200,000 were killed or captured, 91,000 of them surrendering on 31 January 1943. Only a few thousand of them were to see Germany again. By that time the Soviet offensive had been extended westward between the Don and the Donetz and Hitler had been persuaded to order withdrawal from the Caucasus just in time to save Army Group A from being cut off.

The tide turned in other theatres also. Rommel had driven the British Eighth Army back to Tobruk in the early spring, and in June had inflicted a severe defeat on Ritchie, forcing the British back to the El Alamein line, only sixty miles from Alexandria, by the end of the month. But logistics then imposed their harsh constraints. The early days of November saw Rommel's German-Italian army totally defeated by Montgomery, while the Anglo-American armies under Eisenhower landed in French North Africa at the other end of the Mediterranean. There was now no hope that Germany could win the war. Her only hope was to use her conquests as bargaining counters in seeking a settlement, but Hitler and his associates had been responsible for such crimes against humanity, principally against the Jews, that they knew that they themselves could not escape revenge. The army, on the whole, had tried to keep its hands clean of participation in those measures, sometimes protesting, but in general turning a blind eye and content to let the ss do the dirty work.

Whatever the situation, it was not in Hitler's nature to think in such terms, and one of the factors giving him hope that Germany could still win, in spite of the setbacks on land at the end of 1942, was the resounding success that Doenitz's submarines had achieved during the year. As soon as the United States had been brought into the war he concentrated his submarines, of which he then had a total of 250, nearly 100 of which were operational at any one time and to which new ones were being added at the rate of fifteen a month, on the unescorted shipping off the Atlantic coast of America. It gave him a rich picking and by July 1942 he had sunk over three million tons of shipping for a loss of fourteen submarines, only six of them in North American waters. Although Anglo-American countermeasures, including the increasing use of long-range aircraft, became more effective in the second half of the year, forcing Doenitz to shift the principal effort of his 'wolf packs' back into the Central Atlantic, his success continued as the number of operational boats deployed doubled. In August they sank 108 ships totalling over half a million tons, and in November 117 of 700,000 tons.

When he took over from Raeder, Doenitz retained direct command of the submarine fleet, whose operations in the Atlantic in the first six months of 1943 gave him good reason for that hope; but from June onwards there was a dramatic change in his fortunes. Energetic measures had been taken by Britain and the United States: the establishment of a joint command organization, the greatly increased use of long-range aircraft, fitted with short-wave radar and rocket projectiles, and the provision of escort carriers, and the generally improved capability of convoy escorts meant that Doenitz's submarines could seldom afford to move on the surface in daylight. Anglo-American shipping losses, which had been running at over 500,000 tons a month, fell to half that figure in April, when Doenitz lost fifteen submarines, to 200,000 in May, when he lost forty, and 22,000 in June, when he lost seventeen. Thereafter, although the total fleet numbered over 400, the numbers operational declined from a peak of 212 in April to 140 by the end of the year, by which time Anglo-American shipping losses were running at an average of less than 40,000 tons a month compared to over ten times that figure a year before.

The submarine fleet on which Doenitz had placed such high hopes was unable seriously to affect the deployment of American forces to Britain and the Mediterranean or the transfer of equipment and supplies from North America to British, Canadian and US forces in anticipation of operations on the European mainland. His hopes thenceforth rested on the introduction of the *Schnorkel* device and a new class of faster large submarines of 1,600-ton displacement, but the production of both was delayed by the troubles into which German war production fell, including intensive air attack on submarine construction yards. When they were introduced, the bases in France had been lost and it was too late for the submarines to have any significant effect on the war. In spite of severe losses in the last two years of the war, seventy were still at sea when it ended.

The navy laid the blame for the severe reversal of their fortunes on the air force, for failing to provide the submarines with adequate support, both in the form of reconnaissance and of offensive air action to deal with the enemy aircraft which had become the principal threat; for failing to play a full part in the general campaign against enemy shipping, by direct attack and by minelaying, and for failing adequately to defend their bases and construction yards. In fact the *Luftwaffe* did devote effort to all these tasks, but it was never enough to be decisive. The claims of the Eastern Front, of the Mediterranean, of air defence of Germany and the occupied territories generally took priority, and the area of operations in the west, from the northern tip of Norway all the way to the Bay of Biscay, inevitably dissipated the limited forces available. But the criticism was valid. After the failure of Operation ADLERANGRIFF, the *Luftwaffe* High Command never pursued a consistent campaign against shipping within the range of its aircraft, which as far as the Fockewulf 200 was concerned, extended 1,000 miles into the Atlantic, while the Junker 88 and the Heinkel 111 could cover the Bay of Biscay and all of the British Isles except the very northern tip of Scotland. While German submarines sank 4,323,000 tons of shipping in all seas in 1941 and 1942, aircraft sank 973,000, of which less than 400,000 was in 1942, and to which some 50,000 tons over the two years should be added, lost to mines laid by aircraft.

During this period the *Luftwaffe* in the west turned from the offensive to the defensive, although its Chief of Staff Jeschonnek and its

C-in-C Goering only reluctantly accepted the change and, with fatal consequences for the future, half-heartedly. Their complacency was in part due to the *Luftwaffe*'s initial success in this field, although it had devoted very little effort or thought to it before 1941. When the RAF began to bomb Germany in 1941, its results were too small in scale and too inaccurate to cause much concern. Nevertheless the challenge had to be met, little provision having been made to deal with night attacks, to which the British were forced to confine themselves. However, a night-fighter division had been formed in July 1940, under General Josef Kammhuber, with 164 aircraft, mostly Me 110s, who also controlled searchlights and radar, of which there were two types, the early-warning *Freya* and the *Würzburg*, with a range from six to thirty-one miles, which was used to control night-fighters and anti-aircraft guns. A series of zones, each fifty-six miles by twelve, was established, *Helle Nachtjugal* (Henaju) in which the action of guns, searchlights and fighters (only one at a time) could be co-ordinated and *Dunkel Nachtjugal* (Dunaju) zones, mostly along the coast, in which there were no searchlights. Improvements were made in the radars towards the end of 1941, while the number of night-fighters and their efficiency was increased. They had accounted for 422 bombers by November, when twenty-one aircraft out of 169 taking part in a raid on Berlin were shot down, and the tonnage of bombs dropped on Germany in the last quarter of the year fell to 7,600 from 13,550 in the previous quarter; in the first quarter of 1942 it was only 6,753.

The British then turned to area bombing of cities ill-protected by the Kammhuber system – Lübeck, Rostock and Cologne – causing severe damage and casualties. During the whole of 1942 they dropped 50,000 tons in a hundred raids, in seventeen of which they dropped over 500 tons. This was at a time when fighter squadrons were being transferred to the Eastern Front and the Mediterranean, one of the results of which was to put paid to the very promising development of 'intruder' operations. In these Ju 88Cs would mingle with the returning bomber squadrons and attack them over their own bases. 1942 also saw the introduction by the British of their GEE navigation system which enabled them to penetrate a narrow sector of the Kammhuber line in a short period, limiting the number of fighters that it could control against them; but this advantage was soon offset by jamming it and by the introduction of an airborne radar, with which fighters could close to their targets, having been brought within two miles of it by the ground radar. Throughout 1942, Kammhuber, whose command had been redesignated *Fliegerkorps XII*, widened and improved his system, 687 enemy aircraft having been brought down by night-fighters by the end of the year for the loss of 97, with a further 189 damaged; but this represented only 5.6 per cent of British losses, anti-aircraft guns being responsible for the majority. For instance in July and August the RAF reported 696 bombers missing, of which 169 were probably shot down by fighters, 193 by guns and 334 crashed due to unknown causes, while of 1,394 damaged, 153 were by fighters, 941 by guns and the rest by causes other than enemy action. The bomber offensive against Germany was not the only threat with which the *Luftwaffe* had to cope. British light bombers and fighters carried out a large number of attacks against targets in occupied France, in countering which the *Luftwaffe* generally came off the better. The air cover given to the German battleships *Scharnhorst*, *Gneisenau* and *Prinz Eugen* in February and the air battle over the attempted landing at Dieppe in August were notable examples of the success of Sperrle's *Luftflotte 3*. By the end of 1942 the RAF's night bomber force had lost 2,859 aircraft, 3.6 per cent of all the sorties flown, while in certain operations the loss rate was as high as 12 per cent. In daylight raids over the occupied territories 627 bombers (4.2 per cent of sorties) had been accounted for. The disruption caused by the British air offensive to the German economy had been minimal and had probably strengthened national morale. Although the effort devoted to air defence had greatly increased, only Milch was seriously concerned about the future, when the US Air Force would have joined in and the great effort – estimated at one-third of war production facilities – being devoted by Britain to the bomber offensive would have begun to develop its full effect. Impressed by Harris's 1,000 bomber raids, of which there had been three in 1942, Milch pressed for a considerable increase in the fighter defence of the Reich: out of an establishment of 653 it actually had 477, of which 330 were serviceable at the beginning of 1943; but Goering and Jeschonnek made light of his fears.

1941 and 1942 had seen a significant commitment of the *Luftwaffe* to the Mediterranean. Löhr's *Luftflotte 4* supported Operation MARITA in Yugoslavia and Greece with 1,090 aircraft, achieving almost total air supremacy. Soon after Greece had been invaded, Löhr suggested that Student's *Fliegerkorps XI* of airborne troops should be used to capture Crete, and Hitler agreed as the last British troops left Greece. The Air Landing Division could not be moved in time and was replaced by the 5th Mountain Division, already in Greece. The attack, launched on 20 May, supported by 650 aircraft of Richthofen's *Fliegerkorps VIII*, which were almost unopposed, not only rapidly succeeded but in the process inflicted serious losses on the British fleet both when it tried to engage the seaborne follow-up and when it was evacuating the British troops to Egypt. Student's victory had not been a cheap one. Out of a force of 13,000 men, over 5,000 were casualties; 220 aircraft were lost and 148 damaged, 119 of each being Ju 52 transports. These losses discouraged Hitler from embarking again on a major airborne operation, influencing his reluctance to attack Malta.

By that time Rommel had surrounded Tobruk and reached the Egyptian frontier. The logistic supply of his and the Italian forces in Libya depended on shipping from Italy and was vulnerable to aircraft and submarines based in Malta. In the second half of 1941 they sank 280,000 tons of military stores, severely affecting Rommel's operations. In October, before the British launched their offensive which drove Rommel back to where he had started, Hitler decided to reinforce the 400 aircraft of Geisler's *Fliegerkorps X*, responsible for air operations in the eastern Mediterranean, with General Bruno Lörzer's *Fliegerkorps II* and the headquarters of Kesselring's *Luftflotte 2* from the Eastern Front. The former was moved to Sicily to intensify the attack on Malta and the latter to Rome, where Kesselring also assumed the post of C-in-C of all German forces in the area. By the end of April 1942, air attacks delivering 10,000 tons of bombs – half the total dropped on London in the 'Blitz' of 1940 – rendered the island inoperable as a base for either aircraft or submarines and almost completely cut it off from external supply. This made it possible to reinforce Rommel and bring his forces up to strength. In February he drove Ritchie's Eighth Army back to Tobruk and in the first two weeks of June defeated it there. However, air attack alone could not remove Malta altogether as a threat, and Kesselring proposed that it should be captured by Student's *Fliegerkorps XI* and two Italian parachute divisions, their 30,000 men to be followed up by an Italian seaborne force of 70,000. The support of the 500 Ju 52s, each towing a glider, would need all the strength of *Luftflotte 2*, reduced by the return of *Fliegerkorps II* to Russia to 510 aircraft, half of them bombers. Operation HERCULES, as it was named, could not therefore coincide with a major operation in Libya. At the end of April, Hitler and Mussolini, accompanied by Kesselring and Ugo Cavallero, agreed that it would be launched when Rommel had captured Tobruk and reached the Egyptian frontier.

In the intense fighting round Tobruk in the first fortnight of June, General Otto von Waldau's *Fliegerführer Afrika's* (as *Fliegerkorps X* had been renamed) 260 aircraft (of which a high proportion were always unserviceable) were reduced to 100 and, when Tobruk fell on 20 June, Kesselring told Rommel that he could not provide adequate air support beyond the Egyptian frontier. His advance should be halted there, while Operation HERCULES was launched. Rommel, created field marshal, appealed to Hitler and Mussolini, proffering the tempting bait of driving the British out of Egypt. Neither of them were keen on HERCULES, Hitler fearing that Student's *Fliegerkorps XI* would suffer casualties on the scale of Crete and that the Italians would fail to play their part. Mussolini also feared the risks and was sorely tempted to pose as the victor in Egypt. They gave Rommel his head, a decision which, with the advantage of hindsight, can be seen as fatal to their joint venture in the Mediterranean; but the stakes were high and the prize glittering. Rommel did in fact bounce the numerically superior

forces opposing him off, first the frontier defences and then those of
Mersa Matruh, with astonishing ease and speed. It is conceivable that
he could have repeated the process at El Alamein at the beginning
of July but, by a narrow margin, he failed. Like the Russians at Stalingrad, the British were back on their base and the realities of logistics
were all in their favour. They could produce an air effort over the
battlefield far superior to that which *Fliegerführer Afrika* could muster.

In spite of the critical position at Stalingrad, the *Luftwaffe* High
Command found it possible to reinforce Kesselring to meet the threat
posed by the Anglo-American landings in French North Africa, his
total strength of aircraft being raised to 1,220 by the end of the year,
most of them in General Harlinghausen's newly formed *Fliegerführer
Tunisia*, air transport between Sicily and Tunisia having been decisive
in preventing Eisenhower's forces from reaching Tunis in the early
stages of the campaign. From then until the end of the campaign in
North Africa, *Fliegerführer Tunisia*, which absorbed *Fliegerführer
Afrika*, with some 140 combat aircraft, supported by *Fliegerkorps II*
from Sicily and Sardinia, provided a moderately effective degree of
immediate support to General Jürgen von Arnim's forces, but could
inflict little damage either on Anglo-American forces or the air and
shipping support they received.

1943 showed up all the weaknesses in the German war machine:
the lack of any body or staff to consider the strategy of the war as
a whole and exercise authority over its execution, other than the
Führer himself, who suffered from a combination of impetuosity and
vacillation, aggravated by the deteriorating state of his health. The
lack of a clear strategy had resulted in the absence of a clear consistent
plan for organizing the resources of the nation. To set in the balance
against this were both the apparently almost inexhaustible capacity
of the German soldier to endure hardship, to fight with courage and
skill and to improvise, and the hitherto untapped reserves of German
industrial resources, helped by the extraction of material and labour
from the conquered territories. Hitler was almost permanently at loggerheads with his generals, none of whom he trusted, and they were
at odds with each other. Jodl and Zeitzler quarrelled about the authority each wielded, while the Army Group commanders by-passed
Zeitzler and dealt direct with the Führer, who would either give way,
only to be appealed to by a rival, or refuse to give a decision, unless
the request was for a withdrawal when it would be angrily refused.

The near-chaos in the higher command was further complicated by
the appointment of Guderian in March 1943 as Inspector-General of
all *Panzertruppen*. This gave him authority over the organization,
training and equipment of all Panzer troops, including their motorized
infantry and anti-tank units, as well as over development of their weapons including tanks themselves. He was to be independent of OKW

and OKH, responsible solely to the Führer. This further complication of the already confused responsibilities for the army was instigated by the sorry state into which the Panzer arm had fallen. In January 1943, of the German army's total tank strength of about 5,000, only 495 were fit for action on the whole of the Eastern Front, most of them PzKw III and IV, inferior to the Russian KV and T34 tanks. A few Tigers (PzKw VI) were available but none of the new Panther (PzKw V). OKH wished to halt all production of IIIs and IVs and concentrate on Vs and VIs, but this would have severely reduced the tanks available to divisions in 1943. Losses in 1942 had exceeded new production and continued to do so in the first half of 1943.

Guderian backed Jodl in being concerned that the army in the west should be built up to be able to withstand the Anglo-American assault which it seemed certain would take place during the year. Rundstedt's OB West in France had thirty-two divisions with only about 300 tanks. By July it had been brought up to thirty-eight with 350 tanks, compared to the Eastern Front which had 175 with 2,269 tanks and 3.1 million men, 1.3 million being distributed among all the other fronts. At that time the army comprised 243 divisions, but they were 616,000 men short of their full strength. One major source of weakness was Hitler's refusal to reduce the number of divisions in order to keep them up to strength in men and equipment and reduce administrative over-heads. When he did agree to a major transfer of men from the air force to the army, he gave way to Goering's insistence that they should form twenty-two complete *Luftwaffe* field divisions, commanded by *Luftwaffe* officers and NCOs. The result was that, lacking training and experience, they were of limited value. The *Waffen SS* divisions, however, of which there were eleven, were of high quality as fighting material.

It was clear, even to Hitler, that there was no hope of defeating the Soviet Union before facing an invasion in the west. If, however, the latter could be defeated, a settlement with the Soviet Union might be achieved. Guderian and Jodl wished to remain on the defensive in the east, preferably shortening the front in order to make it more secure; but Hitler, backed by Zeitzler and Manstein, commanding Army Group South, preferred an early offensive there in order to deal the Soviet Army a hard blow before it had recovered from its winter offensive and could prepare for the next one. When the ground began to harden after the spring thaw would be the earliest date, and that would be April. Impressed with the way Manstein had pinched out the Soviet salient near Kharkov in the previous year, the salient round Kursk was chosen as the area for the proposed operation, named CITA-DEL. Manstein's proposal was for a withdrawal in the south while formations were transferred to the north of the salient, from which Kluge's Army Group Centre would thrust southwards behind the Russians; but this would have meant abandonment of the Donetz basin and isolation of Kleist's Army Group A in the Crimea, and Hitler insisted on a classic envelopment from both flanks, Field Marshal Walter Model's 9th Army attacking from the north and Hoth's 4th Panzer from the south. Model had serious doubts about his chances of success and kept on asking for postponement to allow his forces to be strengthened. The delay was fatal. Any hope of surprise was lost, and the Russians benefited more from the delay than did the Germans, reinforcing the salient and strengthening its defences, especially with minefields, while making their own preparations for a summer counteroffensive which would outflank the two arms of the pincers.

The result was a battle on a titanic scale, from which the German army never recovered: forty-three divisions, of which seventeen were Panzer, with 1,850 tanks and 530 assault-guns (out of 2,269 and 997 on the whole front) were committed to the attack, launched on 5 July against the salient 100 miles wide and 150 miles deep, protected by six belts of defences, backed by 3,306 tanks and 20,220 guns. After three weeks of intensive fighting it was clear that the operation had failed, and both Model and Hoth were in danger of being encircled themselves. It was called off and Hitler began to transfer divisions to meet the threat of Italian defeat, the Anglo-American invasion of Sicily having started. No respite, however, was to be given to the battle-weary troops on the Eastern Front as the Soviet army launched a series of offensives which forced them back. After the failure at Kursk, Zeitzler wished to construct a fortified defence line, known as Panther, from Narva, west of Leningrad, running through Vitebsk, Gomel and Kiev and thence down the Dnieper to the Sea of Azov, but Hitler refused until 12 August when he allowed work to start, at the same time issuing a stern warning that no withdrawal was allowed without his permission. The Soviet army renewed its offensive in November, chiefly against Army Group Centre, where Busch had recently replaced Kluge, after the latter had complained to Hitler about the serious state of his formations, the soldiers of which 'were beset by a feeling of isolation and neglect' in the face of the masses of Russian infantry. By the end of the year the situation was stabilized more or less on the Panther Line, although Kiev had been lost. Morale at all levels was weakening and relations between Hitler and his generals were at an all-time low.

While the German army was being driven out of the Ukraine, Italy had surrendered. After the fall of Tunis in May, Hitler had set in hand plans to deal with the possibility of defection, hoping to keep Italy in the war by forceful means. He was principally concerned by the threat of an Anglo-American landing in the Balkans, where the Italians provided a large proportion of the forces attempting to hold down the resistance. He feared a link-up between the Anglo-Americans and the Russians in the Danube valley. He built up a force in southern Germany, under Rommel, ready to intervene in Italy, while keeping Kesselring ignorant of his intentions. When, on 26 July, the latter was told of the plan to land an airborne force near Rome, arrest Marshal Pietro Badoglio and reinstate Mussolini, he objected and managed not only to have the plan cancelled but to have his own authority confirmed, displacing Rommel. When the Anglo-American landing took place at Salerno and the Italian government surrendered on 9 September, German troops in Italy and the Balkans disarmed the Italians, who were threatened with being shot if they resisted. Where, as on the island of Cephalonia, they did, they were shot after they had been captured, on the slender grounds that they had committed treason. Hitler's original intention was to hold only northern Italy on the line of the Apennines, but the success of General Heinrich von Vietinghoff's Tenth Army, reinforced from its original seven to a strength of fifteen divisions, in holding the Anglo-American force 100 miles south of Rome in November, caused him to revert to his characteristic strategy of no withdrawal, which in this case paid off.

By 1943 the *Luftwaffe*, which had dominated the skies in the early years of the war, was set on a course of decline. This was primarily due to sheer overstretch, but its own internal troubles aggravated its problems. The campaign in the east undermined all its plans and cherished dictums, which were based on concentrating a balanced force of aircraft on one task at a time, with emphasis on the indirect support of the army and, on a lower priority, the navy. Strategic bombing, directed against the enemy's industrial resources, had not formed part of its strategy, which was tied to the concept of a short war. Milch, Jeschonnek and even Goering himself had strongly opposed Operation BARBAROSSA, but were forced to accept it on the assumption that it would be over by the time winter set in. Milch was sceptical, ordering winter clothing and equipment for the *Luftwaffe*; but the organization, training and production programmes were based on a swift victory.

At first that seemed possible, as the *Luftwaffe* established an even greater degree of air supremacy than it had in the west. Its 2,000 combat aircraft, fewer than had been used there a year before, had by the end of the year destroyed 15,500 aircraft, 3,200 tanks, 57,600 vehicles of all kinds, 2,450 guns, 650 trains (damaging 6,300 others), 1,200 locomotives (damaging an equal number), and cut railway lines in 7,000 places. The vast extent of the area to be covered and the paucity of rewarding targets for indirect support, other than rail traffic and bridges, both of which the Russians were quick to repair, forced the *Luftwaffe* into devoting the greater part of its effort to direct support of the army, for which most of its aircraft were not well suited. Nevertheless by the end of 1941 it had every reason to feel that it had made a major contribution to the apparent success of the campaign, if measured in losses inflicted on the Soviet forces and territory gained.

But the winter severely restricted its activity, a factor of which the Red Army took advantage. By the time it came, the number of combat aircraft in the east had fallen to 1,700,* of which only 43 per cent were serviceable. In 1942 both the Mediterranean and home defence made greater claims, at a time when the aircraft production programme was in disarray. At the heart of the problem lay the basic assumption that the war would be short, that there was no need to make any major changes in the programme that had been planned before it started, nor to impose the sort of restrictions and demands on the workforce and the civilian economy generally that had been adopted in Britain. A forty-hour week, one-shift working and observance of all normal holidays continued. The three armed forces had to compete with each other for the resources, notably steel, devoted to war production, and early in 1941, in preparation for BARBAROSSA, priority had been switched to the army. Two of the new aircraft which formed an important part of the programme, the Me 210 and He 177, were failures, and Udet was proving to be hopelessly incompetent as head of the *Generalluftzugmeisteramt*, responsible for procurement, and thus found himself continuously at loggerheads with Milch. The strain of it drove him to suicide in October 1941, Milch assuming his responsibilities in addition to those of Secretary of State.

While the defence of the homeland and operations in the Mediterranean both increased the demands on the *Luftwaffe* in 1942, the Eastern Front remained first priority to which some 66 per cent of its total strength was devoted. By the middle of June, when the major offensive in the Ukraine, directed towards Stalingrad and the Caucasus, started, that meant 2,750 combat aircraft, of which 1,600 were allotted to the support of the offensive. But by that time the Soviet air force had been increased to three times that number. The demands for support of the army over such a wide front and the improvement in quantity and quality of the opposing air force meant that hardly any effort was available for operations other than direct support of the army, and when Hitler divided effort in the south between the thrust to the Caucasus and the attempt to capture Stalingrad, Richthofen's *Luftflotte* 4 had not the strength to support either adequately, its numbers having fallen by 20 October to 975, of which only 600 were fit for operations.

The *Luftwaffe*'s major test came with the demand for air supply to Paulus's surrounded army. None of the generals, army or air force, believed it could be done, but Goering, already under criticism by

---

* Out of a total *Luftwaffe* combat strength of 3,130 (1,332 bombers, 1,472 fighters, 326 dive-bombers).

Hitler for the failure of the *Luftwaffe* to win the Battle of Britain and to prevent air raids on Germany, as well as for his initial opposition to embarking on the campaign in the east, assured Hitler that his air force could repeat the success they had achieved in supplying two pockets of troops, surrounded by the Russians near Leningrad earlier in the year, the 100,000 men at Demyansk being supplied by air for three months at a cost of 265 aircraft, most of them Ju 52s. Jeschonnek, although fully aware that the task was beyond the *Luftwaffe's* resources, kept quiet for fear of offending his hero, the Führer. In November the total fleet of Ju 52s was 750, of which two-thirds could be made available, although not all immediately. At the beginning of December only 320 were actually deployed with *Fliegerkorps VIII*, while 1,000 would have been required to meet Paulus's needs. Even by using He 111 and FW 200 bombers there was no hope of meeting the target, 3,500 sorties being flown, delivering 6,000 tons of fuel and equipment, at a cost of 488 aircraft, 266 of them Ju 52s.

By the end of 1942 the *Luftwaffe's* future bomber programme was in disarray, and it had to continue to rely on the Ju 88 and He 111 for most of its effort. In any case it was clear that, as the strength of the enemy air forces in all theatres increased and the British bombing offensive was intensified, the principal need was for more and better fighters. Of the latter the Me 109 (of which there were 660) was the mainstay, but the excellent FW 190 (there were 580) was overtaking it. Milch pressed for a dramatic increase in fighter strength as the British bombing offensive against area targets was intensified in 1943. Its attacks on submarine yards achieved little, but the destruction inflicted on cities – Cologne again, the Ruhr, and finally, at the end of July, the four devastating raids on Hamburg – convinced all the senior officers of the *Luftwaffe*, including Goering, that its first priority must be home defence, and that a crash programme of fighter production, and of pilot training to meet it, had to be ordered. But Hitler would have none of it. While castigating the air force for its failure to prevent these attacks, he demanded that terror be met by terror, and that priority should be given to attacks on Britain, for which a special command – *Angriffsführer England* – was created, and that special emphasis should be given to bomber production, particularly of the He 177. In spite of this Milch managed to achieve a significant increase in fighter production, although far short of the 4,000 a month target set for the following year. By July 1943 it had increased to 1,263 from 521 in January, but fell in the second half of the year as a result of the Anglo-American bomber offensive against the aircraft industry. The latter cost the US Army Air Force (USAAF) dear. By the end of October, after its raid on Schweinfurt, in which 60 out of 294 B 17s were shot down, one third of the 8th USAAF's fleet of B 17 and B 24 bombers, operating in daylight without fighter escort, had been put out of action for an average loss in combat of 4 per cent of the German fighters. But the latter had no such success against the British night raids, which suffered an average of 2 per cent loss in aircraft employed against a 3 per cent loss in fighters used against them.

1943 had been a bad year for the *Luftwaffe* on all fronts, losses of all kinds averaging 1,700 aircraft a month. Depressed at a failure for which he could not escape responsibility, although much of the blame was attributable to Goering's idleness, inefficiency and refusal to stand up to Hitler, and disillusioned by the performance of the latter, whom he had originally so much admired, Jeschonnek shot himself on 17 August after the successful British attack on the rocket research station at Peenemünde. He was succeeded by General Korten, an able man who worked more smoothly with others than his predecessor had.

He faced a formidable problem. The build-up of the USAAF in Europe, both in the Mediterranean and in Britain, together with improvement in quality and quantity of both the Soviet and the British air forces, made it impossible for the *Luftwaffe* to expect to establish any form of superiority in the air except over Germany itself, and for the present there only in daylight. That was soon to be challenged by the American long-range Mustang fighter and, later, the British Spitfire XIV. Although highest priority was given to home defence, the largest number of aircraft, in the first half of the year, was committed to the Eastern Front, where 1,150 of its 1,710 combat aircraft were concentrated in the south, largely because of the priority given to keeping the enemy away from the oilfields of Romania, which was only too anxious to get out of the war. But the Russian offensives covered the whole of the 2,000-mile-long front and the *Luftwaffe*'s operations had little effect on them. In the Mediterranean they made practically no impact on the Anglo-American advance to Rome and beyond. Its attempt to affect the preparations for and the subsequent execution of the Anglo-American landing in Normandy in June 1944 had even less effect, which was not surprising in the light of the formidable strength of the Allied air forces based in Britain. Hitler insisted on a renewed so-called 'Baby Blitz' against British cities in the spring with the aim of deterring Allied attacks on German cities. It was an ignominious failure, resulting in the loss of 329 bombers, which had only succeeded in dropping 27 per cent of their total load of 2,812 tons of bombs on the target. It prejudiced the *Luftwaffe*'s ability not only to deliver any effective air attack against the invasion when it came, but also the possibility of carrying out air reconnaissance of the preparations for it. It should have been undertaken at the same time as attack by the V 1 (*Vergeltungswaffe* – retaliation weapon) cruise missile, on which great hopes had been placed. The plan had called for production at a rate of 3,500 a month, so that from the beginning of the year one could be launched at London every twelve minutes. Milch believed that a four-day bombardment would 'be the end of any real life in the city'. But delays in production and deployment, largely caused by Allied air attacks, prevented their use until 15 June, and then only on a limited scale, averaging 120–190 bombs every twenty-four hours, the planned rate falling from one every twenty-six minutes to one every ninety. By the time the launching sites were overrun in September, 8,964 bombs had been fired against London. More were later fired from sites in Holland and from aircraft.* The two V weapons together accounted for one out of every seven people killed and two out of every seven seriously injured by aerial bombardment in Britain throughout the war. Had they been able to start their operations earlier, and had the *Luftwaffe* been able to protect the launching sites, their effect would have been significant.

The Allied air superiority covering the landings themselves and subsequent operations in France was overwhelming. In the first six days they flew 49,000 sorties, while the *Luftwaffe* managed only 400 to 500 a day, the effect of which was hardly noticed by the enemy. In the defence of the Reich they had greater success, inflicting such heavy losses on raids deep into Germany that the British night bombers were diverted to targets in western Germany, France and the Low Countries. In the raids on Berlin and elsewhere between November 1943 and March 1944, involving some 20,000 sorties, the RAF lost 1,047 aircraft shot down and a further 168 damaged beyond repair, culminating in a raid on Nuremberg at the end of the period when 94 out of 795 bombers were shot down, and another 61 damaged beyond repair, a loss rate of 19.4 per cent. This was achieved partly by a considerable increase in the deployment of anti-aircraft guns and in the employment of 'wild boar' tactics, in which the night fighters mingled with the bomber streams.

Although much of the bombing campaign was directed against the aircraft industry, and had a serious effect on planned production, an internal reorganization in March counterbalanced it, achieving a monthly fighter production rate by September of 3,375, an increase since February of 332 per cent. Total production of all types in 1944 was 39,807, an increase of 156 per cent over 1943. Fighter production would have been higher if Hitler and Goering had not still insisted on attempts to double the number of bombers to 2,600, a policy they did not abandon until July. Even then, the defence of the Romanian oilfields remained a high priority. The biggest blow to the *Luftwaffe*

---

* Of the total of 10,942 launched, only 6,725 crossed the English coast and 2,500 reached their target, killing 6,184 people, seriously wounding 17,981 and destroying 23,000 houses. The V 2 ballistic missile, an army weapon, of which 1,054 fell, mostly on London, between September 1944 and March 1945, killed 2,754 and seriously injured 6,523.

came with the switch of Allied bombing priority to the oil industry, the USAAF successfully attacking the nine main synthetic oil plants in May and June, affecting 90 per cent of the *Luftwaffe*'s supply. In May the *Luftwaffe* consumed 195,000 tons of aviation fuel.* In these circumstances the appearance of the jet-engined Me 262 could make little difference: in spite of their high technical performance, the Mustangs learnt to deal with them. By the end of the war the 200 Me 262s in service had destroyed 150 aircraft against a loss of 100.

By 1944 it was clear to all but Hitler that the war could not end in anything that could be called victory, but many of the army generals believed that, if they were allowed to direct strategy and conduct operations in the way they wished, a situation could be created in which a settlement with their enemies could be achieved before Germany itself was ruined. But their chances of doing so, if they existed at all, were destroyed by the strategy of 'no withdrawal' insisted upon by the Führer, by his hanging on to areas he believed to be of political or economic importance, and of interfering in detail with the conduct of operations.

In the east Manstein, commanding Army Group South, was foremost in advocating a significant withdrawal to a shorter front, while creating mobile reserves which could launch a counteroffensive as the Soviet army lumbered forward. In the west Rundstedt favoured the same sort of strategy, although Rommel, believing that the overwhelming power of the Anglo-American air forces precluded mobile operations, did not. Relations between Hitler and the generals became worse and worse, as they pressed for a reorganization of the high command and for a more realistic strategy. Zeitzler, appointed as Chief of the General Staff because he was both a favourite and an admirer of Hitler, became exasperated and confronted his erstwhile hero more boldly than those who had never favoured the jumped-up corporal. Having five times tendered his resignation he reported sick in June and, after the 10 July plot, although he had no part in it, was dismissed from the army and forbidden to wear uniform. He was succeeded by General A. Heusinger and after 10 July by Guderian, who soon found himself protesting as strongly as Zeitzler had, in spite of his unsavoury lickspittle behaviour in the aftermath of the attempted coup.

Insistence on the importance of the Romanian oilfields and of the need to hang on to the occupied Baltic states of Latvia, Lithuania and Estonia had a fatal effect on the campaign in the east. The Soviet

---

* By September monthly output was down to 17,000 tons, 9 per cent of the figure in March and, after a rise to 46,000 in November, it fell away to 17,000 in January 1945 and 1,000 in February, by which time consumption had been forced down to 27,000 tons a month.

offensive started in January 1944 in the south, forcing Manstein back from the Dnieper, and by the end of March it had reached the Carpathians, leaving behind 80,000 Germans in the Crimea, at the same time pushing Army Group North back from Leningrad to the Estonian border. Busch's Army Group Centre, weakened in order to strengthen the flanks, was unable to hold the next main effort when it was attacked in June and was almost totally destroyed, losing 300,000 men, a third of them encircled at Minsk. In the middle of July a major offensive was launched south of the Pripet marshes which brought Russian troops into the outskirts of Warsaw at the end of the month.

Faced with a disastrous situation in both the east and the west, the army suddenly gained a reprieve. It cannot be said with certainty whether that was due to the exhaustion, physically and logistically, of the Soviet armies after a series of major offensives involving advances of 150–200 miles, or to political calculation: that the Poles should be destroyed by the Germans before the Soviet Union swallowed them up, and that the Balkan countries – Romania, Bulgaria and Hungary – should be occupied before they could come under the influence of the Anglo-Americans. Whatever the reasons, the Soviet army stopped in its tracks, while it moved into Romania and Bulgaria. Meanwhile it had occupied the Baltic states, isolating twenty-six German divisions in Courland, which Hitler refused to evacuate. Reduced to twenty-two, they were destined to stay there, making no contribution to the war, until the end.

In spite of the Führer's insistent veto on withdrawal and his demand that 'fortress areas' of surrounded forces should be defended to the last man and round, the German armies in the east reeled back and the long-feared invasion came in the west. From 1942 onwards Hitler's strategy to meet this threat rested on the creation of the 'West Wall', a continuous line of fortifications all the way from Holland to the Spanish border, although, for a long time, his principal anxiety was of a landing in Norway. As 1943 passed without one and the threat of one in 1944 became imminent, Hitler, in theory, gave priority to the west, where Rundstedt was in overall command. Rommel, after his brief commitment in Italy, had been given a special responsibility for the most threatened area in the Low Countries and France, in command of Army Group B, and General J.Blaskowitz, commanding Army Group G, was responsible for everything in France south of the Loire.

As has already been mentioned, Rommel, keenly aware from personal experience of the inhibiting effect of Allied domination of the skies, wanted mobile formations, especially the Panzer divisions, to be stationed within easy reach of the coast, so that the landing forces could be defeated before they could establish a bridgehead capable of being developed for the introduction of more divisions. His superior, Rundstedt, favoured a totally different strategy, wishing to keep the mobile formations concentrated near Paris and to use them, as Manstein had wanted to use his, to deliver a major counterattack when the landing forces had extended themselves. In the event Hitler decided on an unsatisfactory compromise. He supported Rommel in insisting that the invasion must be defeated on the beaches, but split the mobile formations between support to the coastal defences and a central reserve which could not be used without his authority.

All were convinced that the principal landing area would be the Pas de Calais, although the possibility of a landing in Normandy was accepted in March. But the only reinforcements sent to that area between then and June were some anti-aircraft units and one division, bringing the total there to six infantry, four of which were low-grade coastal divisions, and one Panzer. Whether or not the delay on 6 June in authorizing the movement of two reserve Panzer divisions to the landing area was critical, it is difficult to say, as they would undoubtedly have suffered severe casualties from air attack if they had attempted to move in daylight on that day. As it was, even though most of their move was made by night, they both lost heavily from air attack as they approached the battlefield on the 7th. From then on it became a question of which side could build up its strength the more rapidly, the Allies initially hampered by weather and lack of space, the Germans by difficulties of movement caused by air attack and by the pressure being exerted by the Soviet army in the east and the Allies in Italy. Rundstedt wished to withdraw all his forces from south of the Loire, use them to relieve the Panzer divisions in containing the bridgehead, and employ the latter in a major counterattack to drive the Allies back to the beaches. But Hitler refused and insisted on no withdrawals, no transfers from the Pas de Calais, and that four Panzer divisions, two sent from the south of France and two from the east, should be used in an attack on the British sector at the end of June. It never materialized, as it coincided with a major British attack and the Panzer divisions found themselves committed to strengthening the defence to prevent a breakthrough.

Dissatisfied with this, Hitler replaced General L.Freiherr Gehr von Schweppenburg, commander of *Panzergruppe West* and the highest ranking *Waffen ss* officer, by General Hans Eberbach on 1 July. Rundstedt resigned in protest the next day and was succeeded by Kluge. The latter soon found himself making the same sort of representations about the futility of the strategy directed by okw as Rundstedt and Rommel had. When the Americans finally broke out from the western sector of the Normandy bridgehead early in August, Kluge wished to withdraw to the Seine, but instead Hitler ordered the counterattack towards Mortain which sealed the fate of General Paul Hausser's 7th Army. Kluge was less inclined to argue against it for fear of his personal fate. On 20 July Colonel Claus Graf von Stauffenberg had failed in his attempt to dispose of Hitler. Kluge had more than once been approached by conspirators and a group of officers in his headquarters had attempted and failed in a previous bomb plot in March. When he was summoned to Hitler's headquarters as the jaws of Eisenhower's armies closed round the luckless 7th Army south of Falaise, he committed suicide, and was replaced by Model, who also assumed direct command of Army Group B, vacant since Rommel had been wounded in an air attack shortly before the 20 July plot. In the end two field marshals, sixteen generals and one admiral were murdered or forced to commit suicide and many others, including Halder, arrested. Among those later murdered were Rommel and General Fritz Fromm. The latter had been in command of the Reserve Army throughout the war and, as soon as he realized that the plot had failed, arrested the principal conspirators and had them shot; but his previous contact with them was suspected.

Although a considerable number of men escaped from the pocket, the equipment they brought with them was almost all abandoned before they crossed the Seine. No more than 120 armoured vehicles got across, 2,200 having been left behind. Almost half the number of men that had been engaged in Normandy since 6 June had become casualties – 240,000 killed and wounded, 210,000 taken prisoner. Of the fifty divisions employed, only the equivalent of ten survived, and they had practically no equipment.

Any possibility of making a stand on the Seine had disappeared and it proved impossible to stabilize the situation until the British had reached Brussels and Antwerp, and the Americans the easter frontier of France. The astonishing capacity of the German army to improvise, the failure, largely for logistic reasons, of the Allies to maintain the momentum of their advance and a contribution from the *Luftwaffe* of parachute troops and ground crews, made it possible against all the odds for a defence to be established all the way from Holland to the Swiss border which effectively resisted the attempts the Allies made in November to advance from the Meuse to the Rhine, with their eyes on the industrial heartland of the Ruhr.

It was clear to all the generals, although Keitel and Jodl would not admit it, that, having achieved this miracle while the Soviet armies were still over 300 miles from Berlin and had not even entered East Prussia, there was no point in prolonging the war, and that every attempt should be made to bring it to an end, preferably by an arrangement with the Anglo-Americans. But Hitler was now the prey of alternating optimism and pessimism. The optimist in him pretended to believe that a counteroffensive in the west could split the Allies and drive them back to the sea, after which new weapons would swing the balance in his favour. The pessimist welcomed the idea of a supreme sacrifice, a *Gotterdämmerung*, in which he, the German army and the German people would all meet their fate together.

There is some evidence that Hitler was thinking in terms of a counter-offensive in the west as early as the end of August, but it was on 24 October when the order for it was formally issued by OKW. By then the seventy-year-old Rundstedt had been reinstated as C-in-C West, Model remaining in command of Army Group B, which would have to execute the plan. Both thought it impractical. It might succeed in pushing the Americans back to the Meuse, but to expect to reach Antwerp and maintain a salient with its apex there was a fantasy. Attempts to limit the scope of the operation were rejected, and postponement from November to December reluctantly accepted. Twenty-eight divisions, of which nine were Panzer, were committed to an advance in midwinter across the grain of a belt of wooded, hilly country in the face of an enemy who enjoyed a vast superiority of resources and total air supremacy. Launched in the Ardennes on 16 December it met with initial success, helped by the overcast weather and the weakness of the defence in the area; but the final outcome could never have been in doubt. By 22 December Rundstedt knew it had failed and sought permission to withdraw. It was refused, and it was not until 22 January that Hitler resigned himself to abandoning the operation and ordered the return of the SS Panzer divisions to the east, where not only they, but the 600 armoured vehicles and 130,000 men (19,000 of these had been killed) lost in the Ardennes were badly needed.

On 12 January 1945 the Soviet army began the final offensive which was to bring it by the end of the month within fifty miles of Berlin. Once again Hitler rejected the advice of his generals. Guderian had argued for withdrawal of the twenty-two divisions isolated in Courland and the thirty-three, of which eight were Panzer, in East Prussia, liable to be cut off by a Russian advance into Poland. He wished to create a shorter line of defences, the forward line twelve miles in front of the rear lines, which would be the main one, in imitation of the defensive tactics which were successfully used in the First World War, and on a smaller scale in almost every successful German defence, the aim being to force the enemy to use up his effort, in men and ammunition, before he reached the main defence, which by that time could be reinforced from reserves. Not only did Hitler reject all these proposals, but he insisted on ordering a counteroffensive into Hungary, obsessed by the need to retain control of its oilfields. The result was that, of the 103 infantry and 32 Panzer and Panzer-grenadier divisions in the east, only 45 and 16 were left to face the main Soviet effort in the centre aimed towards Berlin. Although they managed to impose some delay on the Soviet advance in February on the Oder-Neisse line, the Russian steamroller rolled on relentlessly until, on 16 April, they launched 193 divisions against the 50 weak German ones, encircling Berlin by the 25th and joining hands with the Americans on the 27th. Three days later Hitler shot himself.

After the failure of the Ardennes offensive it was only a matter of time before the Allies closed up to the Rhine, crossed it and encircled the Ruhr. Once more Hitler's refusal to countenance a withdrawal behind the Rhine led to the loss of 350,000 men in the battles in March between that river and the Meuse. On 23 March the Anglo-American armies crossed to the east bank and by 18 April had surrounded most of Model's twenty-one divisions in the Ruhr, Model himself committing suicide. In that month 1,650,000 German soldiers were taken prisoner in the west, bringing the total since the landings in Normandy to nearly three million. They were more fortunate than their compatriots taken prisoner in the east. On 3 May a formal surrender was requested by Keitel on the orders of Doenitz, whom Hitler on 28 April had nominated as *Reichspresident* when he heard that Heinrich Himmler had made peace overtures to the Western Allies. A number of attempts by Keitel and Jodl to arrange surrender to the Western Allies in all theatres, but not to the Russians, were rejected. The final act of surrender on all fronts was signed by Jodl at Eisenhower's headquarters at Rheims on 7 May and by Keitel and Georgi Zhukov in Berlin the next day.

To what extent should the German armed forces, and particularly the army, be blamed for the disaster which had befallen their country? Could they and should they have prevented it? Although few of their senior officers were active supporters of the Nazi party, they were grateful to Hitler, after he came to power, for removing the restrictions of the Versailles Treaty and restoring their strength. The air force had particular reason to be grateful for the establishment of the *Luftwaffe* as a separate service and for the priority given to it. Most of the officers, and certainly the lower ranks, were also grateful for the contribution Hitler and his party had made in creating employment, restoring the German economy and raising the nation out of the slough of despond into which it had fallen in the 1920s.

The senior officers certainly did not encourage Hitler to embark on military adventures, not even the return of soldiers to the Rhineland. The fact that from then on, through the invasions of Austria, Czechoslovakia, Poland, Norway, France and Russia, Hitler's judgement of the possibility of success proved more accurate than their cautious prognostications weakened their position and their confidence, and that of their subordinates, in their own judgement. Their fatal error was to accept, without much greater protest and opposition, the series of steps by which Hitler drew into his own hands the sole responsibility, not only for the conduct of operations, but also the appointment of senior officers. The fact that the generals were divided among themselves, both within the army and between the three services, and that they could never be certain that they would have the support either of their subordinates or of the public, weighed heavily against them. They had to take account of the fact that Hitler remained a popular Führer to the end. For senior officers in the course of a war to engage in a plot to get rid of, even to murder, the popular and democratically elected head of state was regarded by them, and would have been regarded by the majority of their countrymen, as rank treason. The most likely outcome would have been an internal division within the nation and within its armed forces which, before the Western Allies landed in the continent and were approaching Germany, would have opened the door to a Soviet advance into Eastern Europe. The final insult was the appointment of Himmler, not only to succeed Fromm after the 20 July plot to command the Reserve Army, but as C-in-C Upper Rhine in December 1944 and subsequently of Army Group Vistula.

It is possible that, had the 20 July plot succeeded in killing Hitler, some moves might have been made towards a peaceful settlement, but the Western Allies would not have reneged on their promise to the Soviet Union not to make a separate peace, and the Soviet Union would at least have demanded its pound of flesh. The casualties and destruction of the last nine months of the war might have been avoided, but the overall result is not likely to have been very different and would have involved the occupation of Germany by Soviet, British and American forces.

The army cannot be directly blamed for the atrocities committed against the Jews and the inhabitants of the occupied countries, especially Russia; but there is no doubt that they knew a good deal of what was going on and turned a blind eye. There were notable and honourable exceptions, but there were also dishonourable deeds, unworthy of the soldier's or sailor's honour, to which the German officer corps theoretically attached such high importance. There is no doubt that Russian prisoners of war were harshly dealt with. Of the five million captured, three and a half million died. In the eyes of the world, there was little left of that honour on 8 May 1945.

Whatever criticisms can be made of the behaviour and judgement of the senior officers, there is no doubt that the soldiers fought with a high degree of skill and courage, and endured great hardship and danger with admirable resolution. Nor is there doubt about the achievement of the General Staff in raising such a huge army – 284 divisions at its peak – and keeping it supplied. In 1944, when the Allied bombing offensive was at its height, the production of army equipment rose to 17,800 tanks and assault guns, almost the same number as the USA produced in that year and more than three times as many as Britain, and 40,600 guns, nearly a third of the tanks being Panthers and Tigers, superior in almost every respect to those of their opponents. The German army was a formidable war machine.

1945–86

It was not long after the armistice had been signed before differences arose between the victors, whose forces occupied the whole of the country, those of the Soviet Union also occupying all the territory which lay between Germany and Russia. But on one subject there was no disagreement – that the German armed forces should cease to exist and not be allowed, as in 1919, to rise phoenix-like from the ashes. However, they failed to agree on the terms of a peace treaty, and from the end of 1947 onwards the separation between the zone occupied by the Soviet forces and that occupied by the British and Americans solidified, and the first steps were taken towards the transformation of the two zones into separate states, which eventually would have their own armed forces.

The Berlin blockade, imposed by the Soviet authorities in their occupied zone from March 1948 to May 1949, was a major factor in influencing the British and the Americans to form the North Atlantic alliance, to liberate the Germans in their zones from the restrictions of occupation and to address the thorny problem of how Germany could contribute to her own defence, which the Americans in particular insisted was essential to the defence of Western Europe. Without it, they would not be prepared to commit their own forces. There were formidable obstacles to overcome: the distaste and anxiety of the countries Germany had occupied; and the opposition of just those elements in Germany itself whom the Allies had been at pains to encourage, who were determined to erase the tradition of militarism and of the armed forces as a body apart. In an attempt to overcome external opposition, the French Defence Minister, René Pleven, proposed a European army, controlled by a European Defence Committee, in which no formation above the level of battalion (later changed to a *groupement* of about 15,000 men) should consist of soldiers of the same nation. In spite of strong American pressure, the British refused to participate and, as discussions were prolonged, the French, whose army was embroiled in Indo-China, themselves rejected it; to the great relief of the Germans who, although professing willingness to participate, regarded it as a military fantasy.

The British government in 1954 committed themselves to the permanent stationing in Germany of a significant army and air force, in return for which their allies in the 1948 Brussels Treaty (France, Belgium, the Netherlands and Luxemburg) accepted that the Federal Republic of Germany, as the combined British and American occupied zones had then become, should be welcomed into the Western European Union and the North Atlantic alliance. That welcome was accompanied by conditions: that the army should be limited to twelve divisions and the air force to 1,000 aircraft; that there should be no national chain of command, all the forces being subordinate to the NATO commanders: that the Republic should not manufacture atomic, biological or chemical weapons; and that the forces could not be used in order to achieve the unification of Germany.

By this time the Republic, under the firm guidance of aged Chancellor Konrad Adenauer, had evolved, founded on a Basic Law guaranteeing its constitution as a liberal democracy and devolving considerable powers on its constituent *Länder*. No provision had been made for armed forces. An 'Office for the Commission of the Federal Chancellor for Questions connected with the Increasing of the Allied Troops' was set up in 1950 to consider the question and prepare legislation, headed by Herr Blank. The most influential member was an ex-officer, Count Baudissin, who invented a concept of 'The Citizen in Uniform', subject to civil law, trained on lines of democratic ideology (*Innere Führung*), who could appeal against too Prussian an attitude of his superiors to an independent Ombudsman (*Wehrbeamauftragen*). Few ex-senior officers thought that an efficient army could be created on those lines; but it was that or nothing, and they set about raising one which, with their accustomed efficiency, they did in very quick time. They were now advocates of forces based on voluntary service, and the navy and air force would certainly have preferred that; but it had sinister political undertones, and conscript service was decided upon.

In a remarkably short time the German army achieved its full target and within ten years it was the most efficient, best equipped and best trained contingent of the forces in NATO's central sector, running from the Kiel canal to the Austrian border. The realities of defence economics meant that the air force never reached its permissible total, settling down near the 500 mark. The navy, initially restricted largely to the Baltic, remained the poor relation. The strength of the standing army stabilized at around 400,000, half of whom were conscripts, the figure doubling on mobilization with the recall of reservists, who also formed territorial units, the command of which remained national, in dubious breach of the Brussels Treaty.

The standing army is organized into three corps, one in the north of four Panzer divisions, one of which in war is subordinate to the NATO commander in Denmark. The northern corps is subordinate to the British commander of NATO's Northern Army Group. The central corps of one Panzergrenadier and two Panzer divisions, and the southern, which has one of each and a mountain division, are subordinated in war to the American commander of NATO's Central Army Group. The airborne division, stationed in the south, places one brigade under command of each corps in war. The organization of higher command was clearly a sensitive issue. The heads of the three services, and their joint service co-ordinator, were titled Inspectors and their staffs *Führungstab*. Command in peacetime over all three services, the

Bundeswehr, rests with the Federal Minister of Defence, whose activities are watched over by a Defence Committee of parliament (*Bundestag*), assisted by a Defence Commissioner, to whom any serving member of the forces may appeal. In spite of the reservations of many ex-senior officers about all the restrictions placed on the authority of officers in deference to democratic concepts, the armed forces have developed into highly efficient and well-disciplined bodies. A further sensitive issue was the appointment of German officers to senior NATO commands, with authority over the forces of other nations. Fortunately a good example was set early on by the French, who accepted General Hans Speidel, Rommel's wartime Chief of Staff, who had led the German delegation in the negotiations over the European army, as Commander of the Allied Land Forces Central Europe in place of a French general. When France left the military organization in 1966 the post was merged with that of overall C-in-C, with authority over both land and air forces in NATO's central sector, and became a permanent German appointment.

Efforts have been made without success to acquire other appointments, generally at the expense of the British, although a German deputy to the American Supreme Allied Commander Europe has joined the British deputy and, in the early 1970s, the previous head of the German Air Force, General Johannes Steinhoff, filled the post of Chairman of NATO's Military Committee. By then the armed forces of the Federal Republic of Germany had been fully accepted as a respectable and respected body of armed democratic citizens.

While the armed forces of West Germany had been established to protect a democratic republic, those of East Germany, ironically calling itself the German Democratic Republic, had been built up in a much more Prussian mould, the goose-step providing a reminder of that tradition. By the 1980s their total armed forces amounted to 174,000 men, 94,500 of whom were conscripts. Of these, 120,000 were in the army, the 71,500 conscripts of which served for eighteen months. Those called up into the navy and air force served twice as long. The army produced two tank and four motor rifle divisions; two surface-to-surface missile brigades, two artillery and one anti-aircraft artillery regiment, and eight regiments of anti-aircraft missiles. Its stock of tanks numbered 1,500. The air force, 39,000 men strong, manned 380 combat fixed-wing aircraft and 70 armed helicopters. It is therefore by no means a negligible force.

4

RUSSIA

1900–05

At the dawn of the twentieth century the sprawling giant that was Russia boasted the world's largest army of 1,100,000 men. The nation was struggling to emerge from its almost medieval form into an imperial power which could exert as much influence in the world as its European rivals, Britain, France and Germany. The young Tsar Nicholas II, who succeeded his father Alexander III in 1894, was determined that Russia should not be frustrated in this claim either by Germany in Europe or by Britain elsewhere in the world, as she felt she had been by the Congress of Berlin of 1878, at which she thought she had been cheated of the fruits of her victory over Turkey. The weakness of the Austro-Hungarian, Ottoman and Chinese empires offered opportunities for the extension of her own. Pan-slavism combined with this general desire to rank high in the imperial league had already led to Russian involvement in the Balkans and the war with Turkey. The extension of her empire by land beyond the Urals, supported by the development of railways into Turkestan and Siberia, brought her into potential conflict not only with the British Empire in the Middle East and on the borders of India, but also with the decadent empire of China. Not content with imperial expansion by land, Russia was intent also on challenging Britain as a naval power. It was clearly impossible for her to do so alone, but with a substantial fleet she could play an important part as an ally to any nation with the same aim.

In reaction to the Congress of Berlin, Alexander III, at the beginning of his reign in 1882, had approved a twenty-year plan to build fifteen battleships, ten cruisers and eleven gunboats (later increased to twenty battleships and twenty-four cruisers), which would be split between the Baltic and Black Sea and Far Eastern fleets, the last based at Vladivostok, acquired in 1862, which, like the Baltic Fleet's base at Kronstadt near St Petersburg, was icebound in winter. At that time the Russian navy had only four battleships. The Black Sea Fleet was restricted by the 1841 Convention regarding the passage of foreign warships through the Bosphorus and Dardanelles.

Although the sheer size of the Russian empire and of her army was impressive, the structure of both suffered from severe internal weaknesses. Even before the empire expanded beyond the Urals and the Caspian Sea, it included a large number of different peoples who had never become totally reconciled to their conquest by the Russians, who originally had been confined to the principality of Moscow. In the west the Finns, the Poles and the Ukrainians provided the principal non-Russian elements, the last being split between the Russian and the Austrian empires, and the unfortunate Poles between them both and Germany. In the south the Cossacks and the Muslim Tatars, and to the east and south-east the Tatars, Uzbeks, Georgians and Armenians added to the general hotchpotch. The fissiparous potential of this polyglot group of peoples was undoubtedly a factor in reinforcing the reliance of the Tsar and his advisers on the principle of autocracy, resisting movements in favour of anything which could encourage the autonomy of the separate communities of which the empire was formed. Resistance to pressure in this direction from the educated minority, influenced by developments in Western Europe, led to an increasing emphasis on 'Russification' and intensification of a stultifying centralized bureaucracy, directed from Moscow. This only heightened opposition to it, particularly as it was exceptionally inefficient. These general divisions within the body politic of the Russian empire were aggravated by the combined effects of the abolition of serfdom in 1861 (while slavery was still tolerated in the United States of America) and Russia's belated acceptance of the Industrial Revolution. The combination of the two produced a discontented peasant landless proletariat and a discontented urban population, both those who were at work and demanded better conditions of work and life generally and some means of influencing them, and those who, having drifted to the towns, could find no employment or accommodation.

All this was reflected in the state of the Russian armed forces, of which the army formed the preponderantly larger part. It was in fact the largest army in the world, 1,100,000 men strong, backed by double that number in the reserve in which a man was liable to recall on mobilization for thirteen years after his initial service of four. The army was organized into twenty-nine corps, each of one cavalry and two infantry divisions, in addition to which there were 345,000 Cossacks, 12,000 Caucasian native troops and a national militia of 700,000. Two of the corps were in Siberia and two in Turkestan: the rest were west of the Urals and the Caspian, distributed throughout the length and breadth of the land in garrisons, whose principal function was to act as a reserve to maintain order. Regimental life was akin to that of the British army in the eighteenth century. The commanding officer was allotted a sum of money for the year to cover all the requirements of his regiment, other than its weapons and ammunition. The regiment was regarded by its officers as a business, to be run for their benefit. Most of the soldiers were engaged most of the time in the regimental 'workshops', making their own clothing and equipment or meeting the needs of the officers, who divided their time between filling in endless forms to satisfy the bureaucracy that they were not cheating the system (which they probably were), or in drinking and challenging each other to duels. In the harvest season the soldiers would be hired out as labour. Such training as there was was left to the sergeant-majors and consisted of stereotyped field drills, culminating in a bayonet charge. The Russian soldier kept his bayonet fixed on his rifle at all times, even when firing it on the range, and on active service the scabbard was thrown away. He was trained to fire volleys, standing upright in a line, as his forebears had done in the Crimea and the Napoleonic wars. Primitive as was his training, foreign observers were impressed by the quality of the Russian soldier: his toughness and endurance, his courage and discipline and his sheer physical strength; but they had little good to say of his officers. The latter could be divided into two distinct classes: the Guards, most of the cavalry and the Moscow-based regiments, and all the rest. The former tended to come from the nobility or the rich merchant class who aspired to the same social standing. The majority of the latter had worked their way up through the ranks from a variety of origins, although the commanding officers might have come from the former class, or from a hereditary class of officers, originally of foreign origin, often German, Swedish or Polish. In spite of the poor pay, the social cachet of a commission was sought after. Between 1900 and 1914 40 per cent of all officers between the ranks of subaltern and colonel came from the peasant or lower middle class. Above that rank there was a sharp change.

At the time of the Crimean War, the senior ranks were dominated by the nobility, especially by officers of the Guards, the cavalry and the Moscow-based infantry regiments. There was also an element of what might be called a praetorian guard, senior officers who owed their position directly to the Tsar, not to their noble birth or riches. They were either of the hereditary ex-foreigner type or had themselves, or were descended from those who had, worked their way up from humble origins. They tended to predominate in the infantry, other than the Guards, while the nobility dominated the cavalry and the artillery. The division between the two elements was to become acute after the Russo-Japanese War. Towards the end of the century an increasing number of senior officers were sons of the rich merchant class, keen to establish their social status. The failure of the army's command and staff system in the Crimea had led to major reforms, engineered by the intelligent and able General Count D.A.Milyutin when he became War Minister in 1861. Introduced in 1865, they brought about a decentralization of command to Military Districts, from each of which an operational army was formed. His principal innovation was the integration of a whole number of separate staffs, including the General Staff (*General 'nogo Shtab*) and the artillery and engineer inspectorates into one Main Staff (*Glavnyi Shtab*), responsible to the War Minister. He also reorganized the military academies, intending that their products should displace those who held senior positions in command or on the staff through accidents of birth or influence of wealth. His reforms, which included the introduction of universal conscription in 1873, proved their worth in the war against Turkey in 1877, and were further enhanced by General N.N.Obruchev, when he became Chief of the Main Staff in 1894, his reorganization being approved in 1900 when General A.N.Kuropatkin was War Minister. However, in spite of the attention given in the second half of the nineteenth century to reforms of the command and staff organization and to the training of officers, the actual effect fell far short of the reformers' intentions, and the whole machine creaked when set in motion in 1904.

The Russian navy in 1902 had 65,000 men, conscripted by lot to serve for seven years, followed by three on the reserve. In his seven-year service, the sailor was forbidden to marry, received a pittance of pay, was badly fed and had almost no prospects of promotion. Until Port Arthur became available the navy had no ice-free port in winter. During those months ships were tied up and the crews lived ashore in barracks. The officers were all volunteers, largely from the minor landowning and professional classes, many of them Baltic Germans. The navy lacked the social cachet of the army, although its head was Grand Duke Alexei Alexandrovich.

Since the Russian Empire had expanded to the Pacific in 1860, other nations had been acquiring possessions or concessions, many of them at the expense of the weak Manchu Empire in China. In 1895, only forty years after Commodore Matthew Perry of the US Navy had found an almost medieval regime in control of Japan, the Meiji emperor's army had challenged Chinese authority in Korea, driving them out of both that country and southern Manchuria and acquiring, by the Treaty of Shimonoseki, the Liaotung peninsula and its ice-free Port Arthur, as well as Formosa. This was not to the liking of the European powers, who forced Japan to return the peninsula to China. Within the next few years, to the fury of the Japanese, Russia obtained a long-term lease on Port Arthur and most of its hinterland, while France, Germany and Britain all established naval bases in the area. Russia also obtained permission to build a railway to connect Port Arthur with Mukden, which would link it with the Trans-Siberian railway. The latter was opened in 1901, except for a short stretch round Lake Baikal, not completed until 1905. These naval facilities in the approaches to Peking came in useful at the time of the Boxer rebellion in 1900, when all the nations concerned, including Japan, provided contingents for the International Relief Force.

By that time Japan had become increasingly concerned about Russian activities and interest in Korea and Manchuria, notably that of a timber concession along the Yalu and Tulmen rivers. Considerable effort was being devoted both to developing Port Arthur as a defended base for a major fleet and to the construction of a civilian port at Dalny, immediately to the north of it, while the Russian army in Manchuria, far from being removed, as had been promised after the suppression of the Boxer rebellion, was reinforced. In 1903 Japan proposed an agreement by which she and Russia should recognize the independence of China and Korea, Manchuria as a Russian 'sphere of influence' and Korea as a Japanese one, from which the troops of both would be withdrawn, except for those needed for the protection of the railways. Under the influence of a Far Eastern lobby, which had the ear of the Tsar, the Russians prevaricated, while sending more troops to Manchuria. On 13 January 1904, after six draft agreements had been turned down, the Japanese demanded a speedy reply to the latest proposal. Having received no answer in three weeks, the Japanese ambassador in St Petersburg informed Count V.N.Lamsdorff, the Foreign Minister, that he had been ordered to sever diplomatic relations and leave.

The Russian army in the whole of Eastern Siberia consisted of a corps of one cavalry and two infantry divisions, totalling 84,000 men in 100 infantry battalions, 35 cavalry squadrons, 25 artillery batteries (some 200 guns) and 13 companies of engineers. During the war that was to follow it was reinforced to a strength of 210,000 in seven corps, all, with their equipment and stores, moved the 5,500-mile length of the Trans-Siberian railway. The troops initially available in Manchuria to Lieutenant-General M.I.Zasulich, apart from the garrison of Port Arthur, consisted only of twenty-three squadrons of cavalry, mostly Cossacks, and eight battalions of infantry, increased by March to 5,000 and 15,000 men respectively, supported by sixty guns. For the defence of Port Arthur Lieutenant-General A.M.Stoessel had 40,000 men. Both he and Zasulich came under the command of the Viceroy in the Far East, Admiral E.I.Alexeyev, thought to be an illegitimate son of Tsar Alexander II, as did the Far East Fleet commander Vice-Admiral Starck, both of whom were in Port Arthur, with a fleet of seven battleships, six cruisers, twenty destroyers and ten torpedo-boats. They were ignorant of the fact that the Japanese Admiral Togo had left his base at Sasebo on 6 February, sending ten destroyers to

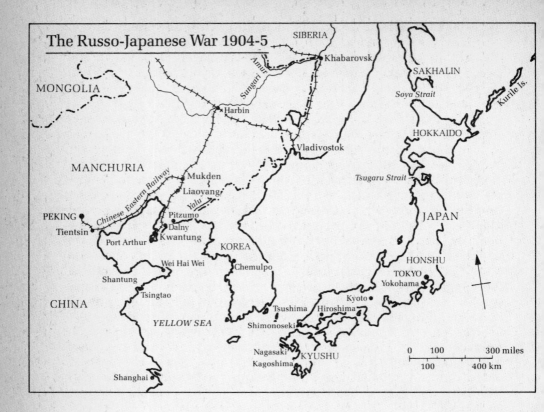

The Russo-Japanese War 1904-5

attack Port Arthur and eight to Dalny where he thought some of the Russian fleet would be. At the same time a force of cruisers and torpedo-boats escorted a landing force of 2,500 men, the advance guard of Major-General Kuroki's 1st Army, to Chemulpo (modern Inchon), the port of Korea's capital Seoul.

At about 11 p.m. on 8 February the ten destroyers were detected by two Russian ones as they approached the entrance to the roadstead in which the fleet lay at anchor, as totally unprepared, except for anti-torpedo nets hung over the side, as were the coast defence guns of the fortress. The Russian destroyers made for the harbour to warn the Admiral and were followed in by Captain Asai's destroyers each firing its two torpedoes. Only three out of the eighteen fired scored hits, disabling two battleships and a cruiser, all of which were grounded as they made for shallow water, the two battleships nearly blocking the harbour mouth. Meanwhile, off Chemulpo, the Russian cruiser *Varyag*, badly damaged in a brave sortie against a superior Japanese cruiser force, scuttled herself. On the following morning Togo appeared with his main fleet of six British-built battleships off Port

Arthur and bombarded the fleet anchorage at a range of 8–9,000 yards. By this time the coast defence guns were in action, and he was not prepared to risk serious loss to his battle fleet in order to press the attack, and no significant damage was done. Togo had to bear in mind that the Russians could reinforce the Far East from their other fleets. He therefore turned his attention to attempts to bottle up the fleet by placing blockships in the harbour entrance, but the Russians sank them all before they reached it.

Russia's strategy was to order Port Arthur to hold out, while the army in Manchuria, commanded by General Kuropatkin, who handed over as Minister of War to General Sakharov, was reinforced. Meanwhile preparations would be made to form a 2nd Pacific Squadron from the Baltic Fleet and also, if diplomatic pressure could persuade other powers to agree, from the Black Sea Fleet. The Franco-British Entente, announced on 8 April, put paid to any hope of that. Meanwhile the energetic and aggressive Admiral S.O.Makarov was sent by rail to Port Arthur to take command of the 1st Pacific Squadron, as the ships at Port Arthur were to be called. The two squadrons together would defeat the Japanese fleet, making it impossible for them to maintain an army of any significant strength in Manchuria or even Korea.

Makarov arrived on 10 March, the day the battleship *Retvisan* was refloated and moved inside the harbour for repair. He immediately invigorated the fleet, reducing the time it took them to move out of the harbour from twenty-four to two and a half hours and putting them through their paces at sea, the first time the battleships did so leading to a collision between three of them. During this period both sides laid mines in the approaches to Port Arthur, and it was partly in reaction to the activities of a Japanese minelaying force that Makarov took part of his fleet to sea on 13 April, Togo having also brought his fleet to the area, hoping to lure Makarov towards it by covering the minelayers with a cruiser force. Makarov flew his flag in the battleship *Petropavlosk*, accompanied by the battleship *Poltava* and three cruisers. Togo had all six battleships and a total of eight cruisers. When Makarov saw this, he turned back to Port Arthur, but soon after he had done so and Togo, disappointed, also turned away, there was a huge explosion as the *Petropavlosk* struck a mine and sank within two minutes, Makarov and 635 of her crew of 715 going down with her. Makarov had a posthumous revenge on 15 May when two of Togo's battleships were sunk by mines, one of his cruisers also being lost that day in a collision with another.

By that time events on land had had serious consequences for the fleet, demoralized by Makarov's death. When Kuropatkin reached Liaoyang, where he established his headquarters, on 27 March, he found that Alexeyev had ordered Zasulich to prevent Kuroki from crossing the River Yalu, the border between Korea and Manchuria. Kuropatkin, although also officially subordinate to Alexeyev, modified this by telling him to fight a delaying action only. By 25 April Kuroki had closed up to the river, and his German-trained army, observed by the British general, Ian Hamilton, showed by its infantry and artillery tactics and its skill in concealment and deception that it was professionally greatly the superior of its stolid, slow opponents, who showed no inclination to conceal or protect their positions or movements. By 1 May Kuroki's army was across the river, having lost 1,000 men, while Zasulich's force was retreating, having suffered 3,000 casualties, of whom 1,400 were dead and 600 taken prisoner.

Four days later General Oku's 2nd Japanese Army began to disembark at Pitzuwo, sixty miles north of Port Arthur, no attempt being made by the Russian fleet to interfere, although the destroyer captains suggested doing so. Alexeyev told them to talk it over with the senior admiral, Vitheft, as he boarded the train to join Kuropatkin at Liaoyang. While Oku began to advance towards Port Arthur and Kuroki north-west from the Yalu towards Liaoyang, General Nozu's 4th Army landed between them and also struck north-west. Before the end of the month Oku had reached the three-mile-wide neck of land at Nanshan connecting the Kwantung peninsula on which Port Arthur and Dalny were situated with the main Liaotung peninsula, thus cutting them off. Its earthwork defences were manned by 10,000 men commanded by Major-General Fock, who had a further 8,000 in reserve. Oku launched his assault with all three of his divisions on 26 May, after a three-hour bombardment. For seven hours Fock's men held the attack, but when, late in the day, Oku renewed it with a concentrated thrust at the western end, Fock failed to employ his reserves in time, and his left flank was overwhelmed by the Osaka division. With Stoessel's permission, Fock withdrew to the outer defences of Port Arthur itself, having lost perhaps 1,000 men, Oku's casualties numbering 4,855 out of a total of 29,500.

Alexeyev now urged Kuropatkin to advance to relieve the siege, which the latter was unwilling to do until he was assured of a numerical superiority over his opponents. A council of war at the Tsar's palace at Tsarskoe Selo ordered him to make the attempt, as a result of which he gave some confusing and irresolute orders to his subordinate, General G.K.von Stackelberg. Before the latter had begun to advance, Oku moved north and on 15 June attacked Stackelberg, who had about 30,000 men at Telissu, forty miles up the railway from Nanshan. The battle lasted for three days, the Russians standing their ground stolidly, while the Japanese made enterprising use of the hilly terrain to outflank and eventually almost encircle their opponents. Leaving behind 2,000 dead, sixteen guns and large quantities of stores and ammunition, Stackelberg's force retreated in disorder, having inflicted 1,000 casualties on Oku's army. Differences between two of Stackelberg's subordinates, Pavel K.Rennenkampf and Aleksandr V.Samsonov, leading to their coming to blows on Mukden railway station, were to have important consequences in East Prussia in August 1914.

Kuropatkin had been badly served by his cavalry, which had failed to keep him or his subordinates informed of the enemy's strength or movements, and had no effect on the battlefield in the face of machine-guns, with which the Japanese were much more liberally equipped than the Russians. A large proportion of them were Cossacks and Uzbeks, enthusiastic at galloping about, uttering war-cries and slashing with their swords, but not well disciplined in routine security and reconnaissance tasks. The Russians were impressed by the courage and high spirits of their opponents, as they were by the speed with which they organized their operations and their skill at deception and concealment, arts of war in which the slow, steady Russians, with their inefficient staff and command arrangements, were sorely deficient.

It was clear to Kuropatkin that he could not resume the attempt to relieve the siege of Port Arthur until his army had been strengthened. Midsummer in any case was a rainy period when movement, except by rail, would be difficult. He felt confident that meanwhile Port Arthur could hold out, although he had little faith in Stoessel, and ordered him to leave by destroyer and hand over to Smirnov, commandant of the fortress. Stoessel, who was senior to Smirnov, and whose real command was the 3rd Siberian Corps, suppressed the telegram, handing over effective command of the infantry (but not the artillery) to one of his divisional commanders, the flamboyant and inspiring Kondratenko. The other was Fock, who, after Nanshan, was not in favour. The army's command arrangements were as Gilbertian as those of the navy.

Immediately after the Battle of Telissu, at a meeting of the Higher Naval Board, the Tsar decided that the Baltic Fleet would be got ready to sail to the Far East, where it was hoped it would arrive by September, and that the relatively junior fifty-five-year-old Rear-Admiral Z.P.Rozhdestvensky should take command. Alexeyev had already ordered Vitheft to sail his fleet, in which the battleships had now been repaired, from the beleaguered Port Arthur and join the squadron of four cruisers at Vladivostok, which rejoiced in the presence of three admirals, all marooned there en route to replace the incompetent and spineless admirals in Port Arthur. On 23 June Vitheft set sail with six battleships, four cruisers and twelve destroyers. Togo was thirty miles away with four, eight and six of the same. Picking up four more cruisers on his way, he sighted Vitheft twenty miles out from his base with only half an hour of daylight left, and cleared decks for action, but Vitheft, fearing a night attack by torpedoes, turned for home. Togo's destroyers pressed their attacks, but failed to score a single hit, the battleship *Sevastopol* striking a mine, but limping home with a list to port. A week before this sorry performance, the Vladivostok squadron had carried out a successful attack on Japanese transports, sinking three of them carrying heavy artillery to the siege of Port Arthur, and causing Togo to divert five cruisers which were not with him on the 23rd.

While Kuropatkin was thinking in terms of playing for time, the aged Marshal Yamagata, Prime Minister and the senior military figure, was fully aware that time was not on his side. He was determined to capture Port Arthur and destroy the fleet there before Kuropatkin could intervene by land or the Russian Baltic Fleet by sea. He now had four armies in Manchuria, Nogi's 3rd Army having taken over the siege of Port Arthur while, from right to left, Kuroki's 1st, Nozu's 4th and Oku's 2nd advanced towards Liaoyang. Hitherto, using cable laid on the seabed, he had attempted to exercise command himself from Hiroshima, but in July he sent General Oyama, Chief of the Army General Staff, to assume the command, which he had exercised six years before against the Chinese.

As Nogi closed in on Port Arthur and brought the harbour under the fire of his heavy artillery, Alexeyev insisted that Vitheft make another attempt to reach Vladivostok before all his ships were disabled or sunk at their moorings. Another motive was to ensure that the fleet was commanded by the more energetic admirals who were waiting at Vladivostok. Recalling his guns from deployment to help the army, Vitheft, in fatalistic mood, and reporting that it was against the advice of himself and all his senior subordinates, set sail on 10 August with six battleships, three armoured and one light cruisers and eight destroyers. Twenty-five miles out he met Togo with four battleships, three armoured and eight 'protected' cruisers, seventeen destroyers and twenty-four torpedo-boats. The latter repeatedly tried to 'cross the T' of Vitheft's fleet, but Vitheft managed to slip behind the rear of Togo's line and continue on his course towards the Tsushima strait. Both sides scored hits on each other's battleships, but it was not until 6.40 p.m. in the evening, when the fleets had been engaged for seven hours, that a stroke of luck tilted the balance decisively in Togo's favour. Vitheft's flagship, the *Tsarevich*, was hit by two 12-inch shells, one killing Vitheft and wounding his principal staff officers, the other hitting the bridge, killing the navigating officer and helmsman and jamming the steering wheel hard to starboard, so that the ship nearly collided with two of the other battleships and threw the whole line into confusion. Togo's fleet closed to 4,000 yards, but failed to sink any of the Russians, who turned back to Port Arthur, leaving the crippled *Tsarevich* with a cruiser and three destroyers to seek refuge in the German port of Kiaochow where they were interned. Although the action was decisive in forcing the Russian fleet back to Port Arthur, it was astonishing that not one ship on either side had been sunk by enemy fire, although Togo's flagship, the *Mikasa*, received twenty-two hits and the *Tsarevich* fifteen from 12-inch guns. None of the Russian ships making their way back to Port Arthur were hit by Japanese torpedoes.

A week after Vitheft had taken his fleet to sea, Nogi launched a major assault on the outer defences of Port Arthur, but after ten days of fighting in which he lost 15,000 men, although he captured two redoubts, the Russian defence stood firm. Oyama realized that an old-fashioned siege would be needed before his principal objective could be gained, and that Nogi's army would not be available to him further north, where he was preparing to bring Kuropatkin to battle at Liaoyang. To defend his main base the latter deployed his army into two groups, the western under General N.P.Zarubayev and the eastern under General Baron A.A.Bilderling, described as an elderly cavalry martinet. Kuroki faced the latter, Oku and Nozu the former, Oyama's cavalry covering the gap of twenty-five miles between the two groups. Kuroki made the first move with an attack on Bilderling's position on 25 August which, by the 27th, had forced the Russians to withdraw in disorder behind the flooded River Tang. Oyama now switched the main effort to the 2nd and 4th Armies, which launched a direct attack on Zarubayev's main defences in the hills south of Liaoyang. After three days of hard fighting, in which Kuropatkin made no use of his reserves, the Japanese gained the crests of the hills overlooking the town. At the same time Kuroki managed to get his troops across the larger River Taitse, threatening the railway line behind it. Fearing encirclement, Kuropatkin ordered withdrawal and left himself by train on 2 September. His attempts to organize a counterattack failed, and his dispirited troops trudged the forty miles northwards towards Mukden in the sultry summer heat, followed up, but not seriously pursued, by Kuroki, while Oyama, with Oku and Nozu, settled down at Liaoyang, capturing significant quantities of weapons and ammunition, having lost 5,000 killed and 18,000 wounded, against the Russian losses of 2,000 and 13,000.

In spite of this defeat, the military balance in Manchuria was turning in Russia's favour. Reinforcements were arriving all the time by rail, while the Japanese, having suffered heavy casualties and used a large amount of ammunition, could not easily or quickly replace either. Under pressure from St Petersburg, Kuropatkin not only decided to launch a counteroffensive, but made his decision public, alerting Oyama to his intention a week before he began a ponderous advance back along the railway on 2 October. When the two armies made contact with each other on 11 October on the River Sha-ho, each decided to make its main effort on its left. Kuropatkin's, on the flat ground east of the railway, was commanded by Stackelberg, his right, in hilly ground to the west, by Bilderling. While the former made some progress, the latter was driven back and appealed to Kuropatkin for help, as he withdrew behind the Sha-ho. Kuropatkin transferred twenty battalions to him from Stackelberg, prejudicing the latter's chances of success. Two days of inconclusive hard fighting followed, after which the two sides dug themselves in, facing each other in a trench system, the pattern of which was to become familiar to European armies ten years later, Oyama seeing his task as being to ensure that Kuropatkin could not move to raise the siege of Port Arthur. The latter succeeded Alexeyev as overall C-in-C, retaining command of the land forces.

At the height of the battle of Sha-ho, Admiral Rozhdestvensky at last set sail with the Baltic Fleet on the first leg of his round the world voyage. The fleet consisted of nine battleships, five cruisers and eight destroyers, two more of each of the latter, unfit to sail at the time, following a few weeks later. The general state of the ships themselves and of their crews, in spite of the long period of preparation, left much to be desired, priority having been given to the Far East and Black Sea Fleets. Coaling en route posed a formidable problem, almost all coaling stations being in the hands of Japan's ally, Britain. Through the good offices of Kaiser Wilhelm, only too anxious to see Russian forces of all kinds committed to the Far East, a contract had been negotiated with the Hamburg-Amerika line to provide coal to be transferred from their ships at sea outside territorial waters, a foretaste of the refuelling at sea which was to be developed in the Second World War, but much more laborious when carried out with coal in lighters and ships' boats in the steaming heat of the tropics.

The fleet was obsessed by the totally irrational fear that Japanese torpedo-boats would be lurking near the narrow exits from the Baltic, and when, thirty miles off course near the Dogger Bank in the North Sea, on the foggy night of 20 October, the First Battleship Division came across first some Danish and then some British fishing trawlers, they opened fire not only on the luckless trawlers, two of which were sunk, but on each other, as well as a Swedish merchant ship and a German trawler. In the international row which followed, Rozhdestvensky, against all the evidence, stoutly maintained that two Japanese torpedo-boats had mingled with the trawlers. The incident brought nothing but disrepute upon the Russian navy generally, whose part in the war hitherto had certainly not been distinguished.

Its reputation was not enhanced by the activities of its Captain Klado, who, although a serving officer, was naval correspondent of an influential newspaper. He had lobbied for Rozhdestvensky's appointment, and accompanied the fleet as far as Portugal, but then returned to Russia and published articles blaming the Dogger Bank incident on deficiencies in training and urging that Rozhdestvensky should be reinforced by all available ships before he reached the Far

East. The result of his agitation was that the fleet, part of which had traversed the Mediterranean and the Suez Canal, but the bulk of which had sailed all round the Cape, waited in the debilitating atmosphere of Madagascar from shortly before Christmas 1904 until March 1905.

During that time the whole strategic, and indeed the political, situation had changed. Nogi had discovered that if he were not to suffer intolerable casualties – 8,000 of his men had been killed and 23,000 wounded between 26 July and 26 October 1904 – he must revert to old-fashioned siege methods of mining and sapping, and that anybody within range of the Russian guns, which included those of the fleet, had to live below ground level. In addition to battle casualties, his losses from sickness were high, a good deal higher than those of the besieged Russian garrison, which by October was beginning to suffer from food shortages. Having been reinforced by 11-inch howitzers from Japan, which did considerable damage to the town itself and the ships in the harbour, Nogi launched a major assault on 26 October, which, after six days of fierce fighting, failed with the loss of 4,000 men. He renewed the assault a month later with a fresh division, and succeeded in capturing important high ground which overlooked the harbour, as a result of which four of the five battleships, two cruisers and several destroyers were sunk, the one remaining battleship, the *Sevastopol*, seeking refuge in deeper water, where she was later scuttled. More significant perhaps than even these losses was the death in action of the robust Kondratenko, leaving Fock in command of the land forces under Stoessel, who had been slightly wounded. By the end of the year Nogi's men had penetrated further and could now bring observed artillery fire onto every area of the fortress. After a council of war on 29 December Stoessel despatched a telegram to the Tsar saying that he could only hold out for a few days, falsely reporting that there was hardly any ammunition left. Without waiting for a reply he opened negotiations with Nogi on 1 January 1905 'for an evacuation of the fortress', to the dismay of his deputy Smirnov and of the senior admiral Wiren, who only had twenty-four hours in which to destroy as much as they could of the ships, the port installations and its defences. The Japanese having refused Stoessel's request that the garrison should not be treated as prisoners of war, surrender terms were signed on 2 January, 24,000 fit men, 15,000 sick and wounded, 3,000 live horses, 600 guns, 34,000 shells, 35,000 rifles and more than 2 million rounds of small arms ammunition falling into their hands. Russian casualties in the 154-day siege had amounted to 6,000 killed and 20,000 wounded: Japanese 14,000 killed and missing, 36,000 wounded and over 30,000 sick.

The news of this blow to Russian pride came at a time of political tension, aggravated by the preoccupation of the Tsar and his ministers with the war. He and they had been little inclined to make any concessions to the different political pressures, represented both by liberal-minded middle- and upper-class movements, who pressed for a greater say in government for the local organizations – the zemstvos – and the socialist movement, split between the Populists and the Social Democrats (who later split into Bolsheviks and Mensheviks), which based its support primarily on the urban working class but also the peasantry. Ironically it was a government-favoured movement, designed to entice workers away from the socialist-dominated trade unions, which sparked off the revolutionary fervour of 1905. It was led in St Petersburg by Father G.A.Gapon. A wave of strikes had already hit the city when the news of the fall of Port Arthur became known. On Sunday 9 January Gapon led a peaceful demonstration to the Winter Palace designed to appeal to the Tsar (who was at Tsarskoe Selo) to meet the strikers' demands. When the leading elements of the 200,000-strong march refused an order by the soldiers guarding the palace to disperse, they were charged by cavalry and then fired upon, over a hundred being killed. The news of these two events naturally cast gloom over Rozhdestvensky and his fleet, sweltering in the heat of Madagascar.

Kuropatkin's reaction to the fall of Port Arthur was to deal a blow to Oyama's forces before Nogi could join him, hoping that Russian familiarity with operating in wintry weather would work in his favour. His forces were now organized into three armies covering a ninety-mile front south of Mukden: the 2nd on the right under General D.K.Grippenberg, the 3rd in the centre under General Baron A.V.Kaulbars, and the 1st on the left under Linievich – Bilderling and Stackelberg being demoted to corps commanders. Rennenkampf's cavalry was east of Linievich and General Mischenko's west of Grippenberg. Operations began with a raid by Mischenko, designed to discover Nogi's whereabouts. It was a dismal failure, his 6,000 horsemen being easily driven off by Japanese infantry. On 26 January Grippenberg launched an assault on the Japanese left wing across the frozen River Han with seven divisions, but could make no headway, and when the Japanese counterattacked, Kuropatkin refused Grippenberg's request for reinforcements and ordered him to withdraw behind the river, bringing the battle to an end, the Russians having suffered 20,000 casualties to the Japanese 9,000. Grippenberg resigned his command and, on return to St Petersburg, laid all the blame on Kuropatkin. Kaulbars took over his army, being himself succeeded by Bilderling.

Kuropatkin did not give up, but he took his time about planning another attack, which was to be a repeat performance on the right wing. But by the time he was ready Oyama had been reinforced, not only by Nogi's 3rd Army, but by a new 5th under Kawamura, Nogi coming up on the left of Oku and Kawamura on the right of Kuroki, opposite Rennenkampf. The Japanese army was now at the peak of its potential strength, Oyama deploying 207,000 men with 1,000 guns and 254 machine-guns against Kuropatkin's 276,000 with 1,200 guns, but only 54 machine-guns. The latter had the advantage in numbers of cavalry, with 16,000 well-mounted men against 7,300 on what Europeans would classify as ponies; but Kuropatkin gained little from it, as they failed miserably to give him any accurate information about the strength and disposition of Oyama's armies.

The Japanese strategy was to deal a mortal blow to the Russian army in Manchuria before Rozhdestvensky's 2nd Pacific Squadron could reach the Yellow Sea and threaten Oyama's communications with Japan. Oyama's operational plan was to encircle Kuropatkin's force by an attack on its eastern flank, followed by one on the west. The attack was launched in a snowstorm on 23 February by Kawamura and led to fierce fighting, causing Kuropatkin to give up any idea of himself attacking from his right and to transfer divisions from Kaulbars on the right to Linievich on the left. From 27 February to 7 March Linievich held attacks by Kawamura, while in the centre Bilderling held repeated attacks by Kurioki, but Nogi was gradually pressing back Kaulbars's right wing. Kuropatkin, fearful of being outflanked there, ordered a withdrawal in the centre, of which all the Japanese army commanders took advantage, Nogi succeeding in cutting the railway north of Mukden, while Kawamura began to outflank Linievich at the other end of the front. Having failed to organize any effective counter-attack, Kuropatkin, almost surrounded, ordered a general withdrawal to Tiehling, forty miles to the north. Although it took place in scenes of total disorder, Oyama's forces had no means of effectively interfering with it, having suffered 70,000 casualties against the Russian 20,000, one third of the latter having been taken prisoner. Twice as many Japanese as Russians had been killed. One casualty of the battle was, not surprisingly, Kuropatkin, who was relieved of his command and succeeded by Linievich.

On the day that Kuropatkin left, Rozhdestvensky, having heard the news of his defeat, at last set sail from Nossi Bé in Madagascar, without waiting for Admiral N.I.Nebogatov's 3rd Squadron of old crocks, despatched through the Suez Canal in response to Klado's complaints. On 8 April the 2nd Squadron was seen off Singapore, trailing the weeds which months in tropical waters had grown on the hulls of the ships, laden in every corner with the soft coal which produced clouds of black smoke from their funnels. Four days later Rozhdestvensky dropped anchor in Cam Ranh bay, off the coast of French Indo-China, and stayed there until Nebogatov caught up with him on 8 May. Six days after that the combined fleet weighed anchor and, after refuelling, headed north for the straits between Japan and Korea, on either side of the island of Tsushima. The alternative was to sail east of Japan and approach Vladivostok through either the Tsugaru or the Soya straits, but both were narrower and more easily mined and would mean a longer voyage with coaling problems and an approach in which the fleet could be more easily shadowed. In any case the whole purpose of the despatch of the two squadrons was to engage and destroy the Japanese fleet in order to cut off their army in Manchuria. Rozhdestvensky organized his fleet into three divisions, the First, led by himself in the Suvorov, of four battleships, the Second, led by Admiral Foelkerzam, of one modern and three older battleships, and the Third by Nebogatov of one old battleship and three coast defence ships. The eight cruisers under Enkvist formed a separate division, and the whole was escorted by forty-five light cruisers and destroyers accompanied by supply ships, whose maximum speed of nine knots dictated the speed of the fleet. Rozhdestvensky does not appear to have discussed with his subordinates any tactical plan for his inevitable encounter with Togo. To make matters worse Foelkerzam died on 23 May but, for fear of depressing morale, his flag was kept flying and nobody, not even Nebogatov himself, was told that he was now the next senior admiral who should take command if anything happened to Rozhdestvensky.

In the misty moonlit early hours of 27 May a Japanese merchant cruiser spotted the lights of one of the three hospital ships accompanying the fleet as it entered the Tsushima strait, and Togo ordered his fleet to sea. His four battleships, six armoured cruisers and light cruisers were still at anchor at Mozampo in Korea, seventy-five miles from Tsushima island, where Admiral Kataoka had four old battleships and four light cruisers. From dawn onwards the Russian fleet was shadowed, and at 10 a.m. Rozhdestvensky ordered his fleet into battle formation, the battleships and armoured cruisers in line ahead with the rest in a separate line ahead to starboard, the whole steaming at eight knots on a north-easterly course. Having lunched at 11 a.m. and drunk a toast on the anniversary of the Tsar's coronation, Rozhdestvensky, in the thickening mist, ordered a change to line abreast, hoping to cross Togo's T when the latter appeared, as he expected, in line ahead. The manœuvre was still being clumsily carried out when he ordered a return to line ahead, just as Togo's fleet was seen at 1.40 p.m. at a range of seven miles, steaming at fourteen knots on an almost parallel opposite course. Togo took the risk of going about and, instead of turning to port and going behind him, which would have exposed his starboard column, Rozhdestvensky turned to starboard, confusing his gunners and being gradually pushed round by Togo's faster fleet. The Suvorov was hit, her helm jammed to starboard and Rozhdestvensky seriously wounded, throwing the powerful First Division into confusion. The Alexander III and the Oslyabya, with the dead Foelkerzam still aboard, were already ablaze, and the latter soon sank, a destroyer hastening to Nebogatov to tell him that he was now in command. By the time he knew this, Togo in a succession of turns closed in on the disordered fleet, sinking in turn the *Alexander III*, *Suvorov* and *Borodino* and seriously damaging three other battleships. While Nebogatov, flying the signal 'Follow me', made for Vladivostok with his own old battleship *Nikolai I*, the damaged *Orel* and the three coast defence ships, Enkvist took his three remaining cruisers in the opposite direction, finishing up at Manila, where they were interned. But Nebogatov was soon cornered by the Japanese, who had meanwhile sunk two older battleships, while two more were scuttled. Only one cruiser and two destroyers reached Vladivostok, thirty-four warships in all being lost at a cost of 4,830 Russian lives: 5,197 were taken prisoner, many of them wounded, and 1,862 interned in neutral ports. The Japanese lost only three torpedo-boats and had 110 sailors killed and 590 wounded. This shattering defeat marked the virtual end of the war, both sides accepting the offer of President Theodore Roosevelt, made on 8 June, to mediate a peace, which was signed on 14 October at Portsmouth, New Hampshire. Russia accepted the 'paramount political, military and economic interests of Japan in Korea' and ceded to her the lease from China of the Liaotung peninsula, including Port Arthur and Dalny, the relevant section of the South Manchurian railway and the southern half of the island of Sakhalin, both agreeing to withdraw their troops from Manchuria.

Japan's train of victories over Russia, on sea and on land, was not primarily attributable to superior equipment, although at sea her battleships had a marginal superiority over the most modern of the Russian, notably in speed (fourteen to eleven knots). On land she was better equipped with machine-guns and a higher proportion of her artillery was of large calibre. Her inferiority in numbers and quality of cavalry proved no handicap, except in pursuit. Both at sea and on land it was the combination of professional skill and efficiency and a high morale, imbued with an offensive spirit, which were the principal causes of her success. The Japanese forces had proved apt pupils of the British Royal Navy and the German army, by whom they had been trained. Russian commanders were slow, unimaginative and cautious, and their staffs inefficient. Infantry tactics were rigid and outmoded, relying more on shock action than on firepower: artillery was generally evenly distributed, ill-concealed and inefficiently handled, while the cavalry, rendered ineffective on the battlefield by modern firepower, was ill-disciplined and ill-trained in the vital task of providing information about the enemy. Both sides had a few balloons for artillery observation and Kuropatkin even had a motorcar. Both navies used primitive radio communications, the Russians being alerted to Togo's move towards Tsushima by listening to Japanese transmissions. The armies of both sides learned through bitter experience the vulnerability of infantry, whether in attack or defence, to artillery and machine-gun fire, and that it must seek protection below ground. The fact that battles were not over in a day, but could last for days on end, also came as a surprise to many. The shadows of 1915–18 had been cast before those who could see them, but most of the military preferred to blame the peculiar pattern of the war on the general inefficiency of the Russian steamroller.

1906–17

Not surprisingly, defeat and demoralization aggravated general discontent with those in authority, and revolutionary sentiments affected the armed forces, the navy in its bases being closely in touch with the urban proletariat and the army with the peasantry. The latter was also affected by being extensively used to suppress civil disorder. Both were affected by mutinies. The first and best known of them was that of the battleship *Potemkin* based at Odessa in the Black Sea at the end of June 1905. It was followed by one at the Baltic naval base of Kronstadt in October and one at Sevastopol, which affected the army also, in November. The general tide of dissatisfaction forced the Tsar to make political concessions, greater power being given to both the local representative assemblies, the *zemstvos*, and a central one, the *Duma*, although the Tsar clung to the principle of autocracy, being particularly incensed when the *Duma* attempted to interfere in the affairs of the armed forces. The latter at least clearly recognized the need for a major shake-up, both among commanders and staffs, and in training in every field, from the highest to the lowest level. In spite of the cost of the war (in some respects because of it), the Russian economy was rapidly expanding, and the defence budget, which normally absorbed a third of government revenue, benefited from it. As a result the navy was able to embark on a significant new construction programme, four 23,000-ton dreadnoughts of Italian design being ordered in 1909, followed by four 32,000-ton ones in 1912, for the Baltic Fleet, and four of the former for the Black Sea Fleet, while each would have four new light cruisers, thirty-six new destroyers for the Baltic and twelve for the Black Sea, with twelve submarines for the former and a small number for the latter, all of modern design, superior to those in service with the Royal Navy at the time. But it was far beyond the capacity of Russian shipyards. Much of the machinery had to be bought from Germany and the rate of construction was so slow – five to seven years for a battleship – that much of the programme had not been completed by 1917. Between 1909 and 1914 Russia spent more on her navy than Germany, which was engaged in the naval race with Britain.

In the first three of these years approximately four times as much was spent on the army, the navy's share increasing to nearly a half by 1914, by which time Russia was spending more on its army than was Germany, producing 114 infantry divisions to the latter's ninety-six and 6,720 mobile guns to Germany's 6,004. The attempt to modernize the army and its methods led to a deepening of the division which ran through the army between the patrician and the 'praetorian guard' elements in the officer corps. The position and power of the Main and General Staffs were at the centre of the quarrel. Defeat by Japan was blamed on the failure of the system for high command and, partly in order to remedy this, a Council of State Defence was established in 1906, assuming much of the responsibility which had been exercised by the Main Staff and the War Ministry. This led to a resurgence of the independent influence of the General Staff as an élite body concerned solely with strategic and operational matters, and also to the appointment of Inspectors-General of the major arms – infantry, cavalry and artillery (the engineers already had one) – who became members of the Council, all of them Grand Dukes. The Chief of the General Staff, to which post the progressive General F.F.Palitsyn was appointed, was made directly responsible to the Tsar, while the Minister for War was responsible to the *Duma* for the army's administration. Similar arrangements were made for the navy. These steps, taken partly to prevent interference by the *Duma* in military affairs, represented a reversal of Milyutin's and Obruchev's achievement in integrating all the staffs; but they were in turn reversed in 1909 when General Sukhomlinov, having intrigued to get rid of the progressive Palitsyn, became War Minister, General A.Z.Mishlaevsky, taking Palitsyn's place as Chief of the General Staff, becoming responsible to him. At the same time the Defence Council was abolished. In the short period it had existed, the Council had been divided over the priority to be allocated between the rebuilding of the navy and the modernization of the army, the Tsar favouring the former and his uncle, the Grand Duke Nicholas, who usually presided over the Council, the latter. Within the army there was a fierce argument between the infantry and the artillery, a legacy of the war against Japan in which the infantry complained that the artillery had failed to give them support, while the gunners stated that the infantry did not understand the true function of artillery, which was reluctant to reduce its eight-gun battery to six because of the reduction in rank of the commander which would follow. The artillery, headed by Grand Duke Sergei Mikhailovich, had become enthusiastic for fortresses manned by heavy guns, and was reluctant to adopt the high-trajectory howitzer for bombardment of field defences, which they maintained should be reduced by infantry. General V.A.Sukhomlinov, a conservative infantryman of the old praetorian guard element, supported the infantry line, as had Palitsyn, against the opposition of the Grand Dukes Sergei and Nicholas. Having established himself as War Minister and downgraded the influence of the General Staff and the Inspectors-General, Sukhomlinov also gained control of promotions and appointments. Thereafter he saw that important commands and posts on the staff were filled by his nominees, while that of the Chief of the General Staff was downgraded by a constant change of occupant. Grand Duke Nicholas's cavalry colleagues were banished to the command of military districts, while the General Staff was greatly expanded, its lower ranks filled by officers of comparatively humble origin.

While the army was riven by this quarrel, the shadow of war was never absent from Europe. Russia's concern for the east had been allayed by the 1907 agreement with Britain over their respective spheres of influence and interest from the Caucasus to Tibet, and by further agreements with Japan and China, the former accepting the absorption of Korea into the Japanese Empire and the latter a Russian protectorate over Outer Mongolia. The powder keg was the Balkans, Russia's Panslav ambitions, her claim to be the protector of the Orthodox Christians and her permanent desire to control the Bosphorus and Dardanelles, all pushing her to take any advantage she could of the weaknesses of the Ottoman and Austro-Hungarian Empires. Nevertheless, the fear that Germany might take advantage of her involvement forced her in 1909 and 1913 to give way to Austrian demands, having failed to erect a Balkan alliance against Austria, based on Serbia, Bulgaria and Turkey. German involvement in Turkey from 1913 onwards made Russia even more sensitive about events in that region, and it was almost inevitable that, when the incident at Sarajevo on 28 June 1914 provoked Austria's demands on Serbia, she was not prepared to give way again.

The artillery argument between the devotees of fortresses and those who wanted more mobile artillery support to the infantry had a decisive impact on Russia's war plans. Before Sukhomlinov's day, they had been strictly defensive as far as Germany was concerned, but influenced by the strengthened alliance with France, they were revised in 1910. Any westward advance from the Polish salient was liable to attack from both flanks, from Germans in East Prussia and Austrians in Galicia. One of those two threats must first be eliminated. To deal with Galicia first would not affect a German attack on France, and Plan 19, compiled by Sukhomlinov's assistant Danilov, aimed to attack East Prussia with four armies, two from the east and two from the south, deploying nineteen army corps, nine being left to contain the Austrians. No more money was to be spent on fortresses, which would be demolished. This aroused the anger not only of the fortress-minded gunners, but also of those who saw Austria, with its Balkan ambitions, as the principal adversary. Two of the Grand Dukes' supporters, the commanders of the military districts of Warsaw and Kiev, led the opposition to Plan 19, and in 1912 the plan was modified to meet two contingencies: Plan G for a German attack against Russia and Plan A for a German attack on France, the latter clearly being the more likely, but the fortresses were retained to meet the former. The major change in Plan A was that the main effort was to be made against Austria, forty-six and a half divisions being allotted to it, twenty-nine and a half in two armies being left for the offensive into East Prussia, which was to take place simultaneously. The change was to have fatal consequences in 1914.

The 'Great Programme', argued about all through 1914 and becoming law in the month that Gavrilo Princip fired his fatal shot, provided for an increase in the annual call-up for three-year service in the army to 585,000, producing a standing army of nearly two million, three times that of Germany, forming 122½ divisions, supported by 8,358 field guns, at last to be organized in six-gun batteries. The number of heavy howitzers in support of each division would be increased from six to twelve – the German divisions had eighteen. After full mobilization of reserves the army would number five million. In spite of its internal divisions, it was in a better state in 1914 than the navy. Few of the new ships of the ambitious programmes of 1909 and 1912 had been completed and, while attention had been concentrated on that programme, the human element had been neglected, the general state of administration, training and morale remaining abysmally low.

German planning was based on the assumption that Russian forces would not be able to launch an offensive until six weeks after the start of mobilization. During that time France would be dealt a mortal blow, and forces could then be transferred to the east before the steamroller got going. The thirteen divisions of Prittwitz's 8th Army should be sufficient to hold East Prussia during that time; but the increase in Russia's peacetime strength and the construction of strategic railways meant that this time could be significantly reduced. Austria's ultimatum to Serbia on 23 July 1914 was rejected on the 26th, and Austria declared war two days later. Russia ordered general mobilization on the 30th, having brought its troops on the Austrian border to a higher state of readiness before that. On the 31st Germany issued an ultimatum to Russia, demanding a reversal of its mobilization within twelve hours, and declared war next day. On 15 August the main body of Rennenkampf's 1st Russian Army crossed the eastern frontier of East Prussia at the northern end of the Masurian Lakes.

Rennenkampf's 1st Army formed part of the north-western 'front', commanded by General Zhilinski. North of him was the 6th Army occupying the Baltic states, which had no part to play in the offensive. Separated from him by sixty miles was Samsonov's 2nd Army, which was to thrust north from Poland to the west of the lakes. It was hoped that this would entrap Prittwitz's army, and so it should have done.

Zhilinski had twenty-nine and a half infantry divisions and a large force of cavalry, out of the total of sixty-seven first line and thirty-one second line infantry and thirty-seven and a half cavalry divisions available to the Supreme Headquarters (*Stavka*) for immediate use against the thirteen German divisions in East Prussia and forty Austrian in Galicia. But not only did the *Stavka* make no attempt to co-ordinate the action of Zhilinski's front with that of General Ivanov's south-western front, but far too high a proportion of the divisions available were held back to man fortresses and to guard the flanks and lines of communication of the 1st and 2nd Armies. The *Stavka* itself was a totally ineffective headquarters. Sukhomlinov had been expected to assume the supreme command, but neither he nor the Tsar, to whom he offered it, was keen to bear the responsibility (and possibly the blame), knowing that whoever held the post would probably be unable to exert his will over the front commanders. Grand Duke Nicholas accepted and was given the nonentity General Yanushkevich as his Chief of Staff, the real power remaining in the hands of Danilov, the Quartermaster-General. An improvised headquarters, with ludicrously inadequate communications, was set up at the railway junction of Baranovichi, eighty miles south-west of Minsk. These factors and inefficiency in the logistic arrangements reduced the force available at the sharp end to only six and a half infantry divisions in 1st Army and nine and a half in the 2nd. Contrary to Prittwitz's intention, his corps commander François attacked Rennenkampf's army before it reached the narrow gap at the north of the lakes, and in a series of battles lasting over five days round Gumbinnen Rennenkampf had the better of the struggle, although at one point he considered withdrawal. Meanwhile Samsonov, delayed by logistic difficulties, at last crossed the southern frontier, just as Prittwitz had decided to withdraw to the Vistula. Suspecting (or even perhaps actually aware of) this, Zhilinski urged him on but, fearful about his flanks, Samsonov allowed his army to get widely separated. His left corps soon became engaged with General Stötz's corps, which had been stationed there, and then with that of François, which had been switched by rail from the north, reinforced by one from France, and at first had some success, while his two corps in the centre penetrated almost to their objective of Allenstein, fifty miles over the border, as did his right-hand corps twenty-five miles away to the east.

An entirely unexpected blow now fell. Two German corps, one led by Mackensen, withdrawing on foot from Gumbinnen, came across the right-hand corps and threw it back in confusion, just as François got round the left flank of the left corps. On 29 August the leading troops of Mackensen and François met behind the whole of Samsonov's army, completing its encirclement. Rennenkampf had done nothing to help him, he and Zhilinski being obsessed with the idea that the

Germans had withdrawn towards Königsberg, to which Rennenkampf was to lay siege. Samsonov shot himself as his army of 100,000 men with 400 guns surrendered, only 10,000 getting away. The Germans christened their victory the Battle of Tannenberg, in revenge for the defeat of the Teutonic Knights there by Poles, Lithuanians and Byelorussians in 1410. Ludendorff was quick to switch the effort of his 8th Army, now eighteen and a half infantry and two cavalry divisions strong, against Rennenkampf. In the battle of the Masurian Lakes which followed, lasting from 7 to 25 September, the Germans at first drove the Russians back behind their frontier, but were then pushed back themselves to the lakes, where the two sides dug in facing each other, the Germans having lost 100,000 of their 250,000 men and captured another 30,000 Russians.

While Zhilinski's north-western front was fighting the Germans, the south-western front, at the headquarters of which at Cholm the commander Ivanov and his chief of staff Alexeyev were bickering about who should be the first to read signals, failed to take advantage of the muddle into which Conrad von Hötzendorf had got the Austro-Hungarian armies. To the dismay of his railway staff, Conrad kept changing his mind about priorities of deployment between Serbia and Galicia and, in the latter, about where to detrain and whether to remain on the defensive or take the offensive before Russian deployment was complete, in order to prevent them from concentrating against the Germans in East Prussia. On the Russian side, there had been two rival plans. Danilov at *Stavka* favoured a plan to make the major effort on the front's left, thrusting west from the Ukraine to the north of the Carpathians, biting into the Austrian right flank, if they moved north. Ivanov and Alexeyev favoured a thrust from their right near Lublin, south-west towards the rail junction of Krakow, severing the Austrian communications with eastern Galicia, where it was assumed their main force would be deployed, as it would have been if Conrad had not changed his mind. Danilov belatedly came round to that view, but no changes were made in the allotment of divisions to the four armies, General Evert's 4th west of Lublin, General Plehve's 5th, on the borders of Poland and the Ukraine and, a hundred miles further south, General Ruzski's 3rd and General Alexei Brusilov's 8th, totalling, by the twenty-fifth day of mobilization, some forty-five infantry and six cavalry divisions, against which Conrad could muster thirty-seven and ten. The Russian strategy lacked finesse. Instead of remaining on the defensive in one area, and concentrating overwhelming force into the offensive in another, it was a strategy, on both fronts and within the south-western, of a general offensive everywhere. Conrad had finally decided to take the offensive and to do so on his left, his 1st and 4th Armies striking north in Poland, while his 3rd Army guarded the right flank against a Russian thrust from the Ukraine. After an inconclusive clash between the sabre-wielding cavalry of the two sides, the Russian 4th and 5th Armies met Conrad's 1st and 4th, and in a confused mêlée were driven back towards Lublin, prompting the *Stavka* to order the 9th Army to move south from the Warsaw area to join them. Meanwhile the western thrust by the 3rd Army from the Ukraine made slow progress, the cautious Ruzski paying scant heed to Ivanov's urgent orders for him to push on and take the Austrians in the flank to relieve the pressure on Plehve's 5th Army, which had been badly knocked about by the Austrian General Auffenberg's 4th, losing 40 per cent of its strength. This Austrian victory was offset by the defeat of its 3rd Army by Ruzski, who failed to exploit his victory and continued his advance with extreme caution. Conrad rashly ordered Auffenberg to turn about and attack him from the north, but by this time all the Russian armies were receiving reinforcements and the revitalized 5th thrust into his rear. By 9 September Conrad was in severe difficulties and, pressed on all sides, withdrew in considerable disorder well west of the River San. Ivanov's troops were too exhausted and their logistics too primitive for him to press Conrad any further west or south. He had lost 250,000 men, 40,000 of them prisoner, and 100 guns, while Conrad had lost almost half his force – 400,000, of whom 100,000 were captured – and 300 guns.

In the operations which took place in the winter of 1914–15, in the area which now marks the frontier between the Soviet Union and Poland, honours were more or less even, the Austrians being saved from defeat by the help given to them by the Germans. In contrast to the Western Front, there was no continuous line of defences on either side, and it was therefore possible for even the largely immobile armies of Russia and Austria to move about as their predecessors had done in previous centuries, the principal difference being that reinforcement and supply were provided by railway. In that the Russians were at a considerable disadvantage. Not only was the railway network on their side less well developed and complicated by the change of gauge as it crossed the border, but their operation of it, although not greatly worse than that of the Austrians, was in stark contrast to the efficiency of the Germans. Rolling stock frequently finished up in the wrong place and major rail movements to switch forces from one sector to another took place at a speed little more rapid than road movement by horse or foot, if adequate roads had existed. This ponderous rail movement was caused in part by the high proportion of cavalry in the Russian army, the movement of both the horses themselves and of their fodder absorbing railway resources and time. When Austrians and Germans attacked Russians in well-prepared defences, they suffered heavy casualties and did not succeed; but Russian defences were seldom well prepared. They were disinclined to dig: it was either too wet or frozen too hard, and one might uncover corpses.

Defence stores, like barbed wire, were in short supply and often stolen by or sold to the local inhabitants. The logistic arrangements of both sides were inefficient, so that, if they did succeed in advancing any distance, they could not exploit success and became vulnerable in turn to a counteroffensive.

After his success against Conrad, Ivanov wished to press on, capture the great fortress at Przemysl on the San and continue an advance to Krakow, even perhaps Budapest itself; but Ruzski, who had succeeded Zhilinski in command of the north-western front, refused to give up forces that would be needed for this, fearing a German thrust from East Prussia. The *Stavka* compromised with a plan for a direct thrust towards Warsaw, Ivanov taking Ruzski's 2nd Army under his wing for that purpose, while withdrawing his left flank east of the San. Movements in preparation for this coincided with an offensive by Ludendorff north of Krakow, designed to help Conrad, and led to confused and inconclusive battles in October and November between the Vistula and the San, in which Mackensen's 9th Army was switched rapidly north to counterattack the Russian 1st and 2nd Armies which had begun their offensive west of Warsaw. A prudent withdrawal to Lödz saved them. This led to a further wrangle between Ruzski who wished to withdraw further, and Ivanov who wished to continue the offensive. The former won the argument, as Ivanov's left flank on the San was threatened by Conrad, attacking from the Carpathians. Fighting died down at the end of the year, the offensive operations of both sides having cancelled each other out.

The argument between Ivanov and Ruzski, the latter supported by Danilov at *Stavka*, continued into the new year, the former arguing that a concentrated attack on Austria could knock her out of the war, the latter that such an attack could not be launched until the Germans had been removed from East Prussia. Once more, there was no firm decision either way, both being allowed to plan offensives, Ruzski with fifteen and a half divisions in East Prussia and twenty-three and a half in northern Poland, while Ivanov, who had received two from Ruzski, had thirty-one divisions in the Carpathians. Ludendorff attempted to forestall their plans with an attack in Poland in January which failed, gas being used for the first time, which blew back into the German lines. At the same time Conrad, with German support, launched a major offensive in the Carpathians which also failed and, when Ivanov launched his offensive in February, Conrad was driven back, Przemysl being left surrounded, finally surrendering in March. Fighting continued in bitter weather for two months, the Austro-Hungarian-German army under Conrad suffering heavy casualties and losing even more men from sickness and desertion. Meanwhile Ludendorff had launched a successful counteroffensive east of the Masurian Lakes which Ruzski held on the line of the Niemen and Narew rivers, after losing nearly 100,000 men, and then pushed back to the Prussian frontier early in March. Grand Duke Nicholas decided not to run the risk of a second Tannenberg, but refused Ruzski's request to shorten his front by a withdrawal in Poland, as this would expose Ivanov's right flank. Pleading exhaustion, Ruzski resigned and was succeeded by Alexeyev. Ivanov hoped that his former Chief of Staff would be sympathetic to his need of reinforcement, but Alexeyev fought as fiercely as Ruzski had done to prevent his front being weakened in order to allow Ivanov to maintain pressure in the Carpathians. As a result the latter called off the offensive on 10 April, pinning hopes for the future on reinforcement with new formations and the build-up of 9th Army in Bukovina on the Dniester.

Ivanov's hopes of reaching Budapest were dashed by the decision, reluctantly taken by Falkenhayn and authorized by the Kaiser on 13 April, to transfer eight divisions from France and plan an offensive, in which Mackensen, with Seeckt as his Chief of Staff, would command the 11th German and 4th Austro-Hungarian Armies. This was launched on 2 May and hit General Radko-Dmitriev's 3rd Russian Army, which faced west and south, north of the Carpathians between Krakow and Przemysl, and after a week's fighting had inflicted severe casualties on it, Radko-Dmitriev making urgent pleas for more artillery ammunition. Ivanov's request for permission to withdraw to the San was refused by the Stavka, which pinned its hopes on a counterattack by the 9th Army, which was in fact successful, but could not avert the inevitable withdrawal to the San. Before defences of any strength could be organized there, Mackensen attacked them, and on 4 June, almost two weeks after Italy had declared war against the Central Powers, his forces recaptured Przemysl. Ivanov withdrew again, this time to the Dniester, his 9th Army having to abandon all their gains. On 22 June the Austrians recaptured Lwow and with it the whole of Galicia, Mackensen's group alone having captured 240,000 prisoners and 224 guns for a loss of 90,000 men.

The disaster on the Carpathian front was blamed on shortage of artillery ammunition and of rifles, although the real cause ran deeper than that. The whole Russian war machine was grossly inefficient: in its tactics and their application; in its staff work, operational and logistic; in its personnel organization and administration; in its procurement arrangements, which were part and parcel of the inefficiency and confusion in its general governmental machine; and in its command organization at all levels, from War Council and Stavka down to the units, where the relations and balance between officer, NCO and soldier were hopelessly unsatisfactory. Fundamentally this can be attributed to the lack of a substantial 'middle' element in Russian society. The result was that, although some nine million men were called up into the army, the potential power of the Russian military machine was never properly translated into effective action on the battlefield. The events of 1915 showed that, in conditions which suited him, the Russian soldier could be effective.

After Mackensen's thrust in the southern sector, the next blow fell unexpectedly in the north, when Ludendorff launched an attack into Courland, into which Alexeyev was reluctant to become drawn; but, when it threatened both Riga and the prestigious fortress of Kovno, he was forced to react. These two offensives left the Russian armies in central Poland in a dangerous salient, for the whole of which Alexeyev was made responsible. He wished to withdraw to the line of the rivers Narew and Bug, but Grand Duke Nicholas at *Stavka* could not bear to give up virtually the whole of Poland with the great fortresses of Osowiec, Novogeorgievsk and Ivangorod, bristling with guns and symbols of Russian rule. But when the Germans launched simultaneous offensives in Courland, northern Poland and Galicia in July, which inflicted heavy losses on the Russian armies, although on Falkenhayn's insistence they were limited in depth, the Grand Duke was forced to change his mind, and a gradual step-by-step withdrawal was put in hand, the Russian armies inflicting significant casualties on the Germans whenever the latter tried to exert pressure without adequate preparation. On 4 August the Germans entered Warsaw and two weeks later after only a short bombardment Novogeorgievsk, which had been ordered to hold out, surrendered. The retreat went on until the end of September, by which time the Russian armies were right back to the Pripet marshes on a line north and south through Baranovichi and Pinsk, the loss of Kovno leading to the transfer of Grand Duke Nicholas to become Viceroy of the Caucasus, the Tsar himself assuming command with Alexeyev as his Chief of Staff. The latter was succeeded by Evert, his command now renamed Western Front, a new Northern Front having been created, commanded by Ruzski. The withdrawal had gone further than necessary and certainly than expected; but the generals, having complained that the shortage of ammunition and rifles, and the cowardice and lack of training (or trainability) of their troops, made it impossible for them to conduct operations successfully, were determined to prove their point. The withdrawal caused dismay to the regional authorities as the army demanded that it should take control of areas deeper into Russia proper, where the population was either evicted or treated in a highhanded, tyrannical fashion, the orders of different military authorities conflicting with each other. The last blow to Russian pride was the fall of Vilna on 18 September, marking the end of the German 1915 offensive; Conrad in the south having, against Falkenhayn's wish, attempted unsuccessfully a major envelopment of Ivanov's left flank. By that time the Germans and Austrians were at the end of long lines of communication in marshy, forested terrain with few roads, while the Russians were back near their supplies in terrain, the defence of which suited their manpower-intensive army. In September Conrad's strength fell from 500,000 to 200,000, a high proportion missing, presumed captured, sick and wounded. Between 1 May and 1 December the Russian armies lost two million men, half of them as prisoners of war, a tribute to the effectiveness of Falkenhayn's insistence on offensives of limited depth and duration, a lesson from which Brusilov was to profit in the following year.

The events of 1915 had major repercussions in the politico-economic field. The *Duma* and all those who were politically opposed to the autocratic state machine, the industrialists who saw their potential contribution to the war effort (and the profits to be derived from it) neglected in favour of government-owned munition factories and foreign suppliers, and the generals at *Stavka* and its subordinate commands, complaining of shortages of guns, shells and rifles, all combined in opposition to the inefficiency of the government's machinery which was blamed for the disastrous failure of the army. The result was a frenzy of suggestions for improvement which initially led to an even greater bureaucratic confusion of committees designed to encourage war industry from the cottage to the large factory. Most of it was soon by-passed by a co-operation between the War Ministry and the major industrial combines, which proved effective, so that by the summer of 1916 the shortages were overcome and by the end of that year Russian war production exceeded that of France. There had been almost as much confusion in making efficient use of the country's vast pool of manpower, caused by the continuation of the large-scale exemptions from call-up that had been the rule before the war, in order to avoid calling up more men than the army could afford. The reserves were therefore ill-balanced and it was easy for peasants to evade call-up, the heaviest burden falling on the urban population, thus affecting the industrial labour force. As time passed the problem of providing trained officers and NCOs became worse. At the beginning of 1916 the army's strength stood at 4,587,145 men, backed by a reserve of 1,545,000. By 1917 the total number called up was 14 million out of a population of almost 180 million, a much lower percentage than in France or Germany. When Zhilinski, Russian representative at French GQG, told Joffre that Alexeyev's front-line strength in January 1916 was 1,693,000, of whom 1,243,000 had rifles, the latter was not impressed. Both the organization for war industry and the unsatisfactory relationship between officers and the rank and file were to make their contributions to the revolution.

After defeating the Russians in 1915, Mackensen in October switched the effort of the Central Powers in the east to crush Serbia, Bulgaria joining in, thus provoking the Entente to deploy a force at Salonika, having encouraged Greece to join them. Russia could not stand idle, and the *Stavka* in November belatedly ordered Ivanov to launch an offensive on the River Strypa in eastern Galicia, which, after several postponements, did not start until 27 December and failed after two weeks with a loss of 50,000 men, blame being once more laid on shell shortage. By March 1916 that excuse could not hold water and, pressed to do something to relieve German pressure on the French, Falkenhayn having transferred a significant number of divisions to the west for his Verdun offensive, which began in February, Alexeyev decided on an offensive north of the Pripet marshes. There Evert's Western Front and the Northern Front, now commanded by Kuropatkin in place of Ruzski, greatly outnumbered both in men and guns the German armies under Ludendorff (Kuropatkin 300,000 to 180,000 and Evert 700,000 to 360,000), while Ivanov, in the South-western Front, had 500,000 men facing about the same number. Alexeyev's aim was to recapture Vilna, Evert making the main attack from the east with Smirnov's 2nd Army, while Kuropatkin drove south-west towards the objective. Smirnov, who had twenty divisions – over 350,000 men in 253 battalions of infantry and 233 squadrons of cavalry supported by 605 light and 282 heavy guns well supplied with ammunition – faced the German 10th Army with four and a half (later built up to seven) infantry divisions, supported initially by 300 guns, later increased to 440. The attack, launched on 18 March, known as the Battle of Lake Naroch, was a total failure, hardly surprising when it was made on a narrow front in a wet area at the time of the first thaw, no effort having been made to achieve surprise. Smirnov lost 100,000 men, 12,000 more dying of frostbite, while the Germans lost 20,000. From then on both Evert and Kuropatkin, and Alexeyev also, were convinced that no attack could succeed unless an even greater artillery bombardment could be provided, on the scale employed in 1916 by the French and British.

Fortunately for Russia there was a general who thought otherwise – Brusilov, the sixty-three-year-old commander of the 8th Army in the South-western Front. Although a cavalryman who had started his service in the Tsar's Corps of Pages, he was a highly practical soldier who paid meticulous attention to detail. After the failure of the winter offensive on the Strypa the Tsar steeled himself to remove Ivanov, who had powerful supporters among both the patrician and praetorian guard lobbies, and replace him with Brusilov, who attended the conference held by Alexeyev on 14 April to decide what the Russian armies should do to take the pressure off the British and French. He proposed a new offensive north of the Pripet, but Kuropatkin and Evert were reluctant to embark on one unless they received reinforcements of heavy artillery. Brusilov however offered to attack without any reinforcement although, unlike on the other fronts, he had no margin of superiority over his opponents. He and an influential body of his staff officers had studied their previous failures and concluded that an attack on a narrow front, in only one sector, preceded by a prolonged bombardment, had no hope of success. The enemy could seal it off by concentrating reserves of artillery and infantry from other areas before it had time to develop. Brusilov proposed a number of simultaneous attacks, each on a much wider front, with only a short preliminary bombardment which would be carefully designed in detail to shatter the enemy's forward defence posts, while the longer-range artillery concentrated on his local reserves. A combination of sapping, mining and well-concealed defence works would make it possible to move artillery and infantry, both the assaulting troops and reserves, close to the enemy so that there would be the minimum delay in exploitation of the assault. Alexeyev's and Brusilov's own subordinate army commanders, with the exception of Sakharov, had little confidence in his proposal, but Alexeyev finally agreed that Brusilov should launch an offensive on these lines as a subsidiary operation, preliminary to the main effort which would be made by Evert and Kuropatkin. All four of his armies would attack, from north to south General Kaledin's 8th, General Sakharov's 11th, General Shcherbachev's 7th and Lechitski's 9th, the last protesting from his sick-bed against what his Front commander and his Chief of Staff had committed him to. Brusilov had no significant margin of superiority in numbers, forty infantry and fifteen cavalry divisions to Conrad's thirty-eight and a half and 11 – 600,000 Russians against 500,000 Austrians and Germans; but the numbers of the latter had been greatly reduced and many of the best Austrian troops had been switched to take part in the Austrian offensive on the Isonzo in Italy, a campaign which was much more popular among all political elements in the Austro-Hungarian empire. Attacks were launched on 4 June, and those by Kaledin and Lechitski were immediately successful. In two days Kaledin reached the River Styr at Lutsk, 50,000 men and seventy-seven guns being captured from Archduke Joseph Ferdinand's demoralized 4th Austrian Army. The pressure exerted by Sakharov and Shcherbachev in the centre pinned down the bulk of the German divisions and attracted Conrad's reserves, while Lechitski in the south trounced the Austrian 7th Army and pushed it south-west away from its German neighbours. By 12 June Brusilov's armies had captured 3,000 officers, 100,000 men, 216 guns and 645 machine-guns; but he had no means of immediately exploiting his victory, having kept his cavalry well to the rear to prevent them cluttering up his logistic supply, even dismounting some to use as infantry. His aim was to strike north-west to Kovel, from the salient Kaledin's 8th Army had created west of Lutsk, threatening the right flank and rear of the forces facing Evert, expecting him to exploit the victory by launching his own offensive. But Evert kept on changing his mind about where to launch it, each change causing further delay, four corps being transferred to Brusilov from his and Kuropatkin's fronts while they found excuses for doing nothing themselves. On 2 July Evert at last attacked near Baranovichi, and after firing off more artillery shells in a week than Brusilov in the whole of his offensive, brought it to a halt when he had lost 80,000 men to the German 16,000 and gained only a few miles. Brusilov, having taken over General Lesh's 3rd Army south of the Pripet from Evert, and reinforced by the newly formed Guards Army commanded by General Bezobrazov, himself took the offensive in July towards Kovel, while Lechitski continued his advance up the valley of the Dniester in the south, causing the Austrians to call off their offensive on the Isonzo and the Germans to exert a tighter control over the Austro-Hungarian armies, installing their own commanders or chiefs of staff to Austrian ones. But there was little scope for the application of Brusilov's meticulously prepared assaults, and the Guards had not been trained in them. The latter's attempt to reach Kovel petered out in the marshes of the Stokhod in August. Only Lechitski's victories still showed promise for exploitation on the borders of Romania, which entered the fray with a declaration of war on Austria-Hungary at the end of August. By this time the Germans had realized that something had to be done to restore the situation in the east. They had been able to contain the British offensive on the Somme without difficulty: the French were clearly exhausted: twenty infantry divisions were transferred from the west and ten redeployed from Ludendorff's forces in the north. Hindenburg was appointed overall commander north of the Dniester, theoretically subordinate to Conrad, who commanded south of it. When Romania's declaration of war led to the replacement of Falkenhayn by Hindenburg, the latter's command in the east was assumed by Prince Leopold of Bavaria.

As has been related in chapter 3, the reaction of the Central Powers was swift, profiting from their ability rapidly to transfer troops by rail from one area to another. The result was a swift revenge, which led to the total collapse of Romania before Russia could do anything effective to help, which she was not much inclined to do in any case. Brusilov's responsibility for the northern part of his sector was transferred to Evert, who was ordered to renew the offensive towards Kovel, while thirty-six infantry and eleven cavalry divisions were transferred slowly from Brusilov's centre to the Russian border with Romania, where they arrived too late to affect events, Mackensen entering Bucharest on 7 December and the Romanian war effort totally collapsing by the end of the year. The encouragement given by the Entente powers to Romania to enter the war was one of their major strategic errors.

Alexeyev hoped to profit from the German commitment to the southern sector by a renewal of the offensive to Kovel; but a succession of attacks, conducted according to Evert's old formula, proved as expensive and fruitless as all his and his predecessor's previous ones. By the time Alexeyev called it off in mid October, the three armies involved, 3rd, Guards and 8th, had suffered heavy casualties and their morale was low. Lechitski's advance into the Carpathians had come to an end a month before and the only bright spot on the horizon before the end of the year was an attack on the Northern Front, where Ruzski had returned to replace Kuropatkin, against Germans withdrawing in Courland, which netted thirty-six guns and 8,000 German prisoners for a few thousand casualties.

Early in 1917 the Russian army, in terms of numbers of men, guns, ammunition and equipment generally, was in a better state than ever before, and its commanders confident that the year could bring decisive victory over the Central Powers. But, if the army in spite of its huge losses was in a reasonably good state, the nation as a whole was not. The basic trouble was that its social and political system could not cope with a major economic transformation which was intensified by the war. Agriculture, or at least the distribution of its products surplus to the peasants' own needs, and rural industry broke down, causing a drift of population to the towns where they could not be fed. The financial and tax policies of the government aggravated the inflation to which war economies were so prone, so that even if food was available, even wage-earners could not afford to buy it. As discontent mounted the Tsar, isolated in his military headquarters and suspected of being under the influence of the Empress's favourite, Grigori Rasputin, was out of touch and losing influence to his opponents. January 1917 saw serious strikes in Petrograd (as St Petersburg had been renamed), Moscow and Kharkov and the crisis became acute as the *Duma* was due to reassemble on 27 February. As the climax of mounting tension, strikes and demonstrations on 8 March in Petrograd were followed by larger ones next day. When the Tsar, from his headquarters, ordered Ivanov, Commandant of the city, to put an end to the disorders immediately, several regiments there mutinied. On 12 March Brusilov supported the appeal of the President of the *Duma*, Mikhail Rodzyanko, to the Tsar that 'a person possessing the confidence of the country' be asked to form a government. The Tsar's reaction was to prorogue the *Duma* until April.

On the same day the mutinous soldiers joined hands with the Soviet of Workers' Deputies to capture the Peter and Paul Fortress in Petrograd and released prisoners there and elsewhere. The Soviet entered into negotiations with a Provisional Committee of the *Duma*, provoking the government to send a telegram to the Tsar asking to be dismissed. The Tsar refused the request and set off for Petrograd by train, but turned back when he learned that the lines there and to Moscow were in the hands of the revolutionaries, and on the evening of

14 March he arrived at Ruzski's headquarters. On that day the Executive Committee of the Soviet had agreed with the Provisional Committee of the *Duma* to support a government formed from the latter, as long as it pursued a political programme which promised a whole host of radical reforms, and that no action would be taken against the mutinous soldiers, who would not be moved from Petrograd. The provisional government under Prince Lvov, with the Octobrist Alexander Guchkov as Minister for War and Alexander Kerensky, from the Soviet, as Minister of Justice, was formed next day. The Tsar abdicated in favour of his brother Grand Duke Michael, who refused the throne, and Russia became a republic.

The new government had to face the question of what to do about the war. It was clearly in a weak position from which to enter negotiations with Germany, which would prejudice its relationship with Britain and France, the argument for continuing the struggle being reinforced as the USA joined the Entente powers in April. However, there could be no question of embarking on any major offensive, which might provoke further mutinies. In any case, the Germans, dominated by the Hindenburg–Ludendorff partnership with its eastern ambitions, were in no mood to make concessions to a Russian government, and their help to Lenin to travel from Switzerland in April to the city which was later to bear his name was designed to undermine it, with eventual dire consequences. Even the Soviet therefore supported continued participation in the Entente's war, although Lenin consistently opposed it. A minor offensive on Evert's front in July, embarked upon as much to occupy the soldiers, who had started grumbling, as for any strategic purpose, soon petered out. A German counteroffensive regained eastern Galicia in August and Riga in September, before the Bolshevik revolution in October sounded the death-knell of the Provisional Government, followed by the armistice negotiated at Brest-Litovsk in November. Estimates of total war casualties vary, but the most reliable is 1,650,000 killed or died of wounds, 3,850,000 wounded and 2,417,000 taken prisoner.[1]

The navy, confined to the Eastern Baltic and the Black Sea, had exerted no influence on the war. The Baltic fleet was initially commanded by Admiral Nicholas Essen, who, as a commander, had been captain in succession of the cruiser *Novik* and the battleship *Sevastopol* in the Russo-Japanese War. Early in the war the navy had to abandon Libau and spent the rest of it keeping the German navy out of the Gulf of Riga and co-operating with British submarines attacking ships carrying iron ore from Sweden to Germany. Essen died in February 1915 at the early age of fifty-four, and was succeeded by the less able Admiral Kanin, who was himself replaced in July 1916 by Admiral A.I. Nepenin, who, in 1914, as Fleet Intelligence Officer, had made the

most significant Russian naval contribution to the war, having personally obtained the code-books of the German cruiser *Magdeburg* when
it grounded in the Gulf of Finland, and forwarded them to the British
Admiralty. The main activity of the Black Sea Fleet, commanded by
Admiral A.A.Eberhardt, was an attempt to impose a blockade of the
Turkish coast in order to interfere with supply of the Turkish army
in the Caucasus and of coal from Zonguldak. Its principal opponents
were the German cruisers *Goeben* and *Breslau*, which had taken refuge
at Constantinople at the beginning of the war. Plans were made for
an amphibious landing at the northern end of the Bosphorus to coincide with the British attempt to force the Dardanelles in April 1915,
but the troops earmarked for it were diverted to reinforce Ivanov in
Galicia. With the completion in that year of the two dreadnoughts,
Imperatriza Maria and *Imperatriza Ekaterina II*, each carrying twelve
12-inch guns and capable of twenty-three knots, Eberhardt could afford
to disregard the *Goeben* and *Breslau*, whose activities were in any
case severely restricted by shortage of coal. The fleet spent more of
its time supporting the Russian army advancing from the Caucasus
into Anatolia in 1916, a force of forty transports being escorted by
the whole fleet in April to help in the capture of Trebizond. In July
Eberhardt was replaced by Admiral Alexander Kolchak, who, as a lieutenant, had distinguished himself in command of a destroyer in the
Russo-Japanese War.* He suffered a severe blow in October when an
explosion in her magazine set fire to the *Imperatriza Maria* at Sevastopol and he had to order her to be scuttled.

   If the navy made no great contribution to the war, it certainly did
to the revolution. Sailors on leave or in shore establishments in Petrograd were among the first to join the rebellious workers, both sailors
and workers at the naval base at Kronstadt soon following suit. Mutinies there spread to Helsingfors, where the bulk of the Baltic Fleet
lay. Nepenin, in common with almost all senior commanders, accepted
the authority of the Provisional Government, but it did not help him.
He and thirty-seven officers at Helsingfors were shot by mutineers,
eighty others sharing their fate at Kronstadt. Ships at the advanced
base at Reval were not affected. In the Black Sea Kolchak was more
fortunate, although he had to accept the establishment of elected
ships' committees, as Admiral A.S.Maximov, styling himself 'The
People's Admiral', had already done in the Baltic, where the fleet
remained in a state bordering on mutiny until the Bolshevik revolution
in October, when the cruiser *Aurora*, sent to Petrograd by the Baltic
Fleet Soviet Committee, made a decisive contribution with a blank
shell bombardment of the Winter Palace. Even after that the fleet

---

* He met his death at the hands of a Bolshevik firing squad in Siberia in 1920, after
taking a leading part in the White campaign against them.

and the Kronstadt base remained a hotbed of anarchy, so much so that as late as 1921 the Bolsheviks themselves, led by Mikhail Tukhachevsky, had to attack the two mutinous battleships *Petropavlovsk* and *Sevastopol*, stuck in the ice at Kronstadt, in order to assert the Bolshevik government's authority over the permanently rebellious sailors.

The German resumption of hostilities against Russia in March 1918 and their advance towards the Crimea posed a threat to the Black Sea Fleet, where Admiral Sablin had replaced Kolchak. After much discussion and internal argument it was sailed, as the Germans entered the town, to Novorossisk, where it became embroiled in the local political turmoil. There Sablin received a secret order from Lenin telling him that, when he received an open message to hand over the fleet to the Germans, he should scuttle it. This led to further internal argument, as a result of which one of the two battleships sailed with five destroyers to Sevastopol, where they were interned together with the ships that had remained there, while the other battleship and nine destroyers were sunk at Novorossisk. At the end of the war they were taken over by Britain and France, but handed back to the Whites in 1919. When the latter were driven out of the Crimea by the Bolsheviks in December 1920, General Baron Peter Wrangel and his followers, 130,000 of them, escaped in three battleships and ten destroyers to Bizerta. The ships remained there, dependent on French charity, until France recognized the Soviet Union in 1924, the last of them being sold for scrap in 1937.

Russian military and naval aviation suffered from heavy reliance on foreign supply. As early as 1869 a Commission on the Use of Aeronautics for Military Purposes had been established, and an army balloon centre had been set up before the end of the century, after the failure in 1894 of a heavier than air machine invented by a naval officer, Mozhaisky. Static balloons were used for observation in the Russo-Japanese War and developments in other countries led to greater interest by both army and navy, inspired by Grand Duke Alexander Mikhailovich, flying-schools being founded in 1910 near St Petersburg and Sevastopol. By 1914 the Russian army had 244 aircraft and 400 more on order. The navy had fewer aircraft in service but almost as many on order, including ten of the remarkable four-engined *Ilia Muromets*, designed by Igor Sikorsky and made by the Russo-Baltic Wagon Factory. Sikorsky biplanes and American Curtiss seaplanes formed the bulk of naval aviation. In 1915 a merchant ship was converted into a seaplane carrier and the first Grigorovich flying-boats built. In the following year the Baltic and Black Sea Fleets each had an aviation division, including dirigibles, which were primarily intended for minelaying.

The army's 244 machines in 1914 were all of French design, 153 Nieuports and ninety-one Farmans. As production in Russia itself increased, partly of foreign (mostly French) designs but also of Russian, Igor Sikorsky being the leader, shortage of engines and spare parts became serious bottlenecks. By 1917 the original seven factories had been increased to eighteen, the monthly rate of production having risen from thirty-seven to 352, by which time the army's total complement numbered 1,039 machines. Shortage of pilots was a constant anxiety for Grand Duke Alexander, head of the Chief Directorate of the Military Aerial Fleet, whose responsibility towards the *Stavka* was ill-defined. Like their allies, the Russians initially thought of military aviation only in terms of reconnaissance and artillery observation, but soon became involved in aerial combat and began to form squadrons specifically for that purpose, which by 1916 were flying in formation, Peter Nesterov, N. Voeodsky and E. N. Krutin being the best known of their 'aces'. The remarkable *Ilia Muromets* bombers were concentrated in a single squadron, commanded by M. V. Shidlovsky, a former naval officer. Its strength never more than twenty-five machines, it carried out 442 raids, dropping 2,000 bombs, between February 1915 and October 1917, penetrating up to 150 miles behind the enemy's front line. It was also extensively used for reconnaissance, 7,000 photographs being taken. Only three machines were lost to enemy action, while their gunners claimed to have shot down forty enemy aircraft. With the advent of the revolution the army's air force disintegrated, officers, who provided the pilots, being driven away from their units or shot, as Shidlovsky was.

5

THE SOVIET UNION

1917–39

The future of the army was one of the most urgent problems facing Lenin and his Bolshevist colleagues when they seized power in November 1917. Hitherto their efforts had been directed towards crippling both the army and the navy. That had been the chief function of the Military Revolutionary Committee, headed by Trotsky, which had become the principal engine-house of the Revolution itself. He was now faced with the dual task of trying to bring the war to an end, and imposing some order on the collection of armed workers and 'politically conscious' soldiers and sailors who had supported it, at the same time as the army and navy were largely disbanding themselves by wholesale desertion. But ending the war was not to prove so easy. The Germans were not impressed by the suggestion of a peace based on 'no annexations and indemnities' and, as Trotsky tried to haggle over the harsh terms imposed, they resumed hostilities in the Baltic states and the Ukraine, recognizing the independence of the latter. In March 1918 Lenin was forced to come to terms and Trotsky, having failed to obtain any help from Russia's allies, who were naturally not in favour of the Soviet making a separate peace, signed the Treaty of Brest-Litovsk, surrendering 400,000 square miles of territory, containing a third of the population. An equally intractable problem was what to do with the nine million men under arms.

A plethora of suggestions was made about whether to form a Socialist Guard, composed entirely of industrial workers, or a militia-type force, or a mixture of revolutionary and ex-Imperial forces. A Demobilization Conference, held at the end of 1917, led the Military Affairs Commissariat to persuade the Third Congress of Soviets in January 1918 to form the Worker Peasant Red Army (RKKA). The revolutionary purists had insisted that it must be based on the voluntary service of politically reliable revolutionaries, but Lenin, facing the imminent threat of renewed German attack, admitted that compulsory service might be needed, that units must be properly disciplined and trained, and that use might be made of the experience and skill of former officers and non-commissioned officers of the Imperial Army. At that point Trotsky could probably call on no more than 50,000 men.

Once the Treaty of Brest-Litovsk had been signed Trotsky's full attention was given to the business of organizing the armed forces, as he assumed the post of People's Commissar for Military Affairs. One of his first acts was the creation of a Supreme Military Soviet, which included ex-Imperial army officers, known as 'military specialists', all ranks having been abolished in favour of 'functions'. The German threat was now replaced by an internal one. The Bolshevik seizure of the *Stavka* by Nikolai Krylenko on 22 November 1917, and the murder of the C-in-C, General Dukhonin, led to a massive desertion of officers and others to join counter-revolutionary movements. The threat from these 'Whites' developed in four areas. In the north, based on the supplies sent to Archangel by the Allies who continued to give him support including some military units, General N.N.Yudenich threatened Petrograd, in co-operation with the Finnish General Mannerheim. In the Baltic states opposition was led by the German General von der Goltz, supported by the Allies. In the south the former C-in-C Alexeyev, later succeeded by General Anton Denikin, commanded the largest force, principally consisting of Don Cossacks, which took over large parts of the Ukraine and southern Russia as the Germans withdrew. In Siberia the core of the force under Kolchak, who attempted to set up a separate state, was the Czech Legion. This consisted of Czech prisoners of war who, having acquired arms, were determined to make their way back to their native land. It was a reflection of the problem facing the foundation of the Red Army that, while demobilizing Russian soldiers, the Supreme Military Soviet sought recruits among their prisoners of war, whom they believed would sympathize with the revolution. When the Czechs preferred a march home to helping the revolution in Russia, Trotsky ordered them to be disarmed, and they promptly joined Kolchak.

Voluntary recruitment soon proved totally inadequate to meet the need to deal with these threats. By 1 April 1918 only 114,678 men had been recruited, to which 40,000 from the old Imperial army were added. Against strong Bolshevik opposition, Lenin and Trotsky introduced universal military training for 'the toiling masses' on 22 April and compulsory military service for the working class at the end of May. Plans were made for an army of twenty-eight first-line divisions, to be followed by two further groups of thirty second-line formations. The initial requirement for 'commanders' was 18,000 and would rise to 55,000. The only way to solve the problem was to make use of officers and NCOs from the Imperial army who were not members of the Bolshevik communist party, nor even sympathizers with it. Commanders, staff officers and specialists from this source were all known as 'military specialists' (*Voenspets*), and many ex-regular NCOs were promoted to fill the function of commander. Their counterparts, originating with the revolutionary forces, were known as 'Red Commanders' (*Kraskomy*). In the end 48,409 ex-officers and 214,000 ex-NCOs were taken on. These measures brought the Red Army's strength by the end of 1918 to over 300,000, although its effective trained strength was well below that. By mid-1919 it had grown to $1\frac{1}{2}$ million and by the end of 1920 officially to a maximum mobilization strength of $5\frac{1}{2}$ million, but an actual effective deployed combat strength of only 159,000, although 1,780,000 were on the payroll and drawing rations.

Trotsky's widespread use of 'military specialists' was hotly opposed by revolutionary purists who wished to see all authority derived from 'cells' of workers, and was also resisted by leaders who had built up their own local armies, like Avtomonov and Kharchenko. However his policy prevailed. An essential element, in order to ensure that there was no deviation from the true political path, was the attachment to commanders at all levels of military commissars (*zampolits*), 'specialist' commanders initially having two. These commissars were also responsible for political indoctrination of the rank and file and formed part of the political administration of the army (PUR, later PURKKA) which had its own hierarchy, separate from that of the operational and administrative command and staff channels. No order was valid without their countersignature. The PUR was entirely separate from the Security Service, the *Cheka* (later OGPU, then NKVD), which inherited the tasks of the Tsar's secret police and the Imperial army's provost service. That element which operated within the armed forces was known as the Special Section (*Osobyi Otdel* – OO).

The Russian Civil War, which lasted from 1918 to 1922, was fought to the accompaniment of intense argument and intrigue about military matters within the Soviet system. The differences on military affairs were inextricably linked to political and personal differences and rivalries. The dominating figure was Lev Trotsky, who, as Commissar for War and President of the Revolutionary Military Council of the Republic (*Revvoensoviet Respubliki*), was served by the All Russian Supreme Staff (*Veroglavshtab*), the first head of which, as C-in-C, was I.I.Vatsetis, an ex-Imperial colonel from Courland. His three subordinates, Boris Shaposhnikov, P.P.Lebedev and A.A.Svechin, were all ex-Imperial army staff officers. Trotsky, Vatsetis and S.I.Aralov formed a triumvirate which, under the general authority of the Supreme Military Soviet, headed by Lenin, raised, organized and commanded the Red Army and conducted the operations of the Civil War. Opposition to them built up from a variety of sources: political rivals, of whom Stalin was the most prominent; front commanders who, like all subordinate commanders, complained of lack of resources and undue interference from above; and ideological purists who objected to Trotsky's reliance on 'specialists' and disagreed with his concept of the purpose both of the armed forces and of war.

Kolchak's capture of Perm, west of the Urals, on 24 December 1918, led to a major shake-up, Josef Stalin, as principal commissar in Petrograd, where he claimed credit for the defeat of Yudenich's attempt to capture the city, insisting on the dismissal of Vatsetis, who was succeeded by S.Kamenev, another ex-Imperial colonel. One of the most important results of this was the appointment of the thirty-five-year-old Mikhail Frunze to command the 4th Army on the Eastern Front. Son of a medical assistant, he became a party member and an active revolutionary in 1904 after leaving the St Petersburg Polytechnic. He had worked as an agitator among the troops on the Russian Western Front during the war against Germany, and as a member of the Yaroslav Military Commissariat had been organizing Red Army units on the Civil War Eastern Front since September 1918. An even younger army commander on that front, who was closely associated with Frunze, was the twenty-six-year-old Mikhail Tukhachevsky. Of aristocratic origin, he had been commissioned into the Semenovsky Guards in 1914 and captured wounded in 1915. He escaped five times, the last being successful, on return from which in October 1917 he joined the Bolshevik revolution. In contrast to these educated men were the ex-NCOS who became army commanders, Vasilii Blyukher and the swashbuckling former sergeant-major of cavalry, Semen Budenny, in whose 1st Cavalry Army served three future Marshals of similar background, Georgi Zhukov, Semen Timoshenko and Konstantin Rokossovsky. Closely associated with Budenny and Stalin was an army commander of different background, the thick-headed thirty-eight-year-old Kliment Voroshilov. A former factory worker from the Ukraine, he had been a party member since 1903 and was City Commissar for Petrograd at the time of the 1917 revolution.

By the end of 1919 the worst of the threats had been removed. Kolchak had been decisively defeated, as had Denikin's thrust towards Moscow. Budenny's cavalry made a decisive contribution to the latter, and Yudenich again failed to reach Petrograd and took refuge in Estonia. With the prospect of an end to the Civil War in sight, argument about the future purpose, organization and training of the Red Army intensified. Trotsky's view was that, as long as the war continued, there was no alternative to reliance on 'specialists' to help command, organize, train and administer a large army, firmly directed from the centre. But once it was over he wished to see a totally different organization, based on a territorial militia, which would not merely be a military organization but would also organize labour and public services and convert the populace into full participants in a socialist political and economic system. Its success would influence other nations to follow its example, thus fulfilling the general desire that the revolution and its armed forces should not be regarded as just Russian, but as international. The opposition to his view was based on many different factors, a strong one being the practical argument that such an army could not be mobilized in time to meet invasion by hostile powers. Political objections were that it could lead to local partisanship and the swamping of the influence of the industrial proletariat by the peasantry. Tukhachevsky and Frunze opposed it as being a purely defensive concept, and an ineffective one at that. They argued that success in the Civil War had resulted from mobile offensive operations, and that the way to convert other countries to socialism was by offensive operations which would destroy the authority of the ruling class and liberate the workers, as had happened in Russia itself.

At the 9th Party Congress in the spring of 1920 Trotsky held his own with difficulty, firmly opposing the concept of offensive operations beyond Russia's borders as impractical and liable to lead to hostilities with Britain and France. One of the major problems of the army was finding and training suitable 'officer' material. It made Trotsky insistent on the need to rely on 'specialists', and intensified the demand of those who disliked that to find an alternative and get rid, once and for all, of the ex-Imperial army officer.

Tukhachevsky's concept was put to the test immediately after the Party Congress. When the Germans withdrew from the Ukraine, the Poles from the west tried to seize areas to which they had laid claim, while the Soviet from the east tried to assert its authority over the Ukrainian national movement, which had been supported by the Germans. Britain tried to establish the eastern frontier of Poland on the 'Curzon' line, but Marshal Josef Pilsudski would have none of it and, after attempts at a negotiated settlement between him and the Soviet had broken down in March 1920, he advanced and occupied Kiev on 6 May. Tukhachevsky, hurriedly transferred from the Caucasus, had been placed in command of the front a week earlier, and by 15 May had some 92,000 men with whom he planned to attack from the north in the area of Vilna and Minsk, while another front under General A. I. Yegorov, to which Budenny's Cavalry Army was transferred, would strike west to the south of Kiev. The operation succeeded in throwing the Poles out of the Ukraine but failed to cut off any substantial element of their forces.

Flushed with victory, Tukhachevsky wished to continue the offensive to capture Warsaw and thus, he hoped, bring the revolution to Poland. Kamenev, at *Glavkom*, the overall command, had misgivings, concerned at the wide gap in the area of the Pripet marshes between Tukhachevsky's proposed thrust, close to the East Prussian border, and Yegorov who was directing his main forces towards Lwow, but he gave his assent. He was also seriously concerned at the threat from Wrangel, who, having succeeded Denikin, had concentrated his forces in the Crimea and advanced north towards the Donetz basin. Having rejected the idea of a peace conference with the Poles and Wrangel on 17 July, the offensive was resumed, and by 6 August Tukhachevsky's leading troops were north-west of Warsaw, as Pilsudski concentrated his forces on either side of Lublin to exploit the gap and strike them in the flank. Belatedly Kamenev, having placed Tukhachevsky in command of the whole front, ordered Yegorov to call off his advance towards Lwow, transfer Timoshenko to deal with Wrangel and order Budenny to move towards Lublin, coming under Tukhachevsky's command. This led to argument, which meant that Budenny failed to divert Pilsudski from his northern thrust which cut off four Red Armies, many of whom were forced to seek refuge in East Prussia. Budenny only extricated his own with difficulty.

By 26 September Pilsudski had regained Grodno and an agreement between the Soviet and Poland was signed on 12 October. That sealed the fate of Wrangel, the operations against whom were under the command of Frunze. In fierce fighting, in which Blyukher's 51st Division played a prominent part, Wrangel was forced back into the Crimea on 16 November, from which his forces were evacuated by sea to Bizerta. Frunze then turned on Nestor Makhno, the leader of the Ukrainian peasant Agrarian-Anarchist Republic, who had helped him to fight Wrangel and, by eliminating his forces, brought the Civil War to an end, except for Eastern Siberia, where it continued until 1922.

With the war virtually at an end and the country in a state of economic and general chaos, argument raged about the future of the armed forces. The unauthorized invasion and occupation of Georgia by Grigori Ordzhonikidze in collusion with Stalin, the Kronstadt naval mutiny, and a crisis over the future of the 1st Cavalry Army, all in March 1921, showed that it was a matter of great urgency, and the 10th Party Congress, although primarily concerned with Lenin's New Economic Policy, addressed itself to the problem. Stalin, Frunze, Voroshilov and several others failed to move Trotsky from his plan for a militia; but it was agreed that for the present it would only be introduced in Moscow, Petrograd and the industrial area of the Urals, while the existing 'regular' army was reduced and modernized. There was considerable debate about the need to abolish the duality of command and the link between the PUR and both local political bodies and 'cells' within the army; but the PUR, in its current form, was to continue at least until a new army of a purer Red hue had been created. The need to respond to the demand for modernization and better training, particularly of commanders, staffs and true specialists, led not only to the establishment of a number of military academies and training schools but also to co-operation with the Germans, who took advantage of it, as has been described in chapter 3, to evade the restrictions of the Treaty of Versailles.

The Treaty of Rapallo in 1922 gave diplomatic sanction to a process which Trotsky had started in 1921. There were political motives also behind it, as it appeared to the Soviet that Germany was ripe for revolution. At this time also the All Russian Supreme Staff, which was concerned with all army matters other than political, and the Field Staff, which had been concerned with operations, were merged into one Red Army Staff. The period 1921 to 1924, ending in the death of Lenin, was one of continuous and active discussion of military matters, intimately bound up with the struggle for power, which included control of the army, between the group centred round Stalin, which included Frunze, Voroshilov, Grigori Zinoviev and L.B.Kamenev (not to be confused with the 'specialist' S.S.Kamenev), while Trotsky was generally supported by Lenin. Hard economic facts imposed a compromise solution to the argument between a national standing army, capable of taking the offensive in mobile operations, and a militia. The outcome was a much smaller 'cadre' army, supplemented by 'territorial' divisions, in which service would be limited to a short annual training period. His opponents, notably Stalin, were determined to get rid of the influence of the ex-Imperial officers and, in the argument about the position of the commissars in the PUR, Frunze, while intent on raising the authority of the commander, was also concerned that the political apparatus should be strengthened and manned by politically reliable men.

After Lenin's death in January 1924 Trotsky ceased to fight hard for his views and gradually lost influence until, a year later, he was replaced as Commissar for War and head of the *Revvoensoviet* by Frunze. In the nine remaining months of his life Frunze expended enormous energy in attempting to reform the armed forces, concentrating on the selection and training of officers at every level and on the vexed question of 'unity of command', that is the commander's relationship to the commissar. A proposal that all commanders must be party members was rejected as impractical. Allied to the officer-training task was that of the production of training manuals, to which he paid great attention, having consistently demanded that the Red Army must have a clear purpose and doctrine which permeated all levels. The solution to the 'unity of command' problem, confirmed by Voroshilov who succeeded him, was that the commissar would have full responsibility for all party matters within the army, but cease to have any confirming responsibility for operational or administrative orders. The sting in the tail of this decision was that the PUR would no longer be responsible to the *Revvoensoviet*, but directly to the Central Committee of the Party, of which Stalin remained the General Secretary. At the time of Frunze's death the 'cadre' army had been reduced to 530,000, producing twenty-six infantry and ten cavalry divisions (also eight independent cavalry brigades) to which thirty-six territorial infantry divisions would be added on mobilization.*

In an important meeting in January 1925, Stalin made it clear that there was no question of the cadre army gradually being absorbed into a militia, as some proposed. However successful revolutionary movements in other countries might be, the Soviet Union had to give first priority to the defence of its own territory, and would have to rely on its own resources for that purpose. If a war broke out, the Soviet Union would not be able to stand aside 'with folded arms' but, he prophetically declared, would be the last to become involved. Frunze developed this into a description of the fundamental characteristics of a future war : it would be a revolutionary class war, not a nationalistic one, and would be dominated by the socio-political and economic elements within all the societies at war : both the 'technical factor' and the mass employed would be important factors : because of their internal social difficulties the capitalist countries might not be able to employ mass and would be forced to rely on 'technical means'. The Soviet Union could not expect to defend itself with a small force, relying principally on technical means, but must be prepared 'in the final resort' to employ the mass of the population and all the resources of the nation.

The decade following Frunze's death was one of dramatic development in the Soviet armed forces, linked to Stalin's crash programme of industrialization, notably in heavy industry. Although Voroshilov had replaced Frunze as Commissar for War, the latter's mantle as an enthusiastic innovator and energetic trainer, organizer and leader was assumed by Tukhachevsky, who was Chief of Staff from 1926 to 1928, when, for reasons that are obscure, he was transferred to command Leningrad District and succeeded by Shaposhnikov. He returned to the centre of affairs in 1931 as deputy to Voroshilov, at the same time as Shaposhnikov, in temporary disgrace, was sent to the Volga Military District, replaced by Yegorov. From 1931 until the 'purge' in 1937 he held the influential post of Chief of Ordnance. He was an enthusiastic supporter of the need to modernize the armed forces, and therefore of the arrangements with the Germans for development and training of aviation at Lipetsk, tanks at Kazan and chemical warfare at Volsk. However, in the Field Service Regulations which he compiled in 1925 and subsequently, he was unable to define clearly the balance of emphasis to be laid on mass, on the one hand, and the 'technical factor' on the other. Any proposal which appeared to be giving too much emphasis to mechanized mobile forces or to aviation, acting other than

---

* To support this force there were some 3,000 guns organized in 59 regiments of artillery. The army was planned to increase by 1926 to 77 infantry (31 cadre, 46 territorial) and 11 cavalry (1 territorial) divisions, supported by 3,718 guns in 77 artillery regiments.

in direct support of the army, both being highly dependent on the better educated element of society, raised the opposition not only of the conservative element, like Budenny, defending his beloved cavalry, and those who saw the long-suffering peasant infantry as Russia's greatest military asset, but also of those who were suspicious of political opposition.

With the preliminary purge of 1924, which did not seriously affect the armed forces, and the exile of Trotsky in 1928, after his bitter criticism of Voroshilov and Stalin, it appeared that the argument about the future organization of the armed forces had been settled. Trotsky warned of the danger that the revolution might be betrayed, as the French Revolution had been, by 'Bonapartism', perpetrated either by the army or by Stalin himself and his immediate entourage. His criticism of Voroshilov was supported in a secret paper to the *Politburo* from a number of senior officers, which did not include Tukhachevsky.

Stalin, Voroshilov, Shaposhnikov and Tukhachevsky were all in favour of developing strong mechanized and air forces, and, although there were those who pressed for a more independent role for both, concepts were developed which divided these between tactical support of infantry attacks, aimed at encircling the enemy, and strategic action in his rear or in wide outflanking movements. In whichever role tanks were to be employed, Tukhachevsky emphasized that they must have the support of artillery. Budenny, supported by Timoshenko and Zhukov, fought to preserve the place of horsed cavalry in the strategic role. The increasing emphasis on the use of mechanized forces in place of cavalry, coming on top of the ill-feeling between Tukhachevsky and Budenny which dated from the Polish campaign and even earlier, was to have fatal consequences for the forces. The air force enthusiasts were disappointed that little emphasis was given to any strategic bombing role, reconnaissance and direct support of the army being given almost total priority.

The general concept of war that had been developed by the 1930s was that war would be preceded by a period of tension in which forces would be mobilized and prepared for major manœuvres, which would terminate in a 'battle for the frontiers'. If this were not quickly concluded, a long war would ensue, in which attrition would play its part alongside that of manœuvre. It would demand the mobilization of the full resources, human and material, of the Soviet Union.

The earliest Soviet tanks were developed from foreign models, the T-26 from a British Vickers light tank, and the BT from a private American venture, the Christie, which was later developed into the highly successful T-34 medium tank, which came into service in 1940 at the same time as the heavy KV-1. Both of the latter carried the same 76.2-mm gun, a much more effective weapon than that carried by any other nation's tanks at that time. From a few hundred tanks in 1925, it was estimated (by the Germans) that the army's total had risen by 1933 to 2,000 and two years later to 10,000, far more than in any other army. The mechanization of agriculture on the collective farms, which had been imposed on a reluctant peasantry, made a notable contribution to the army's ability to man and support this large mechanized force.

As with tanks, Soviet military aviation initially relied heavily on foreign designs and advice. This was particularly the case in respect of aero engines, for which Soviet aircraft design and production continued to rely on foreign supply long after the Union had begun to design and produce its own airframes. Two organizations contributed much to the development of Soviet aviation, the Central Institute of Aerodynamics and Hydrodynamics (TSAGI) and the Association for the Promotion of Defence, Aviation and Chemical Warfare (*Osoaviakhim*), an official organization under the control of the *Revvoensoviet*, designed to encourage popular support for the armed forces, and particularly for their mechanization. TSAGI, headed by N.E.Zhukovski, produced some notable designers: A.N.Tupolev, who specialized in bombers; A.I.Mikoyan in fighters; K.A.Kalinin, a pioneer of the swept-wing; and M.A.Mil, an early auto-gyro designer, who went on to helicopters. In the 1930s aviation received the same sort of official and popular support, as an example of Soviet achievement in the leap into the future, as activity in space was to receive after the Second World War.

As with tanks, progress in numbers was impressive. By 1933, the end of Stalin's First Five-year Plan, the air force (UUS) had a front-line strength of 1,000–1,250 machines, which was doubled by the end of the Second Five-year Plan in 1937. In spite of some remarkable developments the fighters, most of them designed by N.N.Polikarpov, were not equal in performance to the German or British, nor the bombers to the former. Numbers were, however, impressive. By the end of 1937 the Soviet Union was producing over 4,000 aircraft a year, of which half were single-engined fighters, nearly 1,000 twin-engined light bombers and 200 four-engined ones.

The founder of the Red Air Force was General P.I.Baranov, a worker from St Petersburg who joined the party in 1912 and was associated with Frunze. He was killed in an attempted instrument landing in 1933 and succeeded as Chief of the Red Air Force by General Ya.I.Alksnis, a farm labourer's son who had been a party member since 1916 and was a close associate of Tukhachevsky. The Russians were pioneers in the development of parachute troops, two units being formed in 1931, the four-engined TB-3 bomber being used to carry them. In the manoeuvres of 1935 and 1936, 600 paratroopers were dropped, followed up by air-landed troops, greatly impressing the foreign observers. By 1938 the Red Air Force had four parachute 'brigades', each 1,000 men strong.

The navy was the poor relation. After the Kronstadt mutiny in 1921, it was high on the priority for a purge in 1924, when a large number of officers were removed. In order to provide a more politically reliable 'lower deck', sailors were recruited largely from *Komsomol*, the party youth movement. But it was still necessary to retain a significant number of 'specialists' to ensure that the ships, all left over from the wartime fleet, were at least prevented from deteriorating, if not capable of going to sea. It was not until 1929 that a serious effort was made to rebuild the navy, a close liaison being established with that of Germany, which provided designs as well as training. The main emphasis was on submarines as a defensive weapon, a German First World War design and that of a scuttled British submarine being used. Control of the navy was exercised by the *Revvoensoviet* and the Commissar for War, almost as if it were part of the army, until 1938 when a Main Naval Soviet was formed, separate from the Main Military Soviet and directly subordinate to the Defence Committee of the *Politburo*.

By then Stalin's attitude to the navy had changed. Provoked perhaps by the Anglo-German naval treaty of 1935 and the German decision to build pocket battleships, Stalin decided that the navy should no longer be confined to a defensive role, but that it must have an ocean-going fleet of battleships and cruisers. A Northern Fleet, based on Polyarnayy near Murmansk, had already been created and work started on a canal, capable of taking submarines and destroyers, linking it with Leningrad. A small Far East Fleet had also been recreated. The first senior Commissar for the Navy was V.I.Zof, who was succeeded in 1927 by R.A.Muklevich, a party member since 1906 and a textile worker until he was called up into the Imperial navy in 1912. His principal assistant, Commander of the Navy from 1931 until 1937, when both were eliminated in Stalin's purge, was Admiral V.M.Orlov, who had joined the Imperial navy as a cadet in 1916, joining the party in the revolution in 1917. By 1940 the navy was believed to have four battleships, all of First World War vintage, modernized between 1931 and 1937, nine cruisers, four of them of a modern Italian design, thirty-six destroyers, twenty torpedo-boats and 150–175 submarines; and some 700 aircraft, almost all of them seaplanes.

The only active area of operations in which Soviet forces were involved after the Civil War, apart from the Spanish Civil War, was the Far East. After the revolution the Chinese and Japanese took advantage of the inability of the revolutionary regime to establish its authority in Eastern Siberia, until Kolchak's forces had been defeated in 1920. It was not until 1922 that local revolutionaries had decisively scuppered the Whites, sorted out their mutual quarrels, rejected Japanese attempts to suborn them and established the Far Eastern Republic firmly enough to join the Union. By then Outer Mongolia had been cleared of Chinese and established as an independent Soviet Republic, firmly under Moscow control. Under pressure from the USA the Japanese evacuated all the areas of former Russian territory which they had occupied. Blyukher was the principal representative of the Supreme Soviet in the Far East in the activities which culminated in this restoration of Russian authority in the area. In the following year Lenin, while maintaining an official presence with the Chinese government in Peking, decided to give political and military support to both Sun Yat-sen's Kuomintang and Mao Tse-tung's communist rebel movements, Mikhail Borodin being despatched as his political representative and Blyukher as head of a military mission while Sun Yat-sen sent the young Chiang Kai-shek to Russia for discussions. Under the guidance of Blyukher, known in China as Galen, Chiang Kai-shek led the Kuomintang's National Revolutionary Army to victory over the forces of the northern warlords; but soon began to act more independently. In March 1926 he turned against the communists, but in spite of that Stalin continued to support him until August 1927 when, after the final split with Mao Tse-tung, Borodin's and Blyukher's missions were withdrawn. Stalin's failure to support the Chinese communists was based on a deeper motive than that of producing a sympathetic revolution in China. He calculated that a successful revolution that was overtly communist would unite in opposition to it the USA, Britain, France, perhaps Japan and even invite active intervention, with possibly serious consequences for the interests of the Soviet Union in the Far East. However, the success of a revolution which was favourably regarded in America and by western liberal opinion generally, while being anti-imperialist and, it was hoped, favourably disposed towards its Soviet benefactor, would not arouse American opposition and should help to undermine British influence not only in the Far East generally but also in India. Stalin regarded Britain as the arch-imperialist enemy.

In 1928 the Japanese Kwantung Army, having disposed of the dictator of Manchuria, Chang Tso-lin, who showed signs of escaping from their control, encouraged his son to co-operate with Chiang Kai-shek in depriving the Soviet Union of its rights in the Chinese Eastern Railway, the shortcut from the Trans-Siberian railway at Chita in Trans-Baikal to Vladivostok through Harbin. In August 1929 the Special Far Eastern Army was created (ODUA) and Blyukher appointed to command it. Voroshilov hoped that the industrial base could be created in Eastern Siberia to support the army, making it independent of supply from the west. Stalin was reluctant to risk a direct clash with Japan, and it was only after receiving assurances that the latter would not intervene, that in October the Red Army embarked on operations to drive Manchurian forces away from the railway. Four divisions and a cavalry brigade were involved with the support of nine tanks and thirty-nine aircraft. The operations, which included an amphibious landing in the River Amur, were successfully concluded at the end of November. Chinese intervention was prevented by the actions of the Soviet-supported Nationalist General Feng Yü-hsiang. The build-up of the Far Eastern Army linked to the firm Soviet hold on Outer Mongolia, however, caused misgivings in the Japanese Kwantung Army as it prepared to extend its hold over Manchuria. One of the results of this brief campaign, which was to have implications for the future, was the development of a coolness in the relations between Voroshilov and Blyukher, the former jealous of the praise given to the latter for his victory.

When the Kwantung Army sprang into action in the 'Manchurian incident' in September 1931, Stalin was careful to assure the Japanese that he would reciprocate their restraint in 1929; but when Japan refused a non-aggression pact and extended her operations into China proper in the following year, the Far Eastern Army was reinforced and the resurrection of a Soviet Far East Fleet announced. The growing Japanese threat drove Stalin to restore relations with Chiang Kai-shek and forced Voroshilov to conclude that he could no longer rely on transferring forces from the Far East to the west to help meet a threat in Europe.

By 1933 the Far Eastern Army had grown to nine divisions and one and a half cavalry brigades, supported by 300 tanks and 350 aircraft, which included some TB-5 heavy bombers, capable of reaching Japan itself. Extensive defensive works were embarked upon along the long frontier between Manchuria and the Far Eastern Provinces. The conversion of the Trans-Siberian railway to double track was hastened, and Soviet forces were sent to guard the eastern border of Outer Mongolia. The realities of the military situation made it clear that the Soviet Union could no longer rely on the use of the Chinese Eastern Railway and, after twenty-one months of negotiation, Soviet rights were sold to Japan in March 1935. Towards the end of the year there were several direct clashes between Soviet and Japanese troops on the border, and Japanese advances into Inner Mongolia, as Mao Tse-tung made his Long March to the north, threatened Outer Mongolia from another direction. The signature in March 1936 of a pact between the Soviet Union, China and Outer Mongolia, committing them to military co-operation, coincided with a serious border clash with Japanese troops.

There were no really serious developments until 1937, the year in which there were hopes of a *rapprochement* between Chiang Kai-shek and Mao Tse-tung. Sino-Japanese skirmishing developed into full-scale war and Stalin decided to provide material military support to the Kuomintang. The 'Amur Incident' in July 1938, in which Blyukher was forced by Stalin to back down, was immediately followed by a more serious one near Lake Khasan south-west of Vladivostok. Russian occupation of a small hill on a disputed frontier provoked a sharp Japanese reaction which may have been intended to prevent Blyukher from launching an offensive against them, while they embarked on one in southern China. It may also have been designed to test the reaction of the Far Eastern Army which, at the time, was being subjected to a major purge by the NKVD, headed by L.Z.Mekhlis. After an initial Japanese success the Soviet division, under the operational command of General G.M.Shtern, regained the disputed hill, at a considerable cost in casualties, forcing the Japanese, who had committed a division, to acknowledge defeat or raise the stakes. They preferred the former, and a cease-fire was agreed after a fortnight's fighting, on 11 August. Credit for what was greatly exaggerated as a Soviet victory went to Shtern, and at some time within the following three months Blyukher was removed and executed.

The other Soviet operational commitment before the 'great purge', albeit an indirect one, was the Spanish Civil War. Stalin was anxious not to become directly involved and officially supported the French and British policy of non-intervention. However, he could not afford to see communists abandoned, and used neutral shipping to send supplies, including tanks and aircraft, and 'specialists' to man them, as well as commanders and commissars for the International Brigade, the infantry manpower of which was provided by communists and sympathizers from other countries. The whole operation was under the control of the NKVD, headed by Yan Berzin. Kleber (whose real name was Stern) commanded the brigade, the task of which was limited to the defence of Madrid. The 2,000 Russians involved were kept strictly separate from the Spanish communists, who were subjected to as ruthless a purge by the NKVD as it had already begun to execute in the Soviet Union itself.

The fundamental reason for the purge was Stalin's determination that there should be no rival to his own absolute power. To this were added several contributory factors, personal, political and military. August 1936 saw the execution, after a five-day mockery of a trial, of Zinoviev, L.Kamenev and four other party leaders, their statements casting a shadow of suspicion over five more, including Nikolai Bukharin and Karl Radek, the pioneer of co-operation with Germany. Of the military, Tukhachevsky and Blyukher were becoming too prominent and powerful for the liking of Voroshilov and his old crony of the 1st Cavalry Army, Budenny, both of whom were close to Stalin. The latter's dislike and distrust of Tukhachevsky was aggravated by the increasing emphasis being placed on mechanized forces and aviation, with their reliance on specialists : experience in Spain was believed to have shown that they were ineffective unless used in close co-operation with infantry.

The antagonism between the armed forces and the NKVD had been growing for a long time, increased by the tendency of the PURKKA, under Ya.B.Gamarnik, to be more closely associated with Tukhachevsky and the armed forces themselves at all levels and to distance itself from the Party. The reintroduction of the traditional ranks, including that of marshal, in 1935 at the same time as Tukhachevsky succeeded in achieving a high degree of unity of command at the higher levels, freeing the commanders from the control of the commissars, gave further grounds for suspicion by the hard-line NKVD headed by the sinister Nikolai Yezhov, on whom Stalin came increasingly to rely. A division was arising between the younger officers, who had been subjected to Stalin's political propaganda as they passed through the numerous courses of instruction required for their promotion, and the veterans of the Civil War who provided at least 80 per cent of the commanders at the level of division and above and did not necessarily owe much to Stalin. The trial of Radek and four others began on 23 January 1937 and the mere mention of a name, even if it were to exculpate him, as was the case when Radek mentioned Tukhachevsky, was tantamount to listing him for investigation. In May, Yegorov was removed as Chief of the General Staff, the faithful sycophant Shaposhnikov returning to this post, and Tukhachevsky, posted to command the Volga Military District, was arrested by the NKVD before he got there. A host of other moves took place, including that of the tank expert, General I.Khalepsky, to the Commissariat of Posts and Telegraphs. Gamarnik committed suicide before Voroshilov announced, on 11 June, that a special tribunal, which included Budenny, Alksnis and Blyukher, was being set up to try Tukhachevsky and six other senior military figures, all of them with impeccable party credentials. The following day it was announced that they had been found guilty and shot. It is doubtful if in fact the tribunal even met, let alone conducted a trial. One of the principal accusations against Tukhachevsky was that he was in secret collusion with the German *Reichswehr* to overthrow Stalin. There is some disputed evidence that Reinhard Heydrich, head of the SS, colluded with the NKVD in providing faked evidence for this, hoping perhaps for reciprocal help if he needed evidence against German generals.

The purge spread rapidly and by the end of 1938 Yegorov had gone the same way, as had all the members of the tribunal which had condemned Tukhachevsky, except its president, V.V.Ulrikh, Shaposhnikov and Budenny, and only the latter and Voroshilov were left of the five marshals created in 1935. Of the eighty ex-members of the *Revvoensoviet* seventy-five died, as did the inspectors of artillery and mechanized troops, thirteen out of fifteen army district commanders, fifty-seven out of eight-five corps commanders, 110 out of 195 divisional commanders and 220 out of 406 brigade commanders. Apart from Voroshilov and Shaposhnikov, only the veterans of the 1st Cavalry

Army escaped this slaughter : Budenny, Timoshenko, Zhukov and their cronies.

The air force and the navy fared no better. Alksnis, V.V.Khripin and almost all the senior air force generals, as well as several of the leading designers, suffered the fate of their army colleagues, as, in the navy, did Muklevich, Orlov, all the fleet commanders and a high proportion of officers of every rank. Estimates of the total number of officers dismissed vary from 15,000 to 30,000. Allowing for subsequent reinstatement, the most reliable figure[1] appears to be 20–25,000 out of a probable total of 75–80,000 in all three services : 90 per cent of those of general rank and 80 per cent of colonels were involved. It is not surprising that, at the time of the Munich crisis in September 1938 and in the protracted diplomatic manœuvring between then and the Soviet-German non-aggression pact in August 1939, neither the British and French nor the Germans rated highly the capability of the Soviet armed forces.

1939–46

Having by this drastic action removed the internal threat, Stalin feared the external one. That from Germany was obvious, but he relied on a combination of his appeasement policy and the threat from Britain and France towards Hitler to ensure that it did not develop into an actual one. Above all he was intent on not getting involved in a general European war which in 1939 seemed more and more likely. Nevertheless, he was determined to see that the Soviet Union was in a position not only to protect its own territory but to exploit the situation to regain those over which the Russian Empire had held sway before their removal at the end of the First World War : Finland, the Baltic States, Eastern Poland, Galicia, the Bukovina and Bessarabia. As soon as his pact with Germany was signed he began to plan their acquisition, at the same time as he presented Hitler with a large bill for military hardware to help equip the forces which he was rapidly increasing in strength. The expansion, coming on top of the purge, further weakened their quality.

While concerned with developments in Europe, Stalin could not afford to neglect the Far East. Following the Lake Khasan incident, the forces east of Lake Baikal were organized into four armies, totalling twenty-four divisions with 1,900 tanks and 2,000 aircraft, one of them based in Outer Mongolia. An incident on the eastern border of the latter in January 1939, in the area of the Khalkhin-Gol river, developed at the end of May into a small war which by the end of July faced the Russians with the alternatives of commitment to a prolonged defence or conceding victory to the Japanese. Stalin was not prepared to accept either of these and despatched Zhukov with considerable reinforcements to teach the Japanese a lesson. Using three divisions and one of Mongolian cavalry, supported by 500 tanks and the same number of aircraft, Zhukov launched a very carefully organized attack in which the action of all arms, including the air force, was well co-ordinated, on 20 August, the day before Stalin replied favourably to Hitler's offer of a non-aggression pact. By the end of the month the Japanese 6th Army had been driven back behind the Manchurian frontier, an armistice being signed on 16 September, the day before Soviet troops entered Poland, having agreed with Germany a division of that unfortunate country between them, to be closely followed by a move into the Baltic States. Stalin had thus, he thought, cheaply acquired the territories, the preservation of the independence of which had been a stumbling-block in the fruitless negotiations with Britain and France that had still been in progress when the non-aggression pact with Germany was concluded.

Next in line for the restoration of control over the 'lost provinces' was Finland. Free from any fear of German intervention and assured of an ample superiority in numbers, Stalin presented a demand to Field Marshal Carl Gustav von Mannerheim for the transfer of islands in the Gulf of Finland, in return for which Finland would be ceded 183 square miles of barren land north of Lake Ladoga. Mannerheim refused, and on 30 November Soviet forces attacked, the 7th Army on the Karelian Isthmus from Leningrad towards Viipari (Vyborg), the 8th north of Lake Ladoga, the 9th towards the head of the Gulf of Bothnia and the 14th in the extreme north with Petsamo as its principal objective. Only in the last was the offensive successful. Clumsily planned and executed, the infantry plodded forward in massive waves in conditions of extreme cold against well-prepared and concealed defences, held by tough, well-trained and well-equipped Finns, the offensive petered out with heavy losses by the end of December.

Hitherto overall command of the armies involved had been exercised by Voroshilov, the commander of Leningrad Military District, General K.A.Meretskov being responsible for only the attack over the Karelian Isthmus. A North-western Front was now formed with Timoshenko in command, two more armies being allotted to him, the 13th to reinforce the 7th and the 15th the 8th. This gave him an overwhelming superiority which he employed in a carefully prepared offensive launched in mid February, including an attack towards Viipari across the frozen sea, which turned the flank of General Heinrich's gallant Isthmus Army, defending the Mannerheim Line. The Finns, who had appealed in vain to Britain and France for support, were forced to accept Vyacheslav Molotov's harsh peace terms on 12 March.

The armed forces which, at the Party Congress a year before, Voroshilov had described as invincible had made a poor showing. They had been obliged to employ forty-five divisions, with 1,500 tanks, 3,000 aircraft and masses of artillery – 1,200,000 men in all – to crush a small nation with a long, vulnerable frontier, suffering very heavy casualties in doing so : 273,000 dead, an unknown number of wounded and 5,648 taken prisoner, all of whom were sent straight to concentration camps on grounds of treason when released.

Drastic action to remedy the state of affairs was clearly needed. The first step was the replacement of Voroshilov as Defence Commissar by Timoshenko on 6 May, the latter's promotion to the rank of marshal being followed in June by a long list of other promotions and appointments, some to the new rank of colonel-general in imitation of the Germans. Timoshenko's appointment coincided with directives demanding higher standards of discipline and of realistic training under warlike conditions. He did much to restore the balance between the commander and commissar in favour of the former.

While this attempt to improve the standard of the army was being initiated, the Germans had occupied Denmark and Norway and launched their attack on France, Belgium and Holland. Their rapid and spectacular victories caused the gravest concern in Moscow. Stalin had been relying on a prolonged campaign in the west to give him time to mobilize and prepare his forces to meet the situation which it would create, whatever that might be. Not only had that time now disappeared, but Germany's occupation of Norway threatened the security which he thought he had obtained by his subjugation of Finland. The methods by which the German army had gained its victory in France demanded a drastic reversal of the decision taken after the Spanish Civil War to disband mechanized and tank corps and distribute tanks largely in support of infantry formations. In this situation Stalin attempted to follow a compromise policy of preparing his forces for war as rapidly as he could, completing his programme for the return of lost provinces but at the same time professing friendship with Germany and trying not to provoke Hitler into direct action against the Soviet Union. It was not an easy course to steer and it led to a fatal lack of clarity in strategy. The Soviet invasion of Romania, Bessarabia and northern Bukovina in June brought the Red Army to the Danube and was immediately followed by the arrival of a German military mission in Bucharest, whose main interest was to ensure that access to Romanian oil was assured.

In terms of numbers of men and equipment, the Soviet forces by this time looked formidable. By the spring of 1941 the army mustered 170 rifle, thirty-two cavalry divisions and seventy-eight mechanized brigades, of which twenty-five rifle divisions and six mechanized brigades were deployed in the Far East and the Caucasus, and twenty rifle divisions and thirty mechanized brigades held back in the interior. A high proportion of the 125 rifle divisions available for the west, and almost all the cavalry divisions, were held forward in or near the

recently regained 'lost provinces'. A hasty reorganization of the mechanized forces had produced seventeen mechanized corps, one motorized
rifle and two tank divisions. Eight of these corps were allotted to
the South-western Front covering the area south of the Pripet marshes
down to its junction with the Southern Front on the Dniester. There
is much confusion about the organization of Soviet tank formations
in this period, brigades becoming divisions and divisions corps and
vice versa, as well as confusion between tank and mechanized formations. Stalin stated that in June 1941 the Soviet army had 24,000
tanks in sixty divisions or brigades, while the German estimate was
15,000 in thirty-five tank divisions; but by the end of 1941 they had
identified sixty-five. Whatever the correct figure, the great majority
were the light BT tanks, only 967 of the medium T 34s and 508 of
the heavy KV 1s having reached operational units by June 1941.

The air force had been built up to a total front-line strength of something between 10,000 and 15,000 aircraft, of which about 7,500, according to German estimates, were deployed to face them.* Most of the
fighters were the obsolete I 15s and I 16s, and even the new MIG 1,
YAK 1 and LAGG 3 fighters, just coming into service, were not a match
for the German Me 109F and were inferior in performance to the British
Spitfire and Hurricane. Ground-attack was provided by the IL 2 Shtor-
movik, the equivalent of the Ju 87. The PE 2 and DB 3 twin-engined
bombers were more or less as effective as the German Ju 88s and He 111s,
although many obsolescent SB bombers were still in service and Soviet
reconnaissance machines were equally vulnerable to German fighters.
That steady old four-engined war-horse, the TB 3, provided almost
all the long-range transport and heavy bomber force. The strength of
the navy has already been related.

In February 1941 Zhukov became Chief of the General Staff, replacing
Meretskov who had taken over from Shaposhnikov when the latter
fell genuinely ill in August 1940. Zhukov acted as the agent of Timoshenko in exercising general command over five military districts in
the west, which became fronts on the outbreak of war: the Northern,
commanded by General V.S.Popov from Leningrad; the North-western,
by General F.I.Kuznetsov from Riga; the Western, by General D.G.Pav-
lov from Minsk; the South-western, by General M.D.Kirponos from
Kiev; and the Southern, by General I.V.Tyulenev from Odessa.

The recent extension of the Soviet Union's borders into the Baltic
States, Poland and Romania, the ambivalence of Stalin's attitude towards Germany, and the vagueness about operational responsibility for
the whole front resulted in the absence of a clear military plan. The
attention of the Northern Front was primarily directed towards Fin-

---

* 3,000 were fighters, 2–2,500 medium and light bomber and ground attack, 400–500
reconnaissance and 1,500 transport, heavy bomber and other aircraft.

land, and large concentrations of troops were built up in the Western and South-western Front areas, but no clear overall plan issued as to their purpose, either defensive or offensive, all training and doctrine over the past fifteen years having emphasized the former.

The Eastern Front 1941-45

These factors, combined with the fact that responsibility for the security of the frontiers and for intelligence about what was going on beyond them lay with the NKVD and not the army, resulted in the Soviet forces, in spite of clear warnings given to Moscow from several sources including Britain and the USA, being taken completely by surprise when the Germans attacked at dawn on 22 June 1941.

A few hours earlier Timoshenko had issued an order placing all forces on the alert, seven hours later issuing an order in very general terms for the destruction of any forces which crossed the frontier. The Red Air Force, almost all of it caught on the ground, suffered heavily. It was not long before the rapidly advancing German Panzer forces produced a crisis in the North-western and Western Front areas, thrusting through the gaps between their concentrations and enveloping them. Where the Soviet divisions were directly attacked, they fought stubbornly and inflicted heavy casualties on their opponents, but for the most part they were surrounded and cut off from their supplies before they had had a chance to hit back, for they were neither in well-prepared defences nor disposed to launch major counterattacks.

On 23 June a *Stavka*, or High Command, was established to exercise operational command over all the fronts. A week later Stalin, having already combined in himself the chairmanship of the Supreme Soviet with that of General Secretary of the Party, strengthened his power over all governmental, military and administrative organs by the establishment of a State Defence Committee, consisting of himself, Molotov, Voroshilov, Georgi Malenkov and Laurenti Beria, who had succeeded Yezhov as head of the NKVD early in 1939. The members of the *Stavka* were Stalin, Molotov, Marshals Voroshilov, Timoshenko, Budenny and Shaposhnikov and General Zhukov. At the same time three major commands were interposed between the *Stavka* and the fronts, Voroshilov in the north-west, covering the Northern and North-western Fronts, with Andrei Zhdanov as his chief commissar; Timoshenko with Nikolai Bulganin in the west, covering the area of the Western Front; and Budenny with Nikita Khrushchev in the Ukraine, covering the South-western and Southern Fronts. The appointment of these 'super-commissars', who were given the rank of lieutenant-general, was accompanied by the reintroduction of dual command at every level.

The reorganization had little impact on events, Stalin's refusal to abandon Smolensk leading to the loss of 600,000 men and 5,000 tanks. The Western Front commander, Pavlov, was court-martialled and shot for allowing a similar situation to have occurred at Minsk and Bialystok. At the end of the month Zhukov was dismissed as Chief of the General Staff and replaced by Shaposhnikov, for having proposed a withdrawal from Kiev, which he correctly foresaw was threatened with the same fate by the southward move of Guderian's tanks, consequent upon Hitler's critical decision to give priority to the capture of Leningrad, the Ukraine and the Donetz basin over an advance to Moscow. In an attempt to prevent the encirclement of Kiev from the north, the Bryansk Front, commanded by General Andrei Yeremenko, was created, extending Timoshenko's responsibility further south. Budenny was already in trouble, Rundstedt's Southern Army Group having split his command into two, driving a wedge between the

South-western and Southern Fronts as he thrust down the valley of the Bug towards the Lower Dnieper. Voroshilov also was in dire straits as Leeb's Northern Army Group closed up to Leningrad and began to encircle it. Zhukov meanwhile was placed in charge of the Reserve Front, which the *Stavka* began to form from formations gathered in from all over the Union, including the Far East, from which fifteen rifle and three cavalry divisions, 1,700 tanks and 1,500 aircraft were withdrawn before the end of the year. Stalin could do this without anxiety as his spy, Richard Sorge, in the German Embassy in Tokyo, informed him that the Japanese, late in June, had decided to deploy their strength into South-East Asia.

The formation of the Bryansk Front failed to save Kiev, the fight for which ended on 25 September, ten days after its encirclement, with 665,000 prisoners, 884 tanks and 3,178 guns falling into German hands. Before that occurred Timoshenko had replaced his old commander Budenny, who was relegated to the Reserve Front, while Zhukov replaced Voroshilov in besieged Leningrad. In an attempt to raise morale there Voroshilov had taken a step, soon to be followed elsewhere, in designating a militia division, raised from the workers of Leningrad, as a 'Guards' Division. Zhukov put new life into the defences, while the threat to the city changed from one of direct assault to that of severance of its vital land links with Moscow, both locally, in the area of Volkhov south of Lake Ladoga, and nearer Moscow at Kalinin. Timoshenko's successor in command of the Western Front was General Ivan Koniev, and he soon found himself in difficulties as Kluge resumed his advance from Smolensk towards Moscow, with Guderian, released from Kiev, striking north towards it through Orel. The result was another encirclement, between Vyazma and Bryansk, into which fell 673,000 prisoners, 1,242 tanks and 5,412 guns on 7 October. Zhukov was now placed in overall command of the forces which posed a threat to Moscow, with Koniev initially as his deputy, but then as commander of a new Kalinin Front, added to the existing Western and Bryansk Fronts. Zhukov was also given authority over the Reserve Front. Although the threat to Moscow was the most serious one, those to Leningrad and the Donetz basin could not be ignored. Moscow's workers were impressed into militia units and the principal government offices were evacuated to Kuibyshev, 600 miles to the east, Stalin remaining with the *Stavka* in Moscow.

From this period onwards the tide began to turn. The departure of Voroshilov and Budenny from any direct involvement in operations or organization was accompanied by the departure of a large number of other commanders of that vintage and of others whose incompetence had been revealed, as had the ability of their successors, in the harsh test-bed of battle. Zhukov, the rift with Stalin healed, worked closely with Shaposhnikov in adroitly feeding in reinforcements where and when they were needed across the whole front of his seven armies, backed by only some 400 tanks and 1,000 aircraft. The needs of Timoshenko and the Northern Front had to be met, the situation round Volkhov remaining critical until the end of the year, a special Volkhov Front being formed under the command of Meretskov. Shaposhnikov's strategy at this time was to provide Meretskov and Timoshenko with sufficient forces to enable them to launch counteroffensives which would prevent the Germans from reinforcing their Central Army Group's drive on Moscow from the north or the south, while building up Zhukov's command so that it also, while holding the enemy off Moscow, could develop a pincer-like counteroffensive. It required careful judgement and cool heads to balance the conflicting demands and risks. Success was achieved by remarkable feats of improvisation, greatly helped by the problems that the Germans themselves faced, not least in not being suitably equipped or prepared to fight in the severe winter at the end of long and precarious lines of communication. In spite of a loss of two million prisoners on top of all the dead and wounded, manpower was not the principal problem. That was equipment, the possibilities of replacement having been severely limited by the loss of so many of the industrial centres on which production depended.

The first sign that the tide was turning was Timoshenko's re-entry into Rostov at the end of November, when Rundstedt, who had captured it earlier in the month, withdrew to the River Mius, anxious for the safety of his northern flank which the build-up of Timoshenko's forces was threatening. By this time the Germans occupied the whole of the Crimea, except for Sevastopol itself, to which the ships and troops at Odessa had been transferred. The Germans renewed their offensive towards Moscow on 15 November and got within nineteen miles of the city, before giving up as Zhukov's counteroffensive was launched on 6 December in three areas : Koniev between Moscow and Kalinin, General L.A.Govorov in the centre of the Western Front from Tula towards Kaluga, and General Ya.I.Cherevichenko's Bryansk Front towards Orel.

By Christmas these attacks had lifted the threat from Moscow and in the next two months, in a series of hard-fought actions, drove the Germans over a hundred miles away from the capital. In the north Meretskov had driven the Germans back from Volkhov and Tikhvin, preserving a tenuous link with Leningrad, while in the south Timoshenko's offensives, aimed at the recovery of the Donbas by General R.Ya. Malinovsky's Southern and Kharkov by General F.Ya.Kostenko's South-western Fronts, had failed. The traumatic experience of the previous six months, and the need to set priorities for re-equipment, forced the Red Army to abandon its distaste for élitism. Certain divisions were designated 'Guards' divisions, and 'Shock' armies were formed to take the leading part in major offensive operations, a pattern that was to be followed thereafter.

In mid March Shaposhnikov and his assistant General Alexander Vasilevsky tried to persuade Stalin that the correct strategy for the summer months was to remain on the defensive, strengthening the fronts defending Moscow, while building up reserves ready to take the offensive when the Germans had exhausted themselves. Stalin accepted their advice with reservations, lending a receptive ear to Timoshenko's proposal for a major offensive by the Bryansk, South-western and Southern Fronts to cut off the German Army Group South. The result was a compromise; Timoshenko was authorized to launch an offensive with the South-western and Southern Fronts to recapture Kharkov in May, and other limited offensives were authorized elsewhere. The general effect was to undermine Shaposhnikov's strategy of husbanding reserves until later in the year.

On 12 May 1942, with 640,000 men, 1,200 tanks, 13,000 guns and mortars, supported by 926 aircraft, Timoshenko launched his double-pronged attack, five days before Kleist launched his from the south. Stalin initially rejected Vasilevsky's plea that the offensive should be terminated, but soon had to face reality. By the time Timoshenko called it off he had suffered heavy casualties and 200,000 men were marched off into captivity. While Timoshenko was in trouble, the situation deteriorated in the Crimea where the interference of the 'super-commissar' L.Z.Mekhlis in General G.K.Kozlov's command produced confusion from which Manstein profited, and by the end of June Sevastopol was isolated. The news from the north was no better. General A.A.Vlasov's 2nd Shock Army on the Volkhov Front, which had been combined with that of Leningrad, was surrounded, Vlasov himself being captured and exploited by the Germans as leader of a 'Russian Liberation Army'. This blow was followed by a crisis in the centre, when a major German offensive was launched against the Bryansk Front, aimed at Orel, while Stalin, rejecting the evidence of its imminence, was urging General F.I.Golikov to attack. In the midst of this crisis Shaposhnikov's health finally broke down, and Vasilevsky succeeded him as Chief of the General Staff with Zhukov as his team-mate, officially Deputy Defence Commissar to Stalin. The German offensive threatened Voronezh and reserves were rushed to save it. As we have seen from Chapter 3, the pressure in that area slackened, as it was transferred further south, where Timoshenko was struggling back to the Don, hard pressed by Paulus's Sixth German Army.

Stalin reluctantly accepted the necessity for withdrawal, Vasilevsky being determined to avoid disastrous encirclements resulting from a stand fast order. Hitler's decision to make his main effort towards the Caucasus gave the Stavka the breathing-space it needed to stabilize the situation on the line of the Don down to the bend near Stalingrad, from which for historical, political and strategic reasons there was to be no withdrawal. On 24 July Rostov fell, the German capture of the great bridge over the Don opening up the way for List's thrust to the Caucasus, to defend which Budenny had been resurrected to command the North Caucasus Front, covering the whole area. The critical situation provoked Stalin to demand more help from his allies. Convoys of supplies, sailing from Britain in May and June, had met with disaster, and the British government notified Stalin on 17 July that no more convoys would sail to Murmansk, leaving the Persian Gulf as the sole route of supply. On 12 August Churchill arrived in person in Moscow to confirm that the Second Front, which Stalin had been demanding must be opened in 1942 to contain at least forty German divisions in the west, could not be launched until 1943. The failure of the Dieppe Raid a week later confirmed Stalin in his scorn at the feebleness of his allies' efforts, while it confirmed Churchill in his conviction that a full-scale invasion of France was not then possible.

The reorganization to hold Stalingrad led to the break-up of Timoshenko's Front, General V.N.Gordov commanding the Stalingrad Front, facing south along the Don, but not including the city, which was covered by Yeremenko's South-eastern Front, facing west on the line of the Aksai Lakes and the railway to Salisk, near its junction with the Southern Front. In August Yeremenko was given overall responsibility for co-ordinating the action of both Fronts, Golikov taking over the South-eastern, Khrushchev remaining as senior commissar to both. In the desperate fighting which followed, while Paulus's Sixth Army drove into the junction of the two Fronts and forced its way into Stalingrad itself on the west bank of the Volga, Vasilevsky paid constant visits to the South-eastern and Zhukov to the Stalingrad Front; the former was principally concerned to see that General V.I.Chuikov held out on the west bank, the latter with attempts to organize counterattacks from the north to relieve pressure on Chuikov. With the situation in the Caucasus deteriorating and a renewed threat to Leningrad developing on the Volkhov Front, the *Stavka* was faced with major problems of resource allocation. Round Stalingrad a balance had to be struck between strengthening the defences, counterattacks, and the build-up of reserves for a major counteroffensive. From this crisis a general strategy began to emerge. German effort must be sucked into Stalingrad: first, in order both to prevent the application of more resources towards the Caucasus and to preserve the Soviet Union's own communications with the area and, second, to create a situation favourable for a counteroffensive to cut off all the forces Hitler had committed in that direction. Maintaining Chuikov's 62nd Army west of the Volga and counterattacking from Gordov's Stalingrad Front to the north proved expensive in lives and *matériel*, as Paulus's attacks intensified from mid September onward. By then the Soviet forces in the area amounted to sixty-nine divisions, supported by thirty-four tank brigades. West of the Volga Chuikov's 62nd Army had 54,000 men, 900 guns and mortars and 110 tanks in nine divisions on a twenty-mile front, the 64th deploying a further seven divisions on a twelve-mile front to his south. The battle raged for two months, by which time Zhukov and Vasilevsky had completed their preparations for a great counteroffensive, Operation URANUS, to be followed by an even more ambitious Operation SATURN.

During these months a major shake-up had taken place. Unity of command had been restored, the commissars of PURKKA being dressed up as military men, subordinate to the commander. The tendency to distribute the available tanks to support infantry formations was reversed, as the new tank factories began to deliver increasing numbers, particularly of the excellent T 34, and tank divisions, corps and armies were formed. A new Don Front was introduced west of Gordov's Stalingrad Front, commanded by Rokossovsky, while Vatutin took over the South-western Front from Golikov, who was transferred to the Voronezh Front. Yeremenko remained responsible for co-ordinating operations for the defence of Stalingrad. All had star parts to play in the forthcoming drama, of which URANUS was to be the first act.

The great extension of the area occupied by German forces had resulted in their having to rely on allied contingents to man long sectors of the front on which offensive operations were not in progress. Along the line of the Don were Italian, Hungarian and Romanian armies, the latter also south of Stalingrad. Zhukov planned to exploit this by a major pincer movement, the northern by Rokossovsky, the southern by Vatutin, both attacking in sectors held by the Romanians, which would encircle Paulus's Sixth and Hoth's 4th Panzer Armies. This would be followed by Operation SATURN, a wider encircling movement, the right arm of which would be operated by Golikov's Voronezh Front to cut off all German forces in the south, Vatutin's South-western and Malinovsky's Southern Front combining as the left arm to recapture Rostov.

The numerical balance of forces in the whole theatre of war gave the *Stavka* confidence in the prospects of victory. They estimated that of the Axis total strength of 333 divisions, 258 (of which sixty-six were 'satellite') faced the Soviet forces, German strength on the Eastern Front totalling five million. Against them the Red Army, over six million strong, deployed 391 divisions, with 77,734 guns and mortars and 6,956 tanks, supported by 3,254 aircraft. The rifle divisions however were low in strength, the overall average being only 8,800 men, while those on the Don and Stalingrad Fronts were down to about 5,000. Of these, one million men, 13,541 guns, 894 tanks and 1,115 aircraft were involved in Operation URANUS, launched on 19

November, as Paulus's final attack petered out in the heaps of rubble to which Chuikov's exhausted men had been driven on the western bank of the freezing Volga. Four days later the two arms of the pincer met at Kalach, encircling the whole of Paulus's 6th Army and half of Hoth's 4th Panzer, a total of 240,000, although the *Stavka* thought that only some 90,000 were in the pocket. The latter, urged on by Stalin, wished to switch their effort as soon as possible to SATURN, but until the threat of Manstein's relief operation had been removed, and Paulus's force disposed of, the forces involved in that could not be released for the wider encircling movement.

In mid December Stalin reluctantly agreed that 2nd Guards Army, destined to spearhead Golikov's arm of SATURN, should be switched to Rokossovsky for a SMALL SATURN, designed to encircle Manstein's attempt to relieve Paulus. It did not succeed in that task, but did divide one from the other and, under the overall direction of General N. N. Voronov, the supreme artilleryman, Operation KOLTSO, launched on 10 January, sealed Paulus's fate. Golikov opened his offensive against the Hungarians and Italians south of Voronezh four days later, eliminating the 2nd Hungarian and 8th Italian armies and one German Panzer Corps. This was followed by attacks further north by General M. A. Reiter's Bryansk Front towards Kursk and Golikov towards Kharkov, while further north an offensive on the Volkhov Front had restored a rail link with Leningrad.

In the heady atmosphere of this decisive turn of the tide, the most senior officers were promoted to the rank of Marshal, Stalin also assuming that rank. Others, Rokossovsky among them, were promoted to Colonel-General. The *pogon*, the shoulder-board, relic of the Imperial army, was reintroduced, an urgent order being given to Britain for gold braid. At the same time the position of officers was strengthened and discipline tightened.

Success went to Stalin's head and he pressed for offensives everywhere, Rokossovsky's Don Front being renamed Central Front, to be introduced between the Bryansk and Western Fronts in order to carry out a wide scythe-like stroke aimed at Smolensk to cut off all German forces in the north. Had the *Stavka* concentrated its resources at this stage on a decisive south-western thrust towards the Dnieper and the Bug, they could probably have cut off all the German forces in the south, after which an offensive in the centre could have proved decisive. In the event, although Kharkov and Kursk were recaptured, Manstein succeeded in holding the crossings of the Dnieper, covering the German withdrawal from the Caucasus and the Don basin. As the spring thaw slowed operations down, the front stabilized on the rivers Donetz and Mius, where it had been a year before, the Germans having regained Kharkov, Kursk remaining in Soviet hands, forming a salient the size of England, shared by Golikov facing south and Rokossovsky west and north. Zhukov and Vasilevsky recommended to Stalin that the Red Army should go over to the defensive until the Germans had delivered the two-pronged attack to cut off the Kursk salient which the disposition of their armoured forces clearly indicated was their intention. The salient was reinforced and extensive defences constructed in depth, while strong reserves under *Stavka* control were stationed in the rear to cover the approaches to Moscow in a new Steppe Front under Koniev.

The remarkable effort of the redeployed arms industry was now producing guns, tanks and aircraft of high quality in ever-increasing numbers. Allied supplies of these equipments were only marginal, their contribution of lorries being the most significant, providing mobile logistic support to the larger tank formations. In contrast to the trend in the British and American forces, in whose armoured formations the balance between tanks and infantry was shifting towards the latter, the Red Army was moving in the opposite direction. There were now five tank armies, each normally of two tank and one mechanized corps. On the eve of the German attack on the Kursk salient on 4 July, of which Stalin had had accurate information from his invaluable intelligence source 'Lucy' in Switzerland, the Red Army fielded six and a half million men with 103,085 guns and mortars, 9,918 tanks and self-propelled guns, supported by 8,357 combat aircraft. During the period of waiting, practical steps were taken to implement the decision made in the summer of 1942 to proceed with the development of an atomic bomb under the direction of Professors Kurchatov and Kapitza.

Fighting on the flanks of the salient was bitter. Nearly 7,000 tanks in all were engaged, 4,000 of them Soviet, 3,000 German. Air activity was intense and casualties on both sides, in men and vehicles, were heavy. Zhukov and Vasilevsky were despatched, as they had been at Stalingrad, to keep a close eye on how the Fronts were faring, and to see that their needs were met. As usual, Zhukov's main concern was to prepare a counteroffensive. His plan was a major thrust, led by tank armies from the Voronezh, Steppe and South-western Fronts, to Belgorod and Kharkov, cutting behind the German Army Group South's attack on the southern flank of the salient. It was launched on 11 August 1942. Even before that Stalin was pressing for offensives on a wider front to drive the Germans back to and over the Dnieper, while Rokossovsky's renamed Byelorussian Front was directed towards Gomel, with Minsk as a further objective, Marshal Vasili Sokolovsky's Western Front to the north of him towards Mogilev, Orsha and Smolensk, and Yeremenko's Kalinin Front (renamed 1st Baltic) to Vitebsk. The Dnieper was reached in October and by the end of the year, with Kiev recaptured, the leading Red Army troops were a hundred miles beyond it and had isolated the German 17th Army in the Crimea. By that time Rokossovsky had captured Gomel and reached the Dnieper and the Beresina beyond it : Smolensk was in Soviet hands, but not Vitebsk. The onset of winter weather proved abnormally capricious and caused delays and difficulties in deployment.

Stalin was determined that there should be no let-up in the offensive, the Red Army having an advantage over the German in winter warfare, as well as an overall numerical superiority. The General Staff assessed German strength facing them at almost five million men (700,000 from its allies) in 236 divisions.[*] Against them the *Stavka* could field 5,600,000 men in 480 divisions, averaging only 6–7,000 men.[†] Only the smaller calibre field artillery remained integrated into rifle and tank divisions, all the larger calibre guns being concentrated into artillery divisions, so that massive artillery support could be provided for major attacks, launched by 'Shock' armies, usually Guards and Tank. The strategy chosen for the winter offensive involved two major thrusts, the principal one being in the south-west. Three fronts, renamed from north to south 1st (Vatutin), 2nd (Koniev), and 3rd Ukrainian (Malinovsky), were to drive to the lower Bug, and then on to the Dniester, while General F.I.Tolbukhin's 4th Ukrainian mopped up the Crimea. In the centre Rokossovsky was to renew his drive through Bobruisk to Minsk, while Sokolovsky thrust to Orsha and Yeremenko to Vitebsk. In the north General L.A.Govorov's Leningrad, Meretskov's Volkhov and Timoshenko's Baltic Fronts were to destroy the German Army Group North, free Leningrad completely and be prepared to regain the Baltic states.

While the autumn offensive was still in train Stalin travelled by rail to Baku and on by air to Tehran to meet Churchill and Roosevelt. He was in confident mood, looking ahead to the defeat of Hitler and to the installation of governments friendly to the Soviet Union in Eastern Europe. He had no wish to see British or American influence predominate in the countries bordering the Mediterranean and adroitly exploited differences between his allies on the subject, harping on the overriding need to open the Second Front by a cross-Channel invasion. His announcement that he would join in the war against Japan as soon as Germany had been defeated was designed to serve the Soviet Union's own interests and was welcomed by Roosevelt, who foresaw heavy casualties as the Pacific island-hopping campaign approached the Japanese mainland.

The offensive in the Ukraine continued with hardly a pause in exceptionally mild weather which hindered movement, the fiercest fighting taking place round the Korsun salient, between the left of Vatutin's

---

[*] These included 25 Panzer and 18 motorized divisions, with 5,400 tanks and assault guns, 54,000 guns and mortars, supported by 3,000 aircraft.

[†] There were 35 armoured and mechanized corps and 46 tank brigades with 5,628 tanks, and 80 artillery and mortar divisions. The latter were to play an increasingly important part.

1st and the right of Koniev's 2nd Ukrainian Fronts. By late February only partial success had been achieved, and fresh plans were made for a further major effort in March, a new 2nd Byelorussian Front taking over Vatutin's right and all six tank armies being committed to the Ukraine, giving the Red Army a two to one superiority over the Germans. Shortly before the new offensive was launched at the beginning of March, Vatutin was ambushed by Ukrainian partisans of the nationalist UPA and mortally wounded. The UPA fought the Germans as well as the official partisans which the *Stavka* supported and introduced into the Ukraine. Zhukov immediately took over direct command of the 1st Ukrainian Front. In both north and south the Red Army surged forward. In less than two weeks Koniev reached the Dniester : on 10 April Malinovsky entered Odessa, directly threatening Romania, and by the end of the month Tolbukhin had driven the German 17th Army into a tight ring round Sevastopol. Zhukov's thrust in the direction of Lwow turned south in an attempt to encircle General H.V.Hube's 1st Panzer Army at Kamanets Podolskiy, but failed to cut it off entirely.

The northern offensive succeeded in driving the German Army Group North away from Leningrad, clearing the southern shore of the Gulf of Finland and clearing the area between Lakes Ilmen and Peipus. These victories led both Finland and Romania to consider their positions and Hitler to occupy Hungary. Zhukov proposed a further thrust to Lwow and Przemysl, but Stalin held him back while the *Stavka* considered strategy for the summer. The conclusion reached was that the main effort would be made in the centre, in Byelorussia, to open up the direct route into Germany, while deceiving the enemy into thinking that it would be a continuation of the flank thrusts. Previous attacks in the centre had proved expensive and unrewarding, giving rise to criticism by the General Staff of the tactics employed. This time ample resources were to be provided and emphasis laid on careful preparation. The tank armies were not to be let loose until a carefully prepared infantry attack, backed by massive artillery support, had opened the way for them to be launched into deep turning movements which would need careful co-ordination between Fronts. The greatest secrecy was observed about this general intention, knowledge of it initially being confined to only five people, the group being widened, at an important conference on 22 and 23 May to include Front commanders. From north to south they were General I.Kh.Bagramyan (1st Baltic), the recently appointed Jew, General I.D.Chernyakhovsky (3rd Byelorussian), General G.F.Zakharov (2nd Byelorussian) and Rokossovsky (1st Byelorussian). Vasilevsky was to exercise general supervision of the first two, Zhukov of the others. In the south, Koniev took over the 1st Ukrainian, Malinovsky the 2nd and Tolbukhin the 3rd Front, while General I.E.Petrov, who had briefly held command of 2nd Byelorussian, was in reserve with the headquarters of 4th Ukrainian.

The target date for the offensive, codenamed BAGRATION, aimed initially at Vitebsk, Bobruisk and Mogilev, which would open the door for a major encirclement of the German Army Group Centre round Minsk, was 15–20 June. In reply to enquiries from his allies about the contribution which he could make to ensuring that no German forces were transferred to the west to meet the Anglo-American cross-Channel operation, the latest target date of which he was informed was the first week of June, Stalin promised a major offensive, but was non-committal about dates and about the form it would take. Tight security, combined with active deception measures, succeeded in convincing the Germans that the main effort of a summer offensive would be south of the Pripet marshes, towards Lwow and the Carpathians, while a subsidiary effort, in co-operation with the British and Americans, would be launched against northern Norway.

The Soviet railway system was strained to the utmost in effecting the concentration of forces and their supplies, delaying the start of the main offensive until 22 June, third anniversary of their war. Before that a major attack was launched on 10 June against the Finns, which cleared the Karelian isthmus and captured Vyborg on 20 June. The forces of the four Fronts engaged in Operation BAGRATION amounted to one and a quarter million men in 116 divisions.* Within three days Vitebsk had been surrounded, the German garrison surrendering after 20,000 had been killed. By 29 June the same fate befell the forces defending Orsha, Mogilev and Bobruisk, the three German armies involved losing 130,000 dead and 66,000 captured, with 900 tanks and large quantities of transport. On 3 July the leading tank forces of Rokossovsky from the south and Chernyakhovsky from the north met west of Minsk, trapping the German 4th Army, the final suppression of which was left to Zakharov, who completed it on 11 July, two days before Chernyakovsky's tanks, racing westward, captured Vilno. Rokossovsky's effort had been switched south-westward, his progress in that direction bringing into play his powerful left wing, forming one arm of the pincer designed to encircle Brest-Litovsk. On 23 July, Chuikov's 8th Guards Army, secretly transferred from the 3rd Ukrainian Front, entered Lublin and turned north-west towards Warsaw. By then the Germans had lost twenty-five to twenty-eight divisions, totalling some 350,000 men – a bigger loss than that suffered at Stalingrad – and a fresh blow was delivered when Koniev opened his offensive on 12 July.

---

* Including 2,715 tanks, 1,355 self-propelled and 24,000 other guns and mortars, and 2,306 rocket-launchers, supported by 5,327 aircraft as well as the Long Range Bomber Force of 700. Daily, 90–100 trains and 12,000 lorries were needed to keep this large force supplied.

Stalin had wished him to put all his effort into a single thrust aimed at Lwow, but Koniev had persuaded him to agree to a double one, aimed also at Rava-Russkaya, thirty miles further north.* Anticipating the major effort to be directed at Lwow, the Germans had prepared exceptionally strong defences covering it, but had been forced to switch the reserves of Army Group North Ukraine to help save Army Group Centre. For his offensive Koniev had 843,000 men in eighty divisions with 1,614 tanks and self-propelled guns, and 14,000 guns and mortars, supported by 2,806 aircraft. The initial assault made slow progress in heavy rain and casualties among the infantry were exceptionally heavy. Koniev decided to discard the *Stavka* ruling about waiting for a clear gap before launching his tanks and, after ten days of fighting, encircled a major body of enemy between Brody and Lwow, more than 30,000 of them being killed and 17,000 captured. He then decided to try and encircle Lwow itself, but ran into difficulties in doing so. 1st Guards Tank Army on his right was already making good progress further north towards the River San, helped by Rokossovsky's advance on the right. By 24 July he had encircled the city, but it was the 27th before its defences were finally reduced.

Koniev's advance split Army Group North Ukraine into two, 4th Panzer Army withdrawing westward to the Vistula, while the 2nd, with the 1st Hungarian Army, withdrew south-westward into the Carpathians. To enable Koniev to concentrate his effort westward, Petrov's 4th Ukrainian Front was introduced to take over on his left and drive into the mountains. On the day that Lwow fell to Koniev, Rokossovsky succeeded in closing the ring round the eight German divisions in Brest-Litovsk, all that was left of their 2nd and 9th Armies. This encircling movement had not held back the thrust of his left wing towards Warsaw. By 31 July Chuikov was only twelve miles from the Polish capital, annoyed to receive a succession of cautious messages, caused by the *Stavka*'s fear that the considerable concentration of the remaining Panzer formations covering the approach to the city might deliver a sharp counterattack before Rokossovsky could move forward the troops that had encircled Brest. On 1 August the latter authorized Chuikov to cross the Vistula, as General A.I.Radzievsky's 2nd Tank Army on the east bank was ordered to adopt a defensive posture on the eastern outskirts.

On that day General Bor Komorowski ordered the outbreak of active fighting against the Germans by the 'Home Army' (*Armija Krajava* – AK), having been authorized to do so, when he judged the moment ripe, by the Prime Minister of the Polish government-in-exile in London, Mikolajczk, when the latter left England on 26 July en route to Moscow via Tehran. The Soviet Union supported a separate collection

---

* Rokossovsky had won a similar argument with Stalin over the attack on Bobruisk.

of Polish organizations, the political element, known as the Lublin Committee, and the military including both a rival partisan organization within Poland, the *Armija Ludava*, and the 1st Polish Army, commanded by General Berling, serving under Rokossovsky. In addition Soviet partisans had been infiltrated into Poland. Bor Komorowski faced a dilemma. If the AK went into action too soon, it might be defeated by the Germans before the Red Army could liberate German-occupied Poland ; but if it took no action until that happened, it would reinforce the Soviet Union's criticism of its ineffectiveness and allow power and influence to fall into the hands of its Soviet-favoured rivals. With the Red Army in the outskirts of Warsaw he judged that it could only be a matter of days before the Germans were forced to abandon the city, and that AK must strike. It proved to be a fatal decision. Hitler ordered the hardest of SS forces into the city to maintain access to the Vistula, while Model's Army Group Centre counterattacked Rokossovsky's forces between the eastern suburb Praga and Siedlice, fifty miles further east. Appeals to Britain, the USA and the Soviet Union to deliver weapons, arms supplies and parachute troops met with only a very limited response. Stalin fobbed off Mikolajczk with proposals for amalgamation of his organization with that of the Lublin Committee, which he could not accept. In mid September Rokossovsky resumed the offensive, not to try to capture Warsaw but to clear the area east of the Vistula as far as the River Narew, withdrawing those troops that had crossed the Vistula near Warsaw in the face of German attacks. On 2 October the AK's fight in Warsaw came to an end ; 15,000 of its 30,000–40,000 men and 150,000–200,000 out of the one million population of the city were killed in one way or another.

The dilemma facing the Polish underground movement was reflected also in Slovakia, the situation being complicated by the existence of the 1st Czechoslovak army on the side of the Germans, most of which was in eastern Slovakia, helping to defend the Carpathians. While the Chief of Staff of the Slovak army negotiated with the Czechoslovak government-in-exile in London, warning them that a German-Hungarian occupation of the country appeared imminent, partisans, introduced and supported by the Soviet Union, sparked matters off on 26 August by ambushing and killing the head of the German military mission in Romania as he passed through. Colonel Talsky flew to Koniev's headquarters and urged him to launch an offensive through the Carpathians, promising that the Slovak Corps there would change sides. While Koniev was seeking permission to do so from the *Stavka*, the corps was disarmed by the Germans. Ignorant of this, the *Stavka* agreed and Koniev attacked towards the Dukla pass on 8 September with Moskalenko's 8th Army, which included the 1st Czechoslovakian Corps, raised in the Soviet Union. Meanwhile only 16,000 men of the 60,000 strong Slovak army had joined the rising, almost all of them in the centre of the country. Koniev's attack made slow progress, and, although the Soviet Union attempted to support and extend the rising, German countermeasures, executed by some of the ss formations transferred from Warsaw, suppressed it by the end of October, by which time Geneal K.S.Moskalenko had struggled through the mountains to the River Ondava, having suffered 80,000 casualties, 20,000 of them killed, in addition to 6,500 from the Czechoslovak Corps, half its strength.

While Koniev had been battling through the Carpathians and Rokossovsky had been clearing up the area east of the Vistula round Warsaw, Chernyakhovsky's 2nd Byelorussian Front, having surrounded a large body of Germans at Vilno in July, on 17 August reached the frontier of East Prussia. To his north Bagramyan's 1st Baltic Front cut the communications between that area and General Georg Lindemann's Army Group North, which was being forced westward by pressure from Yeremenko's 2nd and General I.I.Maslennikov's 3rd Baltic Fronts. Bagramyan urged the *Stavka* to let him drive north for Riga but, believing that Lindemann was intending to withdraw towards East Prussia, they insisted that he should thrust north-west to Shiaulai to ensure that Lindemann did not threaten Chernyakhovsky's right flank. On the day that Bagramyan captured it, 27 July, Yeremenko also captured Dvinsk (Dangavpils) on the Dvina. Maslennikov having reached Ostrov on 21 July, all three Fronts now headed towards Riga, but the combination of terrain that favoured the defence and the concentration of Army Group North into the Baltic States held them all up.

Govorov's Leningrad Front attempted to clear the southern shore of the Gulf of Finland, west of the River Luga, but after capturing Narva could make no further progress. Further north, Meretskov's Karelian Front followed Govorov's capture of Viborg with an offensive north of Lake Ladoga. By the end of July, Mannerheim's reserves were exhausted and, as he replaced Ryti as President, he opened negotiations with Moscow, which led to an armistice at the end of August. All German forces left the country by 15 September.

Finland was not the only country to have given up. While the main effort had been exerted westward, Malinovsky's 2nd and Tolbukhin's 4th Ukrainian Fronts had had a quiet time. In July they received orders from the *Stavka* to prepare an invasion of Romania, Timoshenko emerging from the shadows as the *Stavka* co-ordinator. Between them they had nearly a hundred divisions with 1,883 tanks and self-propelled guns, and 19,000 guns and mortars, supported by 1,900 aircraft. The plan, as usual, was for a two-pronged pincer, Malinovsky driving south, astride the River Seret, while Tolbukhin struck west and south-westward across the lower reaches of the Dniester to encircle the German and Romanian forces of General Johannes Friessner's Army Group South Ukraine, south of Kishinev. Launched on 19 August, it was an immediate success. King Michael arrested the dictator Marshal Ion

Antonescu and his brother and surrendered unconditionally to the Allies on 23 August. Operations continued for several weeks against the Germans who lost 98,000 prisoners and 100,000 dead before those who escaped, no more than 25,000 according to Soviet sources, had made their way to the west, pursued by Malinovsky who had been ordered to make for Hungary. He was ordered to relieve pressure against the left of Petrov's 4th Ukrainian Front, stuck in the Carpathians, while Tolbukhin pressed on south across the Danube into Bulgaria, the army of which was formally placed under Soviet command on 20 September. Tolbukhin's Chief of Staff, General S.S.Biryuzov, with one corps of two divisions, was left to deal with Bulgaria, while the rest of his Front, including three Bulgarian armies, the left flank of Malinovsky's and Rear-Admiral Gorchkov's Danube Flotilla, pressed westward north and south of the river into northern Yugoslavia.

Tito, none too keen that Belgrade might be liberated without the presence of his troops, was flown on 21 September to Moscow for consultations with Stalin, which did nothing to improve relations between the two. The atmosphere was both more sober and more cordial a few weeks later when Churchill arrived, concerned to settle all the potential disputes between the Allies posed by this irruption of Soviet forces into the Balkans and the consequent withdrawal of the Germans. Without argument Stalin agreed to a mathematical division of spheres of influence: Romania 90 per cent Soviet to 10 per cent British 'in accord with the USA', and the reverse in Greece; 75 to 25 per cent in Bulgaria, and 50–50 in Yugoslavia and Hungary.

However, Poland could not be settled so easily. Mikolajczyk and colleagues, summoned to join the conclave, once more refused to accept the Curzon Line or amalgamation with the Lublin Committee, although Churchill and Stalin accepted the formula of the Curzon Line in the east, with compensation 'to the east of the line of the Oder' in the west. Discussion also took place about participation of the Soviet Union in the war against Japan, once that with Germany was over. Stalin planned to double the Far Eastern Army's strength to sixty divisions, making it possible to thrust through Mongolia, separating the Japanese army in Manchuria from that in China. Even at thirty-six trains a day, the Trans-Siberian railway could not supply such a force and US logistic help would be needed, the bill being set at 3,000 tanks, 5,000 aircraft, 860,410 tons of dry and 200,000 tons of liquid cargo, all to be delivered to the Far East by 30 June 1945.

Tolbukhin's advance towards Belgrade loosened things up on Malinovsky's Front in Transylvania, where the slow progress had provoked Stalin, in reply to the newly promoted marshal's request for more tanks, to say, 'My grandma would know how to fight with tanks. It's time you got a move on'. Löhr's German Army Group F clung on to Belgrade as long as it could in order to be able to withdraw its troops in Serbia and Montenegro. The battle for the city, in which Tito's 1st Proletarian Yugoslavian Division took part, lasted from 14 to 21 October, costing the Germans 15,000 dead and 9,000 prisoners and leaving much of the city in ruins. While it had been going on, all the Ukrainian Fronts further north were on the offensive. Petrov cleared the main Carpathian range and struggled slowly forward to Uzhgorod and Mukachevo in Ruthenia at the end of October, while Malinovsky emerged from the mountainous country of Transylvania into the plains of Hungary from Debrecen southwards. The threat to Hungary set off the familiar pattern of struggle between conflicting political elements, the right-wing dictator Admiral Horthy trying to change sides without falling from the saddle. An armistice delegation arrived in Moscow on 1 October and had to kick their heels for eleven days until it was signed, while Malinovsky fought for Debrecen and outflanked it to the south, heading for Budapest. Debrecen held out until 25 October, by which time the leading troops had reached and crossed the River Tisza.

By this time both Tolbukhin's and Malinovsky's Fronts had been greatly reduced in strength and were at the end of a long and precarious supply line. On 28 October Malinovsky received orders from the *Stavka* to advance immediately with the forces available to him to seize Budapest, while Petrov was ordered to press westward into Czechoslovakia, and Tolbukhin was to reorganize his forces near Belgrade and then drive north up the Danube also to Budapest. Malinovsky's plea for a pause of at least five days was roughly rejected by Stalin with the words, 'I categorically order you to go over to the offensive for Budapest tomorrow'. The result was a failure to capture the dual city, which was reduced to ruins by the bombardment to which it was subjected during the subsequent months. Early in November the whole front south of the Carpathians came to a grinding halt. The attack on Budapest was renewed on 6 January, and it took weeks of bitter house-to-house fighting before it finally fell on 13 February.

The same state of affairs had come about at the other end of the theatre of war, where the Baltic Fronts were trying to eliminate the German Army Group North. They had been reinforced since their concentric thrusts towards Riga had been brought to a halt at the end of July, and the *Stavka* issued orders at the end of August for a fresh offensive to be launched on 14 September. The terrain, interspersed with thick woods and water obstacles of every kind, conferred advantages on the defence: many of the soldiers were new recruits from the liberated areas of Byelorussia, and the *Stavka* directive, attempting to co-ordinate the action of three converging Fronts, imposed a pattern which forced Maslennikov and Yeremenko into costly frontal attacks. Bagramyan's ninety-degree switch in orientation caused him severe movement problems. The combination of these factors and the fierce resistance of the Germans, with no prospect of withdrawal, caused progress to be slow and casualties high. Riga fell to Yeremenko on 13 October and Bagramyan reached the sea on either side of Memel, defended by four German divisions, by the end of the month, by which time Schörner's thirty divisions had withdrawn into the peninsula of Courland between Memel and Riga. There they stayed until the end of the war, the garrison of Memel holding out until January 1945.

This great sweep forward, regaining all the territory lost, except for the Courland pocket, and the prestige it brought to the Red Army and its commanders produced headaches for its two rivals, the Party and the NKVD. There was the problem of the inflated membership of the Party within the armed forces, few of whom had been subjected to the strict supervision needed to guarantee their political soundness; the problem of the population of the areas which had been occupied by the Germans; the need to examine carefully what the attitude of every individual had been, especially if he (or she) were being recruited into the armed forces; the problem of the exposure of members of the armed forces to the experience of entering foreign countries; the problem of handling those who had been prisoners of war and had been liberated by the advance; and, at the heart of the matter, the old problem of 'unity of command': from marshal (a rank rapidly becoming inflated) to corporal, commanders at every level had become accustomed to acting solely on the orders of their military superiors, although military soviets still in theory existed. Relations between the military themselves, their own political organization, the PURKKA, the Party and the NKVD all became tense as operations extended beyond the borders of the Soviet Union and the end of the war was in sight. The armed forces naturally wished to cling to the position and the privileges they had gained through their victories. The Party and the NKVD, as the guardian of its *arcana*, fought to ensure that the purity of dogma was not diluted by sentimental Russian patriotism. Above all there brooded the enigmatic figure of the Caucasian intriguer, Josef Vissarionovich Dzhugashvili, known as Stalin.

The Soviet General Staff estimated that the operations of 1944 had cost the Germans more than one and a half million men, 6,700 tanks, 28,000 guns and mortars and 12,000 aircraft, and that some 130 divisions had been removed from their order of battle. Nevertheless, with their frontage much reduced and fighting for their homeland, it was appreciated that the march to Berlin would not be a walk-over. Although seventy-four German divisions with 1,600 tanks faced the Anglo-American armies in the West, they calculated that Hitler could still muster three million men with 4,000 tanks to face the Soviet forces. After Stalin, Zhukov and Antonov, who had virtually taken Vasilevsky's place as Chief of the General Staff, had discussed future strategy early in November, Rokossovsky, Chernyakhovsky, Koniev and Tolbukhin were called in.

The plan that resulted gave the principal role to the 1st Byelorussian and 1st Ukrainian Fronts, Zhukov to take over the former and Koniev retaining the latter. Rokossovsky was to side-step to command 2nd Byelorussian Front instead of Zakharov. The opening moves were to be made in the south, where it was hoped that the threat to Silesia would draw German reserves away from the direct route to Berlin through Poznan, which would be Zhukov's axis. To protect his northern flank, Rokossovsky, passing north of Warsaw, would thrust towards Marienburg (Malbork) and thence to Danzig (Gdansk) while Chernyakhovsky broke through the eastern defences of East Prussia on the Masurian Lakes and took Königsberg (Kaliningrad).

Stalin himself would act as *Stavka* co-ordinator, Vasilevsky being relegated to the minor role of co-ordinating the action of the Baltic Fronts. The total strength of the Red Army deployed against the Germans would be over six and a half million men in 500 rifle divisions.* To have produced and trained so large an army, and equipped it for the most part from its own resources, which had been subjected to a wholesale removal over hundreds of miles, and to be able to supply and maintain it, was an astonishing achievement. The target date set for the offensive, which the *Stavka* estimated would take forty-five days to reach Berlin, was 15–20 January 1945. When the German offensive in the Ardennes was launched in the second half of December and Churchill appealed to Stalin to help, he was happy to bring it forward by a week, hoping that Hitler's involvement there would prevent transfer of forces to the east.

As the date drew near, Zhukov had crammed 400,000 men and 1,700 armoured vehicles into his fifteen-by-seven-mile bridgehead over the Vistula at Magnuszew, south of Warsaw, Koniev having more room – forty-five by forty miles – in his bridgehead at Sandomierz. Zhukov did his best to deceive General Josef Harpe, commander of Army Group Centre, that his main effort would be round Warsaw, to the occupation of which he allotted Berling's 1st Polish Army. He had persuaded Stalin to allow him to take a more southerly route through Lodz, Koniev's axis being displaced to the south through Breslau (Wroclaw). The latter's deception succeeded in persuading the Germans that his main effort would be directed through Krakow into the heart of the Silesian industrial area, whereas Stalin had given strict orders that Koniev should by-pass it to the north and then surround it to avoid damaging

---

* Grouped into fifty-five all-arms armies, six tank armies and thirteen air armies, with 13,000 tanks and self-propelled guns, 100,000 guns and mortars and 15,000 aircraft. Of this total, Zhukov and Koniev were allotted 163 rifle divisions with 6,500 tanks, 32,143 guns and mortars and 4,772 aircraft, totalling nearly two and a quarter million men, giving them about a five to one superiority over the forces opposing them.

its precious resources. Koniev launched his attack at dawn on 12 January and before the end of the day his tanks passed through the infantry and began their westward drive. Zhukov followed two days later and achieved equally rapid success. At the end of the first week Krakow had been surrounded, Lodz captured by Chuikov's 8th Guards Army, and the Germans withdrawn from Warsaw, prompting Hitler to dismiss Harpe and replace him with Field Marshal Ferdinand Schörner, while making the astonishing decision to send 6th Panzer Army from the Ardennes to Hungary, where the fight for Budapest had been resumed. By the end of January the leading troops of both Fronts had reached the River Oder, but by this time Zhukov was worried about his right flank and Koniev about his left. Zhukov was concerned at the gap which had opened up between his right and Rokossovsky's left, the latter having been ordered by Stalin to switch forces eastward, after he reached the lower Vistula, to help Chernyakhovsky, who was making slow progress, to finish off the German forces in East Prussia. Koniev had meanwhile surrounded the Silesian industrial area and, like Zhukov, had his eyes on Berlin, only forty-eight miles from the latter's forward troops at Küstrin. Both Fronts now needed a breathing-space in which to bring forward their rifle and artillery divisions, with their supplies and ammunition.

At this critical juncture, while the Anglo-American armies were getting ready to clear the area between them and the Rhine, Roosevelt and Churchill flew from Malta to Yalta to meet Stalin, who arrived by train on 4 February. Stalin was concerned to ensure that his allies kept up the pressure in the west to prevent transfers to his front. Arrangements for co-ordination between the Allied forces as they approached each other were discussed.

On 17 February the *Stavka* issued a directive to Malinovsky and Tolbukhin, ordering them to resume the offensive on 15 March, the former towards Vienna and Brno, the latter south of Lake Balaton to Graz, thus cutting off German forces in Yugoslavia and northern Italy. This plan was disrupted by a counterattack by 6th Panzer Army which eliminated Malinovsky's bridgehead over the Hron south of Budapest. The Americans passed on information that a major German counteroffensive was planned north and south of Lake Balaton, and both Front commanders were ordered temporarily to adopt the defensive. The Germans launched their attack on 6 March, but after nine days of heavy fighting it petered out with the loss of over 500 tanks, 300 guns and 40,000 men. It was immediately followed by a riposte from Malinovsky and Tolbukhin which almost succeeded in surrrounding 6th Panzer Army between Budapest and Lake Balaton. By that time the *Stavka* had altered the plan, Tolbukhin's Front being strengthened at the expense of Malinovsky, and his axis redirected to Vienna, which was to be encircled in a joint operation between both.

The Americans had also given warning of an intended German counterattack on Zhukov's right flank from Pomerania. The forces entrapped in East Prussia and Courland were proving difficult to eliminate. Stalin called on Vasilevsky to effect a major reorientation of the Baltic and Byelorussian Fronts involved, as a result of which Bagramyan became responsible for capturing Königsberg and Rokossovsky handed over some of his forces to Chernyakhovsky, who was charged with finishing off East Prussia. Rokossovsky's weakened Front was to turn west to head for Stettin, coming up on Zhukov's right. The combined offensive of all three Fronts started on 10 February and met with little success. Vasilevsky had been recalled to Moscow to his official post of Chief of the General Staff, while Antonov was with Stalin at Yalta, Govorov taking his place as co-ordinator. Vasilevsky was told that he would probably be nominated as commander of the Far Eastern Army when the war with Germany was over. When told this, Vasilevsky asked to be relieved of his post of Chief of the General Staff. Stalin agreed, appointing Antonov in his place, and making Vasilevsky a member of the *Stavka* and a Deputy Defence Commissar, putting him on the same level as Zhukov. Within hours Stalin learned that Chernyakhovsky had died of wounds, and Vasilevsky was sent to command 3rd Byelorussian Front.

The *Stavka* now made a new plan. Rokossovsky and Zhukov together were to drive north through Pomerania to the Baltic, the former to Danzig and Gdynia, the latter to Kolberg (Kolobrzeg) and Stettin. Rokossovsky launched his offensive on 24 February and Zhukov on 1 March, the German Army Group Centre, which had successfully held the defences of the Oder against a series of attacks, astonished to see Zhukov driving north rather than west. Both thrusts were successful, Zhukov reaching the Baltic at Kolberg and Stargard, fifteen miles east of the Oder opposite Stettin, on 5 March, Danzig not being finally cleared until the end of the month. In this period Koniev had been closing up to the Oder and clearing Silesia. By 15 February he had surrounded Breslau, helped by the progress made further south by Petrov, who crossed the Oder and captured Rybnik and Ratibor.

In the last week of March relations between Stalin and his allies, soured by disagreement over Poland, reached an all-time low over contacts made by them in Switzerland with representatives both of Himmler and of the German General Wolff in Italy, interpreted by Stalin and Molotov as attempts to make a separate peace. The atmosphere was improved by a direct communication from General Eisenhower, explaining that, after he had dealt with the German forces in the Ruhr, he intended to make his principal thrust in the direction of Erfurt-Leipzig-Dresden, with a secondary one further south on the axis Regensburg-Linz, and asking for information about Soviet plans. Stalin welcomed the news and said that his forces would make their main effort also in the south to join hands with their allies, making only a secondary effort towards Berlin which 'has lost its former strategic importance'. On the very day, 1 April, he sent his reply, he held a meeting with Zhukov, Koniev, Antonov and the State Defence Committee to finalize plans for the main effort to be launched on Berlin, at which he made General S.M.Shtemenko, Antonov's successor as Chief of the Main Operations Directorate, read out a telegram stating that the Anglo-American forces, headed by Montgomery, were about to launch an operation designed to take Berlin before the Red Army.

As Stalin had told him in November, Zhukov was to have the glory of taking Berlin with two tank and four rifle armies thrusting out from the bridgehead at Küstrin, while subsidiary attacks by two armies each would be made north and south of the main effort, the first task of which would be to clear the defences of the Seelow ridge. His principal opponents would be the German 9th Army. Further south, Koniev, after crossing the Neisse, was to eliminate 4th Panzer Army, between Cottbus and Berlin, and then drive west and north-west to the Elbe, with a subsidiary thrust to Dresden. North of Zhukov, Rokossovsky's 2nd Byelorussian Front would cross the Oder and prevent General Hasso von Manteuffel's 3rd Panzer Army from interfering with Zhukov's operation, driving it up against the Baltic coast, while he also pressed westward to the Elbe. The target date for Zhukov and Koniev was given as 16 April, and 20 April for Rokossovsky, who was faced with severe logistic problems in assembling his force with all that was needed to throw it across the wide lower reaches of the Oder, flanked by swampy terrain.

Stalin saw that the rivalry between Zhukov and Koniev could be exploited to urge both to press their subordinates to the utmost. Rejecting the General Staff's boundary between the two, which excluded Koniev from Berlin itself, he agreed to it only as far west as Lübben, thirty miles south of the city. From that moment on Koniev arranged matters so that his two tank armies, General P.S.Rybalko's 3rd and General D.D.Lelyushenko's 4th Guards Tank, could be switched to Berlin, if Zhukov's frontal assault were delayed. In the latter's initial plan, approved by the *Stavka*, both his tank armies were to envelop Berlin from the north, but he later changed it. General S.I.Bogdanov's 2nd Guards Tank Army would exploit 5th Shock Army's attack round the north of the city, while General H.E.Katukov's 1st Guards Tank Army, initially closely linked with the assault of Chuikov's 8th Guards Army on Seelow, would sweep round the southern outskirts of Berlin to join hands with Bogdanov at Charlottenburg.

The artillery and air support for this offensive by 190 divisions was of unprecedented strength. Zhukov's main attack was to be supported by 8,983 guns ; Koniev's by 7,733. This gave Zhukov 189 guns for each kilometre of front, for which he needed over seven million shells. The

initial artillery plans for the two Fronts however differed greatly.
Koniev opened his assault with a bombardment lasting two and a
quarter hours, while Zhukov, who planned to attack at night, searchlights providing artificial moonlight, limited his preliminary bombardment to ten minutes, followed by a barrage moving forward to a depth
of 4,000 yards. The scale of air support was equally impressive : 7,500
aircraft in four air armies, one of them the Long-Range ADD. Rokossovsky in particular relied on air support, the width of the river and
the restricted area east of it suitable for siting artillery, limiting the
support he could gain from the latter.

In the two weeks between the planning conference and the assault,
a number of significant events occurred. On 5 April the Soviet Union
abrogated its April 1941 treaty of non-aggression with Japan : on 11
April American troops reached the Elbe at Magdeburg and next day
Roosevelt died, succeeded by Vice-President Truman. Vasilevsky
resumed the offensive against Königsberg, which fell after a bitter fight
on 9 April. Petrov's 4th Ukrainian Front was making such slow progress
through the mountains of Slovakia towards Moravska-Ostrava that
he was replaced by Yeremenko and went to join Koniev as Chief of
Staff, replacing Sokolovsky who transferred to the same post with
Zhukov. Tolbukhin was given sole responsibility for the capture of
Vienna, Malinovsky being ordered to drive straight for Bratislava and
on to Brno, threatening the rear of the German forces facing Yeremenko.
The Germans withdrew from Bratislava on 4 April as Malinovsky threatened to surround it, but clung on to Vienna, Tolbukhin's battle for
which lasted from 6 to 13 April, causing grave destruction to the city.

Next day Zhukov launched a preliminary attack on the Seelow
defences, intended as a reconnaissance in force. It failed in its object,
the presence of further defence lines in depth, which were to cause
Zhukov great trouble and heavy casualties over the following week,
not being revealed. The main assault on both Fronts was launched
in the early hours of 16 April, and it soon became clear that Zhukov
had underestimated the task of breaking through the German defences.
His decision to move Katukov's 1st Guards Tank Army immediately
behind Chuikov's 8th Guards caused serious congestion. Koniev, starting from further away, made more rapid progress and on 17 April
received Stalin's permission to turn his tank armies north towards
Berlin, aiming to capture Zossen, the German Army Headquarters, on
the way. Zhukov was given the veiled threat that, if he did not make
more rapid progress, the glory of capturing Berlin would be shared
by Koniev coming up from the south and Rokossovsky from the north.
The threat had its effect. The assault on the Seelow defences was
resumed with great intensity on 18 and 19 April and, on the following
day, Hitler's fifty-sixth birthday, Zhukov's leading troops were into
the north-eastern suburbs of the city.

Both Front commanders urged their men on and on the 21st Zossen was captured, and Koniev relieved of anxiety about the elements of 4th Panzer Army still holding out near Cottbus. His leading tanks were nearing the southern outskirts of Berlin and closing the gap to the west between them and Zhukov's tanks. Berlin was now under air and artillery bombardment from all sides, the perimeter held by the remnants of 9th Army shrinking every day. On 25 April, the day that US and Red Army troops met on the Elbe, the city was firmly encircled and Chuikov's 8th Guards joined hands with Rybalko's 3rd Guards Tank Army in the southern suburbs. Two days before this Rokossovsky, who had crossed the Oder on 20 April and been ordered to join in the encircling operation, was told to revert to his previous task of dealing with Manteuffel's 3rd Panzer Army. The latter, on whom Hitler had been relying for relief, was ordered by General Gotthard Heinrici, commanding what was left of Army Group Vistula, to withdraw to the west. The relentless jaws of the mighty Soviet war machine now closed ever tighter on the unfortunate inhabitants of the city, whose Führer, forbidding surrender although Generals Karl Weidling and Hans Krebs told him there was no hope, committed suicide on the night of 29 April, twelve hours before the Red Army reached the Reichstag.

Operations were held up while Krebs, meeting with Chuikov, tried in vain to negotiate a cease-fire with the Soviet authorities, combined with recognition of a government headed by Joseph Goebbels and Martin Bormann. Stalin insisted on unconditional surrender of all forces to all three allies, and the assault was resumed. Goebbels and Krebs followed Hitler's example, and Weidling, in the presence of Sokolovsky and Chuikov, signed a surrender document on 2 May, effective from 3 p.m. that day. It only covered the area of Berlin. Stalin's immediate aim was to occupy the area of Germany agreed at Yalta as his occupation zone, and the whole of Czechoslovakia. Rokossovsky's and Koniev's tank forces sped as fast as they could to the Elbe, halting only when they met their allies who, in some cases, had advanced into the area due to become the Soviet occupied zone.

The Battle for Berlin had cost the three Fronts involved 304,887 men killed, wounded and missing, 2,156 tanks and self-propelled guns, 1,220 guns and mortars and 527 combat aircraft. They claimed to have destroyed seventy German infantry, twelve Panzer and eleven motorized divisions, capturing 480,000 men, 1,500 tanks and self-propelled guns, 10,000 guns and mortars and a large, uncounted tally of aircraft.

All eyes, including Churchill's, were now on Prague. He urged an unresponsive Truman to press an equally unsympathetic Eisenhower to push an eager Patton to reach the Czech capital before the Soviet army, which Churchill now saw as spreading its red claws all over Europe. After Malinovsky's capture of Brno on 26 April, he and Yeremenko pressed forward to try and encircle Schörner's Army Group

Centre, of which 1st Panzer Army was the most effective formation. On 1 May Stalin ordered Koniev to hand over his forces round Berlin to Zhukov and join in the movement to encircle Army Group Centre. Patton's 3rd US Army had already crossed the Czech frontier, heading for Pilsen.

The date set for the concentrated assault by all three Fronts was 7 May. Eisenhower received a sharp rebuff from Antonov when he sought agreement for Patton to operate east of a previously agreed line, as did General Omar N.Bradley when he visited Koniev on 6 May and proffered help. As rival insurrectionist movements issued appeals from within Prague and the jaws of the trap formed by the three Fronts closed round it, the general surrender was announced on 8 May, Malinovsky's and Koniev's tanks met each other in the early hours of 9 May in the city's outskirts.

The 'Great Patriotic War', as it was later called in Russia, thus came to an end. According to the Soviet authorities the cost to them in human lives, military and civilian, from all causes was twenty million. In comparison to the enormous effort expended and sacrifices made in the war against Germany, Soviet intervention against Japan, finally confirmed at the Potsdam Conference held from 16 July to 2 August 1945, was of little account. Vasilevsky was in overall command, with Malinovsky, Meretskov and General M.A.Purkayev as Front commanders. His forces outnumbered the 750,000-strong Japanese Kwantung Army by three to one. Hostilities were opened on 9 August, the day after the second atom bomb was dropped and, although the Japanese Emperor announced surrender to the Anglo-Americans on 14 August, the Kwantung Army rejected Vasilevsky's ultimatum three days later, fighting on until the Soviet forces took Harbin on 20 August, having already joined hands with Mao Tse-tung's Chinese People's Army. Port Arthur was occupied without resistance on the 23rd.

The Soviet Union had produced, in the face of immense handicaps, the most formidable war machine the world had ever seen. In spite of the break with the past which the Revolution effected, the Soviet army owed much of its strength, as well as some of its weaknesses, to its Imperial predecessor, as reformed by Milyutin between 1861 and 1874. For an army which reached the staggering total of six and a half million men, the Soviet Union itself produced 102,500 armoured vehicles, 490,000 guns and 142,800 military aircraft. Almost all of its original production facilities and the sources of raw material to supply them and the labour force had to be evacuated over hundreds of miles and much of it moved back again. Most of it, especially the tanks and artillery, was of high quality in respect of performance and, in its rugged simplicity, well suited to the harsh physical conditions in which it had to operate, and to the level of training of its operators. To its own production was added the contribution of its allies: 10,000

tanks, 18,700 aircraft and, most significant of all, 427,000 motor vehicles, the last forming two-thirds of the total stock of the Soviet forces
at the end of the war. At the beginning of the war Soviet combat
aircraft were no match for their German counterparts, but the aircraft
designers who had survived the pre-war purge were soon providing
aircraft which, in a combination of quality and overwhelming quantity,
were to drive the *Luftwaffe* from the Russian skies : Lavochkin, with
his LAGG fighter, Yakovlev and Mikoyan with their YAKs and MIGs,
Ilyushin with his IL-Shturmovik close-support aircraft, Petlyakov with
his twin-engined PE 2 bomber and long-range fighter and his four-
engined PE 8, and Tupolev with his twin-engined TU 2 and four-
engined TU 4 bombers. In the early stages of the war production was
concentrated on single-engined fighters, partly for sheer survival,
partly because larger aircraft absorbed more production capacity. The
emphasis was on gaining air superiority in the critical area, where
the Shturmoviks could then be used to give direct support to the
army. As capacity increased and the initial period of acute crisis
passed, greater effort was devoted to offensive bomber operations,
mostly by twin-engined bombers, directly linked to the operational
plans of the army Fronts. Little effort was devoted to long-range strategic operations, partly because production of aircraft for it would have
seriously affected that of other types, partly because few of the influential senior officers of the Soviet army believed in its effectiveness,
but also because, by the time it could have become a possibility, the
British and Americans were fully engaged in it. For the pioneers of
airborne forces, the Red Army made little use of them. Their unfortunate experience in the 1939 campaign against Finland partly explains
it, but their vulnerability on the battlefield, in the air and on the
ground, was the principal cause. Occasions on which they could be
used with any hope of success rarely occurred, and their principal
use became as partisans dropped in comparatively small bodies behind
the German lines. The largest single operation was round Vyazma in
January 1942 when almost all the available parchutists, 8–10,000, were
dropped at night over a period of weeks. The air transport fleet, amalgamated with the long-range bomber force into one long-range air army
(ADD) in the spring of 1942, could theoretically have dropped up to
5,000 men in one operation, but never did so, being used for most
of the time for ferrying troops or supplies. Partisan operations and
their support became the principal role for both the airborne troops
and the air transport fleet in the later stages of the war. The organization of air support of all kinds into air armies allotted to Fronts, which
exercised command over them, proved highly effective, the composition of the air armies being flexible and decided by the *Stavka*
for each major operation. The head of the vvs at the outset of war
was P.F.Zhigarev, who was succeeded by A.A.Novikov in 1942. Both

were members of the *Stavka* and, as Marshals of Aviation, had much the same standing as Voronov, Marshal of Artillery.

The method of command which the Red Army developed proved flexible and effective. The *Stavka*, all of whose decisions and orders required Stalin's approval, decided overall strategy, issuing directives and allotting forces to the Fronts, which had to produce their plans for its approval. The Front commander was then given a fair degree of freedom of decision in implementing the plan, co-ordination between the action of neighbouring Fronts being exercised, on behalf of the *Stavka*, by a senior Marshal. The restrictions imposed on the commander's freedom of action by the political member of his Military Soviet was, as we have seen, reduced in 1943, but as victory approached, and especially as the army entered areas and countries where the political issues were highly charged, their influence was restored. In spite of their inexperience in handling large modern forces, the senior commanders and staffs proved themselves at least the equals of the Germans as strategists and practitioners of what they both called 'the operative art'.

Although casualties in men and equipment were huge, the war was never thought of as one of attrition, but of movement. The huge flat areas of Russia lent themselves to that. The Red Army was never will-ingly for long on the defensive. Bold pincer movements to encircle the enemy were the norm ; Shock Armies forced the breakthrough, Tank Armies exploited it to envelop the enemy, while the Shock Armies followed up to tighten their grip on the ring the tanks had drawn. Rifle Armies mopped up and carried the burden of holding the line where and when offensive operations were not in progress. Frontal assaults were, if possible, avoided, proving very expensive in spite of the Red Army's preponderance in artillery and air support from 1943 onwards. The Russian soldier proved as brave and tough, patiently enduring extremes of hardship and danger, as he had shown himself in Imperial days.

The Soviet navy played an insignificant part in the war. Its Northern Fleet, based at Murmansk, which absorbed the White Sea Fleet at Archangel, consisting of a few destroyers and submarines, did little to help the British convoys which struggled round the north of Norway to bring supplies to its ally. The Baltic Red Banner Fleet, commanded by Admiral V.F.Tributs, was restricted by German domination of the air space and of its access to the Gulf of Finland until late in 1944, by which time it was ice-bound. Only its air arm, built up to some 300–400 aircraft, was able to take an active part in the operations to clear the Baltic coast.

Admiral F.E.Oktyabrsky's Black Sea Fleet did better. Its one old battleship, five cruisers, fifteen destroyers and up to fifty submarines had a clear superiority over the naval forces Germany and Romania could bring to bear. But most of its successes were claimed by its 500-strong air arm, operating twin-engined IL4 bombers, carrying torpedoes and bombs from land bases. German and Romanian shipping sailing to the Crimea suffered severely from their attacks. For obvious reasons the Far East Fleet, based at Vladivostok, played no part until the Soviet Union joined in the war against Japan, and some of its submarines were transferred to the Northern Fleet via the Arctic. The Red Navy's large submarine fleet, the largest one in the world in 1939 with some 250 boats, achieved little, sinking a hundred ships for a loss of eighty submarines, forty in the Baltic and twenty each in the Black Sea and the Arctic.

The Red Army had covered itself in glory. Its achievements appeared to have justified the political and economic regime which produced, controlled and kept it in the field, providing it with huge quantities of modern weapons. The prestige which its victories had conferred on it made it appear as a threat to the other rivals to power, the Party and the NKVD, and Stalin was as suspicious as ever of the danger of it becoming a power base for a rival. The first military figure to incur this suspicion was Zhukov. The hierarchy was reorganized in March 1946, the Defence Commissariat being renamed Ministry of Defence, Bulganin, although a political officer having acquired the rank of Marshal, being appointed Minister with Vasilevsky as Chief of the General Staff, covering both the army and the navy, and Zhukov was named C-in-C of the Ground Forces. Four months later, however, Zhukov was banished to the command of Odessa District.

1947–86

By the end of 1947 the 500 divisions with which the army had finished the war were reduced to 175, but still contained some three million men, a third of whom were stationed outside the Soviet Union borders, ensuring its control of the liberated countries of Eastern Europe. From them, in particular from the Soviet-occupied zone of Germany, the USSR obtained technical personnel and equipment of incalculable value in the development of new equipment for their forces, notably in the design of submarines, rockets, jet aircraft engines and electronic and optical equipment of all kinds, helped also by access to the equipment of her wartime allies. The use of the two US atomic bombs in Japan in 1945 gave added impetus to the Soviet Union's own development in this field, which bore fruit in the explosion of her first atomic bomb in 1949 and a fusion weapon four years later, only a year behind the USA.

Until the death of Stalin in 1953 there was no fundamental change in the outlook and organization of the armed forces. An increasing emphasis was given to air defence in the light of the threat from the

US Strategic Air Force, augmented by Britain's Bomber Command. Hitherto the Soviet Union had had no comprehensive early warning system, and the task of providing one to cover the whole vast area of the Union and the territories its forces occupied in Eastern Europe, linked to an effective method of engaging high-flying aircraft, was a formidable one. The other new field to which emphasis was given was the navy, in particular the submarine arm, to which priority was attached as a means of countering the threat posed by US aircraft-carriers. While the Soviet Union was clearly inferior to the USA in numbers of nuclear weapons and means of delivering them, but also because neither the numbers available nor the current means of delivery appeared to demand any fundamental change from the strategies employed in the recent war – and Soviet political leaders and military men were no different from their former allies in adopting that attitude at the time – emphasis was laid on the principles and methods of warfare which had brought victory.

The death of Stalin was a turning-point. It was already apparent that nuclear weapons were going to become available in far greater numbers than had earlier been envisaged, and could be made much smaller. Progress in the development of rockets (or missiles as they became known) showed great promise as a means of delivery, as it did also as a replacement of the anti-aircraft gun. The Russians had always been interested in rockets and their own development had been greatly helped by the personnel and equipment they had acquired from German test centres at Peenemünde and Heidelage and the factories engaged in production of the V 2.

The five years from Stalin's death to the time when Khrushchev took over as Prime Minister from Bulganin in 1958, combining that post with his own as First Secretary of the Party, was one of flux in military matters. Georgi Malenkov, who had briefly held both posts, hanging on to the premiership until 1955, took the view that the possession of nuclear weapons by both sides cancelled out their influence, and that limited conventional wars were the only ones that needed to be provided for. This pleased neither the old guard nor those who looked to nuclear weapons and rocketry for the future. When Bulganin replaced Malenkov as Prime Minister, Zhukov returned from obscurity to take his place as Minister of Defence, but soon found himself at loggerheads with Khrushchev, as Admiral N.G.Kuznetsov already was. It was much the same argument as was going on in the Ministries of Defence of the Soviet Union's former allies. The political and economic cost in terms of manpower and material resources of maintaining large forces of the Second World War pattern did not seem to be justified by their likely employment. The nuclear weapon, delivered by unmanned missiles, seemed to be both a much more effective and more economic use of defence resources. The large conventional forces then maintained could be reduced in favour of effort devoted to this new form of warfare. Khrushchev, in economic difficulties and anxious to improve the general standard of living, took this line, believing also that the Soviet Union's lead in large ballistic missiles, given dramatic prominence by the launch of the *Sputnik* into space in 1957, made it possible for him to take a more aggressive line in international affairs. Emphasis on the primary part to be played in strategy by the nuclear weapon was reinforced by a prolonged study, carried out between 1953 and 1957, by the Frunze Military Academy and endorsed by the General Staff after two further years of study. Its conclusion, incorporated in a book, said to be the work of the Chief of the General Staff in 1960, Marshal Sokolovsky, was that the nuclear weapon should no longer be considered as supplementary to other arms, but that their actions should be designed to exploit its use.

Zhukov's resistance to these new ideas, combined with his penchant for self-publicity and interference in affairs other than defence, led to his dismissal in 1957, followed by Malinovsky, who was replaced as C-in-C of the Ground Forces by Marshal A.A.Grechko. Kuznetzov had already been replaced as C-in-C of the Navy by Admiral S.G.Gorshkov. In this period there was a major reorganization of the armed forces with the creation, in 1954, of the National Air Defence Command (PVO *Strany*) and, in 1960, the Strategic Rocket Force (RSVN), which took precedence over all the other forces, the ground forces, the Air Defence Command, the navy and the air forces. At the summit of the defence organization was the Defence Council, headed by the Minister of Defence, himself a member of the Council of Ministers, although both Khrushchev and his successor Leonid Brezhnev were recognized as exercising supreme command. The Chief of the General Staff, the C-in-C of the Warsaw Pact Forces, and the Chief of the Main Political Directorate (GLAVPUR) formed, with the Minister, the Main Military Council which exercised direct control over these separate armed forces although, for a time, the ground forces were separately controlled. Khrushchev's attempt to impose severe reductions on the conventional forces contributed to his downfall in 1964, when Brezhnev, who was already President of the Praesidium, became First Secretary of the Party, with Alexei Kosygin as Prime Minister. Grechko succeeded Malinovsky as Minister of Defence, holding the post until his death in 1976, when he was succeeded by Ustinov, whose career had been in arms production.

The dramatic increase in the number of nuclear weapons and their means of delivery on both sides, including the development by the Soviet armed forces of battlefield weapons to support their ground forces, made even the most hard-bitten hawks realize that war between the 'socialist' and 'imperialist' camps would result in disaster for both. The dogma that war between the two systems was inevitable was discarded in Brezhnev's day. Although there was no departure from the concept that, as the two systems were irreconcilable, a war between them could not be limited and must involve the unrestricted use of nuclear weapons, there was a reaction against Khrushchev's attempt to belittle the part played by conventional forces. From 1964 onwards the emphasis was changed from mass to modernization, and it was applied in all fields.

The Soviet navy led the way in equipping its new surface vessels with missiles as their main, if not their sole, armament, adding the cruise missile, of which the German V 1 had been the first operational system, to wire-guided, beam-riding and inertia-controlled surface and submarine weapon systems, the latter also following the US navy in the development of submarine-launched ballistic missiles for nuclear weapon delivery. Under Gorshkov's energetic direction the Soviet navy developed into an ocean-going force with temporary anchorages or bases enabling it to escape from the limitations imposed by the territorial and climatic restrictions on access to the oceans from its bases. By the 1980s it had about 300 submarines of which nearly a third carried ballistic missiles, over 200 major surface ships and 1,200 smaller vessels. Its naval air arm, the majority of which was land-based, but included two small carriers with helicopters and vertical take-off aircraft, had some 800 aircraft. The Soviet navy, after eighty years, had restored the position lost by the Imperial Russian navy to the Japanese Fleet at Tsushima.

The air force, which had never escaped from the army, is not an operational force in its own right. Although it equips, controls, trains and administers all units equipped with aircraft, the units themselves come under the operational command of the other forces : long-range aviation (800 strong) other than naval, under the Strategic Rocket Forces, air defence units (2,700+ strong) covering the homeland under the National Air Defence Command, naval aviation under the navy, and tactical aircraft of all kinds (4,700+), grouped together as 'Front Aviation', under the operational command of Fronts and 'Groups' of Fronts. Only the Air Transport Force might remain directly under operational command of the C-in-C of the Air Forces, but many of its 1,300 fixed-wing aircraft and 2,000 helicopters would be placed under command of army Fronts. The quality of its aircraft has been progressively improved over the years, until they can compete on equal terms with those of the USA, and in some respects are considered to have an advantage. They still bear the names of the famous designers, Mikoyan, Tupolev, Sukhov, Yakovlev, Antonov and Mil.

New missile designs, as delivery vehicles for both nuclear and conventional warheads for the strategic rocket forces, for air defence, for the ground forces and for the navy, have followed each other rapidly, not always with success, some being discarded, some soon replaced by new designs, so that the total arsenal consists of a bewildering array of different types. In the strategic field the Soviet Union has maintained its preference for large missiles, capable of projecting a heavy load into space. The principal step forward in that field, as with missiles of a shorter range, was the development of solid- (or storable liquid-) fuelled rockets, the technique of 'cold launch', in which the rocket is not ignited until it has left the silo, and mobility for missiles of considerable range, the SS 20 being the latest development. The Soviet Union established a superiority in numbers of land-based missiles over the USA in the early 1970s which, with their superior 'throw-weight' and the development of multiple warheads, was only counterbalanced by the US superiority in numbers of submarine-launched ballistic missiles and strategic aircraft, capable of delivering nuclear weapons.

Modernization of the ground forces placed emphasis on mobility. In 1967 conscript service for the ground forces was reduced to two years (remaining at three for the navy and air force), the annual intake of recruits being set at one and a quarter million, providing a total strength of 1,825,000, distributed among four groups of forces in Eastern Europe and sixteen military districts within the Soviet Union. The whole army was gradually mechanized, divisions being either tank or motor rifle, the former having one motor rifle and three tank regiments, a total of 325 tanks : the latter one tank and three motor rifle regiments, but with almost as many tanks – 266. In addition there are seven airborne divisions. The corps was abolished, divisions being subordinated direct to armies. Divisions are classified in three categories : 1, at full strength in peacetime ; 2, at 60–75 per cent strength in men, but with a full complement of equipment ; and 3, at 50 per cent strength and not fully equipped. All the divisions (sixteen tank and seventeen motor rifle) in the four groups outside the USSR in Eastern Europe are Category 1, as are the airborne divisions. Of the rest, approximately one-third is in each category.*

The day of massed attacks by foot-slogging infantry, vulnerable to small-arms fire, has gone. The experience of 1941 and the impossibility of erecting fixed defences to cover the immense frontiers of the Soviet Union and its satellites has wedded the Soviet General Staff to belief in the offensive as the only sound form of defence. As much in the

---

* These are normally distributed as follows : European USSR – 23 tank, 36 motor rifle, 4 airborne : Southern (Caucasus) USSR – 3 tank, 19 motor rifle, 2 airborne : Central (Volga and Urals) USSR – 1 tank, 5 motor rifle ; and east of the Urals 7 tank, 5 motor rifle and 1 airborne. Their equipment consists of approximately 42,000 tanks and the same number of armoured personnel carriers, tracked in the tank and wheeled in the motor rifle divisions ; 20,000 field and heavier guns ; 8–10,000 heavy mortars and rocket launchers, and the same number of anti-tank weapons, and 6–7,000 anti-aircraft guns and missiles.

strategic nuclear as in the conventional field their organization, training and doctrine is based on the need to be able to pre-empt a suspected
intention by an opponent to attack by taking the offensive themselves
and gaining a rapid victory, preferably before the enemy has been able
to employ his nuclear weapons. But the tendency of this to produce
a state of trigger-happiness is balanced by the equally strong belief
that one should not go to war unless the 'correlation of forces' is in
one's favour. By this they mean that, in assessing power, one must
take into account not merely the military forces, but the political,
economic, social and ideological dimensions of power, as well as psychological factors such as morale; and that it is important to take
into account not only the strength of the principal nations and
alliances but also the influence of medium-size and smaller powers,
and the strength of non-aligned countries in the international system.
All their activities are directed towards ensuring that the correlation
of forces, in the world as a whole and in a specific area, is favourable
to them; and, in theory, they will only take action when this is so.
They see military strength as playing a very important part as a political instrument; and to be effective as a political instrument they
believe that it has to be seen as being effective in war. In common
with other military thinkers they see the fundamental function of
politico-military policy or strategy as being to give the political leaders
of the nation the greatest possible freedom of action in both the political and military field. They realize that it is the party which is the
strongest that can determine the limits of action. Their belief in the
offensive, and the importance to it of surprise, is with difficulty reconciled with the attitude of caution induced by the need to ensure that
the correlation of forces is in their favour. On the few occasions on
which the Soviet armed forces have been engaged in an active operation
since 1945, they had every reason to judge that this was the case.
Neither in East Germany in 1953, nor in Hungary in 1956 nor in
Czechoslovakia in 1968 was there any danger of serious resistance or
external intervention. No doubt the same assumption was made over
Afghanistan in 1979, intervention by Iran, possibly supported by the
USA, having been ruled out by the revolution there.

   Within the armed forces the influence of the Party is all-pervasive.
All ranks are subject to the Party's political indoctrination from the
time that they undergo pre-military instruction at school, in the case
of officers perhaps in a full-time military school, to the last day of
their service, the commissar or *zampolit* being responsible at every
level of command for political indoctrination and almost all those
aspects which in western armed forces would come under the heading
of man-management. About two-thirds of all officers, and probably
all senior ones, are members of the Party, as well as a high proportion
of the regular non-commissioned officers. The bulk of the latter, how-

ever, are conscripts, selected on joining as recruits, and although they may have joined the *Komsomol*, the youth branch of the Party, they are unlikely to have qualified as Party members before the end of their conscript service. The political directorate is determined to ensure that the armed forces owe their first allegiance to the Party, not to the nation, which to some of the ever-increasing non-Russian population of the Soviet Union could have serious implications.

There is no doubt about the size of the Soviet armed forces and of the quality of their equipment. It is less easy to judge how well its soldiers, sailors and airmen would perform in practice. By the standard of the instruction they receive, the length of their service, and the traditional fighting qualities they have inherited, there is every reason to suppose that they would prove formidable in combat. Some disaffected officers who have escaped to the west have painted a picture of incompetence, corruption and drunkenness, reminiscent of the worst elements of the Imperial Russian army and navy, but the impression given by those elements with which western military personnel have come into contact is one of a well-trained, well-disciplined and tough fighting force, whose lack of experience of major operations is no greater than that of their potential opponents. The Soviet armed forces of the 1980s undoubtedly represent the most formidable war machine the world has ever seen.

6

THE UNITED STATES OF AMERICA

Pre-1900–22

The start of the twentieth century marked a resurgence in the affairs
of both the United States navy and the army after a thirty-year period
of doldrums following the Civil War. Since at least 1890 the USA
had become the leading industrial nation of the world and was flexing
its muscles, determined that its European rivals should not exclude
its commerce from trading freely in any part of the world and should
not, in pursuit of their own commercial interest, interfere in the affairs
of the American continent.

It was naturally the navy which took the lead in this field. At the
end of the Civil War it had manned 671 warships, mounting 4,610
guns, with 7,000 officers and 51,500 men. By the end of 1867 it had
been reduced to 238, of which 103 were in active commission, manned
by 2,000 officers and 11,900 men; and by 1875 it was down to 147,
a large proportion of which were not in active commission, twenty-six
of them sailing-vessels without auxiliary steam power. The US navy
had not only shrunk in numbers to rank twelfth in the world, but
it was slow to change from sail to steam and from wooden to steel
hulls. The introduction of mechanical propulsion led to a prolonged
contest between 'line' and 'engineer' officers as to their respective
status and responsibilities, particularly in influencing the design of
new craft.

The US Congress kept a firm control over the naval lobby which
sought to build a fleet capable of challenging those of the European
powers. In doing so, it pursued a realistic strategy, based on isolation-
ism from the affairs of Europe; abstention from imperialist colonial
ambitions; maintenance of control over the American continent in
pursuance of the 1823 Monroe declaration; and the vigorous pursuit
of commercial opportunities wherever they presented themselves.
This appeared to require only a small navy, but one that could operate
in small numbers in every ocean, its primary task being to support
US trade and, in the event of active hostilities, to pursue a guerre

de course against that of its opponents. For these tasks large ships, carrying heavy armament and requiring constant supplies of coal, did not seem appropriate, and it is not therefore altogether surprising that the US navy was slow to abandon sail and enter the steam age and the battleship race which characterized the last decades of the nineteenth century.

Nevertheless influences were at work which forced the pace. Naval officers themselves, of whom there was a surplus which could not be employed at sea, were frustrated by the growing obsolescence of the fleet. The establishment of the Naval Academy at Annapolis in 1845 had produced a generation of officers who took their profession seriously and studied with interest and envy developments in other navies. Prominent among these was Alfred Thayer Mahan, who had graduated from Annapolis in 1859 and served at sea in the Civil War. He was among those who founded the US Naval Institute in 1874, pressure from which led to the establishment two years later of the Naval War College, to the staff of which Mahan was appointed, remaining there, twice as President, for most of the rest of his career. His lectures formed the basis of his famous book, published in 1890, *The Influence of Sea Power upon History 1660–1783*, the principal theme of which was that the true function of a navy was to concentrate its resources on the destruction of the enemy's main battle fleet. That gave one command of the seas from which all else, protection of one's own trade and destruction of the enemy's, followed. It was a powerful plea for a reversal of the policy which Congress had imposed on the US navy since the Civil War.

External factors were working in the same direction. European involvement in the affairs of South and Central America, arising out of commercial activities, led not only to the possibility of clashes between them, but also to their intervention in the affairs of these states. Similar rivalries had developed in China and were extending into the Pacific, the growing strength of Japan adding to the problem. With the development in European navies of ocean-going steam-propelled steel-hulled ships, mounting powerful breech-loading rifled guns, the US navy was less and less capable of effectively carrying out even its limited role. At the same time US commercial interests were expanding and were threatened by the extension of trading restrictions, imposed by the imperial regimes of the European powers as they spread their tentacles all over Asia and Africa and into the Americas themselves.

In the latter, the last remnant of European imperialism was Spain, clinging on to Cuba and Puerto Rico. Since the revolt in Cuba in 1873, American sympathy for the rebellion had been influenced by the desire to see this last remnant of European imperialism in America removed, and the US navy became increasingly concerned to acquire bases in the Caribbean from which it could support the Monroe doctrine. Its concern for the need to further US interests in the Pacific generated pressure not only for an adequate fleet for that ocean, drawing its eyes towards the other Spanish colony in the Philippines, but also to the construction of a canal to make it possible for the fleet, traditionally based on the east coast, to be transferred if need be to the west, and *vice versa*.

The decisive shift in US naval policy from *guerre de course* to battle fleet took place with the passage of the Naval Act of 1889, by which Congress authorized the construction of three 'sea-going coast line battleships', six more being authorized in the next few years. In 1895 renewed insurrection in Cuba prompted the preparation of a plan both for an expedition to Cuba, involving the transport of an army of 55,000, and one to the Philippines. The Naval War College, faithful to the tenets of Mahan, resisted this division of resources, particularly as relations with Britain were cool, arising out of a dispute over the frontier between Venezuela and British Guiana. Japanese ambitions in the Pacific also caused concern, after two visits by their warships to Hawaii in 1897.

The climax to these concerns came on 15 February 1898, when the battleship *Maine*, sent to Cuba at the request of the US Consul-General, blew up at its moorings in the harbour of Havana. Two months later the United States was at war with Spain. In spite of the objections of the Naval War College, Commodore Dewey led a force of only six cruisers and gunboats to the Philippines and, sailing through the narrow entrance of Manila Bay on the night of 30 April, destroyed the whole decrepit Spanish squadron there. Meanwhile a Spanish fleet of four armoured cruisers and two destroyers under Admiral Cervera, sailing from the Cape Verde islands, had eluded the two squadrons of the US Atlantic Fleet, Rear-Admiral Schley's, based at Hampton Roads, Virginia, and Rear-Admiral Sampson's at Key West, Florida, and reached Santiago on the south coast of Cuba. There it was blockaded by the combined squadrons, reinforced by the battleship *Oregon* which had sailed round Cape Horn from the west coast. When Cervera attempted a sortie on 3 July his squadron was annihilated; the city surrendered on 16 July, and the war came to an end a month later, followed by Spanish withdrawal from both colonies, over which the United States assumed sovereignty.

The war appeared to have justified Mahan's teaching, and the US navy enthusiastically set about building up a fleet to exercise command of the seas off its eastern seaboard, especially in the Caribbean, and one to assert US influence in the Pacific. It also started putting its own house in order. The long-standing quarrel between 'line' and engineer officers was settled by the passage of the Naval Personnel Act of 1899. A General Board of the Navy was established in 1900, headed by Admiral of the Navy Dewey, and consisting of the chief of the Bureau of Navigation, the president of the Naval War College, the chief of naval intelligence, and eight other senior officers. However, this did not meet the demand of the Naval War College for the establishment of a true naval staff, based on the model of the German General Staff. Officially it was only an advisory board to 'advise the Secretary of the Navy on war plans, bases and naval policy' and to co-ordinate the work of the War College and the Office of Naval Intelligence, the latter founded in 1882. The traditional bureaux fought successfully for their independence, but the Board came to be accepted as representing the voice of the navy.

The early years of the twentieth century were dominated by concern over the Caribbean. There was another crisis over Venezuela in 1902, where German and British intervention to secure their debts raised fears of German ambitions. This was immediately followed by the Panamanian revolt against Colombia which opened the way for the construction of the canal abandoned in 1889 after a disastrous ten-year attempt by a French company. Under the impetus of Theodore Roosevelt, who became President in 1901 and had been Assistant Secretary of the Navy in 1897–8, construction was started in 1908 and completed in 1914. In the early years of his presidency, the navy's principal concern was to acquire bases in the Caribbean, its earlier enthusiasm for acquiring several being tempered by the problems both of cost and of security, and its ambitions bringing it into conflict with the army. It finally settled for Guantanamo in Cuba and Culebra off Puerto Rico, which had also been acquired from Spain.

Up to and during the Russo-Japanese War, concern over the threat from Japan had been allayed by the occupation of Hawaii and the assumption that Japan acted as a counterbalance to Russian ambitions in China which threatened America's Open Door policy. But in 1905 the defeat of the Russian fleet at Tsushima and the return of the Japanese to Port Arthur restored fears of Japanese ambitions in the Pacific, the establishment of a naval base at Subic Bay in the Philippines going some way to providing a counter to them. To emphasize American interest in the area, a major cruise of battleships, known as 'The Great White Fleet', left San Francisco in July 1908, sailing round the world to Hampton Roads by February 1909.

The problem of the disposition of the fleet between the Atlantic and the Pacific, and of where its base or bases should be in the latter, gave rise to keen discussion. Mahan argued that the *rapprochement* with Britain and the increasing involvement of the USA in Asian affairs demanded concentration in the Pacific. This was strongly opposed by east coast vested interests of all kinds, as well as those who were beginning to identify Germany as a major threat. Although the affairs of the unstable countries bordering the Caribbean were of great concern in this period, with revolutions in Mexico in 1910 and in Nicaragua in 1912, from 1907 onwards Japan was seen as the major threat and the war plan to deal with it, codenamed ORANGE, went through several changes. The original version envisaged the Atlantic Fleet sailing east through the Suez canal to Subic Bay to join hands with the Pacific Fleet, whose main base was to be Pearl Harbor in Hawaii. By 1913 the fear of German activity in the Caribbean made President Woodrow Wilson reluctant to envisage transfer of the Atlantic Fleet to the Pacific until the German fleet had either been dealt with or could be definitely left to the attention of others. Until that could be assured, the Pacific Fleet would have to accept the possibility of the loss of the Philippines and Guam, and restrict itself to defensive operations based on the Aleutians, Hawaii and, after it was opened in 1914, the Panama Canal. The dual threat was a powerful argument for setting the target of forty-eight capital ships for the combined fleets, to be achieved by 1920.

The function of the US army in the period between the Civil War and the conflict with Spain was far less clear-cut, the principal source of dissension being the relationship between the regular army and what was vaguely embraced by the term 'militia', and the form which the latter should take. Ever since the American War of Independence, which had been fought and won by an army of militia, the prime task of the regular army had been to protect westward expansion across the continent against the Indians. The Napoleonic wars, culminating in the 1812 war against Britain, had proved the need for it also to provide the nucleus of an army, based on mobilization of the militia, to meet an external threat. The militia's obligation of universal service, although theoretically still in existence in the nineteenth century, had in practice died out and been replaced by volunteer companies on the model of Volunteers in Britain: middle-class and better-off working-class men, who provided their own uniform and equipment. They were absorbed into the state militia and it was they who formed the basis of the armies raised by the states on both sides in the Civil War, led by the products of the Army Academy at West Point, founded by President Thomas Jefferson in 1802.

The immediate aftermath of the Civil War found the army of the Union occupying the states which had formed the Confederacy, acting as the principal agent of the Federal Government in executing its policy of reconstruction. Its task was complicated by a sharp difference between President Andrew Johnson and Congress over reconstruction policy. The former pursued a policy of reconciliation, encouraging the resurrection of state forces and making little attempt to prevent a return to the status quo ante in every field. Congress was determined that the South should not revert to its bad old ways. Caught in between, the army found itself treated with contempt by the people over whom it was supposed to be exercising Federal authority. Once the war was over the obligation to continue service in a state-raised force lapsed, and volunteers, either individually or as a unit, had to engage for further service under Federal authority.

By 1869 the reconstruction task had almost disappeared, and by 1876, the year of Colonel George Custer's last stand against the Sioux Indians at Little Big Horn which, in spite of its disastrous outcome, virtually marked the end of the Indian wars, the total authorized strength of the regular army had been reduced to 27,442, which included two Negro infantry and two cavalry regiments, as well as one thousand Indian scouts.

This period of doldrums, as far as activity and numbers were concerned, coincided with one of significant intellectual activity. Dennis Hart Mahan, father of the naval historian, graduating from West Point in 1824 and returning there as Professor of Engineering only six years later, had an even greater influence over the army than one of his predecessors, after whom his son was christened, Sylvanus Thayer. Over the next forty-one years he inspired cadets to take their profession seriously. One of them was William Tecumseh Sherman who succeeded General Ulysses S. Grant as Commanding General of the army in 1869. Holding that post until 1883, he was instrumental in furthering the higher study of war, the School of Application for Infantry and Cavalry being established at Fort Leavenworth in 1881, which developed into the General Service and Staff College with the encouragement of Colonel Arthur Wagner, who was also responsible for the foundation of the Army War College. Wagner and an even more influential military intellectual, Emory Upton, were firm admirers of the Prussian army. Upton, commandant of cadets and instructor in tactics at West Point, was sent abroad by Sherman primarily to study the British experience in India. As an afterthought he was told to study the European armies on his way home. In his report he recommended that the US army should be modelled on the Prussian army, the regular army to become a cadre for expansion in war by citizen volunteers, and command to be exercised by regulars. He was contemptuous of the value of National Guard units, commanded by their own volunteer officers, and of the confusion caused by the division of responsibility for them between federal and state authorities. Regular units would be stationed in the territorial districts from which, on mobilization, they would draw the volunteers for expansion to wartime strength. He complained that the US army was too much under the control of civilians and recommended the establishment of a General Staff, product of military schools of instruction, which would exercise control over all aspects of the army's organization and activity. Upton died in 1881 before he finished his book *The Military Policy of the United States*. Completed by a friend, it was not published for another twenty years, during which a continuous argument raged between his supporters and the traditionalists, who fought both for the National Guard as the wartime reserve and independence of the separate bureaux, such as the Adjutant-General's and the Quartermaster-General's, each directly responsible to the Secretary of War and with their own unofficial links with Congress.

When Grant became President in 1869 he promised Sherman that the latter would enjoy the same authority as Commanding General as he had wielded, but he did not live up to his word. Throughout the period of his command Sherman was engaged in arguments and struggles for power both with the Secretary of War and with the Adjutant and Quartermaster-General. The issue had still not been satisfactorily settled when the Spanish-American War broke out in 1898.

At the outbreak of that war the army was extremely ill-prepared for action. Its total strength was 2,143 officers and 26,040 men. It was both ill-organized and ill-equipped for modern war. In spite of pressure from progressive officers it was still scattered about in over a hundred posts, few of them holding more than a company or two, although in recent years some regiments had been able to assemble all eight of their companies together for training. The traditional task of fighting the Indians had called for a force consisting almost exclusively of cavalry and infantry. The artillery had concentrated its attention on coast defence, the replacement of the forts guarding ports on the east coast, which had proved vulnerable in the Civil War, having been given a high priority, although it was hard to imagine who would attack them.

The National Guard consisted almost exclusively of infantry. A new rifle, the Danish .300-calibre Krag-Jörgensen, with a five-round magazine, replaced the single-shot Springfield in 1893, and 53,508 were available, as well as 14,895 carbines for the cavalry. There were therefore enough for the regular army, but not for the National Guard. Field artillery was however seriously out of date. Argument started immediately about the method by which the army should be expanded in order to send expeditions to Cuba and the Philippines. The professional officers of the War Department wished to follow Upton's line, to expand the regular army to 104,000 men by filling the ranks of existing companies and expanding regiments to three battalions of twelve companies each with volunteers raised as a federal force, controlled by the regular army. Congress compromised, authorizing an increase in its strength to 64,719, but any militia unit which volunteered as a body would be accepted and states were allowed to raise new units with locally appointed officers. As a sop to the Upton viewpoint, the Secretary of War was authorized to organize volunteer units 'possessing special qualifications' up to a total strength of 3,000. When this turned out to be three regiments of volunteer cavalry, one of which was Colonel Leonard Wood's Rough Riders in which Theodore Roosevelt was to gain valuable publicity, 3,500 engineers were added. Congress also authorized the formation of brigades, divisions and corps.

Acting on the authority provided by Congress, President McKinley called for a total of 125,000 volunteers. V Corps, sent to Cuba under Major-General Shafter, consisted of two infantry divisions and one of dismounted cavalry, manned by 14,412 regulars and 2,465 volunteers, including the Rough Riders and the 2nd Massachusetts and 71st New York National Guard regiments. The only fighting, the attack on Santiago, showed up the inferiority of the American artillery and the Springfield rifles of the National Guard, and the expedition revealed the inefficiency of the army's transport, medical and supply organization. There was chaotic congestion on the railway lines leading to the Florida port of Tampa, from which the expedition sailed, casualties from yellow fever were high and the canned beef provided as rations was almost uneatable.

VIII Corps, which sailed from the west coast to the Philippines, consisted of National Guard regiments, all but one from western states. It saw hardly any fighting against the Spanish, but was soon embroiled in conflict with Emilio Aguinaldo's Tagalog independence movement, involving fighting guerrillas in the jungle, a task for which its soldiers were wholly unprepared. This counter-guerrilla campaign continued and posed a problem when President McKinley, after the armistice with Spain, ordered the discharge of 100,000 volunteers, priority being given to those who had seen action. It proved impossible to apply this to the Philippines, where the National Guard regiments remained until October 1899, to be replaced by regulars. The need to keep soldiers in Cuba, Puerto Rico and the Philippines, to exert American authority in place of Spanish, had persuaded Congress in March of that year to maintain the regular army at a strength of 65,000 and to enlist for a period of twenty-eight months not more than 35,000 volunteers 'to be recruited from the country at large or from localities where their services are needed'.

There was now no doubt in anybody's mind that the army sorely needed modernization. The architect of that was the Secretary of War from 1899 to 1904, the lawyer Elihu Root. Influenced by Upton and the British author Spencer Wilkinson's *The Brain of an Army*, he saw the need for conversion of the post of Commanding General to that of Chief of a General Staff which could both plan in peacetime the measures needed to expand the army rapidly in emergency and exercise control over the execution of those measures. The Army War College would act as the General Staff's 'think-tank'. But he had to move carefully. The current Commanding General, Nelson A. Miles, was not only a stubborn traditionalist, but he had close links with Congress. Cuba and the Philippines were still proving troublesome and, in 1901, Root persuaded Congress to increase the regular army to thirty regiments of infantry and fifteen of cavalry and to reorganize the artillery, hitherto divided between field and fortress batteries, into one artillery corps which nevertheless was heavily dominated by coast defence, 126 companies to thirty batteries of field artillery. With three battalions of engineers it raised the strength of the regular army to 3,820 officers and 84,799 men, and led to the grant of regular commissions not only to volunteers who had shown outstanding ability but also to the product of military colleges other than West Point. A notable entry from the Virginia Military Institute was George C. Marshall.

Other reforms included the rotation of officers between duty on the staff and with their regiments and, in the case of the former, between different branches of the staff. The opposition of Miles prevented the introduction of a General Staff until 1903, Root having to rest content meanwhile with the establishment of a War College Board, similar to the General Board of the Navy. One of his major achievements was the establishment in the same year of the Joint Army and Navy Board to co-ordinate the plans of the two services.

Controversial as these reforms were, they did not attack the central issue: the organization of the reserves by which the US army could be transformed in wartime from a small, long-service, professional force into a national citizens' army. This was resolved by co-operation between Root and Congressman Dick, who represented the interests of the National Guard, resulting in the 1903 Dick Act. This classified all the able-bodied manpower of the states as reserve militia, the actual National Guard units being designated organized militia. The latter would be regarded as a reserve for the regular army and the Federal Government would provide all arms, equipment and military stores for them without charge to the states. To qualify, units had to meet minimum training requirements and be subject to inspection by the regular army, which would provide officers as instructors. In return, National Guard officers could attend regular army schools. When doing so they would draw regular army rates of pay, as would all ranks when their units participated in manœuvres with the regular army. Federal service was limited to nine months, at the end of which a unit could volunteer for further federal service under its own officers. This limit was removed in 1908, when Congress stipulated that the President must call up the Guard before asking for other volunteers, and specify the duration of their service, at the same time recognizing the Guard as a reserve for all wars 'either within or without the territory of the United States', thus removing one of the main planks of its criticism by Upton and his followers. This prompted the army's Judge Advocate General to seek an opinion from the Attorney General as to its legality. The latter took four years to decide that it was unconstitutional to oblige the militia to serve outside the United States. Friction between regular army and National Guard was not removed by the Dick Act, but relations improved and the Guard's standard of training and equipment was significantly raised.

All was not smooth running for the General Staff either. Root had laid down that it should plan and direct, but not operate, and was not to become involved in administration. The result was to bring it into conflict with the Adjutant-General's department which was strengthened by the amalgamation of offices, like the Record and Pension Office, which had previously been independent. The struggle between the Chief of the General Staff and the Adjutant-General was intensified in 1910 when the flamboyant Leonard Wood became Chief and the able and forceful Major-General Frederick Ainsworth was Adjutant-General, both by coincidence having started their careers as army doctors. The power to select and appoint officers for recruiting depots became one of the core issues, and both lobbied Congress for support. Ainsworth finally overstepped the mark and in February 1912, Henry Stimson, Secretary of War, persuaded President Taft that he should be dismissed. Ainsworth's friends in Congress retaliated with a reduction of the General Staff from forty-five to thirty-six officers, with conditions insisting on a minimum regimental service.

The slender resources of the General Staff were greatly assisted by the success of Root's establishment of a whole series of instructional schools, one for each arm, Fort Leavenworth being transformed into the Army School of the Line and throwing off a Staff College. Besides reorganizing the General Staff, Wood, with Stimson's support, tackled the thorny problem of the army's dispersion into a large number of small posts, relic of the Indian wars. He wished to concentrate the army into large stations where units could train together in formations of all arms, but ran up against local vested interests represented in Congress. The civil war in Mexico in 1913 gave him an opportunity, when in response to the President's request for strengthening the army's border patrols he mobilized the 2nd Division in Texas.

Although, in Wood's quarrel with Ainsworth, the latter represented the extreme Uptonian point of view, Wood himself was dissatisfied with the National Guard as the army's sole reserve. Reinforced in his view by the Attorney-General's judgement, he agitated for a separate federal reserve, formed partly by reducing the regulars' term to three years' active duty and adding a three-year reserve liability, and partly by the creation of national militia, in which service would be for two years and officers of which would be provided by expansion of West Point and of military training in colleges. His ideas met with little support from Congress, whose response to the Attorney-General's judgement was to pass The Volunteer Act of 1914, which allowed individual militiamen to volunteer for federal service outside the USA, but stipulated that National Guard units, three-quarters of whose members volunteered, must be preserved intact. The War Department policy which resulted from these pressures was to organize the regular army into infantry divisions and cavalry brigades, which could be employed as an expeditionary force, while a reserve army, based primarily on the National Guard, was being mobilized and given further training. How long the latter should last was a matter of considerable controversy.

Woodrow Wilson's advent to power and the outbreak of war in Europe in 1914 sharpened the debate on these issues. Theodore Roosevelt had prophesied that the USA would not be able to stand aloof from a European war, but Wilson's aim was to ensure that it did, while strengthening his position to act as arbiter and force the participants to change the evil of their ways which resulted in such wars. Agitation by Wood, Root, Stimson and others for universal military training, reinforced by the sinking of the Lusitania in May 1915, forced Wilson reluctantly in July of that year to ask his War and Navy Secretaries to produce a new national security programme. The Secretary of War, Lindley Garrison, sympathized with Wood and company and had already initiated a General Staff study, which recommended expanding the regular army to 142,000 and adding to the National Guard a federal reserve, recruited in three annual increments of 133,000, to a strength of 400,000, to be called the continental army. It would clearly be used before the National Guard. Wilson reluctantly accepted Garrison's plan, but it predictably ran into trouble with supporters of the Guard in Congress, led by James Hay, chairman of the Military Committee of the House of Representatives. Wilson, whose own instincts opposed Garrison's plan, was persuaded by Hay that it would not receive Congressional approval, and switched his support. Hay's rival plan would increase federal responsibility for the Guard, including having a say in the appointment of officers, and required its members to swear an oath to respond with their entire units to a federal call for service anywhere. Garrison resigned, replaced by Newton Baker, who was more acceptable personally to Congress, and Hay's plan, which included an increase in the regular army to 140,000 was approved by the House on 23 March 1916 by 402 votes to 2. When, on the following day, the French cross-channel ship Sussex was torpedoed with the loss of eighty lives, two of them American, the Senate adopted a bill which restored the continental army proposal at a strength of 261,000 and increased the regular army to 250,000.

It was events in Mexico that solved the problem. Anti-American feeling, provoked by Wilson's attempts to influence events there in favour of a democratic solution, led to the arrest of some American sailors on shore leave in Vera Cruz. The US navy retaliated by occupying the city, reinforced by the 2nd Division's 5th Brigade, half of whom

were marines. This accentuated Mexican resentment, one of the contenders for power, Francisco Villa, sending a body of his followers
across the border to shoot up Columbus in New Mexico on 9 March
1916. Baker's reaction was to send a force of 5,000 regulars under Brigadier-General John J.Pershing into Mexico to capture Villa, resulting
in a clash between them and the forces of Venustiano Carranza, whom
the US government recognized as head of the Mexican government.
A truce was patched up, the US agreeing to withdraw Pershing if
Carranza controlled Villa. The latter defiantly crossed the border again
to raid Glen Springs in Texas, bringing the USA and Carranza to the
brink of open war. On 9 May the National Guards of Texas, New Mexico
and Arizona were called out, but they would clearly not suffice if
a full-scale war resulted. In the face of this crisis, the two Houses
of Congress compromised to pass the National Defence Act on 20 May.
It authorized an increase in the peacetime strength of the regular army
to 175,000 over five years.* It adopted Hay's plan for the National
Guard, which would gradually be increased from 100,000 to 400,000,
and it reaffirmed the principle that all able-bodied males between eighteen and forty-five were members of the militia, which was organized
into the National Guard, the Naval Militia and the Unorganized Militia. While the General Staff was increased to fifty-five, including three
generals, its opponents managed to slip in a rider stipulating that
not more than half of the junior officers could be on duty in or near
Washington DC, so that the General Staff branches of the War Department numbered only nineteen officers when the USA joined the Allies
in the war against Germany in April 1917. The National Defence Act
failed to provide all the men needed for the Mexican War, only 158,664
being mobilized. The result convinced Baker that, if the nation became
involved in the wider conflict, conscription would be needed, and
Wilson agreed. The General Staff was set to work in December 1916,
producing a plan which was ready in April and became the Selective
Service Act on 18 May 1917. It authorized the President to bring both
the regular army and the National Guard to the full wartime strength
permitted by the National Defence Act by raising 500,000 men immediately by federal draft, and adding another 500,000 when he thought
it appropriate. Service for all was to be for the duration of the emergency, and the draft was to apply to all male citizens, or friendly aliens
who had declared their intent to become citizens, between the ages

---

* It authorized a wartime establishment of 286,000, forming sixty-five infantry, twenty-
five cavalry, twenty-one field artillery, seven engineer and two mounted engineer
regiments: 263 coast artillery companies, eight aero squadrons and supporting formations.

of eighteen and thirty, later extended to thirty-five. Local boards could defer service for those in industries important to the war effort and those with special family obligations. Volunteers were still accepted, and the prospect of call-up added as a spur, but the draft provided 67 per cent of all those enlisted during the war.

At the time of the act's passage the regular army had thirty-eight infantry, seventeen cavalry, nine field artillery and three engineer regiments, manned by 127,588 men, while the National Guard had 80,446 in federal and 101,174 in state service. In addition the Regular Army Reserve numbered 4,767, the Officers' Reserve 2,000, the Enlisted Reserve Corps of specialists 10,000 and the National Guard Reserve 10,000. The total actually in federal service on 1 April 1917 was 213,557, rising by the end of the war to 3,685,458.

The concern with numbers and how they should be raised was not matched by any great foresight about how they would be provided with weapons and equipment. Only in rifles could this leading industrial nation meet the demand, and that partly because three firms had been producing them for Britain. For almost all other weapons and much military equipment, the US army became dependent on supply from Britain and France direct to Pershing's American Expeditionary Force (AEF) in France. There were enough field guns to equip an army of 220,000, but they were inferior in performance to the French and British and, of the 2,250 guns used by the AEF when fully deployed, only a hundred were American. All its tanks were French.

For a nation that was in the forefront of the development of aviation and motor transport, the army was ill-provided with both and its senior officers reluctant to give either any high priority. Balloons had been used for observation in the Civil War. In 1890 the Signal Corps was given responsibility for them, in 1892 establishing a balloon section which saw active service with one balloon in the attack on Santiago. Experiments were also made with a dirigible, delivered in 1908. After the Wright brothers' successful test of a powered aircraft, the War Department issued a specification in 1907 for a flying machine to carry two men with a combined weight of 350 lb and enough fuel for a flight of 125 miles, a speed of forty mph, and the ability to take off and land in a field without special preparation. The Wright brothers won the contract at a price of $25,000 and delivered their machine on 20 August 1908. Unfortunately it crashed, killing the passenger, Lieutenant Selfridge, and badly injuring Orville Wright who was piloting it. The model was finally accepted in August 1909, the Aeronautical Division of the Signal Corps having been established two years earlier. However its strength remained at one aircraft and one pilot until, in March 1911, Congress appropriated $125,000 for army aeronautics, five more aircraft being ordered immediately. In July 1914 an Army Aviation Section was established from which the 1st Aero Squadron was formed in March 1916 as part of Pershing's force in Mexico. It was equipped with eight Curtiss JN-3s which lacked the power to operate safely at the heights and in the climate and terrain at which the hunt for Villa was carried out. Six JN-3s crashed and the other two were unfit to fly after 540 missions, totalling 346 flying hours.

In April 1917 the Aviation Section, one of whose earliest commanders was Major Billy Mitchell, consisted of 131 officers, almost all pilots, and 1,987 enlisted men, with 2–3,000 aircraft, mostly trainers, of which only fifty-five were fit to fly. By November 1918 the Air Service Army Group, to which Mitchell's command with the AEF had been developed, had 5,000 pilots and observers in forty-five squadrons manning 2,000 aircraft. Almost all his aircraft were of British or French manufacture. The only American-produced combat aircraft was the British-designed DH-4, first test-flown in France on 7 May 1918, three days before the Overman Act removed army aviation from the Signal Corps, establishing it as a separate Air Service US Army under Major-General Kenly as Director of Military Aeronautics. Production was the responsibility of the Army Bureau of Aircraft Production, and the Second Assistant Secretary of War, Ryan, being appointed Director of Air Service to co-ordinate the two. This change was provoked by the serious delays in developing aircraft production on a large scale. The War Department had accepted the advice of America's allies not to get involved in the development and production of fighters, the design of which changed rapidly to meet changing operational requirements, but to concentrate on observation and bomber aircraft and on the production of engines, of which the 12-cylinder Liberty proved the mainstay. The Overman reorganization boosted aircraft production, and by the end of the war US factories in the USA had produced 7,800 trainers, 3,500 combat aircraft, mostly DH-4s, which were coming off the lines at a rate of 260 a week, and 13,500 Liberty engines. The Air Service itself had reached a strength of 11,425.

The size of the army to be sent to France became a controversial issue as soon as war was declared. The General Staff proposed that the regular army should be used to train new formations and that an expeditionary force should not be sent until the strength of the army as a whole had reached $1\frac{1}{2}$ million; but this failed to take account of political and military realities. The Allied effort might have collapsed before the United States could make its presence felt. Four infantry regiments from the Mexican border were formed into the 1st Division and sent, while Pershing, promoted to major-general as the expeditionary force commander, went ahead to set up his headquarters and assess what was required. In July he proposed a target figure of one million men by the end of 1918, forming twenty divisions, which he argued was 'the smallest unit which in modern war will be a complete, well-balanced fighting organization'. The War Department developed his proposal into a plan to provide thirty divisions requiring a total force of 1,372,399 men. The disastrous turn of events in the autumn of 1917 – the failure first of Nivelle's offensive, and then of Haig's, the French mutinies and the Russian revolution – with the prospect of a German strength in 1918 of 250 divisions, against which the British and French would only be able to muster 160, prompted Pershing to raise his sights, after Ludendorff's offensive against the British in March 1918, to three million men in sixty-six divisions by May 1919. After discussion with Foch during the German offensive against the French in June 1918, he raised his target yet again to eighty divisions by April and a hundred by July 1919. The War Department went some way towards his plan by revising its programme to provide fifty-two divisions with 2,350,000 men by the end of 1918, rising to eighty by June 1919, the maximum age for the draft being raised to forty-five.

Total numbers were not the only cause of dissension. The length of the recruit's training before he was considered fit to go to France was another. After much argument, in which Newton Baker took a more balanced view than some of the senior army officers, four months was decided upon. But that was not the end of the story, as Pershing insisted on further intensive training once they got there. A stickler for high standards both of musketry and of general military bearing, he wished to see all his soldiers brought to the standard of the old regular army. The regular 1st Division, hurried to France in July 1917, was subjected to a training programme, including the introduction of battalions to ten-day periods in the front line of quiet sectors of the French front, which delayed its deployment as a division until January 1918, and that also to a similar quiet sector. The 2nd Division, half regular, half marines, the 26th New England National Guard and the 42nd Rainbow, drawn from National Guards of twenty-six different states and the District of Columbia, were all put through the same process.

Pershing was determined that they should not be committed to a major battle until all four could fight together under his command. His resistance to suggestions by the British and French that, in order both to strengthen their dwindling forces and to provide his men with battle experience, units, brigades or divisions should be deployed under British or French command, met with a stubborn refusal. Not until the crisis of March 1918 did Pershing give way, his four divisions being offered to Foch, Pétain tactfully assigned them to a quiet sector of his front. Although Pershing was no doubt motivated by his desire to see the 'minimum force' of twenty divisions employed under his personal command in a decisive operation which would be the climax of the war, he also genuinely, and probably rightly, feared that his troops would not give of their best under British or French command. Not only might they be affected by the dispiriting atmosphere of trench warfare and the general cynicism which experience of it over the previous years had implanted in his allies' armies, but their motivation to prove the American soldier in battle, fighting for an American cause, would be weakened. He sought for an opportunity to demonstrate that the American soldier, using his rifle and bayonet, could take the offensive and restore mobility on foot to a battlefield in which soldiers had ceased to believe that was possible. Tactically he was being unrealistic; but as far as the factor of morale was concerned, his judgement was sound. The organization of his force, however, was not attuned to his tactical ideas. All formations, from platoon up to division, were so large as to pose problems of control to the few and inexperienced officers in command and to the small staffs they were given.

A compromise was reached in June. Faced by arguments of shipping limitations on the build-up of his strength, Pershing reluctantly agreed to priority being accorded to infantry and machine-gun units at the expense of complete divisions. When the German offensive against the French opened on the Chemin des Dames in the last week of May, capturing Soissons and threatening Paris, Foch demanded that it be prolonged indefinitely. In a tense conference at Versailles on 1–2 June, Pershing negotiated an agreement by which, in return for a temporary continuation of the policy he had accepted a month before, priority would be reversed to bring over the other elements, combat and support, to restore balance to his army, which on 10 August he officially designated as the US First Army. It then consisted of three American corps of fourteen divisions, 550,000 American and 110,000 French soldiers with 3,010 guns (1,329 manned by the French), 267 light tanks (all of French manufacture, 113 manned by the French), supported by 1,400 aircraft under Colonel Mitchell, which included the British Major-General Trenchard's Independent Bomber Force, and a French air division of 600 aircraft. The first major action in which US divisions took part was the counterattack in June in the Château-Thierry sector, in which Major-General Hunter Liggett's 1st Corps showed that American soldiers were full of fight, and the Marine Brigade of the 2nd Division suffered heavy casualties in the struggle for Belleau Wood.

From the earliest days Pershing had sought a sector in which his army could operate as independently as possible, taking the offensive in an area where a decisive result could be achieved in terrain which was not honeycombed with the detritus of previous prolonged trench warfare. His eyes turned towards Metz, the important industrial area of Briey-Longwy beyond it, and the railway linking it to Sedan and Mézières. Operating on Pétain's right would suit his logistic arrangements, dependent on French Atlantic ports. As his forces increased in size he saw himself taking over responsibility for the whole front from Metz to the Swiss border, making use also of Marseilles as a supply port. As soon as he could free 1st Corps from the Château-

Thierry sector, he wished to concentrate his forces, the first phase of his offensive being to eliminate the St Mihiel salient between the Meuse south of Verdun and the Moselle south of Metz.

At the meeting on 24 July between Foch, Haig, Pétain and Pershing, Foch agreed, seeing it as one of a series of limited offensives designed to free lateral railways which would be useful for a winter build-up to a subsequent major offensive in the spring of 1919. But the success of Haig's offensive, starting on 8 August, prompted Foch to agree to the former's suggestion that Pershing's offensive should be directed north-westward towards Mézières, as the right hook of a major envelopment of the Hindenburg Line, of which Haig would provide the left arm of the pincer. On 30 August Foch visited Pershing and proposed that, after a minor operation to take out the St Mihiel salient, Pershing's 1st Army should be switched to west of the Forest of Argonne for an attack towards Mézières. A Franco-American Army, under a French commander, would operate on its right in the more difficult terrain between the forest and the Meuse, a French general being attached to Pershing as an adviser. Pershing protested vigorously. After setting out all the arguments why he should attack north-eastward from St Mihiel and then develop operations further south, he said that he would accept Foch's strategic direction provided that American troops were kept together under American command. 'Our officers and soldiers alike', he wrote to Foch, 'are, after one experience, no longer willing to be incorporated in other armies.'

The argument was resolved at a meeting between Pershing, Foch and Pétain on 2 September. The 1st US Army was to attack towards Mézières on Pétain's right, between the Forest of Argonne and the Meuse. Foch wished to start this offensive on 20 September, reluctantly agreeing that Pershing should eliminate the St Mihiel salient beforehand in order to remove its potential threat to his right rear. It left little time for a switch of forces from St Mihiel to the Meuse-Argonne sector and meant that the attack in the latter had to be led by inexperienced divisions. The St Mihiel attack was launched on 12 September, Liggett's 1st Corps on the right bearing the brunt, Major-General Joseph T.Dickman's 4th moving up on the left as Liggett advanced. The Germans began to withdraw as soon as Liggett attacked, but could not make a clean break and lost heavily as they retired in some confusion.

Both Germans and Americans suffered problems from congested roads, but the former managed to get most of their forces out of the trap, the other side of which was held by the French, leaving 15,000 prisoners and 443 guns behind. Pershing's casualties, almost all in 1st Corps, were less than 8,000. Some smart staff work, in which Colonel George Marshall played a prominent part, effected a rapid redeployment of 400,000 men from the St Mihiel to the Meuse-Argonne sector, where the attack was launched on 26 September. Foch had told Pershing not to restrict his advance by the need to keep in line with Gouraud's 4th French Army on his left, but Pershing's orders to his own three corps, Liggett's 1st on the left, Major-General George H.Cameron's 5th in the centre and Major-General Robert L.Bullard's 3rd on the right, linked the action of the outer corps to Cameron's, which he had expected to progress faster. The reverse was the case, and the whole operation slowed down after the capture of the initial objective, and thereafter lost impetus. After the first two weeks Pershing reorganized his army, sacking several commanders including Cameron. On 12 October he formed a 2nd Army, commanded by Bullard, to take over the inactive area east of the Meuse and handed over 1st Army to Liggett, whose corps was taken over by Dickman. Major-General John L.Hines took Bullard's 3rd Corps and Major-General Charles P.Summerall Cameron's 5th. Pershing himself formed an Army Group Headquarters to exercise overall command of both armies.

When the offensive was resumed on 14 October it again met with little success and Liggett decided to halt and carry out a thorough preparation for a major thrust in the centre of his sector. This was launched on 1 November, by which time the progress being made elsewhere on the front and events in Germany itself led to the weakening of resistance. As Pershing tried to switch his effort north-eastward to his old target of Metz, the war came to an end. In the event, it would probably have made no difference to the outcome if Foch had allowed him to drive in that direction from St Mihiel. Pershing's total casualties had been 224,000, of whom 50,000 were killed in an army that by November 1918 numbered 1,200,000. His men had learned the hard way that skill with the rifle and courage with the bayonet were not enough to turn the enemy out of his trenches and impose mobile warfare. They suffered as a result of their commander's lack of enthusiasm for the tank.

While Pershing had been struggling against allied pressure to keep his army together and then gaining the laurels for its success when it was finally deployed in action, battles of another kind had been waged in Washington DC. There the fight between the Chief of the General Staff and the bureaux continued. The age of the two first wartime incumbents of the post, Hugh Scott and Tasker Bliss, and the game of musical chairs they played with each other, one going off on prolonged tours while the other acted for him, meant that they exerted little authority and firm direction over the complex business of the army's rapid expansion and deployment overseas.

The remarkable independent authority granted by the Wilson administration to Pershing, who assumed that all the War Department had to do was to meet his demands, weakened the position of the General Staff and the authority its Chief exercised. The first signs of trouble appeared on the supply front, where the responsibility for procurement and distribution of all forms of equipment and supply was spread over a number of different bodies. Not until Major-General Goethals, who had been responsible for the construction of the Panama Canal, was brought back from retirement to replace Henry Sharpe as Quartermaster-General at the end of 1917, and Major-General Peyton C.March, Pershing's head of artillery, returned from France in March 1918 to replace Biddle as Chief of the General Staff, did matters improve. When March was confirmed in his post and promoted General in May, he made Goethals his deputy and succeeded in establishing the authority of the General Staff across the whole board. The bureaux were defeated and the army's house at last put in order, co-operating effectively with the War Industries Board, which had evolved out of a number of previous attempts to establish a coherent and efficient organization to control procurement for all war purposes.

When Bernard Baruch became its chairman in March 1918, the board at last became an effective and efficient organization. Differences between March and Pershing caused the former a number of problems and each suspected the other of attempting to encroach upon their respective spheres of power. There is no doubt that March made a highly significant contribution to the efficiency of the machine which supplied Pershing with both his soldiers and all their requirements, and he resented the fact that, when the war ended, Pershing received all the glory, with the rank of General of the Army, while he reverted to his permanent rank of major-general. He served on as Chief of Staff, tackling with equal efficiency the demobilization of three and a half million soldiers and the reorganization of the army into its peacetime organization, until his retirement in 1921.

The navy contributed little to the defeat of Germany. It had not seen that as its principal purpose. Its total casualties were only 893 killed or lost at sea and 819 wounded, excluding the Marine Corps which lost 2,461 killed and 9,520 wounded. When war broke out in 1914 the concentration of the British and German fleets, watching each other in the North Sea, removed the threat that either might interfere with American interests elsewhere. The Japanese seizure of German possessions in the Pacific accentuated the navy's concern about that ocean. However, the sinking of the *Lusitania* in May 1915 placed a different complexion on the naval outlook, and President Wilson became a big navy man overnight. The result was the Naval Act of 1916 which authorized a programme leading to a fleet of sixty capital ships by 1925 and 146 other warships, including sixty-seven submarines; ten battleships and six battle-cruisers were to be built in the first three years. This programme was not embarked upon because the General Board of the Navy envisaged participation in the war against

Germany, but because it feared that Japan would form an alliance with whichever power won the war, and that either a German-Japanese or an Anglo-Japanese alliance would attempt to exclude the United States from large areas of world trade. Wilson's motives in supporting this ambitious programme were different. He looked to a strong navy as an influence which he could use, first to bring pressure on the belligerents to end the war, and while it was still in progress not to interfere with neutral rights, and after it to adopt an international organization which would prevent a recurrence. His decision to join the conflict in April 1917 was due more to his realization that he could not achieve these ends, as long as the USA stood apart from the struggle, than it was to take revenge on Germany for the adoption of unrestricted submarine warfare or for her attempts to intervene in the affairs of Mexico.

When that happened Admiral Benson, first holder of the new office of Chief of Naval Operations, resisted suggestions that the naval programme should be changed into one based primarily on the needs of anti-submarine warfare, as recommended by Admiral Sims, the officer sent to London to maintain liaison with the British Admiralty. Wilson backed Sims, and forced Benson to depart from Mahan's insistence that the battle fleet should remain concentrated, by detaching a squadron of dreadnoughts under Rear-Admiral Rodman to join Beatty's Grand Fleet in the North Sea. But even before the war ended American naval suspicion of British intentions was revived when Sir Eric Geddes, British Navy Minister (First Lord of the Admiralty), visited Washington to try to persuade his American colleague that the US navy should continue to maintain a strong anti-submarine capability after the war instead of reverting to the capital ship programme of the 1916 Naval Act. The navy was determined not only to maintain the principle enshrined in that act, that the US navy should be second to none, but that it should maintain a fleet superior to that of Britain. 'In interpreting the basic naval policy [as stated in May 1918],' wrote the US navy's planning section in London, 'we should consider for the present the British Navy as the maximum possible force which we must be prepared to meet. We should always have in mind the possibility of the co-operation of the British and Japanese navies.'

The Navy Department's proposals for 1919 called for ten battleships and six battle-cruisers beyond those already authorized. Wilson supported them, believing that it would strengthen his hand at the peace conference. There he hoped to persuade the participants to agree to a major programme of disarmament, naval parity with Britain being the basis from which the naval element of the programme should start. The position of Japan was of even greater concern to the Navy Department, her demand for recognition of a sphere of influence in China's Shantung province being added to that of a mandate over the Marshall,

Caroline and Mariana islands. The general strategy the department adopted was to maintain a defensive capability against Britain's Royal Navy in the Atlantic and an offensive one against Japan in the Pacific, with 682,000 tons of capital ships in the former and 846,000 in the latter. The navy's proposals fell by the wayside when Wilson failed to persuade the Senate to ratify the Versailles Treaty. Congress refused the General Board's modest proposal for an interim programme for construction of two battleships, one battle-cruiser and twenty-five smaller vessels to start in 1920.

Opposition to naval expansion was confirmed by Wilson's successor, Warren Harding, who reverted to the traditional Republican policy of trying to keep aloof from the potentially troublesome world beyond America. One of his first acts was to summon a nine-nation conference to reduce world tensions by a limitation of naval power. Britain was anxious to avoid another battleship race, which its ailing economy could ill afford, and accepted an agreement which maintained a 5 : 5 : 3 ratio in capital ships between herself, the USA and Japan, a ten-year ban on capital ship construction, a limitation on total naval tonnage and on the size and age of capital ships, and a veto on fortification of bases in the Pacific, Singapore and Pearl Harbor being excluded. The Treaty, signed by the USA, Britain, Japan, France and Italy, also terminated the Anglo-Japanese Alliance and bound the signatories to respect each others' possessions in eastern Asia.

1922–45

Although the General Board had initially opposed any agreement, the US navy gained considerably from it. By scrapping its older ships, completing two battleships under construction and converting two others to aircraft-carriers, and diverting effort from increases in numbers to modernization and to the creation of a balanced fleet, with greater emphasis on aircraft and submarines, the navy not only greatly improved its quality but was forced by circumstances to begin a move away from its devotion to the battleship. The restriction on bases in the Pacific accelerated the change from coal to oil and added emphasis to the submarine and the aircraft-carrier as offensive elements of the fleet. Radio communication was another field in which important advances were made, the navy taking the lead in sponsoring and co-ordinating the civil radio industry, helping to shake it free from dependence on foreign patents and protecting it from foreign competition. Development of aircraft-carriers evaded the restrictions of the Washington Treaty by their classification as experimental vessels, and was stimulated by the fear that an independent air force, following the British pattern, might be established which would deprive the navy of its own air arm, the existence of which was confirmed by Congress in 1926 and developed over the next fifteen years by the Morrow Board, the navy's inventory of aircraft being increased from 351 to 1,000. From 1923 to 1929 the fleet's three aircraft-carriers, converted two from battle-cruisers and one from a collier, took part in fleet manœuvres, their task being regarded primarily as reconnaissance. From 1929 onwards they were often used as an independent carrier group with the primary task of strike action against enemy ships and aircraft. In 1934 the first aircraft carrier, designed specifically as such, was commissioned, the *Ranger* of 17,500 tons. Throughout this period argument raged over the vulnerability of the battleship to air attack, following General Mitchell's demonstrations in 1921 and 1923, when his aircraft sank three old German warships and three obsolete American ones.

In spite of these improvements, the general political climate was unfavourable to the navy's ambitions throughout the Republican administrations of Warren Harding, Calvin Coolidge and Herbert Hoover. The international climate laid emphasis on disarmament negotiations, economic constraints and the general feeling that air power had made the battleship obsolete. To counter it the navy embarked on a major public relations exercise, emphasizing the benefits which a strong navy conferred on opportunities for international trade and the protection of US commerce from foreign interference, and inferring that opponents of a strong navy were unpatriotic. To carry conviction there had to be an enemy. Britain was represented as a serious rival, but the principal villain was Japan. President Hoover was a firm opponent of the navy's persistent pressure for an active interventionist policy in the western Pacific, in which a US naval presence, based on the Philippines, played an important part. He had the army on his side. It doubted the viability of the revised Plan ORANGE, which envisaged the army holding out there against Japanese attack until the US Pacific Fleet could arrive and defeat the Japanese navy in a major naval battle. The navy pursued a chicken-and-egg type of strategic argument over the islands, emphasizing the importance of retaining a base there in order to protect US interests in eastern Asia, and then justifying a large fleet in order to maintain its security.

The tide turned decisively in favour of the navy with the advent to power in 1933 of Franklin Roosevelt, who had been Assistant Secretary of the Navy during the First World War. Not only was he a passionate devotee of everything maritime, but he was also a strong supporter of Chiang Kai-shek's faction in China. By-passing Congress, he allocated $238 million in 1933 from National Recovery Administration funds to the construction of thirty-two naval vessels over three years. In 1934 he pushed the Vinson-Trannell Act through Congress, authorizing an eight-year programme of 102 ships, which would bring the navy up to the full limits of the Washington Treaty. As a result, by 1937 the navy had fifteen battleships, three aircraft-carriers, seventeen heavy and ten light cruisers, 196 destroyers, eighty-one submarines and 1,259 aircraft, manned by 113,617 sailors and 18,223 marines. Under construction were three carriers, one heavy and nine light cruisers, forty-one destroyers and fifteen submarines. Almost all of the active fleet was in the Pacific. Only three old battleships, one cruiser, eight destroyers and six submarines were kept on the east coast, a carrier and two destroyers being sent to European waters to evacuate American citizens from Spain in the Civil War. In January 1936 Japan notified the other signatories that she would no longer observe the restrictions of the Washington Treaty which expired at the end of the year. On the outbreak of full-scale hostilities between Japan and China in 1937, Roosevelt refused to reinforce the Asiatic Fleet in the Western Pacific, except with marines, even after the gunboat *Panay* was sunk by Japanese aircraft off Nanking on 12 December. Yet another revision of Plan ORANGE once again led to arguments between the navy and the army. The former proposed an offensive to capture Truk in the Japanese Carolines, which the latter considered reckless, being reluctant to provide the 20,000 men needed to add to the marines for the planned amphibious assaults. The *Panay* incident changed the political climate into a more favourable one for new construction. Congress did not oppose Roosevelt's January 1938 proposal, originating with the Chief of Naval Operations, Leahy, for three more battleships, two carriers, nine light cruisers, twenty-three destroyers and nine submarines over the next two years, the navy's inventory of aircraft was to be increased to 3,000, to the fury of the army air corps, who got nothing out of the bill. They wanted money for their B-17 bomber, which had a range of over 700 miles, while the navy tried to get the army to restrict its air operations to 100 miles from the coast. In a rival demonstration of naval air power, Admiral King, commanding the carrier group, carried out an impressive air attack on Pearl Harbor in the fleet's spring manœuvres.

While the two services were vying with each other in the exercise of air power over the sea, the strategic pendulum began to swing back to the Atlantic, with concern about Axis activities in Central and South America. After the Czechoslovak crisis in September 1938, the Joint Army-Navy Board called for a revision of plans on the assumption that the USA might find itself fighting the Axis alliance of Germany, Italy and Japan, while Britain, France and the Netherlands remained neutral as long as their possessions in the western hemisphere were not attacked. Roosevelt's reaction was to switch emphasis to the expansion of the army air corps, to which he gave the task, which the navy regarded as its own, of protecting any part of the Americas and acting as a deterrent to any hostile intervention in the western hemisphere. Congress delivered a further blow to the navy by refusing to approve funds for the development of the base near Manila, on the grounds that the Philippines had been promised independence by 1946, and the State Department objected to the construction of one on Guam as provocative to Japan. The result of the Joint Board's study was a recommendation that priority should be given to the defence of the Caribbean and Panama Canal. A defensive strategy would have to be pursued in the Pacific, abandoning if necessary Guam and the Philippines, a total reversal of the navy's proposals for the revision of Plan ORANGE.

The immediate result of this reversion to the strategy of two main battle fleets was a request by Leahy for the funds to construct two 45,000-ton battleships, with two more and a cruiser in 1940–41. The Joint Board's planners began work on five RAINBOW plans, covering a series of different eventualities, the highest priority accorded to the unilateral defence of America north of 10° latitude (the Panama canal), the next to war in the Western Pacific with or without allies, then the unilateral defence of all the American continent and the eastern Atlantic, and last 'a war in Europe in association with Britain and France, whereby US forces would be sent into the eastern Atlantic and to Europe and/or Africa in order to effect the defeat of Germany, or Italy, or both'.

The first recommendation to emerge from the planners' study was a request for base facilities, most of them airfields, in the British Caribbean islands, Brazil, Venezuela, Ecuador and Central America. The new CNO, Admiral Stark, assessed that he had just enough capital ships to mount an offensive in the Atlantic and stand on the defensive in the Pacific. He was short of cruisers and submarines and on the borderline for destroyers. Although he had enough carriers, bearing in mind plans to convert nine merchantmen to that role, he was deficient in patrol aircraft. The fleet as a whole was undermanned at 78 per cent of established strength, and the manpower budgeted for up to 1941 fell short of the requirement of 136,000 by 9,000 men. He put forward ambitious proposals for a 25 per cent increase in tonnage, battleships for the first time forming no part.

With the outbreak of war in Europe in September 1939, planning activity became frantic, especially in regard to the RAINBOW plans for the Pacific. The traumatic events of May 1940 naturally intensified anxiety, as the German conquest of France and the Netherlands and the threat of an invasion of Britain foreshadowed the possible seizure by Japan of those countries' possessions in the western Pacific. The need for allies suddenly became urgent, and suggestions were made for the use by the US navy of the British base at Singapore. The naval C-in-C in the Pacific, Admiral Richardson, still wedded to the concept of an offensive strike to Truk, after fleet manœuvres near Hawaii was ordered to keep his fleet at Pearl Harbor instead of returning to west coast ports 'because', Stark told him, 'of the deterrent effect which it is thought your presence may have on the Japs going into the East Indies'. The possibility of both the French and British fleets falling into German hands prompted Stark, with the agreement of his army colleague, Marshall, to recommend that the bulk of the fleet be transferred to the Caribbean and that the RAINBOW plan for unilateral continental defence be given first priority. Preparations were put in hand to seize, if necessary, British, French, Dutch and Danish possessions in the western hemisphere. At the same time Stark proposed a 70 per cent increase in naval tonnage, which would give the navy, over and above the 358 warships in service and 130 under construction, another seven battleships, eighteen carriers, twenty-seven cruisers, 115 destroyers and forty-three submarines. Congress gave its approval on 20 July.

A month before that, when France accepted German armistice terms, Stark and Marshall also proposed that military aid to Britain should cease. Roosevelt overruled both that proposal and the transfer of the major part of the fleet to the Atlantic. It was to remain at Pearl Harbor, as the British had requested. Only if the French fleet were to fall into German hands would it be transferred to the Atlantic. Naval co-operation with Britain was to be stepped up, and the fifty over-age US destroyers, which Churchill had requested on 15 May, were to be transferred to the Royal Navy. Stark's objections were overcome by the grant of base facilities in British Caribbean islands, British Guiana and Newfoundland. Several British requests for the dispatch of a US naval force to Singapore were refused, but instead an embargo was imposed on the supply of aviation fuel and high-grade scrap metal to Japan.

Richardson, the US C-in-C Pacific, was deeply unhappy about keeping his fleet at Pearl Harbor. He argued that its presence there and any naval activity, combined with embargoes, further west was more likely to be provocative than to act as a deterrent; that his fleet could not be maintained in a state of war readiness there, but could be brought to such a state more quickly and securely in its home ports on the west coast. His persistent opposition to Roosevelt's policy led to his replacement by Admiral Kimmel, who was to bear the brunt of the consequences of failure to accept Richardson's advice. Fresh decisions could not be taken until after the presidential election on 5 November 1940. Stark produced a memorandum, posing four strategic options: (A) to concentrate on hemisphere defence in both oceans; (B) to remain on the defensive in the Atlantic, while launching a full offensive against Japan, in co-operation with the British, Dutch and Chinese; (C) to restrict action to military support of the British in Europe, and to them, the Dutch and the Chinese in the Far East; (D) to plan for an eventual strong offensive in the Atlantic as an ally of the British, while remaining on the defensive in the Pacific. He opted for D, Plan DOG. Germany could not be defeated, however much weakened by bombing and blockade, without major operations on land, which Britain could not launch without American manpower. 'In addition to sending naval assistance,' he wrote, '[the USA] would also need to send large air and land forces to Europe or Africa, or both, and to participate strongly in this land offensive.' Marshall was in broad agreement, but attached less importance to the Far East.

The period between the two world wars saw the US army at its lowest ebb. Anti-militarism and financial parsimony combined to keep it in a national backwater, lacking both resources and a sense of purpose. Almost as soon as the First World War was over the old argument about organization came to the fore again. By 30 June 1919, 128,436 officers and 2,608,218 enlisted men had been discharged, leaving only 130,000 to meet all the regular army's commitments, including a token occupation force in Germany. General March and his General Staff assumed that the army should plan to be able to mobilize a similar army some day in the future and proposed a permanent regular army of 500,000, which could be expanded by recall of reserves to produce a field army of five corps, with the National Guard providing an additional reserve which would have lower priority. An army of that size could not be provided without universal military service, the length of which the General Staff planners recommended should be eleven months, but which March was prepared to reduce to three. This was a reversion to Upton's concept, and Congress was no more inclined to accept this Prussian model than it had been before. It preferred the ideas put forward by Colonel John McAuley Palmer. He proposed a small regular army which would itself provide a force that was immediately available and would train a citizen army, based on universal service in the National Guard, which could be mobilized for a major emergency. His views were backed by Pershing, who, in spite of his insistence on thorough training, had been impressed by the performance of his National Guard divisions in the Meuse-Argonne battles. Palmer's concept lay behind the 1920 National Defense Act, although Congress rejected any form of universal military service.

The regular army was limited to 280,000 men, wartime expansion depending primarily on the National Guard and secondarily on an Organized Reserve, the counterpart of the 1917 National Army. Officers were to be provided, especially for the Organized Reserve, by expansion of the Reserve Officers Training Corps in colleges and Citizen Military Training Camps. The old territorial division of the country into military departments was to be changed for one which provided for three army and nine corps headquarters, each corps containing one regular, two National Guard and three Organized Reserve divisions. The regular division would be responsible for training the others. The Act authorized the four combat arms, infantry, cavalry, coast artillery and field artillery, and administrative headquarters in the War Department, responsible to the General Staff, on an equal footing with the traditional bureaux. The Air and Chemical Warfare Services and March's invention, the Finance Department, were given the same status.

Unfortunately, Congress and the Executive did not provide the resources to back the Act. By 1927 the army's strength was cut to 118,750 and the National Guard seldom reached half its authorized strength of 435,000. No way was found by which to recruit enlisted men for the Organized Reserve, which throughout the 1920s consisted of an Officers' Reserve of some 100,000. A 1933 amendment to the National Defense Act gave the National Guard a dual status, both as the militia of the states and as a permanent reserve component of the regular army. The improved authority of the General Staff, which March had succeeded in establishing, was confirmed by a board headed by General Harbord, Chief of Staff and Supply to Pershing, who succeeded March as Chief of the Army Staff in 1921. It was reorganized on the French model that Pershing had adopted in the AEF into five branches: G-1, Personnel; G-2, Intelligence; G-3, Operations and Training; G-4, Supply; and the War Plans Division.

This last division was not so much concerned with strategic planning as with the problems of mobilization. When General Douglas C. MacArthur became Chief of the Army Staff in 1930 he found this unrealistic; the General Staff still favoured Uptonian solutions which regarded the regular army as a framework which would be expanded into the mobilized army. He wanted the regular army to be able to provide an 'Instant Readiness Force', which in an emergency would be commanded by himself, converting the War Plans Division of the General Staff into a GHQ field forces. The three army headquarters provided for in the 1920 National Defense Act had never been created and the nine corps headquarters had become solely administrative organizations. He divided the country into four army areas, the senior corps commander in the area being designated as the army commander. He also devised a complicated scheme which allotted every regular officer to a specific post on mobilization.

The General Staff's obsession with plans to mobilize manpower led it, as before 1914, to neglect the problems of equipment and supply. MacArthur tried to remedy this by initiating a series of six-year programmes to cover research, development and production of new equipment. The 1920 Act had charged the office of the Assistant Secretary of the War Department with responsibility for keeping alive the experience of the First World War in this field. A Planning Branch was set up for the purpose, a valuable by-product of which was the Army Industrial College. In spite of the creation in 1922 of the Army and

Navy Munitions Board, the navy took little interest in the subject, assuming that it would fight the war with what it possessed at the outset. Although little equipment was actually produced, assessments were made of the industrial capacity and raw material requirements of a major war.

The nation's pride in its industrial enterprise and its leadership in the automotive age was not reflected in the army's equipment. Pershing had been unenthusiastic about tanks, unlike his former aide, Major George S.Patton, who had commanded the AEF's tank training centre in France and the 304th Tank Brigade at St Mihiel, while Major Dwight D.Eisenhower had been in charge of the tank training centre in the USA. The army's post-war stock of 1,115 tanks was almost all of light 6-ton copies of the French Renault, and when the 1920 Act allotted all tanks to the infantry, little interest was taken in them, Patton returning to the horsed cavalry. A visit by the War Secretary, Dwight F.Davis, to the British Experimental Mechanized Force in 1927 forced the reluctant Summerall, who had replaced Pershing as Chief of Army Staff, to create a similar force at Camp Meade in Maryland. Fortunately the General Staff officer responsible for the project, Major Adna. R.Chaffee, was an imaginative cavalry officer who saw the tank not just as an infantry support weapon or a replacement for horsed cavalry, but as a new combat arm which could restore mobility to the battlefield. Unfortunately, he was given little support of any kind, and the army missed the opportunity to exploit the inventiveness of J.Walter Christie, whose ingenious suspension was adopted and developed by the British and, more successfully, in their T-34 by the Russians.

The record in the field of small arms and artillery was little better. Although the Garand semi-automatic rifle was developed by the Ordnance Department, the army clung to the 1903 Springfield, as it did to the 75-mm field gun, copied from the French, while the .50-inch heavy machine-gun remained the only anti-tank and anti-aircraft weapon. This sorry state of affairs was principally due to the refusal of Congress to authorize new equipment as long as First World War stock remained serviceable. Not until 1935, when General Malin Craig replaced MacArthur as Chief of Army Staff, was any real progress made. He realized that the army would never be re-equipped if it waited for development to be authorized, and that the best course was to purchase already developed foreign equipment. Acquisition of the Swedish Bofors 40-mm anti-aircraft and the German 37-mm anti-tank guns were the first results of his policy.

The development of the army air service was affected by controversy about its proper role. Billy Mitchell, an admirer of Trenchard, wanted an independent force, but the army's view prevailed and the 1920 Act authorized the air service as a combatant arm of the army with a strength of 1,516 officers, 16,000 enlisted men and 2,500 flying cadets. It was to be headed by a major-general, with a brigadier-general as assistant, and was to be responsible for controlling its own research, development and procurement, as well as all personnel and training matters. Its tactical squadrons were, however, to operate as integral elements of divisions and corps under control of army commanders. The Assistant Chief was Mitchell who, after his successful demonstrations of the ability of aircraft to sink ships, was court-martialled for accusing the senior officers of the army and navy of 'incompetence, criminal negligence, and almost treasonable administration of the National Defense', following the loss of the navy's dirigible *Shenandoah* in 1925.

Inquiries into the future of the air service followed each other thick and fast, one of the most influential being the 1925 Morrow Board. It rejected both the suggestion of a Department of Defense responsible for the navy, the army and an independent air force and also the creation of the latter, although a further advance in the air service's status was established by the 1926 Act, naming it the Army Air Corps, resurrecting the post of Assistant Secretary of War responsible for 'military aeronautics', and insisting that the corps was represented in all branches of the General Staff. It also authorized a five-year programme to bring the corps up to the strength authorized in 1920, with 1,800 serviceable aircraft. A further Board, headed by the Army Deputy Chief of Staff, Major-General Hugh A.Drum, accepted the Air Corps's 1933 proposal for a GHQ Air Force of 1,800 aircraft, which would both provide support for the army and undertake strategic operations directly against targets in the enemy's territory. Its commanding general would be subordinate to the Chief of the Army Staff in peacetime and the commander of the field forces in war. He would exercise command for all purposes, other than administration, over the air corps units in each corps area in the USA.

This division of responsibility between the Chief of the Air Corps in the War Department, the Commanding General GHQ Air Force and the corps area commanders led to endless bickering, being partially resolved by the merger of the posts of Chief of the Air Corps and CG GHQ Air Force. By that time the Air Corps had won its fight to justify the long-range B-17 Flying Fortress bomber. The mission of the Air Corps, defined in 1935, stated that it 'would operate as an arm of the mobile army both in the conduct of operations over the land in support of land operations and in the conduct of air operations over the sea in direct defense of the coast, and that, under some conditions, it would conduct air operations in support of or in lieu of naval operations'.

Roosevelt's switch of emphasis in rearmament from the navy to the air at the beginning of 1939 resulted in Congress authorizing in April the procurement of 3,251 aircraft and an expansion of the Air Corps to 3,203 officers, 45,000 men and 5,500 aircraft. This would provide twenty-four tactical groups to be ready for action by June 1941. The events of 1940 led to a succession of revisions of this target. By the autumn of 1941 it had risen to eighty-four groups, with 7,800 combat aircraft, manned by 400,000 officers and men. By the time of the attack on Pearl Harbor seventy tactical groups had been formed, of which twenty-five were fighters, fourteen heavy, nine medium and five light bomber craft, and eleven reconnaissance and six transport planes, but few had combat-worthy aircraft or were at full strength. In May 1940 Roosevelt called for an increase in production from the current 2,000 aircraft a year to 50,000, and between July 1940 and December 1941, 22,077 military aircraft were built in the USA, of which 9,932 went to the Air Corps, 4,034 to the US navy and 6,756 to Britain.

The higher organization was not satisfactorily sorted out until Henry Stimson became Secretary of War in March 1941. Robert A.Lovett was appointed Assistant Secretary of War for Air and General Henry H. Arnold was appointed Chief of Army Air Forces, remaining a Deputy to the Army Chief of Staff, Marshall, and responsible both for Air Force Combat Command, as GHQ Air Force had been renamed, and for the Air Corps, the former headed by Lieutenant-General Emmons and the latter by Major-General Bret.

On 1 September 1939, the day that Hitler invaded Poland, the fifty-nine-year-old General George C.Marshall succeeded Craig as Chief of the Army Staff. A product not of West Point but of the private Virginia Military Institute, he was recognized as a superb staff officer, and had served on Pershing's staff in the AEF, becoming his aide-de-camp after the war. After three years in China, he held a series of routine administrative staff posts, the only significant one being his period on the staff of the infantry school at Fort Benning, when he spotted promising officers like Eisenhower, Bradley, Mark W.Clark and Matthew B.Ridgway. If it had not been for pressure from Pershing, he would probably have been considered too old at fifty-six for promotion to Brigadier-General in 1936 and might not have been accepted by Craig as head of the War Plans Division of the General Staff two years later, designated to succeed Embick as deputy chief of staff in 1939. It was certainly Pershing's pressure which persuaded Roosevelt to promote him above more senior generals to succeed Craig.

A week later Roosevelt declared a limited national emergency, authorizing an increase in the strength of the regular army from 210,000 to 227,000 (still 53,000 short of that allowed by the 1920 Act) and an increment of 235,000 to the National Guard. Marshall's first steps were to activate an army headquarters and several corps headquarters, capable of exercising command in the field, and to reduce the division from four brigades (regiments) to three, making it possible to form five active divisions. After the fall of France in June 1940, Congress, going beyond Roosevelt's request, authorized a regular army strength of 375,000. Plans for industrial mobilization were co-ordinated by an Advisory Commission for National Defense, headed by William Knudson from General Motors. When he asked the War Department for planning figures, they postulated a combat army of one million by October 1941, two million by 1 January 1942 and four million by 1 April of that year, and stated a requirement for an annual production of 36,000 aircraft. These figures could clearly not be achieved without conscription, but neither Marshall nor Roosevelt dared to ask for it, their final reaction being to seek mobilization of the National Guard. The inequity of the latter without the former prompted Congress, on 16 September 1940, to authorize selective service for one year, having three weeks before authorized the President to call up the National Guard and other reserves for that period. To exercise control over the whole training machine GHQ Field Forces had been activated in July. Although Marshall officially held the post of Commanding General as well as that of Chief of the Army Staff, the Chief of Staff, General Lesley J.McNair, actually exercised command over the four field armies. Lack of adequate training camps, facilities and equipment meant that by December 1941 the General Staff's plans had not been fully implemented.

For many years the US navy had been intercepting high-level Japanese radio traffic, knowledge of which made Secretary of State Cordell Hull sceptical of the peaceful professions of Japan's Ambassador Nomura. This scepticism was confirmed by interception of a signal on 14 July 1941 which said that

> next on our schedule [after Indo-China] is sending ultimatum to Nether-lands Indies ... in the seizing of Singapore the navy will play the principal part ... Army will need only one division to seize Singapore and two divisions to seize Netherlands Indies ... with air forces based on Canton, Spratly Islands, Palau, Singora in Thailand, Portuguese Timor and Indo-China and with submarine fleets in mandates, Hainan and Indo-China we will crush British-American military power.

When this was followed by the deployment of Japanese aircraft to bases in southern French Indo-China, all Japanese assets in the United States were frozen, which effectively invoked an embargo on export of oil to them. MacArthur, still C-in-C of the Philippine Army, was appointed US C-in-C Far East, and several squadrons of Army Air Corps B-17 bombers were sent to join him there, the highest priority being given to the reinforcement and fortification of Manila.

Admiral Harold R.Stark disapproved of these measures, which he thought might provoke the Japanese to seize the oilfields of the Dutch

East Indies. He had no intention of reinforcing the Asiatic Fleet and wished to play for time in the Pacific, while he dealt with the problems of the Atlantic. On 1 September the US navy had assumed responsibility for escort to convoys sailing to Britain in the western Atlantic. After three US navy destroyers had been torpedoed while engaged on that task, Congress authorized the arming of American merchantmen, and Admiral Ernest J.King, C-in-C of the Atlantic Fleet, assumed responsibility for preventing surface raiders from passing through the Denmark Strait between Greenland and Iceland.

When General Tojo, leader of the pro-war faction in Japan, succeeded Prince Konoye Fumimaro as Prime Minister on 18 October, Roosevelt seems to have made up his mind that war was inevitable. In the first week of November Admiral Yamamoto ordered his ships to sail from the Inland Sea to the Kuriles, from which, under cover of fog banks, his carriers could stealthily approach the Hawaiian Islands to deliver the devastating attacks which, in spite of some intelligence indications, caught the US forces there, navy and army, by surprise on 7 December 1941. Out of eight battleships in Pearl Harbor, seven were sunk and the other seriously damaged; ten other warships were lost; 394 navy and army aircraft were destroyed and 159 damaged; casualties amounted to 2,403 killed and 1,178 wounded – all for the loss by the Japanese of only twenty-nine aircraft.

Fortunately, not only were fuel tanks not hit, but the three US aircraft-carriers were not there, two of them engaged in ferrying aircraft to Guam and Midway, the third in its home port at San Diego. Humiliating as this blow was to the nation as a whole and the navy in particular, it proved a blessing in disguise. First, it brought the nation into the war with Germany and Japan with full public support. Hitherto Roosevelt had appeared ambivalent. It seemed that, by Lend-Lease and other measures, he hoped to provide sufficient support to Britain and exiled governments like the Dutch to enable them to resist, although perhaps not defeat, the Germans. The latter's attack on the Soviet Union made that policy more credible. At the same time he appeared to wish to escape from the uncertainties of that policy and bring the great power of the United States to bear on a situation which might persist for a long time if that were not done. His fear that he could not persuade the American people, and the Congress which represented it, to participate directly in the war, led him to appear to favour the first policy. Pearl Harbor and Hitler's subsequent declaration of war on the USA resolved his dilemma. The blessing for the navy was that it forced it, willy-nilly, to abandon the battleship as the core of the fleet and replace it with the aircraft-carrier. The immediate problem was the threat to the Philippines and the link with allies which Japan's action had joined to the USA – Britain, the Netherlands, Australia, New Zealand and China.

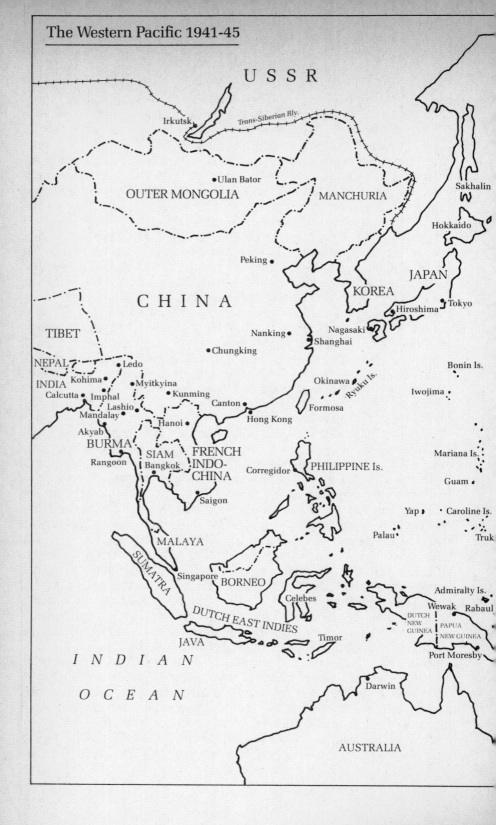

The Western Pacific 1941-45

U S S R

Irkutsk

Trans-Siberian Rly.

Sakhalin

Hokkaido

Ulan Bator

OUTER MONGOLIA

MANCHURIA

Peking

JAPAN

KOREA

Hiroshima

Tokyo

CHINA

Nagasaki

Shanghai

TIBET

Nanking

Chungking

Bonin Is.

NEPAL

Ledo

Okinawa

Ryuku Is.

Iwojima

INDIA

Kohima

Myitkyina

Calcutta

Imphal

Kunming

Formosa

Mandalay

Lashio

Canton

Akyab

Hanoi

Hong Kong

BURMA

Mariana Is.

SIAM

FRENCH

INDO-

Rangoon

Bangkok

CHINA

Corregidor

PHILIPPINE Is.

Guam

Saigon

Yap

Caroline Is.

Palau

Truk

MALAYA

SUMATRA

Admiralty Is.

Singapore

BORNEO

Wewak

Rabaul

Celebes

DUTCH

NEW

GUINEA

PAPUA

DUTCH EAST INDIES

NEW GUINEA

JAVA

Timor

Port Moresby

I N D I A N

O C E A N

Darwin

AUSTRALIA

ALASKA

Kurile Is.

Aleutian Is.

P A C I F I C   O C E A N

• Midway

Hawaii

Pearl Harbor

• Wake

Marshall Is.

Gilbert Is.

Ellice Is.

Solomon Is.

Guadalcanal

C O R A L

S E A

New Hebrides

Fiji Is.

0 200   1000 miles

500   1000 km

The fact that it was now an alliance war was quickly brought home by the visit to Washington of Churchill with two of his chiefs of staff (General Sir Alan Brooke had only just taken over from Sir John Dill as CIGS). The conference, codenamed ARCADIA, confirmed the priority, agreed in previous unofficial staff conversations, to be given to the defeat of Germany. It also revealed a difference about the method by which that was to be achieved which was to persist throughout the war and engender American suspicions about British strategy. Not only did the British appear to Marshall and Stark to put too much faith in weakening Germany by air bombardment, but their concept of land operations was one of 'encirclement': the main American effort to be made in the Mediterranean, the first step in which would be the capture of French North Africa, joining hands with a British advance across Libya. Marshall had grave doubts about this strategy, which he saw as a diversion of effort for the sake of British imperial interests. He wished to concentrate American strength as soon as possible against the principal enemy, the German army, and the quickest and shortest route to that was via Britain and across the Channel. This was important to ensure not only that the Soviet Union did not give up before America's war effort was deployed, but that too much of the latter was not devoted to the war against Japan. It was also important to prevent the idea taking hold that the war could be won by the navy and the air force without a major army contribution.

Two organizational decisions were taken, one contributing nothing, the other a great deal to the conduct of the war. The first was the creation of an allied ABDA (American-British-Dutch-Australian) Command, with Wavell, then C-in-C India, as Supreme Commander, covering the area from Burma to the Philippines; the other was the establishment of the Combined Chiefs of Staff Committee in Washington, to which Dill was appointed as the resident representative of the British Chiefs of Staff. This not only effected permanent liaison with the latter, but it forced the American authorities, in order not to be at a disadvantage in discussion with the British, to put their own house in order. The result was a formal recognition in March 1942 of the US Joint Chiefs of Staff, replacing the former Joint Army-Navy Board, Marshall and Stark being joined by Arnold, thus recognizing de facto that the Army Air Force had become virtually a separate service, and by Admiral Leahy, the President's representative, as chairman. One of the results of Pearl Harbor had been the establishment of the post of C-in-C US Fleet, held by King, in addition to the Chief of Naval Operations, but a few months' experience of this unsatisfactory dual authority led to the merger of the two posts, King remaining and Stark being sent to London to liaise with the British. In practice everything depended on agreement between Marshall and King, and both fully appreciated the importance of reaching it. The relation between the committee and the President was never straightforward, Roosevelt being averse to formal procedures and preferring to deal direct with Marshall or King in person. The British Chiefs of Staff agreed that their US colleagues should have principal responsibility in the Pacific and the British in the Middle and Far East, except for China. Overall responsibility for Europe was left vague, that for the anti-submarine war in the Atlantic being split between Britain for the eastern and the USA for the western half, in which Doenitz concentrated the effort of his submarine fleet in the first half of the year.

In the Pacific Admiral Chester W.Nimitz replaced Admiral Husband E.Kimmel and had to face the unpleasant reality that for the moment there was little he could do to affect the string of Japanese victories which brought them in March 1942 to New Guinea, only 500 miles from Australia. By then they had occupied Malaya, Singapore, Burma, the Dutch East Indies and all of the Philippines except for the peninsula of Bataan in Manila Bay, where MacArthur, taken by surprise with less excuse than Kimmel at Pearl Harbor, was holding out, furious that no attempt was being made to send a force to relieve him. It would have been folly to attempt it. Marshall and King were in agreement that the sound foundations on which to base a strategy in the Pacific were Hawaii and Australia. MacArthur was therefore ordered to leave Corregidor and head a new South West Pacific Command, based in Australia. Responsibility for the war in the Pacific thereafter was split between himself, reporting to Marshall, and Nimitz, reporting to King. A number of conflicting influences affected the direction of the war in that area. One basic one was the overall priority between it and the war against Germany. MacArthur never moved from his view that Asia was more important to the USA than Europe, and Roosevelt's interest in China reinforced it; but not even King, who was tempted, and sometimes threatened to do so, diverged from the basic priority agreed at the ARCADIA conference.

The other source of strategic divergence was between those who wished to give priority to a direct thrust across the Central Pacific aimed as far as possible directly at Japan, and those who saw Japan's conquests as her Achilles' heel and wished to combine regaining lost territory, notably the Philippines, with attacking her where she was weakest. The navy had an unexpected supporter of the former strategy in the Army Air Force, which was interested in acquiring bases from which its bombers, B-17s and later B-29s, could bomb Japan's homeland. They first attacked Tokyo in April 1942, General James H.Doolittle's bombers operating from one of Admiral William F.Halsey's carriers, the *Hornet*. Force of circumstances dictated that initially the latter should have priority, operations in the Central Pacific, resulting in the battles of the Coral Sea and Midway in May and June 1942, being designed to divert Japanese effort from their southern thrust into the Solomon Islands. After Midway, operational command was divided into three main forces, not two, Nimitz's command being split between a Central Pacific Force, commanded first by Halsey and then Admiral Raymond A.Spruance, and a South Pacific Force, originally commanded by Admiral Robert L.Ghormley, which operated in close co-operation with MacArthur, but not under his command.

During this period Marshall pressed for MacArthur to be given overall command in the Pacific. King's suggestion that it should be given to Nimitz effectively countered it. The latter's initiative in committing the marines under Ghormley to Guadalcanal ensured that the US was fully committed to a major campaign in the Pacific before that against Germany drained all its resources away to Europe. The prolonged struggle there which continued until February 1943 proved to be the turning-point which decided the Pacific War, inflicting losses on the Japanese, particularly in ships and aircraft, from which they never recovered. Ghormley's failure to provide full support to the forces ashore in the early stages led to his replacement by Halsey. As MacArthur fought his island-hopping war northwards towards the Philippines in 1943 and 1944, Nimitz inclined to support Halsey and MacArthur in giving greater emphasis to the recapture of the Philippines than to a direct thrust across the Central Pacific, but was firmly brought into line by King. Both he and Marshall were concerned at the ever-increasing demands of the Pacific War on manpower and material, not least in landing-craft and shipping of all kinds, which were proving to be the limiting factor in the timing of operations in Europe.

Marshall's concern on that issue had been one of his motives for proposing, on a visit to Britain in April 1942, that a cross-Channel operation should be planned for September of that year to secure a lodgement area in Normandy, if it appeared that either Germany or the Soviet Union were about to collapse on the Eastern Front. Known as SLEDGEHAMMER, it would be succeeded in 1943 by a larger operation, known as ROUND UP, in which the Allied armies would be deployed into Northern France, defeat the German army in the west and advance to Germany itself. King supported Marshall, although he realized that the amphibious capability for SLEDGEHAMMER could not be made available. He preferred to let the realities speak for themselves rather than side with the British against Marshall; but emphasis on the need for amphibious operations at an early date in Europe, coming on top of the demands of the Pacific, forced him to allot a priority to the construction of landing-craft at the expense of that for fast aircraft-carriers, essential for the Pacific, and destroyer escorts, equally important for the Atlantic, if the US forces, destined for Europe, and their supplies, were to be safely transported across that ocean.

The British never considered SLEDGEHAMMER a practical proposition but, like King, expected the facts of life to kill it, hoping at the same time to convince their allies that GYMNAST, an invasion of French North Africa, should take its place. Marshall and King opposed the latter, believing that, whatever happened to SLEDGEHAMMER, GYMNAST would lead to a drain of effort into the Mediterranean, which would undermine their strategy of a direct blow at Germany through a cross-Channel operation. If the latter could not be launched in 1942, it must be in 1943. By-passing the Combined Chiefs of Staff, Churchill, on a visit to Washington in June, persuaded Roosevelt that SLEDGE-HAMMER could not be mounted in 1942 and to accept GYMNAST. The President was primarily motivated by the desire, as important for political as for military reasons, to see American soldiers in action against Germany before the year was out. The British defeat at the other end of the North African coast, signalled by Rommel's capture of Tobruk while Churchill was in Washington, reinforced his decision, although it seemed likely to remove one of the principal attractions of GYMNAST, a link-up between the forces engaged in it and the British 8th Army. The US Chiefs of Staff were partially mollified by an assurance from their British colleagues that they remained fully committed to the priority of BOLERO, the build-up of US land and air forces in Britain in preparation for ROUND UP.

Marshall was as eager as Roosevelt to get the army that he was building up into action as soon as possible. He did not want to see any support given to the concept, put forward by Arnold's staff as Air War Plans Division 1 in August, which suggested that the economic and social life of Germany was so severely strained by the war in Russia that concentration of effort on strategic bombardment of Germany could render unnecessary a land offensive, which in any case could not be launched before 1945. Marshall had already had to fend off a suggestion by Roosevelt, made before Pearl Harbor, that because of difficulties over equipment production the army's target strength should be reduced. He had been forced to plan for the deactivation of eighteen National Guard divisions in February 1942. Pearl Harbor changed all that.

The army's eventual target was for 213 divisions to be reached by June 1944, which would need a total army strength of 8.8 million men. The immediate target was seventy-one divisions by the end of 1942, double the number formed in December 1941. The task of forming and training this huge army lay in the capable hands of General McNair at GHQ Army Ground Forces. His policy was to form new divisions by making existing ones throw off a cadre, and to streamline the division itself to contain only what it needed to advance against average resistance. Anything above that, required for exceptional circumstances, either in attack or defence, should be in army reserve and organized in battalion-sized units. There should be no permanent allotment of troops to corps, and the minimum number of headquarters. In spite of his devotion to the concept of a lean, streamlined army, the increasing demand for manpower for support units of all kinds, combat and logistic, forced a drastic reduction in the army's target. The eventual total reached eighty-nine divisions and the strength in combat units approximated two million, hardly changing from the end of 1942 until the end of the war, when the army's total strength was 8,291,336. Although two million men entered the army during 1943, only 365,000 went to the combat arms, most of them as replacements for casualties.

The achievement of McNair and his subordinates in raising and training such a large army in so short a time was a remarkable one. They owed much both to the quality of the officers of the regular army, for the most part products of West Point, and to the training they had received. They provided the essential command element at the higher level, which on the whole performed well.

Manpower was not the only factor restricting the number of divisions. Equipment also imposed limits. The needs of Britain and Russia under Lend-Lease had to be met at the expense of the US armed forces. Only those divisions under orders to move to active theatres overseas received their full scale: those under training in the USA frequently found themselves deprived of equipment that had already been issued in order to equip the former or to provide replacements for active theatres. In general the army's policy was to concentrate on quantity, if necessary at the expense of seeking higher quality in performance. That was certainly the case as far as tanks were concerned, the M 3 light (Stuart) and the M 4 medium (Sherman), although reliable, were significantly inferior to German tanks that opposed them from 1943 onwards. In artillery, however, the record was better, the principal field gun, the 105-mm howitzer, proving to be both effective and reliable, as were the 155-mm medium gun, and the heavy 8-inch and 240-mm pieces. The Browning .300 automatic rifle and light machine guns also proved successful. The record in anti-tank weapons was no more impressive than that in tanks, and by the time that the bulk of the army was deployed in action, the enemy air threat had diminished to a degree that did not test the adequacy of the army's anti-aircraft defence.

Responsibility for the supply of the army was centred in the office of the Chief of the Services of Supply, General Brehan B. Somervell, a post created by the reorganization of March 1942, which also, in establishing the Army Air Forces, gave them responsibility for their own supply. The reorganization was the fruit of a study by Major-General Joseph T. McNarney, one of the principal recommendations of which was the creation of an Operations and Plans Division within the General Staff. It became Marshall's planning and executive staff, and resolved the division of responsibility between the General Staff and GHQ. Thenceforth there were four distinct areas of responsibility, Marshall as Chief of Staff of the Army exercising direct control of plans and operations, intelligence and general organization; McNair, as Commanding General Army Ground Forces, raising and training the army; Somervell, as Chief of Supply Services, equipping and supplying it; and Arnold, responsible for all aspects of the activity of the Army Air Forces.

Roosevelt's dislike of creating machinery dominated by the services, which might appropriate to itself decisions which he wished to keep in his own hands, led him to reject the advice, given to him in 1939 by Bernard Baruch, who had successfully headed the 1918 War Industries Board, to create a similar organization. The result was a free-for-all and a failure to co-ordinate the demands, not only of the navy and army, but of the Allies and the civilian economy. The ARCADIA conference set production targets of 60,000 combat aircraft for 1942 and 125,000 for 1943: for tanks of 45,000 and 75,000, and for merchant shipping 6 million and 10 million tons. But the machinery to achieve these was not created, the peacetime procedure being followed of requests to Congress and orders being placed as soon as its authority was given. Steel soon became a bottleneck, but it was not the only one. In September 1942 the Production Executive Committee was set up to co-ordinate the demands of the military with the needs of the civilian economy, followed in November by the Controlled Materials Plan, and in the spring of 1943 by the Office of War Mobilization. By the end of the war the army had received 96,000 tanks, 61,000 field guns, 7 million rifles, 1.2 million radio sets and 2.3 million trucks, and the Army Air Force acquired 100,000 aircraft.

On 24 July 1942 the US Chiefs of Staff finally accepted that here would be no cross-Channel operation that year and gave the go-ahead to GYMNAST, renamed TORCH. Lieutenant-General Dwight D.Eisenhower, former head of the Operations and Plans Division of the General Staff, who had been based in London since June as commander of the US Army in the European theatre of operations, was appointed to overall command, with Major-General Mark Clark as his deputy. The initial landing force was split into three: Eastern, landing round Algiers, of 10,000 British and 10,000 US troops under the command of the British Lieutenant-General Anderson; Central, round Oran, of 18,500 US troops, under the US Major-General Lloyd R.Fredendall; and Western, of 24,500 US troops, under Major-General Patton, landing round Casablanca in Morocco, the first two sailing from the United Kingdom, escorted by the Royal Navy, the last direct from the United States by the US navy. The participation of the British was to be kept as secret as possible in the hope that the Vichy French forces in North Africa would not oppose an American expedition. The force was to be built up to a total of seven US and six British divisions.

The landings took place on 8 November in an atmosphere bordering on *opéra bouffe*. The subsequent development of operations proved disappointing, German reaction being swift and effective; but for the US army and its air force the campaign was an invaluable proving ground, revealing the bad and the good points of both their commanders and their training. Nevertheless it confirmed all Marshall's fears, shared by King. These fears were intensified by the proposals made by the British at the SYMBOL conference at Casablanca in January 1943. Convinced that ROUND UP would not be a practical proposition in 1943, the British suggested that the clearance of North Africa, still far from complete with Eisenhower's troops bogged down in Tunisia and Montgomery not yet in Tripoli, should be followed by the capture of Sicily. The principal aim would be to detach Italy from the side of Germany, which would have repercussions in the Balkans, where the Germans relied on Italian troops to keep that troublesome area suppressed. It would force Hitler to commit more troops to his southern flank and might even persuade Turkey to abandon its neutrality. Marshall and King fought hard for a decision that there should be no further commitment to the Mediterranean after the clearance of Tunisia, while the British fought for limitations to be imposed on operations in the Pacific in order to free more landing-craft for Europe. Both agreed that strategic bombing of Germany should be intensified. The result was a compromise. The US Chiefs of Staff agreed that an invasion of Sicily should follow the clearance of Tunisia, provided that there was no commitment to any further Mediterranean ventures, and that an amphibious operation should be launched against southern Burma. Differences of opinion about the feasibility of ROUND UP were covered up by the formula that in Britain there would be 'the assembly of the strongest possible force... in constant readiness to re-enter the Continent as soon as German resistance is weakened to the required extent', and about the Pacific that 'operations must be kept within such limits as will not, in the opinion of the Combined Chiefs of Staff, jeopardize the capacity of the United Nations [as the alliance was then officially called] to take advantage of any favourable opportunity that may present itself for the decisive defeat of Germany in 1943'. The press conference at the end of SYMBOL was notable for Roosevelt's statement, made apparently without consultation with anyone, that the war against Germany would be pursued until her 'unconditional surrender'.

On his return from Casablanca, King pressed for an early offensive in the Central Pacific to the Marshall and Caroline Islands, but met opposition not only from Marshall and Arnold, but also from Nimitz who preferred to put his main effort into Halsey's operations. Arnold interpreted SYMBOL as justifying the removal of bomber support from

MacArthur and Halsey, but protests from both prevented it. Tunisia was not cleared until early in May 1943. By then it was clear that there was no hope of launching ROUND UP in that year, and Marshall agreed that King should go ahead with his Central Pacific operation while MacArthur and Halsey launched an offensive north from the Solomons, by-passing the well-defended Rabaul. In a meeting with the British in May, the US Chiefs of Staff insisted that they commit themselves to a cross-Channel operation, renamed OVERLORD. In return the Joint Chiefs accepted that, after Sicily, 'pressure would be maintained against Italy', leaving it to Eisenhower to suggest how that should be done. The amphibious operation against southern Burma was dropped, the British, pressed by General Joseph W. Stilwell who was present, reluctantly committing themselves to an operation in Upper Burma to reopen the road to China.

The forces involved in Operation HUSKY, the landing in Sicily in July, were more or less evenly divided between British and US troops, the former slightly larger. The six US divisions, three of which, with one parachute regimental combat team (the equivalent of a British brigade), were involved in the initial assault, were formed into the 7th US Army, commanded by Patton, who had one subordinate corps under Bradley. Clark had already been appointed to command 5th US Army which had the unglamorous task of occupying Morocco. The British contribution of about the same size was provided by Montgomery's 8th Army, and the campaign took on the semblance of a contest between him and Patton, umpired by Alexander, the Germans escaping across the Straits of Messina while the two bickered. The invasion of Sicily which began on 10 July succeeded in its main aim of provoking the defection of Italy from the Axis, the first step being Mussolini's dismissal and replacement by Badoglio on 25 July.

This convinced Eisenhower that the mainland should be his next objective, and he was supported by the Joint Chiefs, strongly by Arnold, who wished to deploy his bombers to the airfields round Foggia, from which they could attack the Romanian oil installations at Ploesti. Not surprisingly, Eisenhower's recommendation was warmly endorsed by the British, whose ambitions in Italy led to a clash with their American colleagues at the QUADRANT conference which started in Quebec on 15 August. At that stage Brooke assumed that, on the assumption that Italy surrendered, entry to the mainland would not involve severe fighting and would open up the possibility of establishing a front in northern Italy which would draw German forces away both from France and from the Eastern Front. This would be of significant help to OVERLORD. He did not share Churchill's wider ambition to open up a front in the Balkans, but the latter made the Joint Chiefs intensely suspicious of anything which would divert effort from executing OVERLORD in the first half of 1944. Brooke's request to keep in the Mediterranean seven divisions, due to be transferred to Britain, and landing-craft due to be sent to the Far East, created a tense atmosphere at QUAD-RANT. As on many previous occasions, Dill, whom Marshall liked and respected, helped to find a compromise. The British committed themselves firmly to OVERLORD not later than May 1944 and agreed to the release of the seven divisions, while the Americans accepted that 'unremitting pressure' would be maintained on the German forces in Italy. The British also agreed that 'the full resources of both countries would be applied to the war against Japan when Germany had been defeated'. A future source of argument was provided by mention of an operation in southern France to assist OVERLORD. Discussion of Far Eastern and Pacific strategy had at times been acrimonious, the Americans considering that the British forces in India were making little contribution to the war, Roosevelt in particular demanding action to open the road to China and generally give more support to Chaing Kai-shek. Arnold, who had exaggerated notions of the effect on Japan of basing bombers on Chinese airfields, supported him.

The landing at Salerno on 9 September, which coincided with Italy's surrender, was the first test of Clark as a commander, and it was a touch-and-go affair. His 5th Army was half British and the American contribution to the Italian campaign was throughout to be smaller than that of its allies. In the initial stages there were nine US divisions out of nineteen in both the Fifth US and the 8th British armies. At the time of the fall of Rome in June 1944 it was reduced to seven out of twenty-five, falling to five out of twenty in August, increasing to six out of nineteen in December with the arrival of the 92nd (Negro) Division, and seven out of seventeen in April 1945, with that of the 10th US Mountain Division. Brooke's assumption that entry into Italy would not involve serious fighting, although true in the case of Montgomery's crossing to the toe, was far from being the case; the long hard slog up the country's terrain proved to be more akin to the battles of the First World War.

The long delay before the cross-Channel operation was mounted allowed King not only to put more effort into the Pacific but to divert construction from landing-craft to destroyers and thus contribute to the anti-submarine war in the Atlantic. An equally important contribution was the allocation of long-range aircraft to the anti-submarine role, which King managed to prise out of the hands of the Army Air Force. The tide fortunately turned in the Battle of the Atlantic before the major transfer of troops and their supplies across the ocean to Europe was called for.

One of the agreements of QUADRANT was that the overall commander of OVERLORD would be American, and it was assumed that Marshall would be the President's choice. King pressed for American command of all the naval forces engaged, but Churchill adamantly and successfully opposed that. Between QUADRANT and SEXTANT, the conference in Cairo in the last week of November, attended by Chiang Kai-shek, two other American suggestions about the higher command were rejected. One was that Marshall, as it was assumed it would be, should exercise command over both North West Europe and the Mediterranean, while perhaps still officially remaining Chief of Staff of the Army, and Eisenhower acting as his deputy in Washington, and that, in addition to the Combined Chiefs of Staff, there should be an 'United Chiefs of Staff' to include the Soviet Union and China.

SEXTANT was largely taken up with attempts, for the sake of China, to commit Mountbatten's South East Asia Command to both a land offensive in Upper Burma and an amphibious assault on the Andaman Islands in the Bay of Bengal. King complained that, having provided ships for the Atlantic to replace British ones sent to Mountbatten, the latter was making little use of them. SEXTANT was immediately followed by EUREKA in Tehran, where the Russians replaced the Chinese. Stalin lined up with Roosevelt in opposing any further commitment to Mediterranean operations other than a landing in southern France, codenamed ANVIL, which should precede OVERLORD by a month. Brooke was strongly opposed to this, as he foresaw that it would result in pressure on the Germans in Italy being removed just at the time when it could contribute most to tying down forces which might otherwise be employed against OVERLORD. On their return to Cairo, the Combined Chiefs of Staff agreed that, in return for British acceptance of ANVIL, 'on as big a scale as landing-craft permit', without commitment to a firm date other than that 'for planning purposes', it would coincide with OVERLORD 'during May'. The amphibious operation in the Bay of Bengal would be cancelled and the landing-craft due to be sent to Mountbatten for it would be retained in the Mediterranean. The thorny issue of the competition between the Pacific, the Mediterranean and OVERLORD for landing-craft was largely resolved by an assurance from King that he could increase construction sufficiently to meet all the requirements approved by the CCS. Agreement on the surface concealed the continuing suspicion of the Americans that the British were not fully behind OVERLORD and preferred Balkan adventures; and, on the part of the British, complaints that opportunities were being missed in the Mediterranean.

Roosevelt decided that he could not spare Marshall, on whose sage advice and calm, strong character he had come to rely, to command OVERLORD and chose Eisenhower, informing Stalin at Tehran, Churchill on return to Cairo and Eisenhower himself when he stopped at Tunis on his way back. The latter had already recommended Bradley to command the US army involved in the OVERLORD assault, a choice Marshall heartily endorsed. At that stage he had not been nominated to command the army group which would be established after the initial landings, all the forces employed in which would be under the command of Montgomery's 21st Army Group. General Jacob L. Devers, who had succeeded Eisenhower in Britain, and McNair were considered, but by the end of the year Marshall had settled on Bradley, General Courtney H. Hodges being appointed his deputy, due to take command of 1st Army when 12th Army Group was formed. Patton was out of the running as a result of an incident in Sicily, but was reinstated as commander of 3rd Army, which would join Bradley in the follow-up. The disappointed Clark was left to soldier on in Italy, beating his head up against the mountain wall at Cassino.

Although there was now no doubt that the maximum effort was being applied to OVERLORD, argument between the Allies flared up again over ANVIL. The landing at Anzio in January 1944, intended to loosen up the front for an advance to Rome, after which US and French divisions would be released for ANVIL, resulted instead in an additional commitment, particularly of landing-craft. The British tried to have ANVIL cancelled and to get the CCS to order General Sir Henry Maitland Wilson, Eisenhower's successor in the Mediterranean, to capture Rome by the first week in June and to press north thereafter, but the Americans insisted on a commitment to ANVIL, promising to provide more shipping and landing-craft from the Pacific. Eisenhower accepted that the operation could not precede or even coincide with OVERLORD, but still wished to see it executed, as he had his eye on Marseilles as a route by which his forces, once in France, could be reinforced. The British thought that they had won the argument when Wilson was instructed 'to make plans for the best possible use of the amphibious lift remaining to you, either in support of operations in Italy or in order to take advantage of opportunities arising in the South of France or elsewhere' and that this was not to detract from his main task, to 'launch, as early as possible, an all-out offensive in Italy'.

During this period the British Chiefs of Staff, opposed by Churchill, wished to plan a major participation in MacArthur's campaign in the South West Pacific, but King opposed it. He was himself involved in fighting for priority for his Central Pacific thrust over MacArthur's desire to liberate the Philippines, and had no desire to see the latter's priority reinforced or to be involved in the complication of co-ordination with a British fleet. Roosevelt, King and Churchill were all united in pressing the British to concentrate on regaining Burma to help China, and thereafter recapturing Malaya and Singapore.

The great operation towards which the whole American war effort was intended to be designed was at last about to take place. An Anglo-American staff, headed by the British General Morgan, had been established in April 1943 to begin work on an outline plan for it. The first problem was the choice of the area for landing, the critical factors being the radius of air cover, about 150 miles from airfields in the south of England; the capacity of beaches to receive landing-craft and of exits from them to accept a rapid flow of vehicles; and the strength of the German defences. It was the last of these which ruled out the Pas-de-Calais, although it was favoured by some as providing the shortest distance from England and the most direct route to Germany. It soon became clear that the area of Normandy east of the Cotentin peninsula was the best choice and that the initial operation should be designed to capture Caen, Bayeux and St Lô and be rapidly developed to secure Falaise and Cherbourg. It should be combined with a deception plan to persuade the Germans that the main landings would be made in the Pas-de-Calais. Morgan had to base his planning on the limits of landing-craft laid down by the CCS, which restricted the landings on the first day to three divisions. Eisenhower and Montgomery each claimed credit for being the first to recognize that this number was too small, enforcing too narrow a landing area. Whatever the validity of their claims, the important factor was that they were agreed and persuaded the CCS to provide the additional landing-craft, although it meant postponing the assault until early June and reducing the size of the assault force for ANVIL, renamed DRAGOON. The British tried once more to cancel it, but Eisenhower insisted on its retention, even though it could not be launched until at least a month after OVERLORD. His arguments for it were that it would obviate the need for him to guard his right flank once he was ashore in France and advancing towards Germany; that it would provide an important means of entry of US reinforcements, in addition to the ports of Brittany and Normandy; and that it would offer the opportunity for participation in the liberation of their country to the French forces in the Mediterranean, for which the USA had provided significant quantities of equipment.

The command of land and naval forces for the actual invasion had been settled, the former given to Montgomery (although Eisenhower would have preferred Alexander) and the latter to the British Admiral Sir Bertram Ramsay, the naval forces supporting the operation amounting to 7,000 vessels, 1,213 being combat ships of which 79 per cent were supplied by the British and Canadian navies. The difficulties arose over the command and employment of the air forces. Eisenhower demanded that all the air forces engaged in the preparation for and execution of OVERLORD should be under his command. This was resisted by the strategic bombing forces, the US under General Carl A.Spaatz, recently transferred from the Mediterranean, the British under Harris. The tactical air forces directly supporting the operation were commanded by the British Air Marshal Sir Trafford Leigh-Mallory, who had the advantage of also commanding the RAF's Fighter Command; but Spaatz and Harris were not prepared to take orders from

him. After much argument, it was agreed that they would be directly
subordinate to Eisenhower, advised by his British deputy, Air Marshal
Tedder. The argument about employment of the strategic bombers
in preparation for the invasion was even more intense. The bomber
chiefs believed that continuation of their attacks on targets within
Germany itself would serve Eisenhower's interests best, but Eisenhower's own airmen, basing their argument on a study carried out
by Professor Zuckerman, pressed for a campaign against transportation
targets in France which would limit the ability of the Germans to
move troops towards the invasion area and supply them there. Eisenhower prevailed, 195,200 aircraft sorties dropping 195,400 tons of
bombs, 101,200 by the USAAF, which lost 1,251 aircraft in the process,
753 from the strategic 8th US Air Force.

It was natural that, in view of all the problems it posed, it was
the initial landings and the development of a beachhead, expanding
to a lodgement area, which it was hoped would include Falaise, St
Lô and Cherbourg, which dominated the thinking of all concerned,
particularly at the level of Montgomery who would be in command
of all the land forces ashore in that phase. The actual assault was
to be carried out, in the western sector, at the corner of the Cotentin
peninsula and the Bay of the Seine, by Bradley's 1st US Army with
two airborne and two infantry divisions; in the eastern sector, between
them and the mouth of the River Orne, by General Sir Miles Dempsey's
2nd British Army with one airborne and three infantry (one Canadian)
divisions. In the subsequent two weeks both armies would be reinforced to a strength of twelve divisions each. Thereafter the British
and Canadian armies would hardly increase at all, while the American
forces would be continually reinforced, both from Britain and direct
from the United States, the first major step being to introduce Patton's
3rd US Army, strong in armoured divisions. At that stage the US
12th Army Group would be formed with Bradley in command, Hodges
taking over 1st Army, and Eisenhower, bringing his Supreme Headquarters Allied Expeditionary Force over from England, would assume
overall command of the land forces, to which Devers's 6th Army Group
from DRAGOON would be added. He would continue, in addition, to
act as Supreme Commander over land and air forces and coastal naval
ones. He resisted suggestions for the appointment of a land force commander, subordinate to him, with authority over all the army groups.

At that stage it was generally envisaged that the initial phase would
be followed by a fairly even gradual advance up to the Seine, pivoting
on Dempsey's 2nd and later General Henry Crerar's 1st Canadian Army
on the left, while Patton pushed out to capture the ports of Brittany
before turning east for Paris. It was estimated that this phase of the
campaign would last up to the end of August. Beyond the Seine, Eisenhower expected his forces to divide into two major thrusts, one led

by Montgomery, to which priority would be given until Antwerp had been secured, aimed to envelop the northern end of the German Siegfried Line covering the approaches to the Ruhr, the other by Bradley, generally aimed at the area of Metz, which would envelop the southern end. If these major enveloping moves did not cut off the German forces west of the Rhine, further enveloping movements would be executed north and south of the Ruhr. Eisenhower saw this as the best way of bringing to bear the maximum force, which would be increasingly American, to fulfil his directive from the CCS to 'enter the Continent of Europe and, in conjunction with the other Allied Nationals, undertake operations aimed at the heart of Germany and the destruction of her Armed Forces'. True to the Clausewitzian principles of US army teaching, he was clear in his own mind that destruction of the enemy's armed forces, not occupation of any particular territory for political reasons, was the overriding priority.

Exhilaration at the success of the Normandy landings was dissipated in the following weeks as the British struggled vainly to capture Caen, and Bradley, after capturing Cherbourg, could make no progress towards a break-out round St Lô. The frustration of the airmen, unable to deploy aircraft from England to newly-constructed airfields in France; of those Americans still held in Britain, notably Patton, until there was room for their deployment in France, and of Eisenhower and his staff, sitting in England bombarded by V 1s and feeling unable to influence events, led to criticism of Montgomery, who was accused of failing to push Dempsey hard enough to break out beyond Caen as they maintained he had undertaken to do. The British were accused of hanging back, leaving the Americans to do most of the fighting. Both criticisms were unfair. There is no doubt that Montgomery had wished, if he could, to capture Caen on the first day and extend the beachhead as rapidly as possible at least as far as Falaise; but not at the expense of the cohesion and solidarity of the area already captured, nor if the cost of doing so, in terms of casualties, would prejudice his ability to develop the general strategy of the campaign on the lines that he, and Eisenhower also, had originally intended. When operations to capture Caen, and extend the British 2nd Army's sector south and east of it, progressed much more slowly and at greater cost than had originally been hoped, as also had those of Bradley's Army in clearing the Cotentin peninsula and capturing St Lô, Montgomery continued to try to push out towards Falaise. But by this time, in late June and all of July, his primary aim was to tie down the bulk of the German Panzer forces in Dempsey's area, while Bradley thrust towards St Lô and developed operations into a break-out from there.

When, at last, that break-out came at the end of July, giving the recently-arrived Patton his chance, the situation was very different from that which had originally been envisaged would then be the case.

The Germans had committed larger forces to fighting as far forward as possible in Normandy than had been expected, instead of carrying out a phased withdrawal to the Seine, as the planners had envisaged. This change from what had been forecast was accentuated when Hitler ordered Kluge to launch the counterattack towards Mortain which was so disastrous to his 5th Panzer and 7th Armies. When the Battle of Normandy finished with the closure of the Falaise gap on 21 August, the Battle of France was also virtually over. DRAGOON had taken place six days before in spite of last-minute attempts by the British to cancel it after Alexander's capture of Rome (or Clark's as the latter would have claimed it) on 4 June. Instead of a second phase of the campaign, forcing the Germans gradually back eastwards from the Seine, operations to clear France of German forces involved little more than an unopposed pursuit. Paradoxically, therefore, the slow progress made in the two months that followed the Normandy landings worked to the advantage of the Allies, and particularly of France, whose liberation was swifter and involved less destruction, beyond the battlefields of Normandy, than had been feared.

Although Patton had himself arrived in Normandy on 7 July, it was not until 1 August that his army became operational. On that date Bradley's 12th Army Group was formed, Hodges assuming command of 1st Army. Eisenhower realized that, at such a crucial stage of the Battle of Normandy, it would be unwise, indeed impractical, for him to attempt to take control of the intense land/air battle for which his headquarters and staff, even if they could be quickly transported and established as a functioning operational organ, were unsuited. He therefore left the overall control of the battle to deal with Kluge's counterstroke to Montgomery, although he personally moved to France and sat at Bradley's elbow.

Ten days after the end of the Battle of Normandy on 1 September, Eisenhower's headquarters became operational on the continent. By then he was still dependent on supply across the beaches and on Cherbourg, which was only just beginning to operate fully, he had twenty US, twelve British, one French and one Polish division. Six more US divisions, three of them airborne, were in reserve in England. Away to the south, Devers had landed three US divisions, which were followed up by seven French. Eisenhower's air support was provided by 1,720 light and medium bombers and 5,000 fighter/fighter bombers, and he could call on 4,035 heavy bombers and 2,000 transport aircraft. From the moment he assumed command, he found himself assailed by Montgomery with suggestions that an overall land force commander be appointed and that Bradley's 12th Army Group should be aligned with his 21st in a single thrust north of the Ardennes. Montgomery's argument was that logistic restraints made it impossible to develop the full effort planned for both of the arms of the envelopment manœuvre which Eisenhower had envisaged, and that, if logistic priority were given to a force of forty divisions in the northern thrust, a bridgehead over the Rhine could be established from which mobile operations in the plain north of the Ruhr could be developed. Overall command of that force would have to be exercised by one man, whom he naturally saw as himself; but, at a subsequent stage of the argument, he offered to serve under Bradley, if the latter were given the responsibility. Eisenhower gave way to the extent of ordering Bradley to direct Hodges to advance in parallel with Dempsey's 2nd Army as far as Belgium, Patton's advance towards Metz being limited to effecting a junction with Devers near Dijon. He gave way even further when Montgomery proposed an ambitious airborne operation to secure a bridgehead over the Rhine at Nijmegen and Arnhem. Meanwhile Patton, who had disregarded attempts to restrict his thrust towards the Saar, was protesting loudly at the priority given to the northern thrust. After the failure of the airborne landings at Arnhem, although the bridge at Nijmegen was firmly secured, Eisenhower had to face the realities of the logistic situation. The ports of Brittany as well as Le Havre were still being held by isolated German garrisons, as well as the lower reaches of the Scheldt, so that Antwerp could not be used. Hopes of enveloping German forces west of the Rhine had faded, and from then on progress depended on tough infantry fighting in wooded or wet terrain in winter weather, straining resources both of manpower, after the high rate of casualties experienced in Normandy, and of artillery ammunition.

Montgomery, whom Brooke tried to restrain although he supported his views, continued his sniping, reverting to the need for an overall land force commander and to give priority to the northern sector, on the grounds that it was necessary to force mobile warfare on the Germans, who were short of oil fuel, and that only the area north of the Ruhr was suitable for that. Matters were brought to a head after the Germans themselves forced mobile warfare on Eisenhower in the Ardennes in December. Montgomery's tactless claim that he had saved the day, and the eruption of anger from Bradley and Patton to which it gave rise, following on a renewed attempt by Montgomery to raise the same old questions all over again, arguing that the Ardennes setback proved his case, almost led to a demand by Eisenhower for his removal. He had already received strong support from Marshall in a signal saying that 'under no circumstances make any concessions whatsoever. You not only have our complete confidence, but there would be a terrific resentment in the country following such an action [acceptance of Montgomery's proposals]'. A major inter-allied row was averted by Montgomery's tactful chief of staff, Major-General Francis de Guingand, who fortunately was on good terms with Eisenhower's General Walter Bedell-Smith.

The plan for the final offensive into Germany itself was presented to Churchill, Roosevelt and the Combined Chiefs, on their way to Yalta, in Malta on 1 February 1945. Eisenhower argued that he must first close up to the Rhine throughout its length as, until he had done that, he could not economize forces in sectors on which he did not plan to develop his main thrusts. The latter would be developed north and south of the Ruhr, with the intention of enveloping it. The US 9th Army, commanded by General William H. Simpson, would be placed under Montgomery's command to strengthen the northern thrust. Brooke continued to complain that neither thrust was strong enough and that, because Bradley faced more difficult terrain, unsuitable for mobile operations, greater weight should be applied to the northern arm of the pincer. Rather than continue the argument, at a time when the preponderance of American strength inevitably weakened British influence, he grudgingly accepted the plan. The forces lined up for the final offensive consisted of sixty US, thirteen British, seven French, three Canadian and one Polish division, the twelve divisions of the US 9th Army being under command of Montgomery's 21st Army Group until after the crossing of the Rhine in the last week of March. Eisenhower's plan worked smoothly, although Anglo-American argument flared up once more as a result of his direct communication to Stalin, notifying the latter that he had no intention of driving towards Berlin, but would direct his forces to the ports of Bremen, Hamburg and the Baltic coast in the north, and to link up with the Russians in the general area of Leipzig.

It was with justified satisfaction that Eisenhower received Doenitz's emissary, Admiral Friedeburg, who arrived with Jodl at his headquarters at Rheims on 5 May. Bedell-Smith and Jodl signed the surrender document at 2.41 a.m. on 7 May 1945. Three days earlier the German forces in Italy had surrendered to Clark, whose frustration at being relegated to what most Americans considered a backwater was intensified by the removal from his 5th Army of two US and four French divisions after the fall of Rome, and he had been only partially compensated by succeeding Alexander as the Army Group commander in November 1944. Alexander replaced Wilson as Supreme Commander in Mediterranean, the latter moving to Washington to take the place of Dill, who had died. The last winter of the war had been a trying time for the Allied troops in Italy, stuck in the mountains south of the Po valley, and morale in both US and British divisions had sunk low. In the final offensive which cleared the valley, Clark commanded seventeen divisions; seven US and one Brazilian in 5th US Army, three British, three Indian, two Polish and one New Zealand in General Sir Richard McCreery's British 8th.

The war with Japan had yet to be finished, and Harry S. Truman, who succeeded Roosevelt as President on the latter's death on 12 April

1945, was concerned about maintaining popular support for the war which many thought would become tougher as US forces drew closer to Japan itself. The defeats which Chiang Kai-shek's forces suffered at the hands of the Japanese in 1944 removed all hope of any help from that quarter and encouraged the Joint Chiefs to press for Russian entry into the war, once Germany was defeated. At Yalta they received the assurance they sought, conditional on the delivery of supplies by the US to eastern Siberia. In July 1944, at a meeting at Pearl Harbor, Roosevelt had personally come down in favour of MacArthur's plan, which King and Nimitz opposed, to liberate the whole of the Philippines after the southern island, Mindanao, had been secured. King wanted the next step to be Formosa, from which a base on the Chinese mainland could be secured for US naval and air forces. Nimitz and Halsey favoured the capture of a base in the Central Philippines as a stepping-stone to Formosa on the way to Okinawa and Japan itself. On his way to invade the Carolines, Halsey found Japanese opposition east of the Philippines so weak that he suggested that MacArthur should go directly for Leyte, cutting out Mindanao. The Combined Chiefs, in session at the OCTAGON conference in Quebec, authorized the change of plan, MacArthur's agreement having been obtained. The result was the decisive naval battle of Leyte Gulf at the end of October.

It was followed by fierce fighting in Leyte itself, to which MacArthur was forced to deploy 250,000 men to defeat the 70,000 Japanese, 50,000 of whom were killed. MacArthur moved on to the main island of Luzon in January 1945, landing in the Lingayen gulf. Although he reached Manila and Clark Field on 2 February, it was another month before he had cleared the city and recaptured Corregidor. The resources expended on the Philippines and the unlikelihood of being able to establish a base on the Chinese mainland made it out of the question to attempt to occupy Formosa. Iwo Jima in the Kazan Islands and Okinawa in the Ryukus were to be the stepping-stones to Kyushu, the southern island of Japan proper. Iwo Jima, five miles long and half as wide, was attacked by three marine divisions in February 1945 and took a month to capture, at a cost of 6,821 killed and 17,000 wounded. For the attack on Okinawa, launched on 1 April, General Buckner's newly formed 10th Army had three marine and four infantry divisions totalling 180,000 men supported by 110 warships. The initial landings were hardly opposed, but thereafter the 120,000 strong garrison fought fiercely. The battle raged until 22 June, when the Japanese General and his staff having committed suicide, all that was left of the garrison, 7,400 men, surrendered. Buckner and 12,500 of his men had been killed.

During the battle for both of these islands heavy losses had been inflicted on landing-craft by Japanese aircraft in suicide (*kamikaze*) attacks. Their fleet, reduced to one battleship, one cruiser and eight destroyers, had also ventured on a suicide mission during the battle for Okinawa, facing Admiral Spruance's twenty-seven aircraft-carriers, ten battleships and nine heavy cruisers. The tough fighting for these islands and the heavy casualties involved appeared to be a foretaste of what might be expected in the next step, a landing in Kyushu.

It is not surprising therefore that Marshall and King, discussing with Truman the possible use of the atomic bomb, due to be tested the day before he was to begin conferring at Potsdam with Stalin and Churchill on 17 July, pressed for its use as an alternative to a direct assault. They opposed a warning demonstration, which would use up one of the only three available. Truman had no doubts, nor had his Allies, and the first was dropped from a B-29 bomber on Hiroshima on 6 August, immediately killing 71,000 people. The second bomb dropped three days later on Nagasaki killed 80,000. The Soviet Union entered the war, and the Japanese Emperor immediately started nego-tiations for surrender, publicly announcing the fact on 14 August. The formal surrender to MacArthur took place on the battleship *Missouri* on 2 September.

That swift conclusion to America's greatest military effort was inter-preted by the advocates of strategic bombing as justification of their claim that it could bring the enemy to his knees without the need for an expensive campaign to defeat his armies. However, other air operations of the war failed to support their thesis, at least when they were analysed after the war by the official US Strategic Bombing Survey and other authoritative accounts. The major effort made by the USAAF in that field was by the US 8th Air Force, based in England, commanded first by Eaker and then by Doolittle. Its attempt to bomb in daylight without fighter escort cost it dear and limited its effect. Its attacks, and those of the British Bomber Command, made little impact either on industrial production or national morale in Germany until after the diversion of the effort of both to support OVERLORD. In January 1943 General Ira C.Eaker had only 413 bombers to add to Harris's 515: by March 1945 Doolittle had 5,027 and Harris 1,609. In 1944 the efforts of the 8th Air Force were augmented by those of the 15th, based in Italy, the operations of both being co-ordinated by Spaatz's US Strategic Air Forces in Europe. Of the total of 2,770,540 tons of bombs dropped by both the USAAF and the RAF on Germany, only 17 per cent fell before 1 January and 28 per cent before 1 July 1944. Their main contribution to victory was to force the Germans to concentrate on fighter defence and thus to lose the air war against the overwhelming strength of the Allied fighters.

Strategic bombing achieved even less in the war against Japan. Only in the final stages was it possible to maintain sufficiently strong attacks to have a significant effect on Japanese industry, its loss of access to raw materials as a consequence of the operations of Nimitz and

MacArthur being far more serious. From Washington, the 20th Air Force, directly subordinate to the Joint Chiefs, commanded the 20th Bomber Command in China and the 21st, based in the Marianas. The former achieved little and the latter did not develop its full potential until its commander, General Curtis Le May, received a third wing of B-29 bombers in February 1945. The first major result was the fire-raid on Tokyo on 10 March.

Far the most significant contribution was made by the tactical air forces in support of the army, although the initial operations of the 12th Air Force in Operation TORCH were not impressive. Fortunately the senior officers of the USAAF in that theatre were less averse to profiting from the experience of the British than their soldier collea-gues. Spaatz, appointed in January 1943 as Eisenhower's air adviser, worked smoothly with Tedder and even with Air Marshal Sir Arthur Coningham, after the British Desert Air Force, supporting Mont-gomery's 8th Army, had joined forces with Eisenhower. Tedder was instrumental in welding the Allied air forces into one, and in the pro-cess the USAAF profited from the hard-earned, often painful experience of the RAF in developing procedures and concepts for the direct and indirect support of army operations, and of naval ones also, while never losing sight of the importance to both of establishing general air supremacy, air force fighting air force.

This educative process paid high dividends when most of the princi-pal actors, Tedder, Spaatz, Doolittle (who commanded the North-West African Strategic Air Force) and Coningham moved with Eisenhower to North-West Europe. Spaatz and Doolittle realized that there was no hard and fast dividing line between the operations of tactical and strategic air forces: each could help the other; although the use of strategic heavy bombers in direct support of army operations remained controversial to the end, the attacks on Cassino and in the area of Caen and St Lô raising the issue most keenly. The army accused the air force of inaccuracy (General McNair was killed in the bombardment near St Lô); the airmen accused the soldiers of failure to exploit the devastation they had created on the ground. There is no doubt what-ever that the total domination of the skies over France by the Allied air forces made an essential contribution to Eisenhower's campaign in North-West Europe, especially in Normandy.

In the Pacific, operations had been heavily dependent on carrier sup-port, and the lack of it was keenly felt when, for good naval reasons, the carriers had to be moved away, the reason why the US Marine Corps had developed its own air arm, although never used in that way until the battle for Okinawa. Support of MacArthur's forces by the USAAF, commanded by General Kenny, was provided by the 5th Air Force for the US 6th Army and the 13th for the US 8th Army; each air force, as in the case of the 12th when it started in operations in North Africa, was divided into air defence, bombardment, air support and air service commands.

Equally significant was the USAAF's contribution to air transport, in which its resources eclipsed those of its allies. The great workhorse, the Douglas DC-3 Dakota, was the mainstay of all airlift operations, combat or logistic, and the USAAF dominated air transport operations, whether combat airborne drops or the routine flights of the world-wide Military Air Transport Service, forerunner of post-war civil aviation. The contribution of the US aircraft industry to the war was decisive. Apart from naval aircraft, the Boeing B-17, B-24 and B-29 (Flying Fortress) bombers equipped the US Strategic Air Forces, the Douglas DB-7 (Boston), Martin B-26 (Marauder), and the North American B-25 (Mitchell) provided the medium and light bombers; the Curtiss P-40 C (Tomahawk) and P-40 F (Kittyhawk) and Republic P-47 (Thunderbolt), the fighter-bombers. The Lockheed P-38 (Lightning) and North American P-51 (Mustang) made a significant contribution as long-range fighters, and the Consolidated PBY-5 (Catalina) flying-boat and B-24 (Liberator) long-range general purpose aircraft made essential contributions in the maritime field. In addition to producing 100,000 aircraft to meet the needs of the US navy and USAAF, significant numbers were provided to its allies through Lend-Lease, which overall absorbed 16 per cent of US war production. In tanks it was even higher: 38 per cent.

The Joint Chiefs had envisaged a major redeployment of the army and the USAAF from Europe to the Pacific (which would have raised major issues over the exercise of command) and, when the war with Japan was over, an orderly demobilization based on a points system. The sudden end of the war upset their carefully laid plans, and political pressure, including that from servicemen themselves, led to the immediate discharge of all men who had served for two years. The army, including the USAAF, was reduced from eight and a quarter million to two million by June 1946 and to one million a year later. The navy, whose 3.4 million men manned 1,200 ships in the summer of 1945, transported most of the men released in a major shipping operation, named MAGIC CARPET. Its casualties, excluding the Marine Corps, amounted to 36,950 killed and 37,778 wounded. The Marine Corps lost 19,606 killed and 67,207 wounded. At its peak strength, the navy had manned twenty-three battleships: ten had been built during the war and two were lost in action.* Between 1945 and 1950, the navy's strength fell to 375,000, manning 237 major combatant ships and 4,300

---

* The equivalent figures for other types were: strike carriers, 28, 27 and 5; escort carriers, 71, 77 and 6; cruisers, 72, 47 and 10; destroyers, frigates and escorts, 436, 485 and 11; submarines, 232, 217 and 45; landing ships (excluding landing-craft, manned by the army and marines), 2,116, 2,531 and 79.

aircraft. The Marine Corps was reduced from 475,000 to 75,000, most of the reductions occurring in the first two years.

1945–86

One of the first issues to be tackled was the future status of the Army Air Force and the higher organization for defence. In spite of fierce opposition from the navy, strongly supported by the Secretary for the Navy, James Forrestal, the Army Air Force won its long battle for recognition as a separate service in the National Security Act of 1947. The act resulted from a compromise between the navy and the army, both fighting against proposals for a unified and integrated defence establishment. The 'National Military Establishment', which the act created, preserved the Joint Chiefs of Staff Committee as it had evolved during the war and, although it established the post of Secretary for Defense, a National Security Council, the Central Intelligence Agency and a National Security Resources Board, it preserved the Army (renamed from War) and Navy Departments as independent executive departments, with their own Secretaries and Chiefs of Staff, and created them for the Air Force. The Defense Secretary had almost no staff, civil or military, to impose unified solutions, and could act as little more than a co-ordinator. Forrestal, the first holder of the post, who had fought as hard as anyone against all forms of unification, soon discovered that he was almost powerless in the fierce struggle that immediately developed for the dwindling sums of money the Truman administration and Congress were prepared to devote to defence. The fiercest struggle was between the fledgling air force and the navy, both claiming priority as the most effective deliverers of the atomic bomb. Neither thought that large armies had any future. The Defense Secretary's position was strengthened by an amendment to the National Defense Act in 1949 which replaced the National Military Establishment by the Department of Defense and downgraded the army, navy and air force to 'military departments' within it. The Defense Secretary was given his own staff as head of an executive department and a Chairman of the Joint Chiefs of Staff appointed, Bradley being the first incumbent, having succeeded Eisenhower as Chief of the Army Staff in 1948, the latter having replaced Marshall in November 1945.

The army, which was reduced to a strength of ten divisions, was still concerned with occupation duties in Europe and Japan. Devotees of Uptonian solutions for its reserve organization gave way to those who favoured preservation of the National Guard, from which an Air National Guard was formed. The Army Guard was to consist of twenty-nine divisions, two of them armoured, and an appropriate number of combat and logistic support units. It was never at full strength, but by June 1950 its 4,597 units contained 324,761 men. At the same time the Organized Reserve held 68,785 officers and 117,756 men in 10,629 active training units and another 390,961 outside them.

Marshall had called the seventy-one-year-old Brigadier-General Palmer out of retirement to advise on the future organization of the reserves, and both of them put forward a proposal for universal military service to succeed the Selective Service Act which Congress had brought to an end in 1947. Although President Truman gave his support, Congress rejected it at the time, but reversed its decision in 1951, during the Korean War, by passing the Universal Military Training and Service Act which decreed that all male citizens between the ages of eighteen and a half and twenty-six who joined or were inducted into the armed forces also incurred an obligation to military reserve service. That was not what Marshall and Palmer had in mind. They wanted to create a national resource of young men, trained in the basic military skills, available for service in the National Guard or Organized Reserve. In the Army Department itself there was a limited reversion to the old independence of the bureaux, the OPD losing the central control it had exercised during the war. Emphasis was theoretically laid on decentralization to commands and technical services, the function of the General Staff being defined as to 'plan, direct, co-ordinate and supervise'.

The army's planning for future mobilization took place in an atmosphere which threw doubt on the need for it, but it was rescued by the formation of NATO, and by the Korean War, which broke out on 25 June 1950. The latter was a test of the new National Security machinery, from which it did not emerge with great credit. If National Security policy had been well co-ordinated, either the war would never have taken place or the armed forces would have been prepared for it. At first it was thought that air and naval support to the South Korean army would suffice, but the latter's precipitate retreat forced the rapid despatch to Korea of the one cavalry and two of the three infantry divisions acting as occupation troops in Japan, never having thought for a moment that they would be called upon to fight.

Together with the 58,000 men of the South Korean army and a reinforcement of 5,000 marines, General Walton Walker's 8th Army, under the overall command of MacArthur in Japan, reached a strength of 47,000 to face a North Korean army of 70,000. While Walker held on to the southern tip of the peninsula round Pusan, MacArthur, overriding the advice of all the experts, including that of the Joint Chiefs, landed the 1st Marine Division at Inchon, followed up by the 7th Infantry, on 15 September, the two divisions forming 10th Corps under the command of his Chief of Staff, General Almond, who remained directly subordinate to him. Walker's men joined up with them eleven days later, and on the 28th MacArthur entered the ruins of the capital Seoul with Syngman Rhee. Victory went to his head. He disregarded orders from the Joint Chiefs to limit any action north of the frontier on the 38th parallel to South Korean forces 'unless there were indications that Russian or Chinese forces were entering the country'. Having assured Truman in October that the war would be over by Thanksgiving Day and 8th Army back in Japan by Christmas, and dismissing the possibility of Chinese intervention, although they had publicly threatened it if US troops moved north of the 38th parallel, he rapidly changed his tune when it became clear that they were massing a large army west of the Yalu. The Joint Chiefs now told MacArthur to clear the area up to the Yalu, but forbade him to bomb the bridges or targets beyond the river, which the Soviet Union announced would provoke retaliation by them. At the end of November the Chinese attacked, by-passing Walker's and Almond's troops (the latter transferred to the east coast) by pushing through the hills while the US and South Korean troops clung to the few roads. By that time MacArthur had 205,000 troops facing 300,000 Chinese. After four days they were in full retreat and MacArthur decided to make a clean break back to the 38th parallel; but he could not hold there, the line finally being stabilized in January forty to seventy-five miles south of it, Seoul, Inchon and the important airfield of Kimpo being abandoned.

Walker was killed in a traffic accident on Christmas Eve and was succeeded by Ridgway whose impact was soon felt. Taking Almond's corps under command also, he pulled his army, now 365,000 strong, together, imposed sterner discipline and began a series of well-co-ordinated deliberate attacks to drive Lin Piao's army of twenty-one Chinese and twelve North Korean divisions, 485,000 strong, back to the 38th parallel. While these pedestrian, costly, but successful operations were in progress, MacArthur, to Ridgway's dismay, was publicly asserting that the war could not be won unless it were extended to the mainland of China. As he persisted in that line, in spite of warnings, and lobbied Republican support for it, Truman finally lost patience, dismissed him and replaced him with Ridgway, handing over 8th Army to General James Van Fleet. The latter repulsed a Chinese counteroffensive and by mid June had driven them north of the parallel. He wanted to press on with amphibious landings, but Truman had had enough.

The war was unpopular, having cost 80,000 casualties, including 12,000 dead and 10,000 missing, and negotiations were opened for a cease-fire. It did not end the war, however, which lasted for another two years, developing into a replica of the trench warfare of 1915–18. The 8th Army's strength rose to 768,000 men, with sixteen divisions in the front line (four US, eleven ROK and one British Commonwealth) and with three more US divisions and one ROK in reserve.

The final armistice agreement was signed in July 1953, after Eisenhower had replaced Truman as President and made threatening noises about the possible use of atomic weapons, and Stalin had died. The war cost the lives of nearly 30,000 Americans and 3,143 other UN servicemen: 107,000 of the former and 15,700 of the latter were wounded or missing. South Korean military casualties were estimated at about 400,000 killed and an equal number wounded or missing, civilian casualties being of the same order.

In no quarter was there any enthusiasm for the war, which had no parallel in American experience. It was neither an 'Indian' type war which hardly affected anyone but the regular forces, nor was it a patriotic war for a noble cause, for which all the resources of the nation could be mustered to achieve victory as soon as possible. The navy disliked it. They regarded the Pacific as their preserve and the only major naval action, the landing at Inchon, was embarked upon against their advice. Their marines became stuck on shore, used as ordinary infantry, and the cost of the war, which the Department of Defense tried to meet as far as possible within normal appropriations, enforced restrictions on their planned new construction programme. The fledgling air force disliked it, as they were forbidden to use their heavy bombers in the strategic role which they believed in or to attack the enemy's air force beyond the Yalu. To the army it posed several problems. Its performance in the early stages did it little credit, and throughout it imposed a severe strain on morale, much of it originating from the problems of manning. Before the war, to encourage regular recruitment, discipline had been loosened and the general standard of welfare raised in an attempt to reduce the contrast between civil and military life. Expectations were raised which could not, once men were deployed to Korea, be fulfilled, attempts to do so having an adverse effect on combat ability and greatly increasing the manpower devoted to support services.

The inequity and apparent irrationality of the army's method of filling its ranks was the principal problem. Initially the divisions in Korea were brought up to strength by drafting men from other formations and Congress revived the Selective Service Act. This not only brought in men to refill the places thus left vacant, but stimulated enlistments to avoid the uncertainty of call-up. The act also authorized the President to call up all reserves, including the National Guard, for twenty-one months' service on active federal duty. Four National Guard divisions were initially called up, their ranks being filled with draftees, and four more when the Chinese entered the war, two of which were sent not to Korea but to Germany. A further four were called up to act as training formations.

In June 1951 Congress extended the Selective Service Act at the same time as it passed the Universal Military Training and Service Act. The passage of the latter was influenced by the patent unfairness of the way that the call-up operated. Not only were there numerous grounds for exemption, but only a small proportion of the 2,834,000 men called up were sent to Korea. In order to try to reduce the arbitrariness of the system, a policy of rapid rotation of individuals was introduced which had a deleterious effect on the morale of units. In all, twenty divisions were formed of which one Marine and eight Army Corps served in Korea.

Reaction to the Korean War took the form of a determination that the most powerful nation in the world should not find itself again fighting an old-fashioned infantry war with no hope of a clear victory. The answer seemed to be atomic weapons, and all three services turned to them, as it became clear that they could be produced in greater numbers and in smaller size than had previously been envisaged. They were encouraged to do so by President Eisenhower's approval, in October 1953, of a National Security Council paper, authorizing the Joint Chiefs to plan on using nuclear weapons, tactical as well as strategic, whenever their use was desirable from a military standpoint. The air force had originally expected that it alone would be responsible for delivering the weapon, using its new B-36 bomber, the range and effectiveness of which could be extended by air-breathing cruise missiles, developed from the German V 1. The navy fought against the B-36, developing aircraft of its own to deliver nuclear weapons. At first they were too large and heavy to station permanently on aircraft-carriers, although they could land back on them; but the development of smaller weapons, larger carriers and more powerful aircraft strengthened the navy's case, reinforced by basing the 6th Fleet in the Mediterranean to provide nuclear support for NATO's southern flank.

The army was not going to be left out of the game. It had taken under its wing the German team, led by Werner von Braun, which had designed the V 2 at Peenemünde. In 1954 the air force, disappointed at progress with its cruise missiles and becoming aware that technical advances in the development both of ballistic missiles and of the fusion nuclear warhead made the former a leading contender for intercontinental delivery, itself turned to their development. The Soviet feat of putting *Sputnik* into space in 1957 intensified the rivalry between its *Atlas* and *Thor* missiles and the army's *Jupiter*, which placed the first American satellite into space in 1958. Meanwhile the navy had also entered the missile race. It also began with a cruise missile, the *Regulus*, first fired from a submarine in 1953; but the navy gave it up in 1958 in favour of *Polaris*, the development of which started in 1955 and led to the first underwater launch in July 1960 from the submarine *George Washington* which became operational with sixteen missiles in November of that year. By that time the army had handed over its *Jupiter* to the air force and was concentrating on the development of shorter-range systems, missiles and guns, for the delivery of nuclear warheads which, although much smaller, produced yields higher than those used at Hiroshima and Nagasaki.

The navy's enthusiasm for things nuclear was not restricted to weapons. Led by the formidable Captain, later Admiral, Rickover, it adopted nuclear power for propulsion of both submarines and surface ships with enthusiasm. The result, however, was to increase the size and cost of new ships, tending to reduce the number which the navy could afford, the number of major combatant ships being maintained at the expense of smaller work-horse vessels. The number of carriers rose from twenty-six in 1960 to thirty-one in 1968, dropping to twenty-one by 1978; cruisers rose from fourteen to thirty-four, then down to twenty-six; and destroyers fell from 226 to sixty-two in the same year. By 1978 the navy's conventionally-powered submarines were reduced to ten, while nuclear-powered ones increased from ten to 109, forty-one of them ballistic-missile boats.

This emphasis on nuclear systems did not go unchallenged. Not surprisingly, it was from the army that the challenge came, its loud and clear voice being that of General Maxwell D.Taylor, wartime commander of 101st Airborne Division, when he succeeded Ridgway as Chief of Staff of the Army in 1955. He argued against the concept of massive nuclear retaliation to any infringement of either the Iron or the Bamboo curtain. He put forward a policy of 'flexible response', by which he meant that the US government should be able to choose its response to a challenge anywhere in the world appropriate to the form which that challenge took – political, economic, diplomatic or military – and that the military element of its response should be adjusted to the military nature of the challenge. Concentration on nuclear methods of military response had reduced the capability of the United States to take less drastic action. His views did not find favour until John F.Kennedy became President in 1961 and Robert S.McNamara was appointed Secretary for Defense. Khrushchev's threatening attitude over Berlin, and the fact that the Soviet Union could not only deliver nuclear weapons directly at targets in the USA, but also respond with shorter-range ones to their use on the battlefield, led to a swing of the pendulum back towards conventional combat strength, reinforced in 1962 by the Cuban missile crisis.

The army reverted to a more traditional type of division after experimenting with a 'pentomic' organization, like the Pentagon a five-sided affair, which was expected to survive nuclear weapon attack and deliver short-range ones in every direction. McNamara's aim was to have sufficient military strength on every level so that the United States could respond appropriately without immediate resort to nuclear arms. While trying to persuade America's NATO allies to move in the same direction, he planned to increase the regular army's active divisions from eleven to sixteen, requiring an increase in manpower from 875,000 to 1,000,000 men, and to back them up with six high-priority reserve divisions. The USAF's troop-lifting capacity would be increased fourfold to enable it to transport two divisions rapidly to Europe, where their heavy equipment would be stockpiled. After the Korean War the Eisenhower administration reorganized the army's reserves into a Ready, Standby and Retired Reserve. The Ready Reserve was designed to fill the ranks of regular army, National Guard and Army Reserve units called to active duty in a limited emergency or the initial stages of general mobilization. The President could call up one million men without further Congressional authority, which was needed for call-up of the other reserves. By the Reserve Forces Act of 1955 the ceiling of the Ready Reserve was increased to 2,900,000 men.

The general result of these measures was to swell the ranks of the reserves with men who were unlikely to be of much value in combat, and the need for whom was doubtful. McNamara set about cutting them down and adjusting their organization to likely needs. The National Guard was reduced to 467,000 men, in twenty-three combat divisions, and the Army Reserve to 375,000, producing six combat and thirteen training divisions, both also forming support units of various kinds. Six of the reserve divisions would be categorized as high-priority, to be maintained at 75 per cent of wartime strength. His stated aim was to have the capability to fight two fairly large limited wars simultaneously or to effect a rapid reinforcement of the US 7th Army in Germany.

Emphasis on conventional warfare capability, and the need to transport forces rapidly overseas, led to an expansion of the US air force, which had already grown into a formidable organization. While encouraging improvements in its capability for support of the army, both in tactical and transport forces, McNamara struggled to restrain its voracious appetite for means of delivery of nuclear weapons. The introduction of ballistic missiles, *Titan* succeeding *Atlas*, with *Minuteman* on the horizon, did not, in its own view, call for a reduction in its bomber fleet, now dependent on the B-52. The argument was produced that a triad of delivery systems was needed: submarine-launched and land-based ballistic missiles and cruise missiles launched from aircraft. Soviet missile-sites, the number of which was later found to be exaggerated, were added to the target list.

By the mid 1960s the USAF was a vast organization, absorbing the largest slice of the nation's defence budget. Its Strategic, Tactical, Air Defense, Pacific and European Air Commands each had three subordinate air forces; the Pacific Air Command's based in Japan, the Philippines and eventually Vietnam; the European in Britain, Germany and Spain. Strategic, Tactical and Air Defense Commands were based in the USA, Continental Command being concerned with the Air National Guard and all reserves, and Military Airlift Command operating world-wide. Southern Command, based in the Panama Canal Zone, provided airlift to meet contingency plans and supervised the training of Latin American air forces.

The navy also benefited from the new look. Many new ships came into service in this period while numbers of all types were kept at a reasonably high level. In 1961, thirty carriers, twenty-eight cruisers, 213 destroyers and forty-nine amphibious ships were in service, and the number of nuclear-powered submarines had risen to forty (twenty-one ballistic-missile), while eighty-five conventional-powered boats were still active. Emphasis on flexible response also favoured the US Marine Corps, which could field three divisions with their own integrated air support.

The test of the change of emphasis to preparation for limited conventional war was not long in coming. President Kennedy inherited a crisis in Laos, which appeared to be falling under communist control, and a commitment to support the shaky government of Ngo Dinh Diem in Vietnam, south of the 17th parallel, fixed as a temporary frontier with Ho Chi Minh's government in the north at the 1954 Geneva Conference. Kennedy, McNamara and General Taylor, whom Kennedy had brought out of retirement into the White House as Military Representative of the President (in 1962 appointing him Chairman of the Joint Chiefs), all hoped to limit military support to provision of arms, a training mission and, if necessary, the introduction of Special Forces of which Kennedy was a keen supporter. The army had created them in the 1950s on the pattern of the British Special Air Service, which had been resurrected in 1956 in Malaya to train and keep contact with the aboriginal jungle tribes, enlisting their support against Chin Peng's communist guerrillas. All were keen to give American military support a low profile, in order to avert both local antagonism and domestic opposition, and to encourage the South Vietnamese to make a greater effort themselves. Step by reluctant step Kennedy's administration became further committed, fearing that, if it did not do so, South Vietnam, and the whole of South East Asia after it, would fall into the hands of the communists.

In November 1961 there were 948 US servicemen in South Vietnam; in January 1962, 2,600; in June, 5,500. When McNamara and Taylor paid a visit in September 1963, before the coup in which President Diem was killed, they suggested withdrawal of 1,000 by the end of the year and an end to the commitment in 1965, when they expected that the South Vietnamese Army would have been built up to a strength to enable it to shoulder the burden alone. Visiting again three months later, when the coup had, if anything, made things worse, McNamara took a gloomy view, as did General William C. Westmoreland on his arrival a few weeks later as deputy to General Paul D. Harkins, head of the Military Advisory Group, whom he was to succeed in June 1964, by which time the US presence had increased to 23,000.

Responding to the argument that, unless the USA demonstrated its willingness to give active support, the South Vietnamese would not themselves make sufficient effort, and rather than let US troops become directly involved in operations, President Johnson, who had succeeded on Kennedy's death in November 1963, was persuaded to initiate a bombing campaign against the North, known as ROLLING THUNDER, using the Gulf of Tonkin incident in August 1964 to obtain Congressional approval. Opposition to it came from several quarters: from those who thought it would have no effect on the North's intervention in the South; those who thought it would provoke greater intervention; and those who objected to the concept of graduated pressure and believed that only an all-out onslaught was worth attempting.

Westmoreland was concerned about the security of the air force's bases in Vietnam and it was ostensibly for that purpose that, in the spring of 1965, he asked for additional troops, which brought his strength up to 82,000 US and 7,250 troops from South Korea, Australia and New Zealand by the middle of that year. These forces, however, were to be restricted to operating within fifty miles of the enclaves surrounding the bases. Westmoreland chafed under this restriction, which precluded him from planning major offensive operations, without which he saw no hope of defeating the combined forces of the Viet Cong and Giap's North Vietnamese army, which embarked on an offensive of their own in June. He requested an increase from seventeen to forty-four battalions, which would bring his total up to 175,000, stating that this would achieve no more than preventing the situation from deteriorating. To support an offensive in 1966, which could lead to victory in 1968, he would need a further 100,000. He was sent the helicopter-equipped Airmobile Cavalry Division which brought his strength up to 125,000 in August 1965, but he was soon asking for more troops, as more of Giap's army penetrated into the south. In February 1966 he requested an increase from seventy-one to 102 battalions, seventy-nine US and twenty-three Allied, which would involve a rise in the troop level from the 235,000 he had at the time to 429,000. McNamara jibbed at the latter, which could probably not be achieved without calling up the reserves, and recommended to the President limits of 367,000 for 1966 and 395,000 by mid 1967.

In the early years the army had filled the ranks of units sent to Vietnam by transfers from regular units elsewhere, including from Germany, and when the Selective Service draft was reintroduced, posting draftees to take their place; but these increases in the force in Vietnam compelled the army to use draftees there also. All the complaints that had been made in the Korean War about the inequities of the Selective Service system were repeated, aggravated by the fact that none of the reserves, National Guard or other, were mobilized. As in Korea, the attempt was made to spread the load more fairly by frequent rotation with its adverse effect on both military skill and morale, made worse by the thoroughly unsatisfactory nature of the war, the terrain and the attitude of its inhabitants. The threat of an offensive by Giap to cut off the forces, mostly Marine, immediately south of the 17th parallel, provoked Westmoreland to raise his sights again to 542,000 for 1967. After much discussion, Johnson agreed to 469,000 by mid 1968, still refusing to mobilize the reserves and hoping that ROLLING THUNDER would at last produce some results.

In 1967, encouraged by what seemed to him an improvement in the general political and military situation in the south, Westmoreland proposed to launch a major offensive in 1968 into the sanctuaries his opponents enjoyed in Laos and Cambodia, combined with an amphibious assault north of the 17th parallel. For these he would need another 200,000 men, bring his total to 670,000, and producing over twelve divisions on the ground, four more than he had already. Realizing that he was unlikely to get agreement to this, he lowered his target to 80,500 ($2\frac{1}{3}$ divisions), bringing him to 550,000, which he realized could probably not be done without a reserve call-up. He was granted 525,000 but refused permission to carry the war into Laos, Cambodia and the north. The air force at last was given permission to attack all the fifty-seven targets the Joint Chiefs had listed, including ones near Hanoi and Haiphong.

After the Tet offensive in January 1968, in which the Viet Cong and North Vietnamese lost 37,000 killed and 6,000 captured, Westmoreland renewed his proposal for a counteroffensive, requesting also that all restrictions on air attacks should be removed. For that he would need another 206,000 men. Johnson jibbed at the proposal and, seeing no way out, announced that he would not stand for another term of office. Westmoreland left in June 1968 to succeed General Harold R. Johnson as Chief of Staff of the Army, his place in Vietnam being taken by General Creighton W. Abrams.

President Richard Nixon, although authorizing extension of the war into Cambodia, began to run the show down, preferring to pursue the path of negotiation instead of victory. With no high cards in his or his Secretary of State Henry Kissinger's hand to play, he had virtually to accept the terms demanded by the other side, which soon abrogated them anyway. The war, which failed to achieve its aim, was the longest ever waged by the US and undoubtedly the least popular. It cost 46,397 American lives in action, 13,000 of them regular soldiers, marines and airmen, and another 10,340 dead from other causes. It had demonstrated that even the most powerful nation in the world could not dictate the limits of a limited war. It could impose limits on itself, but not on the enemy. The conflict had shown that, if a nation does not want to devote resources to a war sufficient to achieve its aim, it had better not become engaged in it at all. Gingerly dipping a toe in and then stepping gradually into the water is fatal: better a high dive and rapid total immersion.

The tonnage of bombs dropped by the USAF in the Vietnam War exceeded that it had dropped in the Second World War. At its peak in 1969, 6,431 aircraft of all types were engaged, fighter/bomber sorties in that year amounting to about 200,000. Aircraft losses in that year amounted to 466 fixed-wing and 1,048 helicopters, the figures for the whole campaign totalling 5,248 fixed-wing and 4,252 helicopters, aircrew deaths amounting to 6,160, of which 3,559 were due to hostile action. These figures include US navy, marine corps and army aviation. The significant developments of the war, from the point of view of air forces, were the introduction of both the television and laser-guided 'smart' weapons and the electronic countermeasures to the Soviet-supplied surface-to-air missiles employed by North Vietnam.

The Vietnam War was a test of the organization not only of the army but also of the joint service command which had been modified since the Korean War. At that time the arrangement, agreed at Forrestal's conference at Key West with the Joint Chiefs in 1948, was that command of an overseas expedition should be exercised through the Chief of Staff of the service which had the primary interest. In Korea it was the army, and Collins had not been successful in imposing his will on MacArthur, for which he could not be entirely blamed. President Eisenhower, fresh from the Joint Service post of Supreme Allied Commander Europe, and with his wartime experience behind him, tried to eliminate the single-service Chiefs of Staff from the command chain, which he proposed should run from him through the Defense Secretary to unified commanders, the Joint Chiefs acting solely as military advisers to the Defense Secretary. Congress, strongly lobbied by ex-Chiefs, watered it down to the same organization in principle, but with the Joint Chiefs, as a committee, acting as the agents of the Defense Secretary in the execution of command, their number having been increased to five with the addition of the Commandant of the Marine Corps. This was the organization in force during the Vietnam War.

Westmoreland commanded the Military Assistance Command Vietnam, which comprised all the US armed forces actually based in Vietnam. It excluded both the naval forces of the 7th Fleet at sea and the USAF forces which bombed the north. The former, as was Westmoreland himself, were under the command of the joint service C-in-C Pacific, an admiral based in Hawaii, who received his orders from the Joint Chiefs. In practice Westmoreland, on important issues, dealt with General Earle G. Wheeler, Chairman of the Joint Chiefs, keeping CINCPAC, for most of the time Admiral U.S. Grant Sharp, informed. Logistic responsibility, on the other hand, was primarily dealt with through single-service channels from the Pentagon, through CINCPAC's single-service subordinates, CINCARPAC and CINCAIRPAC. Because Wheeler, Sharp and Westmoreland were determined to make the system work, it did, although it creaked. It probably would not have done so if prima donnas, like MacArthur, had occupied their posts. Command of the B-52s which bombed North Vietnam was exercised by USAF's Strategic Air Command, directly responsible to the Joint Chiefs, but had to be closely co-ordinated with all other air operations through CINCPAC and the USAF's 7th Air Force under Westmoreland's command in Vietnam. Only the sophistication of modern communications made such a complex command work at all. It never worked smoothly.

In the aftermath of Vietnam all three services – all four if the Marines were accepted as a separate service, as they increasingly tended to be – felt frustrated and dissatisfied, the army most of all. Its reputation sullied by apparent failure and a partial breakdown of morale in the later years, when the soldiers knew that the nation did not support the cause for which they were fighting and that the administration did not expect their campaign to succeed, the army laid the blame partly on the news media and partly on the administration's whizz-kids who thought they could win wars by clever psychological tricks like graduated pressure. Large sectors of the public, including the pro-testing students, blamed the army. Both the army and the air force tended to believe that, if only the restraints on their action had been removed, they could have successfully defeated the forces of North Vietnam, and that, if that had been achieved, a viable anti-communist regime, capable of standing on its own feet, could have been estab-lished, at least in the south: a doubtful proposition. The navy felt that valuable national effort had been wasted in a fruitless campaign, which proved that the nation should not get entangled in land cam-paigns in foreign parts; that strategic air bombing achieved nothing; and that the best method for the United States to employ to project its power overseas was by naval strength, centred on the aircraft-carrier, supplemented by the amphibious capability of the US Marine Corps. The costly war in Vietnam had seriously interfered with its new Rickover-inspired construction programme.

While the war had been in progress, the nuclear arms race had acceler-ated, responding to what was perceived as a 'missile gap' in favour of the Soviet Union. In the 1960s the USA had a clear lead in numbers of missiles and warheads, but the balance began to change in the 1970s. In 1963 the USA had 229 intercontinental ballistic missiles (ICBMS) and 105 of shorter range based in Europė, which could reach targets in the Soviet Union, to the Soviet Union's forty-four capable of reach-ing targets in the USA; 144 submarine-launched ballistic missiles (SLBMS) to the Soviet Union's ninety-seven, and 1,300 long-range bombers to 155+. In 1970 the ICBM balance was 1,054 to 1,300, SLBMS 656 to 240, and long-range bombers 520 to 140, the USAF's B-47s being phased out and the B-52 fleet halved. The balance in warheads or bombs was 4,000 to 1,800, the greater yield of many of the Soviet warheads producing a megatonnage balance of 4,300 to 3,100.

Having with limited success tried to restrict the USAF's insatiable appetite for offensive nuclear systems, McNamara also had to struggle against pressure for the development of an anti-ballistic missile (ABM) system, the Soviet Union having already clearly begun to develop one. He believed that it could not provide effective defence of the USA, but would only provoke the Soviet Union to increase their offensive capability. He managed to stall the issue, and it was left to Nixon's Republican administration to settle it. Instead, he authorized the development of multiple independently targeted re-entry vehicles (MIRVs) as a counter to Soviet ABM systems.

Kissinger, as National Security Adviser to the President, saw the danger of never-ending escalation in the nuclear arms race, and realized that the security of both the USA and the USSR would be enhanced if it could be frozen. With the first Strategic Arms Limitation Talks (SALT 1) agreement and the ABM Treaty of October 1972, he appeared to have achieved his aim, restricting ABM systems to one area, either the capital or an ICBM field, setting a limit on the number and siting of launchers and radars, and limiting the number of ICBM and SLBM launchers to those existing or under construction at the time – ICBMs: US 1,054, USSR 1,608; SLBMs: US 710 on forty-four submarines, USSR 950 on sixty-two submarines. The stability of these agreements, which excluded long-range bombers, was undermined by the introduction of MIRVs by both sides, by the development by the Soviet Union of missiles of greater 'throw-weight', and later by the American proposal to develop a comprehensive anti-ballistic missile system, the Strategic Defence Initiative (SDI), claiming that the Soviet Union had already embarked on that course. The number of warheads on both sides inexorably multiplied until the strategic balance over a decade later stood at: ICBMs, USA 1,045 with 2,145 warheads (1,375 megatons) to USSR 1,398 with 5,678 (5,481 megatons); SLBMs, USA 568 with 5,152 (333 megatons) to USSR 977 with 2,857 (914 megatons); long-range bombers, USA 272 with 2,570 (1,745 megatons) to USSR 145 with 290 (290 megatons). But the total of 9,867 US warheads was only a proportion of the whole US nuclear armoury which, counting all the naval, army and tactical air force weapons, had by that time reached an estimated 26,000.

The immediate aftermath of the Vietnam War was not a climate favourable to defence expenditure, and President Carter's Democrat administration, which succeeded Gerald Ford's Republican one in 1977, tried to hold it down against the pressure of the military/industrial complex, which was particularly powerful in the aerospace industry with its developing investment in electronics. As an ex-navy man himself, he was kinder to his former service than he was to the army or the air force, the latter failing to obtain approval for a new bomber to succeed the ageing B-52. President Reagan came to power in 1981 on a platform which accused Carter of having neglected the nation's defences and, in his first term at least, extracted from Congress sufficient funds to enable him to remedy the defects which he maintained had been caused by the lean years following the Vietnam War. The army's fears that it would be unable to recruit enough soldiers without the draft, or that its ranks could only be filled with a high proportion of society's underprivileged, including the coloured community, proved unfounded, perhaps because of the general level of unemployment, although senior officers continued to be concerned about the level of drug addiction. The regular army's authorized strength in 1985 was 780,648, of whom 75,500 were women. The army had ten divisions in the USA, four in Germany and one in South Korea; the manpower deployed in Europe amounted to 217,100; and in the Pacific, excluding Hawaii, where one division less one brigade was stationed, 32,300, almost all in South Korea. The combat forces stationed in the USA were divided between an initial reinforcement for Europe of seven divisions and an airborne corps of two divisions, one airborne and one air assault, envisaged as a Rapid Deployment Force, the Persian Gulf being considered the most likely deployment area. Army aviation consisted of sixty attack and 200 transport and general utility helicopters: its artillery amounting to 5,250 guns, mostly of 155-mm calibre, and its stock of tanks to 13,423, a large proportion of which were the out-of-date M48 and M60s, in the process of being replaced by the M1 *Abrams* and M2 *Bradley* tanks. The 7th Army in Germany was supported by fifty-four Pershing IA and fifty-four Pershing II ballistic missiles. The National Guard had an authorized strength of 379,000, organized into eight combat divisions, and the Army Reserve of 212,000 was organized into twelve training divisions. From these two a Selected Reserve Force of three divisions could be mobilized within five weeks.

In the same year the US navy numbered 568,781, of whom 48,200 were women. Its strategic nuclear strike force comprised six *Ohio* class submarines, each with twenty-four Trident missiles, twelve *Franklin* class (sixteen Tridents) and nineteen *Lafayette* class (sixteen Poseidons). It also had four cruise-missile and ninety-five attack submarines, all nuclear-powered. Its offensive strength lay in its four nuclear-powered and nine other aircraft-carriers, each normally carrying an air-wing of seventy to ninety-five aircraft, two F14 fighter squadrons, three F/A eighteen attack squadrons and one squadron each of electronic counter-measure and airborne early warning aircraft. Its thirteen carrier wings totalled 1,350 fixed-wing combat aircraft and 160 combat helicopters. It still kept two battleships in service, each mounting *Harpoon* and *Tomahawk* missiles. Twenty-nine cruisers (nine nuclear-powered), sixty-eight destroyers and 101 frigates were distributed among the three fleets, the 2nd in the Atlantic, 6th in the Mediterranean (two carriers and twelve other surface vessels), and the 7th, split between the Western Pacific (three carriers, twenty-three other surface vessels, six amphibious ships, twenty submarines) and the Indian Ocean, where a carrier battle group was normally deployed. The navy's sixty-one amphibious ships and fifty-four landing-craft supported the 198,241 Marines (9,200 women) in three divisions, each with its own air wing. The Marine Corps's total aircraft strength was 350 and its tank strength 716.

The USAF's authorized strength in 1985 was 603,898, of whom 67,500 were women, manning 3,700 combat aircraft and 1,023 ICBMs, most of the latter *Minuteman*, *Titan* having been almost entirely phased out. The Strategic Air Force manned 241 B-52s, fifty-six F1–11s, 646 tankers, forty-one command and twenty-four reconnaissance aircraft. The Tactical Air Force contained thirty-six fighter (F15 and 16), forty-eight fighter/ground-attack (F4, F1–11 and A10) and three reconnaissance squadrons. The Transport Force operated seventeen strategic (four C5 and thirteen C141) and fourteen tactical (C130) squadrons. The Air National Guard numbers 107,900, manning 1,020 combat aircraft in ten interceptor and thirty-six fighter/ground-attack squadrons. Overseas the USAF in Europe manned 725 combat aircraft, 309 in Britain, 324 in Germany, the remainder in the Netherlands, Spain, Greece and Turkey. In the Pacific, Guam was operated as a staging-post; ninety aircraft were based in Japan, the same number in South Korea and sixty in the Philippines. Thirty-two ground-launched *Tomahawk* cruise-missiles were based in Britain and sixteen in Belgium.

The considerable peacetime strength of the US armed forces, especially that of the US navy and the US air force, and the permanent stationing of significant numbers of all three services overseas, represents a fundamental change in US politico-military policy since the beginning of the century. The threat of Soviet-based communism in Europe, Asia, Central and Southern America, and to a lesser extent in Africa south of the Sahara, is seen as a threat to the US itself. Isolation from economic, political and military influences elsewhere in the world is no longer a viable policy for a nation, the economic and political tentacles of which reach out to every corner of the world, although isolationist voices still find ready echoes. Dependence on Gulf oil and links with Israel have been two strong influences drawing the US into involvement in the volatile affairs of the Middle East. Historical links with China and Japan draw the nation equally into the affairs of the Far East, while both proximity and strong economic links reinforce attempts to maintain the Monroe doctrine in Central and South America. Although in terms of numbers of men the US armed forces are by no means the largest in the world, their material strength can be compared only with that of the Soviet Union. The hope that those two giants, diametrically opposed in their political and military outlook, will avoid a military clash, rests on the mutual deterrence of their nuclear arsenals which, if unleashed, could come near to destroying life on earth, at least human life in the northern hemisphere.

The sheer size of the American defence establishment; the ramifications of its involvement in the military-industrial complex, and the huge staffs, incorporating a bewildering variety of technical specialists, which have proliferated, have produced a US war machine which contrasts starkly with the small élite professional cadres which nurtured the art of war and then practised it in the two world wars. As a result, the US armed forces have come under criticism from elements of the nation's well-developed academic defence establishment. They have been accused of neglecting the study of war and the art of training men (and now women) to fight in favour of reliance on the latest form of weapon technology, in support of which huge sums are spent by a military bureaucracy, serving its own vested interest. The human material has been neglected in pursuit of technology, the latest example of which is the Strategic Defense Initiative. The cost of this technical race aggravates the rivalry between the three armed forces, which is itself partly a cause of the race.

Whatever may be the validity of this criticism, there is no doubt that the sheer size of the machine and its technical complexity do pose severe problems; and that the impression is often given that the men who operate it – at all levels – have become obsessed by their own individual advancement within it. But, as in all nations at all times, the armed forces reflect the nation itself. Emphasis on Big Business has been a feature of American life for a long time, and the defence establishment of the United States of America has now become the Biggest Business of all.

7

JAPAN

1900–22

When Japan in June 1900 provided the largest contingent, 10,000 out of 20,111,* of the force despatched from Tientsin for the relief of the legations at Peking besieged by the Boxers, the Imperial Japanese army and navy were less than thirty years old. They were the creations of the Meiji imperial regime which had succeeded the *bakufu*,† the government of the Tokugawa shogunate in 1868. In 1192 the *tenno*, or Emperor, Go-Toba, created the post of *sei-i-tai-shogun*, 'commander-in-chief in charge of suppression of the barbarians', and since the battle of Sekigakara in 1600 the post had been held by the *daimyo*, a chief of the powerful Tokugawa clan, whose capital was Edo, renamed Tokyo in 1868 when the young *tenno* Meiji moved there from Kyoto. Up to that time the country was divided into 271 clans, the *daimyo* of which, while owing feudal and spiritual allegiance to the *tenno*, were practically subordinate in a loose fashion to the *shogun*, whose *bakufu* exercised a monopoly of all dealings with foreigners, including the possession of ocean-going ships. Each *daimyo* maintained his own army, drawn from the *samurai* warrior class who alone were entitled to bear arms. In 1868 there were some two million *samurai* out of a total population of thirty-two million, and within them there were higher and lower classes.

The contacts between western nations and Japan in the mid nineteenth century, of which the American Commodore Perry's arrival in Tokyo bay in 1853 is the best known, led to a clash between the shogunate, which pursued a policy of reluctantly granting concessions to foreign traders, and the powerful clans of Satsuma on the island of Kyushu, and Choshu, on the opposite side of the Shimonoseki strait on the main island of Honshu. Both these clans contained a high pro-

---

* Russia provided 4,000, Britain 3,000, USA 2,000, France 800, Germany 200, Austria 58, Italy 53
† Literally 'tent government'.

portion of samurai and, being the closest to the mainland – Korea was only 120 miles away across the Tsushima strait – were keenly aware of the indignities which China was suffering at the hands of Europeans. They themselves had direct experience of it. As a result of an incident near Yokohama in 1862, in which an Englishman had been killed for insulting the Satsuma daimyo, a British squadron had bombarded Kagoshima, among the defenders of which was the fifteen-year-old samurai Togo Heihachiro and the twenty-two-year-old Oyama Iwao, both destined to gain renown, the first at the head of the navy, the second of the army, in the Russo-Japanese War. They enlisted the help of a leading samurai, a veritable giant of a man in every sense of the word, Saigo Takamori, to found a secret society with the aim of expelling all foreigners and overthrowing the shogunate. The pressure they exerted was influential in persuading the Emperor in the following year to declare a policy of joi, expulsion of foreigners, as part of which the Shimonoseki straits were closed to all foreign shipping. In retaliation an international squadron, led by the US navy, bombarded the Choshu capital of Shimonoseki in September 1864. The Choshu reacted as the Satsuma had done, forsaking their rivalry with the latter in favour of a joint plot against the bakufu. The leading figures in their clan were Omura Masujiro and Yamagata Aritomo. Yamagata was to become the acknowledged father of the Imperial Japanese army, surviving until 1922; Omura had been assassinated in 1869 by a fanatical samurai who objected to the western-style reforms he was introducing into the newly formed Imperial army.

That act illustrated the dichotomy in the outlook of those who engineered the overthrow of the shogunate and the so-called 'restoration' of the imperial regime. Their opportunity came with the death of the young shogun Iemochi in 1866, followed by that of the Emperor Komei in the following year. Iemochi was succeeded by his weak guardian Keiki, and Komei by his fifteen-year-old son Mutsuhito, who, as the Emperor Meiji, was to rule for forty-five years and preside over the astonishing transformation of Japan from a medieval country, almost isolated from the rest of the world, into a modern international power. The instigators of the revolution, however, were fundamentally conservative. They were not motivated by any admiration for European civilization as such, but by an intense desire to preserve traditional Japanese values. They saw what was happening in China, where European nations, using modern weapons, were imposing their trade, and with it their alien culture, on an ancient civilization, from which that of Japan was derived, and humiliating the Chinese in the process. They saw with realistic clarity that, unless Japan herself could develop armed forces on European lines, she might suffer the same fate, and that she could not muster the power to do so without a fundamental change in the way affairs were conducted. The Choshu leaders had already accepted this in raising a force known as *kiheitai*, or surprise troops, which employed peasants as well as *samurai*, equipped with modern weapons bought in Shanghai from the British and Dutch. The Kishu clan took the same step, buying weapons from Germany and engaging the services of a German sergeant, Karl Köppen. In reaction to this threat to its own authority, the shogunate, encouraged by the French ambassador, Léon Roches, made overtures to France in 1866 for help in developing a modern army of its own. Nothing much had happened in response to that when the *shogun* was forced to hand over his powers to the young Emperor in 1867, finally resigning in the following year, when the new imperial regime was established, abolishing the feudal status of the *daimyo*, whose territory was transformed into provinces, of which they themselves for the most part became the governors or prefects, appointed by the Emperor. The latter's 'government' was formed by a coalition of the leading personalities of the revolution, later to be described as the *genro* or elder statesmen, the principal bureaucrats of the shogunate and the officials of the Emperor's court.

Omura, given the task of raising an imperial army, was greatly influenced by the scholar Nishi Amane, who had studied military matters in Holland. At first he had to rely on the existing *samurai* bands, supplemented by the officers who had been trained for the shogunate by the French, whose mission was retained and expanded. When Omura was assassinated in 1869, Yamagata took his place, returning from a tour of Europe where he had been greatly impressed by the German army. After the French defeat in the Franco-Prussian War, opinion swung in favour of the Germans, and in 1872 Yamagata persuaded the Emperor and his ministers to introduce conscription for all males on reaching the age of twenty, to serve two years on active duty, followed by four in the reserve. His ideal was an army of 240,000 in peace, based on 40,000 for each of the six main garrisons, which would be expanded in war to 400,000; but financial realities forced him to accept a peacetime target of 40,000, expandable to 70,000 in war. When the measure was announced in 1873, it actually provided for only 31,860 men in peace and 46,350 in war. Conscription introduced some major changes into Japanese life, eating meat instead of fish being one of them. The most significant was the downgrading of both the actual and the symbolic importance of the sword, which the *samurai* fought fiercely for the right to retain.

All these measures, political and military, were deeply resented by many of the *samurai*, although all the leaders of the revolution were to come from that class and many were to benefit materially from the development of the *zaibatsu*, the great commercial empires of Mitsui, Mitsubishi, Sumitomo and Yasuda. Not only were the *samurai* deprived of their monopoly in bearing arms, but the great majority, particularly of the lower class, found themselves without an occupation, pensioned off with a pittance. One way of employing them, backed by Saigo Takamori, was to organize an expedition to Korea, the king of which had refused to recognize the new regime, and settle unemployed *samurai* there. Yamagata and the majority of the Emperor's advisers, notably the influential Ito Hirobumi and Inouye Kaoru, both, like Yamagata, Choshu men, opposed this. They emphasized that Japan must build up its own strength first before it could risk an adventure beyond its islands, which would almost certainly involve reaction by other powers, particularly Russia.

Saigo left the government and returned to Satsuma where, in 1877, he led the most serious of the revolts that erupted over this issue, testing the loyalty and effectiveness of Yamagata's new Imperial army. It was significant that several prominent Satsuma colleagues of Saigo fought on the government's side against him, Oyama skilfully employing the guns bought from Krupp at Kagoshima, where the wounded Saigo committed suicide. Also under Yamagata's command at that decisive battle was a twenty-nine-year-old Choshu officer, later to gain renown in the Russo-Japanese War, Nogi Maresuke. Attacked by a stronger force of Satsuma *samurai* and forced to retreat, he lost the regimental standard, presented by the Emperor. Ashamed, he decided to commit *seppuku* but was prevented from doing so by his men. When Yamagata insisted that honour demanded it, the Emperor reprieved him. Thereafter Nogi sought death on the battlefield in vain, finally retrieving his honour by disembowelling himself on 13 September 1913 as the first gun fired to mark the departure of the hearse, carrying the Emperor's coffin from the gates of the palace in Tokyo. The *samurai* might feel that they had been rejected and humiliated, but their military code of behaviour, *bushido*, was preserved and fostered in the Imperial army of which they were the founders.

While the Choshu dominated that army in its early years, the Satsuma were to prevail in the navy, formed in 1873. When the *bakufu*'s ban on the possession of ocean-going ships was lifted in 1854, the shogunate assumed control of the private fleets which, in spite of it, the *daimyo* had acquired. The nucleus of a navy therefore existed at the time of the restoration; and when Queen Victoria's son the Duke of Edinburgh, a serving naval officer, arrived promptly to pay his respects to the new Emperor in 1868, arrangements were made for the British Royal Navy to train it. One of the first cadets to go to England to train at Greenwich and Portsmouth was the twenty-two-year-old Togo. He arrived in 1871, the year in which the first orders were placed for warships in Britain. The Emperor personally contributed a large sum and 10 per cent was deducted from the salary of every government employee to pay for them; an additional source was the indemnity of 500,000 Mexican dollars which was exacted from

China when an expedition was sent in 1874 to occupy the Ryukyu islands and Formosa, in retaliation for the primitive inhabitants of the former having been said to have eaten some shipwrecked Japanese.

The defeat of the Satsuma rebellion did not put an end to Japanese ambitions in Korea, whose weak king was regarded by China as a tributary. The Japanese government became more and more anxious as representatives of European countries, especially Russia, began to establish links with this primitive 'hermit kingdom' as it was known. Attempts by the powerful Queen Min to encourage these contacts were opposed by her father-in-law, Prince Taewongun, who led a revolt against her in 1882, in which the Japanese legation and soldiers training the Koreans were attacked. The royal party successfully appealed to China for help, Li Hung-chang, Viceroy of Chihli-Li, the province surrounding Peking, leading an expedition, carried in six gunboats, to Chemulpo (later to be known as Inchon). After he had captured Taewongun and removed him to China, the Japanese minister returned with a battalion of Japanese soldiers, transported in the cruiser *Amagi*, commanded by Lieutenant-Commander Togo, and enforced an indemnity of half a million yen, the right to station troops to protect his embassy and for Japanese to trade and travel freely. At the same time the Chinese, commanded by Yuan Shih-kai, strengthened their position there.

Sino-Japanese rivalry intensified, the Japanese becoming more aggressive after the French had sunk the Chinese fleet without warning at Foochow in August as a preliminary to their expulsion of the Chinese from Indo-China. As both sides strengthened their position in support of rival Korean factions, a serious incident occurred in December, leading to a direct clash between Japanese and Chinese soldiers, both sides reinforcing their garrisons. But the Japanese government had no desire to find itself at war with China and reached an agreement with Li Hung-chang, confirmed in the Treaty of Tientsin in April 1885, which recognized Japanese rights in Korea. China, however, did not abandon its claim to suzerainty, and Yuan, as Chinese pro-consul in Seoul, became the virtual ruler of Korea, the King being a mere puppet in his hands. In the face of this, Japan switched its support to secret societies opposed to the King and the Chinese. Matters came to a head in June 1894 when in response to an appeal from the King for help, Chinese troops were moved up to the Yalu river and landed at Chemulpo, supported by warships, Li Hung-chang's notification of this action to Ito referring to Korea as a tributary. Ito replied that Japan did not recognize Chinese suzerainty, and despatched a brigade of marines and soldiers to Seoul, at the same time mobilizing three divisions. By August a Japanese army of 10,000 was established in Korea and it clashed with the Chinese at Pingyang on 15 September, gaining a clear victory. So did the Japanese navy two days later at sea off the mouth of the Yalu, Togo's cruiser having already sunk a Chinese ship transporting troops.

Under the overall command of Oyama, the First Army crossed the Yalu into Manchuria, while the Second Army landed on the southern coast of the Liao-tung peninsula and captured Port Arthur with 26,000 men on 21 November in a professionally conducted assault. Part of the army then moved north to join the First Army, which inflicted a crushing defeat on the Chinese at Tien-chuang-tai on the Liao river on 9 March 1895. The navy had already landed a force to occupy Wei-hai-wei at the northern end of the Shantung peninsula. The Chinese sued for an armistice, and on 10 April a peace treaty was signed at Shimonoseki, by which China recognized the independence of Korea, ceded to Japan Formosa (which Japan already occupied), the nearby Pescadores Islands and, of great significance, the Liao-tung peninsula, and agreed to pay a war indemnity of 200 million taels, the equivalent of over £25 million. Ten days later, before the treaty had been ratified, Russia, France and Germany presented a joint note to Japan (known as the Triple Intervention) urging that she forego her claim to the Liao-tung peninsula on the grounds that it would be prejudicial to the lasting peace of the area. Reluctantly swallowing his pride the Emperor, on the advice of Ito and Inouye, who realized that Japan was in no position to challenge those three powers, 'yielded to the dictates of magnanimity, and accepted the advice of the three powers'. In compensation 39 million taels (£5 million) was added to the indemnity. Most of it was to be spent on building up the navy.

The impressive performance of the Imperial army owed much to the training it had received from the Germans. After the Franco-Prussian War admirers of the German army, like Yamagata, gained ground. The turning-point was the 1877 Satsuma rebellion which revealed a number of serious weaknesses in the army, for which the Japanese high command blamed the French. Their mission left in 1880 although two officers remained as instructors at the military academy, which had been founded on the model of St Cyr in 1875. It was not until 1885, after Oyama had led a mission to Europe and the USA, weighing the merits of the armies he saw, that the forty-three-year-old Rhinelander Major Meckel arrived to teach at the army staff college, founded in 1882, four years after the establishment of a General Staff. His task was officially restricted to instruction at the staff college; but, with Japanese approval, he extended it to advice to the General Staff. In the three years he spent there he was largely responsible for major reforms, which included the conversion of garrisons to divisions of all arms, and for mobilization preparations, including all the logistic and administrative measures and troops needed to support an army in the field. He was also instrumental in the development of the post of Inspector-General, independent of both the Army Minister and the Chief of the General Staff. When the latter post had been created in 1878, Yamagata insisted that its occupant should be responsible directly to the Emperor and not the Army Minister. On Meckel's recommendation the Inspector-General was given equal status. He became responsible not merely for inspection, but for all training and education, as well as personnel matters.

The constitution, introduced by imperial rescript in 1890, owed much to the Prussian model, of which Ito and Yamagata were admirers. It stipulated that the appointment of the Army and Navy Ministers must be approved by their respective General Staffs; in 1900 it was decreed that they must be serving officers, although from 1913 to 1936 that was altered to allow retired officers to hold the post. It meant that the army or the navy could bring a cabinet down, if either of them refused to nominate or approve a minister, as the navy did in 1914 as a result of disagreement about the naval budget, and the army did on several subsequent occasions. At a lower level, the duties and rights of soldiers were defined in two important rescripts, the *Gunjin kunkai* (Admonition to Soldiers) in 1878 and the *Gunjin chokugo* (Imperial Rescript to Soldiers and Sailors), issued on the authority of the Emperor. The special military police unit, the *Kempeitai*, set up in 1881 to enforce them, later extended its activities into the civilian field. The rescripts laid strong emphasis on soldierly qualities, reverence for the Emperor and national patriotism, as did a similar Imperial rescript on education disseminated to all schools.

The military rescripts stressed the need for soldiers and sailors to refrain from taking part in politics. From 1890 to 1945 they were strictly forbidden to attend political meetings (except in the course of their duty – probably to break them up) and were denied the vote. But this did not prevent generals and admirals from indulging in political activity. Indeed, for much of the time, they dominated it. In the early years political influence was largely in the hands of men who were territorial magnates as well as military leaders, a situation typical of eastern Asia, surviving today in Thailand.

Developments in the army had their parallels in the navy, although the latter was being trained by the British. In 1876 a naval academy was formed, upgraded in 1888 to a college when it moved from Tokyo to the island of Etajima. From then until 1938 there were always two British naval officers there as instructors, strong emphasis being laid on physical fitness. Meckel was joined in 1886 by Captain von Blankenburg and, when he left in 1888, he was himself replaced by Major von Wildenbruck. Tired of continuing tension between French and German officers, the Japanese decided to get rid of both in 1890, but re-employed a German major until 1895. An authoritative source summed up the influence of ten years of German advice as follows:

What in 1885 had been a poorly organized and directed defence establishment, dependent on garrisons for its supplies, without a mobilization or operational plan, had by 1890 become a flexible organization of six divisions, led by a general staff corps that was capable of preparing large manœuvres or sending armies to the Asiatic continent. In the process the army had lost much of its French character and assumed a German outlook. Theory and appearance were replaced by practical knowledge and realistic performance.[1]

German influence was reinforced by the support Germany gave to the revision of the 'Unequal Treaties' in the Treaty of Tokyo in 1882, but her co-operation with Russia and France in the 'Triple Intervention' undermined it. Germany's determination not to be left out of the process of 'carving up China' was further emphasized in 1897 when, after the murder of two German missionaries, she forced China to grant her a ninety-nine-year lease of Kiao-chao (modern Tsingtao) with other concessions in the Shantung peninsula. With the Japanese still in Wei-hai-wei and the British showing an interest in Ta-lien-wan on the Liao-tung peninsula, Russian warships were despatched to Port Arthur in December, forcing China to agree in March 1898 to a twenty-five-year lease of the port and its hinterland, with rights to erect defences. Britain's response was to acquire a lease of territory on the mainland opposite Hong Kong and of Wei-hai-wei, as long as the Russians were at Port Arthur. The French chimed in with a lease of Kuang-chou-wan.

These foreign encroachments onto the mainland of China provoked the xenophobic movement known as the Boxer rebellion, to counter the threat of which to the foreign legations a small force of American, French, English, Russian and Japanese soldiers was sent to Peking in June 1900, followed by an international force of 2,000 men, fifty-four of whom were Japanese, commanded by the aged British Admiral Seymour. Before the force had got far, the Boxers occupied Peking and laid siege to the embassies, with the apparent approval of the Chinese Dowager Empress and her authorities. A larger force was clearly needed, and the Japanese Prime Minister Katsura Taro and Oyama, the Chief of Staff, prepared to send two divisions, totalling 30,000 men; but, in the face of opposition from both the Russians and the Germans, reduced their contingent to 10,000. At that figure it was almost equal to the total of all the other contingents under the command of the Russian General Linievich, and did most of the fighting before the siege was raised on 14 August. Thereafter the international force was built up to a strength of 70,000, under the command of the German Field Marshal Waldersee, the Russians withdrawing their force to concentrate on the occupation of Manchuria where, in addition to incidents on the Amur river, several attacks had been made on the Chinese Eastern railway either side of Mukden and on the Imperial railway linking it with Tientsin and Peking. By the end of October 1900 Russia was occupying virtually the whole of Manchuria, and in February 1901 presented a draft treaty which, although recognizing Chinese suzerainty, would have transferred real power in Manchuria to Russia. This naturally gave rise to grave concern in Japan, which was successful in mobilizing the opposition of the governors of other Chinese provinces and the European powers involved (although Germany played a double game) to stiffen Li Hung-chang's refusal to accept it. Pressure was also exerted by the USA for the maintenance of the Open Door policy in China, i.e. that no foreign nation should be given any rights to the exclusion of other powers. Punishment of the rebels and a large indemnity were demanded, but no cession of territory. Legation guards in Peking were to total 2,000, with 6,000 more to be stationed at Tientsin and elsewhere to protect the railway. The agreement did not cover the thorny problem of Russian troops in Manchuria. That was the subject of a treaty signed in April 1902, by which Russia agreed to withdraw her troops from Manchuria (except Port Arthur) in stages over eighteen months, handing back territory to China as she did so. The first stage was completed in October, when part of Mukden province was evacuated.

These events brought Japan and Britain together. The latter, engaged in the Boer War, felt isolated in the international field, Russia and France having concluded an alliance in 1894 and Kaiser Wilhelm II having already been in collusion with his cousin Tsar Nicholas II over events in the Far East. Japan had been devoting much of the indemnity obtained from China in 1895 to purchase warships from Britain and develop her navy, while Britain was seeking means of reducing her naval commitments far from European waters. Prime Minister Ito, a Germanophile who sought a diplomatic solution to potential conflict with Russia, was succeeded in 1901 by General Katsura, a Choshu protégé of Yamagata, a Russophobe. He backed the initiative taken by Hayashi Tadasu, ambassador in London, to negotiate an alliance, cashing in on British anxiety about the Russian threat to India, as did his Foreign Minister Komura, all of them, except Hayashi, military men. Yamagata, holding the influential post of President of the Privy Council, supported them. Only Ito and Inouye among the genro opposed it. Secrecy in the negotiations avoided attempts by other powers to block it. Announced on 30 January 1902, the two governments 'activated solely by a desire to maintain the status quo and general peace in the Far East, being moreover especially interested in maintaining the independence and territorial integrity of the Empire of China and the Empire of Korea, and in securing opportunities in those countries for the commerce and industry of all nations' contracted, if either of them became involved in war with a third power in defence of its interests in the area, to remain neutral and do its best to prevent other powers from joining in hostilities against its ally: but if any other power or powers did so, to come to its ally's help and take part in the war. The agreement was to last for five years initially.

Reassured by this alliance, Japan was prepared to take a tougher line with the Russians over events in Manchuria and Korea. In spite of her agreement with China, concluded three months after the Anglo-Japanese alliance, Russia failed to implement the second stage in 1903, and presented seven demands affecting Mongolia and Manchuria which China, backed by Britain, the USA and Japan, refused to accept. Russian activity in Korea, notably in the area of the River Yalu, also caused concern. Since 1895 Japan had extended its activity in Korea, almost to the extent of treating it as a conquered territory, driving the King into the hands of the Russians. The more powerful Queen had been killed by Japanese soldiers in 1895. It was fears that Russia would penetrate Korea in the same way as she had established her position in Manchuria that finally convinced the Japanese authorities that they must be prepared to fight her. A series of meetings took place in May and June 1903, before and after a visit to Japan of the Russian General Kuropatkin. A public demonstration by the anti-Russian Black Dragon Society was followed by several meetings of the genro. Yamagata, Katsura and Komura persuaded Ito that, if Russia penetrated into Korea, there was no alternative to war. Concerned at the cautious attitude of Oyama, Chief of the General Staff, and his deputy Tamura, a meeting of relatively junior officers of the army and navy General Staffs brought pressure to bear on their seniors, which influenced the decisive meeting on 23 June between the Emperor, Prime Minister Katsura, Army Minister Terauchi, Navy Minister Yama-moto and Foreign Minister Komura. They agreed that any cession of Korean territory to Russia was unacceptable, and that, if negotiations to prevent it failed, war was inevitable.

The result was a proposal, made in July, to the Russians of a treaty by which Japan would recognize Manchuria as a Russian sphere of influence, in which she had a right to protect her own interests, while the same applied to Japanese rights in Korea. Neither side would inter-fere with the predominant position of the other in these spheres of influence, including that of the Japanese to extend the railway in Korea into Manchuria. In the negotiations which followed and were finally broken off in February 1904, Russia tried to limit the treaty to Korea, maintaining that Manchuria was none of Japan's business. Japan was prepared to accept that, if Russia agreed in return to be excluded from Korea. From October 1903 onwards, when the bellicose Choshu General Kodama was appointed Vice-Chief of the General Staff, both sides were clearly preparing for war.

By then the Japanese army could field thirteen divisions with a simi-lar number of kobi (reserve) brigades for line of communication duties, two cavalry and two artillery brigades, some 200,000 men, supported by 666 guns. A further 400,000 trained men were available to replace casualties. During the war four more divisions and four *kobi* brigades were added. This army was provided by conscript service to which a man was called up at the age of twenty (although liable from seventeen to forty). After three years in the active army, the conscript served a further 4⅓ in the First Reserve, the two together forming the Standing Army of about 380,000 men. A further five years was spent in the Second Reserve, both First and Second Reserves being liable to call-up for periodic training. In addition there was a Conscript Reserve consisting of all those who, although liable to conscription and medically fit, had for one reason or another escaped service in the Standing Army. Service in this reserve lasted for 7⅓ years and produced 50,000 men. Thereafter reservists of all kinds, until they reached the age of forty, were liable to call-up in the National Army. Altogether this produced some 850,000 men who had received military training. In addition, on the basis of the 1898 population of 46 million, 4¼ million untrained men were liable for service. The infantry were armed with a Japanese version of the 2.56-inch calibre Mauser rifle, the cavalry with a shortened version as well as sabres. At the beginning of the war they had no machine guns, but before the end of 1904 ten French Hotchkiss guns were supplied to each division, increased in March 1905 to fourteen. The artillery was based on the Arisaka 75-mm field gun, firing a 13-lb shell to a range of 5,000 yards. Half of the divisions had the mountain version of shorter range. The artillery brigades were equipped with a 120-mm howitzer. Both were inferior to European artillery of the same period, and were replaced by Krupp designs in 1905. The infantry were trained to close up to the enemy under cover of artillery fire, which was decentralized to a low level, moving rapidly and making the greatest possible use of natural features. Emphasis was laid on night operations. Major Meckel's emphasis on mass shock action and the use of the bayonet had been modified in theory to a greater stress on firepower, but assaults were made in close formation with great vigour and determination.

The Imperial Japanese navy had been built up since 1895 into a formidable force, headed by six first-class battleships and six armoured cruisers. The oldest battleships, the *Fuji* and *Yashima*, were launched in England in 1896 and were then the fastest and most heavily armoured battleships in the world. They were refitted in 1901 with sixteen rapid-firing 12-inch guns. Two years later they were followed by the *Asahi* and *Shikishima*, each mounting four 12-inch and fourteen 6-inch guns and carrying five torpedo-tubes. In 1900 the flagship *Mikas* was launched, carrying the same armament. She was the largest and most powerful battleship in the world, 400 feet long with a top speed of eighteen knots and even more heavily armoured than her predecessors.

In addition the navy had eleven second-class and seven third-class protected cruisers, nineteen destroyers capable of thirty knots, fifty-eight first-class torpedo boats, capable of twenty-two to twenty-nine knots, and twenty-seven second-class, varying in top speed from eighteen to twenty knots. Twenty-six old cruisers, sloops and gunboats were too out of date to take part in fleet actions, but were useful for escort and bombardment support to landing operations. Although Japanese shipyards were capable of building some types of naval vessel, they could not produce armoured ones. The bulk of the fleet was concentrated at Sasebo on the western coast of Kyushu, with a detachment at the island of Tsushima and one protected cruiser at Chemulpo in Korea.

The events of the war have been described in Chapter 4. As the size of the army deployed in Manchuria increased and the operations of the three armies, Kuroki's First, Oku's Second and Nozu's Fourth, all converging on Liao-yang, with Nogi's Third besieging Port Arthur, could no longer be effectively commanded from Japan, the problem arose as to who should be appointed C-in-C in Manchuria. The choice was affected by the rivalry between Satsuma and Choshu. Oyama, Kuroki, Nozu and Kawamura were all Satsuma and Oku came from an island close to Kyushu. Nogi's wife and his Chief of Staff were also Satsuma. Yamagata, the principal Choshu military figure, had a clear claim to the command, but not only was he Prime Minister, he was also not in very good health and was considered by the leading Satsuma genro to be too obsessed with detail. A compromise was reached by which Oyama was appointed C-in-C in July 1904, with the forceful Choshu Kodama as his chief of staff, while Yamagata took over Oyama's responsibilities as Chief of the Army General Staff as well as remaining Prime Minister.

The attitude of mind of the old-fashioned *samurai* in the Japanese army is illustrated by the stories about General Nogi. After serving as Governor of Formosa, following the Sino-Japanese War in which he had fought with conspicuous courage, he retired to a life of farming and poetry. When he heard that his eldest son had been killed before he arrived, he replied, 'I am glad he died so splendidly. It was the highest honour he could have.' When his second and favourite son was killed carrying a message in the closing stages of the assault on Port Arthur in November 1904, having satisfied himself that he had completed his task, he said, 'I often wonder how I could apologize to His Majesty and to the people for having killed so many of my men. But now that my son has been killed....' And in a letter to Terauchi, the army minister, when the siege was over and the total casualties had risen to 57,780 killed and wounded and 33,769 had died from other causes, he wrote:

The feeling I have at this moment is solely one of anguish and humiliation that I have expended so many lives, so much ammunition and such a long time upon an unaccomplished task. I have no excuse to offer to my sovereign and to my countrymen for this unscientific, unstrategical combat of brute force. ... I thank you heartily for your kind condolences on the deaths of my sons and I beg you to forgive my long display of military unskilfulness.

He also wrote a poem:

> His Majesty's millions conquer the strong foe.
> Field battles and sieges result in mountains of corpses.
> How can I, in shame, face their fathers?
> Songs of triumph today, but how many have returned?[2]

One who was singing a song of triumph was Lieutenant Tojo Hideki, celebrating both the fall of Port Arthur and his twenty-first birthday. It is difficult to imagine him, forty years later, adopting such a humble and sensitive attitude.

The Battle of Mukden at the end of February 1905 saw the Japanese army at the peak of its strength, Oyama deploying 207,000 men supported by 1,000 guns in five armies, against Kuropatkin's 276,000. In spite of his great victory, his failure to prevent the Russian army from escaping to the north was held against him, although the blame should more fairly be laid at the door of his subordinates, who did not always obey their orders and allowed the momentum of their operations to die away once Mukden was reached. There could however be no doubt that it was a decisive victory which, with that of the navy at Tsushima three months later, was instrumental in bringing the war to a successful conclusion. Given the primitive communications of the time, it was a remarkable feat on the part of Oyama's Chief of Staff, Kodama, to have been able to maintain control of such a large force in carefully planned and executed operations, involving several armies manœuvring separately. Few armies were to achieve it as well in the First World War that followed nine years later.

In the command of his armies Oyama had not enjoyed the facility available to the admiral who, as a fifteen-year-old, had fought by his side in the defence of Kagoshima against the British forty-three years before. Togo had the advantage of wireless telegraphy as a means of communication between his ships, and it served him well in keeping him informed of the movements of the Russian fleet as it approached the straits of Tsushima in the early hours of 27 May. His sets were copied from a Marconi design. His opponents also had wireless, German Slaby-Arco sets, but seldom used them and, when they did, could never get them to work. Although the British Royal Navy used wireless in the Boer War, Tsushima was the first occasion when it was used in a naval battle. Togo had left his main base at Sasebo in February, concentrating his fleet in a deep sheltered bay at Masan near Pusan at the south-eastern tip of Korea. He took the gamble of leaving Japan's eastern shores unguarded, assuming that Rozhdestvensky would not take that route to either the Tsugaru or the Soya (La Pérouse) straits. His brilliant and unconventional operations officer, Commander Akiyama, had prepared seven plans from which he could choose, depending on how the Russian admiral handled his barnacle-encrusted ships. By 22 May Togo began to be anxious, as he had had no news of the Russian fleet since it had passed between the northern Philippines and Formosa almost a week before. If he waited too long and Rozhdestvensky did sail east of Japan to the other straits, he would be too late to intercept him before he reached Vladivostok. He heaved a sigh of relief on 25 May when he received information that Rozhdestvensky had sent six supply ships back to Shanghai. If he had taken the other route, he would have kept them with him. At 3.30 a.m. on 27 May Togo received a wireless message from the *Shinano Maru*, an armed merchant cruiser, that the enemy had been sighted 'in square 203' heading for the eastern channel.

At 5 a.m. Togo, in the *Mikasa*,* led his fleet out to sea, followed by the battleships *Shikishima*, *Fuji*, *Asahi* (with Captain Pakenham RN aboard as an observer) and the armoured cruisers *Kasuga* and *Nisshin*. Vice-Admiral Kamimura (who had been fishing when the fleet was ordered to sea and whose independent action in the battle was to be of great value to Togo) followed in his flagship *Izumo*, leading the Second Division of the battleships *Azuma*, *Tokiwa*, *Yagumo* and *Iwate*. The Third Division of four old battleships under Rear-Admiral Kataoka and the Fourth of cruisers under Rear-Admiral Dewa, were already deployed forward at Tsushima island itself, covering the straits on either side. The action was certainly one of the decisive battles of history, revealing the Imperial Japanese navy at every level as having mastered in a remarkably short period the technicalities of naval tactics in the mechanical age. One of the midshipmen in the cruiser *Nisshin*, escorting Togo's flagship, who was wounded when his ship was hit, was the twenty-year-old Yamamoto, who thirty-five years later was to find himself occupying Togo's exalted post. He was no relative of the Navy Minister, Yamamoto Gonnohyoe, a Satsuma, but, born Gombei, was adopted into the latter's clan on the death in 1913 of his father, a schoolmaster from Nagaoka in northern Honshu. While Yamamoto Gonnohyoe was Navy Minister from 1895 to 1905, the chief of the naval General Staff was another Satsuma, Admiral Ito Sukeyuki.

With the end of the war the admirals and generals, particularly Togo and Nogi, were the heroes of the hour, and the standing of military

---

* Built at Barrow-in-Furness and fitted out on Tyneside.

men in the affairs of state was higher than ever. Between 1900 and 1945 the Prime Minister was either a general or an admiral in twenty-five out of the forty-six years. But the victory of an Asian country over a European power also encouraged both xenophobic movements like the Amur River (or Black Dragon) Society, which had close links with the army, and anti-colonial socialist movements, which were basically anti-military. Only three years after the end of the war a parade of socialists and anarcho-syndicalists took place in Tokyo. In 1911, a year before the death of the Emperor Meiji, the police severely repressed these movements when a plot against his life was unearthed. One of the results of the war was a tightening of Japan's grip on Korea, where Ito had been appointed Resident-General. When he presented demands in 1907, which would virtually have turned Korea into a Japanese colony, the King abdicated in favour of his son Yi. At that time Yamagata favoured outright annexation, which Ito opposed, and the issue precipitated a cabinet crisis leading to the replacement of Prince Saionji as Prime Minister by General Katsura, Yamagata's protégé. Ito resigned in 1909, and his murder in the same year by a Korean at the railway station at Harbin gave Tokyo a pretext for annexation, which was confirmed by treaty in August 1910. Yi was pensioned off, and General Terauchi took over the reins of government of the country, given the inappropriate new name of *Chosen* – 'land of morning calm'. It was to be ruled by the Japanese with an iron hand for the next thirty-five years.

The strength of the army's political influence, even when the head of government was a civilian, was shown in 1912. Saionji had resumed his post as Prime Minister in August 1911 and was in office when the new Emperor, weak both physically and mentally, succeeded to the throne as Emperor Taisho in July 1912. The army, having assumed fresh responsibilities in Korea as well as in the Liao-tung peninsula, pressed for an increase in its strength from the seventeen divisions which had been raised during the war to nineteen. Yamagata backed this demand, but Saionji, concerned at the financial situation, opposed it. Outvoted in the cabinet, the Army Minister, General Uehara, submitted his resignation directly to the Emperor, forcing Saionji and his cabinet to resign when the army refused to nominate a successor. General Katsura, who returned to replace him, authorized the increase in army strength, but found himself opposed not only by the Diet, the lower house of parliament, when it met in December, but also by the navy who disliked him personally and regarded him as an inflexible Choshu partisan. Riots, suppressed by the army, followed, inflaming feeling between Choshu and Satsuma. The outcome favoured the latter, Katsura being forced to resign in favour of Admiral Yamamoto Gonnohyoe, the political influence of the Satsuma having been strengthened by an alliance with the *Seiyukai* party, which Saionji had led since the death of Ito Hirobumi, backed financially by the Mitsui *zaibatsu*, rival of Mitsubishi who backed Katsura's Choshu-based *Rikken Doshi-kai* party, later renamed *Kenseikai* (Constitutional) and later *Minseito* (Democratic) party. In very rough terms the *Seiyukai* were regarded as Liberals and the *Kenseikai* as Conservatives. Yamamoto's administration did not last long, being forced to resign in April 1914 over a scandal involving naval officers accused of accepting bribes from a German firm, which provoked public protests, troops being called out to suppress them. He was succeeded by the moderate civilian Okuma, who was expected to be neutral in the Satsuma–Choshu rivalry; but he quickly allied himself with Yamagata, who supported him until the incident of the 'Twenty-One Demands' on China in 1915.

On the outbreak of war in Europe in 1914, Japan delivered an ultimatum to Germany to hand over her lease of Kiao-chou, which she promised at some time to return to China. Having received no reply she declared war on 23 August 1914, occupying the leased territory and German island possessions in the Pacific before the end of the year. The siege of the port of Tsing-tao on the Shantung peninsula saw the first use of aircraft by the Japanese forces. With their forces in both the Liao-tung and the Shantung peninsulas and the European powers preoccupied with their war in Europe, the Japanese government felt itself to be in a very favourable position to extract far-reaching concessions from the feeble Chinese government, of which Yuan Shih-kai was the *de facto* ruler.

Egged on by the military attaché in Peking, the Foreign Minister, Kato Komei, persuaded Okuma and his cabinet to agree to the presentation of twenty-one demands in five groups, which included transfer of all German concessions and rights, the extension of Japanese rights in Manchuria and, in the fifth group, Chinese agreement to the appointment of Japanese political, financial and military advisers to the central government, to the participation of Japanese police in the administration of the large cities and to Chinese arms purchases being restricted to Japan. Not surprisingly, these demands aroused strong anti-Japanese feelings in China and stern protests from the USA, while Britain tried to persuade the Japanese government to tone them down. The more extreme demands in the fifth group were dropped, but China was forced to sign two humiliating treaties, transferring all German concessions to Japan and extending the latter's rights not only in Manchuria but in Inner Mongolia also.

It was ironic that the opportunity, in the absence of the rival European powers, to establish a satisfactory relationship between Japan and China, and between both of them and the USA, was lost by an administration headed not by the military but by the civilian political leaders of liberal outlook, Kato having created a good impression as ambassador in London before he became Foreign Minister. Indeed

Yamagata was dismayed at the anti-Japanese feeling aroused and withdrew his support from Okuma. The result was the latter's replacement in November 1916 by Yamagata's Choshu protégé General Terauchi who, as Governor-General of Korea, had cultivated good relations with Yuan Shih-kai. When the latter died in June 1916, he sent a personal emissary to Peking to try and influence the northern war-lords who exercised power. When he became Prime Minister he continued this policy by offering financial help, personal and public, to them.

He was able to do so as Japan was profiting substantially from the orders placed on her industry by the British and French, especially for shipping. The economy prospered, led by the *zaibatsu*: gold reserves, held by the government and the Bank of Japan, increased a hundredfold, and the country became more self-sufficient industrially. But the prosperity did not reach down to the lower levels of society, who suffered from inflation and a shortage of rice, Korea having become an important source of food. There were serious riots in September 1918, provoked by the price of rice, which forced the resignation of Terauchi's administration. Earlier in the year Japan had joined the Allied expeditionary force in Siberia, supporting Kolchak's White Russian campaign against the Bolsheviks. Her motive was to ensure that neither the Bolsheviks nor any of the European powers should establish their influence in the area of the Amur valley in Eastern Siberia. She kept her troops there long after others had withdrawn, remaining in the Amur river area until 1922 and in North Sakhalin until 1925. This expedition was unpopular at home and had a deleterious effect on the Japanese army. The example of the behaviour of both the Bolsheviks and the White Russians gave the Japanese officer a poor opinion of Europeans and encouraged brutality in many forms.

The Treaty of Versailles awarded to Japan, as League of Nations mandates, the former German Pacific Islands north of the Equator, which she had occupied in 1914 – the Caroline, Marshall, Mariana and Ladrone Islands; but, as mandates, they could not be fortified. It was followed, in 1921, by the Washington Conference which considered both the general naval situation and the problems of the Far East. In the face of opposition from the naval General Staff, the government had to accept the ratio for the replacement tonnage of its capital ships of 525,000 tons each for the USA and Britain to 315,000 for Japan, having stood out for a ratio of 10:10:7. If no agreement had been reached Japan, like Britain, would have found herself involved in a warship building competition with the USA which she could not afford, and which would have resulted in a worse ratio. In compensation, it was also agreed that the status quo would be maintained in regard to fortifying bases in the Pacific, excluding Hawaii and Singapore, which was clearly to her advantage.

Pressure from the USA and Canada replaced the Anglo-Japanese alliance with a Quadruple Agreement between the USA, Japan, Britain and France that disputes in the Pacific, which had not been settled by diplomatic means or involved aggressive action by some other power, would be referred to a joint conference of the four. A Nine-Power Treaty was signed, binding the signatories to respect the integrity and sovereignty of China and apply the Open Door policy to trade and industry in that country. Japan retained her rights in the Liao-tung peninsula, including basing the Kwantung army there, and some concessions in the Shantung peninsula, but she withdrew from the leased territory of Kiao-chou and accepted the abolition of the Twenty-One Demands. However the USA, Britain, France and Japan retained the right to station troops at Tientsin and Peking and on the railway between the two.

1922–41

The year 1922 saw not only the signature of these treaties, but the deaths of two of the last of the genro, Okuma and Yamagata, the latter aged eighty-four, leaving Saionji as the sole survivor. Until his death Yamagata acted as a stabilizing influence both within the armed forces and between them, the politicians and the palace, where the young Hirohito acted as regent for his mentally deficient father until the latter's death in 1926. Yamagata's death released two floods of frustration: one was that of the civilian politicians of the Diet against the military. The expense of the Siberian expedition, in which the army had taken matters entirely into its own hands, dislike of the repressive actions of the army during the rice riots in 1918 and a general resentment against both the expenditure and the arrogance of the military, who continued to act as if they were still the honoured heroes of the Russo-Japanese War, produced a reaction. With neither Russia nor China posing a military threat, the politicians of both parties in the Diet pressed for reductions in defence expenditure, resulting in a cut in the army's strength of 50,000.

This reduction aggravated the frustration, which had built up within the army among the middle-ranking officers who had graduated at the staff college, against the domination of Satsuma and Choshu, particularly the latter. It was a period which saw a proliferation of patriotic societies, with which army and navy officers, especially the former, had links. Both already had societies of their own, the oldest of which, founded in 1895, was the Butshukai (The Society of Military Virtues). The Zaigo Gunjinkai (Ex-servicemen's Association), founded in 1910 by Generals Terauchi and Tanaka, was one of the most influential. Of considerable significance was the 40,000 strong Kodogikai (The Imperial Way), founded in 1918 by Seiyukai politicians, of which General Araki, although a serving officer, was the director. It organized the *Kodo-ha* movement, popular among junior middle-rank officers influenced by the writings of Ikki Kita, who advocated a return to a primarily agricultural community owing direct allegiance to the Emperor, the problem of over-population being solved by emigration to Manchuria. He was opposed to all Western influences. *Kodo-ha* supporters were anti-establishment, political, military and commercial, and favoured a radical policy very similar, in its National Socialism, to that advocated by Hitler in the same period.

Balancing them, with strong support at the lieutenant-colonel level on the staff, was retired Major-General Kitu's *Kaikakai* (The Expansion Society), founded in 1924, and a highly influential and secret small group of a hundred army officers, founded in 1930, the *Sakurakai* (The Society of the Cherry). Its leading figure was Major-General Tatekawa, although almost all the members were captains, majors and lieutenant-colonels from the General Staff, the military academies, the *kempei* and the troops stationed in Tokyo. The society was dedicated to the use of the armed forces, if necessary, in order to achieve its aim of using Korea and Manchuria as the solution of Japan's economic problems.

Tension between the radical and the pragmatic wings of the movements arose over attempts to modernize the army which, in terms of training and equipment, had stagnated since 1905. In 1924 General Ugaki, although not himself a Choshu, was nominated with their support as Army Minister. He had, however, his own non-Choshu supporters in the *Sakurakai*. He was determined to modernize the army and decided to reduce its twenty-one divisions to seventeen and the length of conscript service from three years to two, using the money thus saved to develop new weapons, such as tanks and aircraft, as well as introducing a large number of other reforms. His proposals coincided with anxieties about a *rapprochement* between Russia and China, threatening the Japanese position in Manchuria. With anti-American feeling, arising from the 1924 US Immigration Act, following so closely on the Washington Conference, and resentment against the British for cancelling their alliance, the Japanese, and especially the armed forces, felt that once more they were being hemmed in.

These anxieties strengthened conservative opposition to Ugaki's reforms, led by Marshal Uehara, around whom also all the anti-Choshu factions rallied. Ugaki succeeded in forcing the retirement of many of the generals in his clique but could not touch Uehara himself. His principal protégé, General Muto, was removed from his post of Vice-Chief of the General Staff and sent to command the Kwantung Army in Manchuria. When Ugaki was out of office for a short time in 1928, Uehara managed to have Muto appointed Inspector-General, but failed in the following year to have him appointed Chief of the General Staff, Ugaki, having returned as Army Minister, appointing a less well qualified general from his own clique. This divided even further the senior officers of the army, many of whom, supported by junior officers, resented the abolition of regiments in the 1925 reforms. Regiments had strong territorial affiliations, and officers and soldiers whose regiments were disbanded felt that they had been deprived both of their homes and of the status they held in their own locality. These discontents were aggravated by inflation, reductions in pay and the generally depressed state of both the economy and the armed forces. The army's assistance to the population in the terrible earthquake of 1923 had raised its status in the estimation of the public for a time. So did the transformation of officers made redundant in the 1922 and 1925 reforms into schoolteachers as a deliberate policy of countering anti-militarism and reviving bushido. But throughout the 1920s and 1930s the army was riven by factional intrigues, policies at the top often appearing to be dictated by political pressure from the groups of middle- and junior-ranking officers.

The Kodo-ha supported and were supported by General Araki, who in 1925 became head of the operations section of the General Staff and backed Uehara in his resistance to Ugaki's reforms. He was opposed by what was known as the Tosei-ha (Control) group, prominent among who was Major-General Tojo. Although opposed to civilian political influence and as much in favour of 'expansionism' as the Kodo-ha, they were pragmatic and more inclined to compromise with the establishment in order to achieve the same basic aim. Personal rivalries played a significant part in the faction fighting between the two, and their expansionism diverged, the Kodo-ha giving priority to action against Russia, the Tosei-ha against China.

Both factions presented a united front towards civilian politicians, against whom a series of plots and attempted coups d'état were planned in the early 1930s. A military coup to seize power in March 1931, after an attempt on the life of Prime Minister Kamagushi Minseito in the previous November, following his disagreement with the navy, failed when Ugaki refused to be involved in the conspirator's proposal to appoint him head of a military government. In 1932, after the murder of a prominent politician and the chairman of the Mitsui company, the seventy-five-year-old Prime Minister, Inukai Ki, was assassinated by followers of Ikki Kita. The army then warned the surviving genro, Saionji, that the appointment of another civilian party administration would only lead to further violence, as a result of which Admiral Saito was appointed, with a predominantly non-party Cabinet, Araki remaining as Army Minister. This advancement of his own career by exploiting the division within the army led to the dismissal of Araki and several of his faction in 1934. General Hayashi Senjuro, supported by the Tosei-ha, replaced him as Minister; the other leading Kodo-ha general, Mazaki Jinzaburo, also was removed from his post as Inspector-General. In revenge for the latter, Nagata Tetsuzan, one of the principal *Tosei-ha* generals who had been instrumental in the removal of *Kodo-ha* supporters, was murdered by a *Kodo-ha* lieutenant-colonel. While he was on trial, on 26 February 1936, *Kodo-ha* officers, with 1,400 troops of the First Division in Tokyo, tried to seize power. They killed Takahashi Korekiyo the Finance Minister, General Watanabe the Inspector-General and Admiral Saito Makoto, who had handed over the premiership to Admiral Okada, who escaped. Saionji also evaded them, and the Emperor's Grand Chamberlain, Admiral Suzuki, was seriously wounded. After three days the rebellion was suppressed on the orders of the Emperor, and the *Kodo-ha* with it. From then on the *Tosei-ha* group controlled the army and virtually controlled the government as well.

The struggle between different army factions and between the armed forces and the political parties was closely linked to events in Manchuria and China. In 1928, when the Chinese warlord of Manchuria, Chang Tso-lin, was assassinated in Mukden, Japanese officers of the Kwantung Army were suspected of being involved, as indeed they were, in blowing up his railway carriage intending it to be the prelude to the occupation of Mukden. Although the plot was not supported by the senior officers of the Kwantung Army, and therefore failed, the attempt by the Prime Minister, General Tanaka, to have the officers tried and punished was frustrated by the army authorities in Tokyo, to the dismay of the Emperor. In September 1931, however, the climate was more favourable to the plotters. In the intervening period Chang Tso-lin's son, Chang Hsueh-liang, had allied himself to Chiang Kai-shek's Kuomintang movement and took a series of measures aimed at Japanese commercial interests in Manchuria. The officers of the Kwantung Army received support from both the *Kodo-ha* and the *Tosei-ha* groups in Tokyo, and notably from the Army Minister in 1931, General Minami. On the pretext of an attempt by Chinese soldiers to sabotage the Japanese South Manchurian Railway near Mukden (to which Chang Hsueh-liang was building a competitor), the Kwantung Army occupied Mukden on 19 September and, reinforced from Korea, expanded their operations to occupy almost all of Manchuria, not only without any orders from Tokyo to do so, but in blatant opposition to the policy of the cabinet, headed by the civilian Wakatsuki Reijiro and his Foreign Minister Shidehara Kijuro, against whom an officers' plot to seize power by force was detected in October, the ringleaders being let off with a reprimand. The feebleness of international reaction, neither the Americans nor the Russians being prepared to support any measures to reverse the situation, reinforced the determination of both the army and the navy to override the anxieties of civilian politicians to press ahead with a rearmament programme and an aggressive policy towards China designed to counter the growing influence of both the Kuomintang and the communists, both of whom pursued anti-Japanese policies as part of their general campaign to get rid of humiliating concessions to foreigners.

The navy had already taken a stand in 1930 over the rejection by the government, headed by the civilian Hamaguchi Yuko, of their General Staff's advice to reject the proposal of the London Disarmament Conference in that year to agree to a ratio of cruisers vis-à-vis the US navy which they regarded as unacceptable. While the Navy Minister, Admiral Takarabe, their nominee, was heading the delegation in London, Hamaguchi assumed the portfolio of Navy Minister in Tokyo himself and, in that capacity, advised the willing Emperor to accept the proposal. This provoked a constitutional crisis, seen as a test between the civilian politicians and the military, in which the victory, a very temporary one, went to Hamaguchi, who was shot and seriously wounded by a right-wing supporter in November. After a gallant attempt to resume office in April 1931, he handed it over to Wakatsuki and died before the year was out.

After the Mukden incident, the navy's determination to match the increasing aggressiveness of the army was reinforced. Early in 1932, following incidents arising from anti-Japanese agitation, the Japanese naval garrison of the international settlement in Shanghai clashed with troops of the Chinese Nationalist 19th Route Army, in the course of which an aircraft-carrier was used to bomb the suburb of Chapei, causing severe civilian casualties. When Admiral Saito was Prime Minister, not only did the government announce its withdrawal from the League of Nations, which had passed a resolution condemning the occupation of Manchuria, but it gave notice to the other signatories of its intention to abrogate the Washington Treaty's limitations on naval construction.

Since the development of her ship-building industry as a result of Allied orders in the First World War, Japan designed and built her own naval vessels. The attempts of her naval constructors to produce warships which openly conformed with but secretly evaded the Washington Treaty limitations had not proved successful, and they advised the naval General Staff that only by abrogating these limitations could the Imperial Japanese navy be built up to a position from which it could successfully challenge its obvious rival, the US navy. The decision was taken to build a new class of giant battleships, the Yamato class, of 64,000 ton displacement, while older battleships were rebuilt with emphasis on an increase in speed. One of the arguments for the Yamato class was that they would be superior to any battleship which could pass through the Panama canal.

The general strategy of the naval staff was to try to repeat Tsushima. The US fleet would be lured across the Pacific to Japanese waters, attacked by submarines and aircraft-carriers on the way, so that it would be reduced to a strength which the Japanese battleship fleet could successfully challenge in its home waters. The navy already had three aircraft-carriers and one under construction, and plans for four more were made, as well as significant new production programmes of cruisers, destroyers and submarines. An excellent torpedo, leaving no trace on the surface, had already been developed. The navy was not beset with the rival factions which divided the army, although there was certainly an element of *gekokujo*, assumption of authority by junior officers over their seniors, which favoured an aggressive policy directed against the United States.

The Kwantung Army and the navy were not the only forces to take an aggressive line. The Japanese North China Army, based in the Peking-Tientsin area, was not going to be left out of it, and its commander, General Tada, tried to negotiate with the northern war-lords that his troops should occupy the province of Hopei, ostensibly to keep out the forces of the Kuomintang and the communists. At the same time the Kwantung Army was extending its influence into Inner Mongolia and the Tokyo government was trying to come to an agreement with Chiang Kai-shek, by which he would be given Japanese support in return for agreeing to receive no help from any other power. Meanwhile the Japanese attempted to achieve international recognition of Manchuria, renamed Manchukuo, as an independent country under the nominal leadership of Pú-Yi, the last remaining scion of the Manchu dynasty which had ruled China from Peking. In fact he was a mere puppet of the Kwantung Army. Japanese occupation of Manchuria and penetration into Inner Mongolia led to a deterioration in relations with the Soviet Union which, in March 1936, concluded a pact of mutual assistance with Outer Mongolia, allowing for the presence of Soviet troops in that country. Discussions about boundary disputes were broken off in November after eight months of argument, immediately followed by the conclusion of the German-Japanese Anti-Comintern Pact, the secret clauses of which bound each party to do nothing to help the USSR if it were in dispute with the other.

The pact had been opposed by the Foreign Ministry but favoured by the army which, with the co-operation of the navy, persuaded the Inner Cabinet to adopt four 'Basic Principles of National Policy'. They were that

> Japan must strive to correct the aggressive policies of the great Powers and to realize the spirit of the Imperial Way by a consistent policy of overseas expansion: must complete her defensive armament to secure the position of her Empire as the stabilizing power in East Asia: should strive to eradicate the menace of the USSR in the north in order to stabilize Japan-Manchukuo national defence and to promote sound economic development: should also be prepared against Britain and the USA, and attempt to bring about economic development by close collaboration between Japan, Manchukuo and China, paying attention to

friendly relations with other Powers; and further her plan to promote social and economic development in the South Seas and, without rousing other Powers, should attempt to extend her strength by moderate and peaceful means.

In 1937 the army General Staff came to the conclusion that a war in Europe would take place before long and that, in view of the increasing power of the Soviet Union and the growing national feeling in China, Japan should observe a cautious strategy, building up her strength and avoiding provocation of other powers; but the Kwantung Army, of which Tojo was the chief of staff, took a different view. Chiang Kai-shek and Mao Tse-tung had agreed to co-operate in an anti-Japanese campaign and he believed that they would be supported by the Soviet Union. This could be countered by the maintenance of a buffer state in Inner Mongolia, combined with strong action against the Kuomintang, then based on Nanking. The purges in the USSR made it a propitious time to seize the initiative.

The year 1937 was one of political jockeying in Tokyo. In reaction to criticism of the army by a liberal member of the Diet, Terauchi, the Army Minister, demanded the dissolution of parliament, and when the Prime Minister, Hirota Koki, who hitherto had been subservient to the army's demands, refused, he resigned, bringing the cabinet down with him. The venerable Saionji recommended that General Ugaki should head a new administration, but the army would not provide him with a minister,* and the choice fell on the unsuitable General Hayashi, who only lasted four months, both political parties combining against his dictatorial ways and campaigning against him in an election to the Diet held in April. He was succeeded by the universally respected Prince Konoye, at forty-six much younger than any of his predecessors, who had been President of the House of Peers since 1933. He soon found that he could not control the army. On 7 July an incident occurred at Lukouchiao (Marco Polo Bridge) on the railway leading south from Peking to Hankow. The Japanese North China Army had insisted on acquiring land there to build barracks and an airfield. By thus getting control of both the Peking-Tientsin and the Peking-Hankow railway lines, they could exert a stranglehold on the capital of the weak Chinese Central Government. Occupation of the area by Japanese troops was resisted by those of the Central Government, and when the latter appeared to be prepared to negotiate some agreement to keep the peace, Chiang Kai-shek accused Japan, justifiably, of having engineered the incident which, if the government gave way, would

---

* In 1936 the army and navy had co-operated in forcing a return to the previous article of the constitution which stipulated that the army and navy ministers must be serving, not retired, officers.

prove a repetition of Mukden, China suffering the same fate as Manchuria. Without reference to Konoye, the tough *Tosei-ha* Army Minister, General Sugiyama, ordered reinforcements from the Kwantung Army and Korea to be sent to the North China Army. Konoye, faced with the certainty of Sugiyama's resignation and the refusal of the army to appoint a successor, and therefore the fall of his government if he attempted to countermand Sugiyama's orders, as the Emperor wished him to do, acquiesced. The die was cast which led to war in China and later against the USA.

From then on, hostilities between Japan and both the Central Government and the Kuomintang spread. In August hostilities extended to the naval garrison at Shanghai, where the Japanese accused the Chinese of stepping up military measures in the neutral zone, in breach of the agreement made after the 1932 incident. As both sides took military precautions their forces clashed on 13 August, Chinese aircraft bombing the Japanese cruiser *Izumo*, causing heavy loss of civilian life in the international settlement. Next day General Matsui was appointed to command an expeditionary force of two divisions to be sent to Shanghai, built up by October to five.

A week later the Sino-Soviet Non-Aggression Pact was signed. China appealed to the League of Nations, which condemned Japan's action as a breach both of the Nine-Power Treaty and the 1928 Kellogg Pact and proposed a meeting of the signatories of the former, which Japan refused to attend. The conference had an unfortunate effect on both the Japanese and Chinese. It steeled the determination of the former, both because they felt isolated – even their ally Germany disapproved of their action – and because they saw that no other power was prepared to take effective measures against them; and it encouraged the Chinese in hopes of receiving foreign help and in refusing to seek a settlement with Japan.

In the remaining months of the year Matsui's force drove Chiang Kai-shek's troops back up the Yangtse valley, capturing his base at Nanking, where the Japanese troops indulged in acts of indiscipline and brutality which would never have been tolerated by Japanese generals in the Russo-Japanese War. In the course of these operations naval aircraft bombed and sank the US navy's river gunboat *Panay* and the army's artillery tried in vain to sink the British *Ladybird*, both actions contrary to orders from Tokyo to avoid antagonizing foreigners. Matsui and two of his divisional commanders were recalled as a result.

Prince Konoye was now anxious to bring hostilities in China to an end. With the support of the army General Staff, negotiations had been opened with Chiang Kai-shek, through the German ambassador in Tokyo, before the fall of Nanking, about terms which the leaders of the Kuomintang, notably Wang Ching-wei, advised him to accept. But after Nanking fell, pressure from Matsui's Central China Army forced Hirota, Konoye's Foreign Minister, to stiffen the terms to include the stipulation that China should receive no help from other powers and join Japan and Manchukuo in an anti-communist league. Konoye's peace overtures to Chiang Kai-shek were complicated by the Japanese North China Army's establishment of a puppet régime in their area of operations, which they expected would eventually, with their support, extend its authority over the whole country as both their operations and those of the Central Army defeated the Kuomintang. In the early months of 1938 the Japanese forces in China were reduced from sixteen to ten divisions by withdrawal of the divisions formed from reservists but, after the North China Army had suffered a defeat in Shantung province, they were sent back and helped to inflict revenge on the Chinese near Hsuchow on the Peking-Nanking railway, where the Chinese army barely managed to escape destruction and prevent the Japanese from reaching Hankow by blowing a gap in the banks of the Yellow River, causing floods which had catastrophic consequences for the peasantry of the region.

Konoye was anxious to bring the war in China to an end not only for its own sake but because, as long as it went on, he knew that he could not control the army. Hoping to resurrect the *Kodo-ha* to counter the influence of the *Tosei-ha*, he brought Araki back as Minister of Education and replaced the hard-line Sugiyama as Army Minister by the younger Itagaki, with Tojo as Vice-Minister, although both were *Tosei-ha* generals. Ugaki was made Foreign Minister. His attempt to revive *Kodo-ha* failed. His position was weakened by the Kwantung Army, when it provoked clashes with Soviet forces in July and August on the Amur river and near Lake Khasan, south of Vladivostok.* This resulted in the Emperor telling Konoye, Itagaki and the Chief of the General Staff, 'From now on, you may not move one soldier without my command'. It had little effect. Konoye's and Ugaki's attempts to come to some agreement with the Kuomintang were opposed by all the Japanese armies in China, reinforced in their confidence of defeating Chiang Kai-shek by the peaceful settlement of the border disputes with the Soviet Union. Frustrated by their opposition and by the failure of Konoye to back him up, Ugaki resigned as Foreign Minister at the end of September, after which Konoye gave in to the army's demand that policy towards China should be determined by an Asia Development (or China Affairs) Board, dominated by the military. The capture of Hankow on 21 October and of Canton by a naval expedition in the same month, forcing Chiang Kai-shek to transfer his base to Chungking in the extreme south-west, strengthened the hand of the hard-liners and was followed by a statement of policy which would have reduced China to the condition of a vassal state in a Japanese

* See p. 219.

Empire, which by that time had deployed thirty-five divisions, containing over a million men, in China. When Chiang Kai-shek rejected the Japanese terms at the end of the year, his Prime Minister, Wang Ching-wei, deserted him, making his way through Indo-China to Shanghai, where he co-operated with the Japanese in establishing a rival regime, based at Nanking.

Meanwhile Konoye, who had wished to resign earlier, gave up the struggle. He had been faced with serious divisions of opinion, not only about China but also over pressure from the army to respond favourably to a proposal for a military alliance with Germany and Italy directed against the Soviet Union. The civilian politicians, including Togo, the Japanese ambassador in Berlin, and the navy were opposed to it, fearing that it might involve awkward commitments which could bring Japan into conflict not only with the USSR, but also Britain and perhaps the USA. The proposal was strongly backed by General Oshima, the military attaché in Berlin, who negotiated with Ribbentrop, the German Foreign Minister, behind the back of the ambassador, referring matters directly to the army General Staff. Konoye's successor, Hiranuma Kiichiro, continued negotiations on the issue, under strong pressure from Arita Hachiro, his Foreign Minister, and Admiral Yonai, the Navy Minister, not to agree to anything which might antagonize the USA and Britain. Hiranuma tried to get Germany and Italy to accept a pact which did not openly commit Japan to action against the Soviet Union, but was accompanied by secret clauses, suggested by Itagaki on behalf of the army, which gave that assurance. Her prospective allies were not prepared to accept that solution and nothing had been agreed when, to Japan's dismay, the Soviet-German non-aggression pact was announced in August 1939. That slap in the face coincided with the defeat of the Japanese 6th Army on the borders of Manchuria and Outer Mongolia by Soviet forces under Zhukov,* causing the resignation of Hiranuma's government.

The choice of the politicians as his successor was Hirota, but the army refused to accept him, or any other civilian, and an administration was formed with a retired general of limited political or military experience, Abe, as its head, General Hata, former commander of the North China Army as War Minister, Admiral Yoshida as Navy Minister, and Admiral Nomura Foreign Minister. Abe's first cabinet meeting after the outbreak of war in Europe on 2 September 1939 resulted in the statement: 'Faced by the European War which has just broken out, Japan intends not to be involved in it. She will concentrate her efforts on a settlement of the China affair.' Abe did not last long and was succeeded by Yonai who himself only held office until July 1940, when he was forced out by the army through the usual device of insist-

* See p. 223.

ing on the resignation of the Army Minister, General Hata, and refusing to appoint a successor.

Hitler's startling victories reinforced the arguments of the army in favour of an alliance with the victors, which neither the political parties nor the navy felt any longer able to oppose. The former went virtually into voluntary liquidation, and Konoye returned to head a 'national' party to 'assist the Imperial Throne'. In September Japan joined the Berlin-Rome Axis, and in the same month Britain, under pressure from Japan and to the anger of the USA, agreed to close the Burma road to Chungking. France was forced to allow Japanese troops to enter Indo-China in order to prevent supplies reaching the Kuomintang from Haiphong. Both countries withdrew their troops from Shanghai, together with the few left at Tientsin since most had been withdrawn in September 1939. With the Netherlands occupied by Germany, the situation of the Dutch East Indies was critical, dependent on British or American naval support.

Konoye's Foreign Minister was the hot-headed civilian Matsuoka Yosuke, a convinced Germanophile, and his Army Minister Tojo. Yoshida remained as Navy Minister. Konoye was prepared to accept the Tripartite Axis Pact as he hoped that, through the good offices of the Germans, a Soviet-Japanese non-aggression pact could be achieved and also a settlement with Chungking, although the prospects of the latter were reduced by increasing Japanese support of Wang Ching-wei's regime. Matsuoka's attempt to persuade Germany to include the Soviet Union in the Axis fell on increasingly stony ground, as the relations between Stalin and Hitler cooled and the USSR stepped up its support of Chiang Kai-shek. All that was achieved was a Pact of Neutrality, signed in April 1941, which bound both Japan and the Soviet Union to remain neutral if either were involved in war. It was observed by both sides to their mutual advantage until August 1945. The pact was followed by an agreement on Soviet-Japanese trade and on German-Japanese trade via the Trans-Siberian railway, both signed only eleven days before Germany attacked the Soviet Union on 22 June 1941, another slap in the face for Japan from her ally. Matsuoka wanted Japan to join in, but he was opposed by Tojo, who knew that the army was ill-equipped to fight the Soviet armed forces, and by all the other members of the cabinet, which resigned on 16 July in order to rid itself of Matsuoka.

On 2 July an Imperial Conference* had been held which decided that the Russians should be told that the Neutrality Pact would be observed. However, if the war went in Germany's favour, Japan would

---

* An Imperial Conference was normally preceded by a Liaison Conference, a joint meeting of the Chiefs of Staff and the Inner Cabinet. Its result was reported to the Emperor in an Imperial Conference, at which he traditionally remained silent.

intervene 'to secure stability in the northern regions'. The conference also decided to prosecute the plans for an advance into Indo-China and Siam, at the risk of war with Britain and the USA, which was promptly acted upon by the occupation of bases in southern Indo-China, which Admiral Decoux, the Vichy French Governor-General, was helpless to resist. Konoye appears to have acquiesced in this as a sop to the activists in order to avoid fighting either the Soviet Union or the British by an attack on Hong Kong and Malaya. Matsuoka was replaced by Admiral Toyoda.

Konoye was taken aback by the sharp reaction of the USA, Britain and the Netherlands authorities who imposed an economic embargo on Japan and began to make clear preparations for war in the Pacific. He was just as keen as Tojo on the establishment of a Japanese-dominated Greater East Asia Co-Prosperity Sphere, but he believed it could be achieved without fighting either Britain or the USA. The former was certain to be defeated by Germany, unless the United States came to her help, and he thought that the USA would not become involved in war against either Germany or Japan. Tojo and the generals, and by now many of the admirals, urged on by their subordinates, saw the situation, with the Russian threat removed, as presenting a golden opportunity, and did not believe that the US armed forces, even if their nation went to war, would fight hard. Japan could not afford to wait and allow economic sanctions, particularly interruption of the supply of oil, to exert a gradual stranglehold, which would undermine the Japanese effort in China and South-East Asia.

Discussions between Japan and the USA had started in April, after Admiral Nomura had been appointed ambassador in Washington. The Japanese proposed that the USA should persuade Chiang Kai-shek to make peace terms which included recognition of the independence of Manchukuo and agreement to merge his regime with that of Wang Ching-wei. In return Japan would not demand any indemnities or territorial concessions or large-scale immigration, and would observe the Open Door policy in respect of trade. Japan and the USA would guarantee the independence of the Philippines, and the USA would make a gold loan and co-operate in the supply of vital raw materials to Japan, and would 'avoid aggressive alliances'. If Chiang Kai-shek rejected the proposals, the USA would withdraw their support from him. Japanese troops would remain in Inner Mongolia and part of North China 'in furtherance of Sino-Japanese co-operative defence against communistic activities', but would be withdrawn from other areas of China over a period of two to three years. These proposals were clearly unacceptable to the USA, and not solely because the assistance given to Britain under Lend-Lease was interpreted by Japan as tantamount to an 'aggressive alliance'.

American counterproposals, which included withdrawal from China and from the Tripartite Axis Pact, leaving Sino-Japanese co-operation against the communists for later discussion, were equally unacceptable to the Japanese cabinet. After the American freezing of Japanese assets at the end of July, Konoye made further proposals. Japan was to agree not to advance beyond Indo-China, and withdraw her troops from there once peace with Chiang Kai-shek had been secured. She would guarantee the neutrality of the Philippines. In return the USA should discontinue their military preparations in the Pacific, lift economic sanctions, mediate with Chiang Kai-shek and recognize Japan's special position in Indo-China. He proposed a meeting with Roosevelt to discuss them, to which he would bring senior army and navy authorities as proof that they were involved in whatever was agreed. Roosevelt favoured a meeting, but his Secretary of State, Cordell Hull, successfully opposed it, fearing a form of Munich settlement, and demanded further clarification. Konoye was becoming desperate as pressure from the pro-Axis and pro-war factions intensified, both arguing that the Americans were just playing for time while economic sanctions and their preparations for war gained strength. After receipt of Roosevelt's reply on 4 September, a full scale Imperial Conference on 6 September confirmed the decisions of that of 2 July, deciding that preparations for war against the USA, Britain and the Netherlands should be completed by the end of October. Diplomatic negotiations would continue, but 'if by the early part of October there is no reasonable hope of having our demands agreed to ... we will immediately make up our minds to get ready for war against America (and England and Holland)'.

Further discussions between the two countries centred on the key issues of Japanese withdrawal from China and from the Axis. Konoye and Toyoda tried to get Hull to agree to a compromise wording which evaded them, and again pressed for a meeting with Roosevelt, holding out the prospect of greater flexibility; but Hull did not trust a vague agreement, which he was certain they would interpret in their own way. When he returned a dusty answer to their proposal on 20 October, Tojo demanded that the decision to go to war should be taken rather than give way to a demand for withdrawal of Japanese troops from all of China. Admiral Oikawa, the Navy Minister, was lukewarm. He wished to avoid war but, having given the 2 July and 6 September Conference decisions his approval, was not prepared to reverse his position and left the decision to Konoye who, after consulting the Emperor, resigned on 16 October in favour of Tojo, arguing that the Americans would know that, if an agreement was reached, he would be able to implement it, having the army behind him. To add credibility to this, Tojo remained Army Minister and also assumed the portfolio of the Home Ministry. The pro-Russian Togo was appointed Foreign Minister, and both the Soviet Union and the USA were assured that Japan had no aggressive intentions towards the USSR. The Emperor told Tojo that he was no longer bound by the decisions of the Imperial Conference of 6 September, but Nomura, constantly asking to be relieved of his difficult post, forecast no likelihood of change in the American position.

Both the army, headed by Generals Tojo and Sugiyama, and the navy, by Admirals Shimada and Nagano, opposed further procrastination. They were confident of victory, reckoning that oil stocks would last for two years, long before which the capture of the Dutch East Indies would have solved the problem. Tojo warned Nomura that Japan was not prepared to make any further concessions, and informed him that a final decision would be made at an Imperial Conference on 5 November. The conference decided that, if further discusssion with the USA had not led to a satisfactory agreement by 25 November, Japan would go to war with the USA and Britain, asking her Axis partners to join her, but making clear that Japan would not only not engage in war against the Soviet Union but would try to mediate a peace between the USSR and Germany, so that a direct link between Germany and Japan could be restored.

Nomura was told to make a new approach in Washington, called Proposal A. Japan would agree to 'non-discrimination' in trade in the Pacific area, including China, if the principle were applied to the whole world: she could not change her position in regard to the Tripartite Axis Pact, but wished to prevent the European war from spreading to the Pacific. Once a Sino-Japanese peace had been concluded, 'some' of the Japanese troops in China would remain 'for a suitable interval' in North China, Inner Mongolia and on Hainan island. If Nomura were pressed about the 'suitable interval', he was to answer 'vaguely' that it meant about twenty-five years. Japan would respect the territorial integrity of French Indo-China and remove their troops once affairs with China were satisfactorily concluded, but that pledge should not be included in a formal agreement with the USA. If Proposal A were not immediately accepted by the USA, an alternative Proposal B should be offered, in which the two governments would agree not to invade any area of South-East Asia or the South Seas, except French Indo-China; and would co-operate in securing the raw materials they needed from the Dutch East Indies, economic sanctions having been removed. The USA would provide Japan with the oil she needed and refrain from any activity which would hamper Japan's efforts in making peace with China. If necessary, a general promise by Japan to evacuate her troops from China after such a peace could be added, as could the first two provisions of the previous proposal.

The US government deciphered Nomura's instructions, including a message from Togo that 'because of various circumstances, it is absolutely necessary that all arrangements for the signing of this agreement be completed by the 25th of this month', as a result of which the

US Secretary of State warned his colleagues that a Japanese attack might come 'anywhere at any time'. After further discussions about Proposal A, Nomura was instructed to present Proposal B on 20 November, the deadline being extended to 29 November, adding 'after that things are automatically going to happen'. The initial American response was to propose a so-called *modus vivendi*, to which Chiang Kai-shek and the powerful China lobby in the US raised strong objections, supported mildly by Churchill. This pressure persuaded Roosevelt and Hull to propose a general settlement, which they had intended should follow the *modus vivendi*, but without the latter. Under it, Japan was to withdraw all her armed forces from China and Indo-China, recognize no Chinese government other than Chiang Kai-shek's and surrender all her extra-territorial rights. She was to interpret the Tripartite Axis Pact in a way that would not conflict with peace in the Pacific area. In return the USA would conclude a commercial treaty, removing economic sanctions and establishing a dollar–yen stabilization fund. These terms were presented to Nomura and Kurusu Saburo, a professional diplomat sent to help him, on 26 November, with little, if any, expectation that they would be accepted. In extensive discussions in Tokyo, in which Konoye, Kido Koin and other civilians, urged on by Nomura and Kurusu from Washington, argued for acceptance against Tojo's determination to force the issue, for which he had the backing of the navy, Tojo prevailed. At the Imperial Conference on 1 December, following a belligerent public speech by Tojo which alarmed the US government, the Emperor observed his traditional silence, as the conference confirmed the decision of the preceding 'Liaison' Conference to go to war. The government's rejection of the American proposal was cabled to Nomura on 6 December, with instructions to deliver it at 1.00 p.m. 7 December Washington time, which would be just twenty minutes before the planned attack on Pearl Harbor (7.50 a.m. Honolulu time). The Japanese armed forces thus sealed their own fate.

Like many others who had led their countries into war, Tojo believed that a favourable decision could be reached quickly, brushing aside the warning that General Hata had given at the final Imperial Conference about the consequences of a long war. He was confident that the general state and strength of the Imperial Japanese army and navy could achieve swift victory, particularly as long as the potential threat from the Soviet Union was removed by her war with Germany. He had some sound reasons for that, as events were to prove; but to imagine that the USA, once committed to war, would be content to accept Japan's victories as a *fait accompli* was gravely to misjudge the world's most economically powerful nation.

The rearmament programme, which had been initiated by Admiral Saito's government in 1932, had borne fruit. The navy had ten battleships, with three of the Yamato class under construction. Two were of the Nagato class, built in 1920–21, the first battleships to carry 16-inch guns, and both they and the eight others, which all dated from the First World War, had been extensively modernized, although it had not been possible to thicken their armour. The fleet contained eight aircraft-carriers; one, the *Ryujo*, was originally built to appear to conform to the Washington Treaty and therefore, at 10,000 tons and carrying only forty-eight aircraft, was too small to be really effective. Of the others, two were recently constructed 25,000-ton vessels, each carrying eighty-four aircraft; two converted from a battleship and a battle-cruiser, two of the modern but unsatisfactory Hiryu class (15,900 and 17,300 tons); and an old one, the *Hosho*, used for training.* In spite of its lead in torpedo design, the navy's weakest link was its submarine arm. Design had been influenced by the Germans and aimed at long endurance, which was seldom achieved. Training concentrated on support to the fleet, in reconnaissance and attack on warships, to the neglect of operations against merchant shipping. The key to victory, indeed for survival, was an early and devastating blow delivered against the US Navy's Pacific Fleet. Nobody saw that more clearly than Admiral Yamamoto, C-in-C of the Combined Fleet, but he had difficulty in persuading Nagano, chief of the naval General Staff, that a direct attack by Admiral Nagumo's First Air Fleet on the US fleet's base at Pearl Harbor would succeed and that the risk of the loss of perhaps three carriers was acceptable. It was not until the beginning of November, using the threat that he would take personal command of the carrier fleet himself, that he finally persuaded Nagano that none of the other operations planned, for which the army was prepared to run the risk of war – against the Philippines, Malaya and Borneo, least of all against the Dutch East Indies – could be undertaken until the US fleet had been dealt with. The old strategy of luring the US fleet towards home waters could not be applied.

The army was confident of success, but was entirely dependent on the navy to transport it to its objectives and keep it there. The General Staff reckoned that, although the bulk of the army would be required to remain in China, eleven divisions could be made available for operations elsewhere. That should suffice against the weak resistance expected, and, in any case, it was doubtful if more than that could be transported and kept supplied. Although the defeat of the Sixth Army by Soviet forces on the Khalkin-Gol in 1939 had brought home

---

* In addition two fast oilers and six liners were available for conversion as light fleet carriers. The cruiser fleet included eighteen heavy, of which ten had been built since 1932, and twenty light, all completed in the 1920s, with one heavy and nine light under construction. Of the 108 destroyers, fifty-nine dated from before 1932 and forty-three more were under construction.

to the army how ill-equipped it was to face modern forces, well-equipped with tanks, air forces and artillery, it did not expect to have to meet that sort of opposition. It had found its armament perfectly adequate against the Chinese, and its generals had unbounded faith, not without justification, in the fighting spirit of the Japanese soldier to compensate for the lack of sophistication of his equipment. His rifle was a development from the original Mauser, used in the Russo-Japanese War, and he was not given a sub-machine-gun. The machine-gun was a copy of the Czech ZB 26, as was the British Bren gun. Most of the artillery was of equally ancient derivation, developed from the original Krupp 75-mm and employed in small numbers in direct support of the infantry. Some more modern field artillery, copied from the French Schneider, was restricted to Manchuria.

Although outmoded by the standards of other armies, the field artillery, especially the pack mountain guns, being light and easily transported, proved very suitable to the conditions in which the army was to fight in South-East Asia, but less effective in providing concentrated defensive fire in the later phases of the war. Medium artillery of 150-mm calibre, which was not available in great quantity, was developed from an original Japanese design of 1915. Anti-tank artillery included a copy of the German 37-mm, and a 47-mm gun developed from the Russian 45-mm. The standard anti-aircraft gun was a 75-mm, which could theoretically fire a 14-lb shell to 29,000 feet, but was in fact not effective above 20,000. It was adequate in 1941, but unable to deal with US aircraft by 1944, when it was superseded by a new model.

The army had begun to show an interest in tanks at the end of the First World War, but continued to regard them as primarily an infantry support arm. Experiments with heavy designs in the 1920s had not been successful and experience in Manchuria in the 1930s had fostered an enthusiasm for light tanks. The result was the Type 95, with an excellent 114 hp diesel engine, which, armed with a 37-mm gun, was a successful design within the limitations imposed by a weight of 6½ tons. It went into quantity production in 1936. A year later the medium Type 97, a four-man 15-ton tank with armour 25-mm thick, was produced with a 57-mm gun, a low-velocity weapon designed primarily to fire high-explosive in support of infantry. The army's total tank strength was about 1,500, of which half were deployed to Manchuria and China. The conditions in which the Japanese army fought in South-East Asia were not generally favourable to tanks and their number and quality was not a decisive factor.

The arm in which both the navy and the army reposed great confidence was the air, and they had good reasons for doing so. The army and navy air forces – there was no separate independent air force – were of the highest quality, not only in terms of aircraft, but also of crews, most of whom had considerable operational experience over

China. The army bought its first aircraft, a French Farman, in 1910, but neither it nor the navy took a serious interest in aviation until 1919, when they bought some surplus British and French Sopwiths, Nieuports and Spads. By 1921 the Nakajima Company had acquired licences to produce both British Gloster and French Nieuport aircraft and their Bristol and Hispano-Suiza engines. In the same year Mitsubishi engaged the services of Herbert Smith of Sopwiths to design a carrier-borne fighter for the navy, and later also a reconnaissance aircraft and a torpedo-bomber. Kawasaki manufactured French Salmson bombers under licence and employed the German, Richard Vogt, in 1923, as their chief designer to build an army bomber, powered by a BMW engine. By the end of the 1920s Japan was designing and producing all her own aircraft for both services, other firms joining the original three, notably Tachikawa, Aichi and Kawanishi, the last specializing in seaplanes and flying-boats.

By 1941 both the navy and the army had sizeable fleets of modern aircraft, in quality equal or superior to the American and British. The navy had 2,029 aircraft, of which 1,380 were first-line and operational. They included the famous Mitsubishi A6M Type O fighter, known as 'Zero', the same firm's G3M bomber ('Nell') and Nakajima's B5N bomber ('Kate') which inflicted most of the damage at Pearl Harbor. The army air force had about 1,600 aircraft, organized in some fifty air regiments (*sentais*) of about forty-five fighters or thirty bombers and a number of independent squadrons (*chutais*). They included the Nakajima Ki-43 *Hayabusa* fighter ('Oscar') and the same firm's Ki-21 ('Sally') and Ki-49 ('Helen') bombers. The quality of both the Japanese air forces was gravely underestimated by the Americans and the British, partly because of their success in preserving secrecy, partly due to confusion about the complicated method of designating types of aircraft, and partly to the arrogant belief that, while Japanese industry was adept at copying other people's designs, it could not develop effective aircraft of its own. The events of December 1941 and January 1942 were to prove that underestimate to be fatally mistaken.

1941–86

Admiral Yamamoto ordered Nagumo's task force to sail from the Inland Sea to Tankan Bay in the remote Kuriles early in November. It consisted of six carriers, escorted by a light cruiser and nine destroyers, a support force of two battleships and two heavy cruisers and a reconnaissance force of three submarines, supported by a fleet train of eight tankers and supply ships. In thick fog the force sailed from the Kuriles on 26 November to refuel in the deserted space of the northern Pacific where, on 2 December, Nagumo received the codeword ordering the attack. Refuelling next day, he set sail at twenty-six knots, still covered by fog, for the planned launching position 275 miles north of Pearl Harbor. In spite of anxiety about the weather, the first wave of 183 aircraft, led by Commander Mitsuo Fuchida, took off at 6.15 a.m. Hawaiian time on 7 December, delivering their attack at 7.55 a.m. with the results that have already been recorded.*

The attack on the US fleet was part of the wider plan which included the seizure of Hong Kong and the invasion of Thailand, Malaya and the Philippines, to be followed by the occupation of Singapore, Sumatra and Borneo. A defensive perimeter was to be established at sea on the line of the Kuriles, through Wake Island, the Marshall Islands and the Dutch East Indies, and on land on the Burma-India frontier. The navy was responsible for the Pacific operations, including the capture of the outer ring of islands, to include the Bismarck Archipelago and New Guinea, for all of which their marines were supplemented by one army regiment. They were also responsible for the capture of Hong Kong, although the attack from the mainland was the responsibility of the Central China Army. The invasion of the Philippines, Thailand, Malaya, Singapore and the Dutch East Indies was the responsibility of General Terauchi's Southern Army, whose headquarters was at Saigon, supported by Vice-Admiral Kondo's Southern Fleet.†

All the planned operations were accomplished by the forecast date of early April 1942, at a remarkably low cost in terms both of equipment and of human casualties. If the original plan had been adhered to, Japan would have consolidated her gains there and begun negotiations with the USA, but the ease of her victories whetted her appetite for further conquest, her leaders being affected by the temptation of all military men to believe that a further extension of space gained would add to security by removing potential threats. The navy was sensitive to the fact that their attack on Pearl Harbor had missed the aircraft carriers and wished to lure the US fleet westward to force a fleet action. Yamamoto argued for an attack on Midway which would both force the US navy to react with its carriers and also, when acquired, would provide a more satisfactory defensive perimeter. The naval staff

---

* See p. 291.
† Consisting of two old battleships, two light fleet carriers, eleven heavy and seven light cruisers, fifty-two destroyers and eighteen submarines. Terauchi had ten divisions: two allotted to 15th Army for Thailand and Burma; four, supported by one air division, to 25th Army for Malaya, North Borneo, Singapore and thereafter northern Sumatra; two with one air division, to the Philippines; and one to 16th Army for the Dutch East Indies, to be reinforced by one transferred from Hong Kong and one from Manila when they had been taken. One was held in reserve for the internal security of French Indo-China, twenty-two divisions in five armies remained in China, thirteen, with one air division, in the Kwantung Army in Manchuria, two in Korea and five, with one air division, in Japan.

opposed him until, on 18 April, US Army Air Force bombers, launched from a carrier, bombed Tokyo.

Before the attack on Midway could be carried out, the extension of operations beyond the Philippines to the Solomon Islands and New Guinea led to the Battle of the Coral Sea at the beginning of May, in which the navy suffered its first rebuff. A light fleet carrier was sunk, one of the heavy carriers, the *Shokaku*, was so badly damaged that it had to return to Japan and its sister ship lost so many aircraft that it was not available for the attack on Midway a month later. That did, as Yamamoto intended, lure the US navy into action, but not with the results he intended. He lost all four of the carriers in Nagumo's First Carrier Strike Force, and the fact that he retained a superiority over the US Admiral Nimitz in surface vessels counted for nothing. It was the turning-point of the war in the Pacific.

From then on, over-extended by their thrust towards Australia, the navy, and therefore the army with it, was forced onto the defensive as the American fleet, and particularly its carrier force, gathered strength. After Midway, the test of strength came round Guadalcanal in the Solomons and in Papua. Fighting on land and sea continued for six months, that in Papua costing the army 13,000 battle casualties, of whom only thirty-eight were taken prisoner, and the struggle for Guadalcanal 25,000, of whom 8,000 died from hunger or disease. Twenty-four ships and over six hundred aircraft were lost by the time that the evacuation of the remaining 12,000 was successfuly effected in February 1943.

A dispute between the navy and the army then arose about the strategy which should follow. Both agreed that there should be no voluntary withdrawals, but argued about priorities. The navy was principally concerned to safeguard its major base at Truk in the Carolines and gave priority to defence of the Bismarck and Solomon archipelagos, a thousand miles to the south. The army was less interested in archipelagos than in clinging on to the coast of New Guinea, as an outpost, with New Georgia in the Solomons, to keep enemy aircraft away from the Philippines. The centre of their interest was Rabaul, at the northern tip of New Britain in the Bismarcks, from which its 40,000 strong 17th Army in the Solomons and 55,000 strong 18th in New Guinea were commanded, supported by 410 land-based aircraft. They argued that these outposts could be held, if ten to fifteen divisions and 850 aircraft could be redeployed there from the forty divisions in China and Manchukuo, none of which were engaged in active operations. Twelve were facing Chiang Kai-shek's theoretically huge army and about the same number in the north were keeping an eye on Mao Tse-tung's, but neither of their opponents showed any sign of engaging in serious offensive action.

In spite of the signs from Stalingrad, El Alamein and Algiers that the tide of war had turned against Germany and Italy, Tojo and his associates remained confident that they could consolidate the Greater East Asia Co-Prosperity Sphere by establishing pro-Japanese regimes in all the countries they had occupied, resting their hopes on Wang Ching-wei in China. Given time, the Sphere could provide the resources required to maintain Japan's military strength to a degree enabling her to negotiate an agreement with the USA which would preserve her gains. Provided that the Japanese soldiers and sailors continued to fight with the unquenchable spirit of *bushido* and loyalty to the *tenno* which they had already shown, the effete Americans and British would prefer compromise to the prospect of casualties on the scale which operations, such as those at Guadalcanal, had inflicted on them. They clung to the hope that somehow the war between Germany and the Soviet Union could be brought to an end, allowing the former, with Italy, to concentrate her strength against Britain and the USA, which would then seek peace in the Pacific.

The course of the war in 1943 proved that their strategy was unrealistic. The defence of New Georgia cost the army 2,500 casualties out of a garrison of 10,000 and seventeen warships, MacArthur's casualties being less than half that and Halsey's a third. By mid September it was clear that hanging on to the Bismarck Archipelago was counter-productive. Meanwhile Nimitz in the Central Pacific was building up a formidable fleet, which the Japanese, in spite of all their efforts to build new aircraft-carriers and aircraft for them, would be hard put to it to deal with. An ambitious programme of new construction had been drawn up on the outbreak of war, revised after Midway to exclude battleships and concentrate on carriers, twenty of which were ordered, two being completed and three more launched in 1944, the keels of the remaining fifteen not having been laid down. Of the whole 1942 programme, only 31 destroyers, 122 escorts and 40 submarines, eleven of which were designed solely for cargo, were complete before the end of the war.

Reviewing the war situation in September 1943, Imperial General Headquarters, presided over by Tojo, concluded that the 'zone of absolute national defence' ran from Burma through Malaya and the Dutch East Indies to New Guinea and thence northward through the Marshall and Caroline Islands to the Kuriles. The Bismarck Archipelago would be held for six months, while the defences of that ring were strengthened. One of those who might have challenged this unrealistic strategy would have been Yamamoto, but he was dead, shot down by American aircraft on 18 April on a visit to Rabaul, as a result of the interception of a radio message about his movements.

The contrast between the American and the Japanese situations at this time was extreme. The former were churning out ships, combat and transport, in vast quantities, and also aircraft and crews to fly them, while the Japanese, for all their industrial skills, had banked on a short, sharp war and leapt from one expedient to another, the army and navy staffs drawing up totally unrealistic production programmes, while constantly changing the technical requirements, and neglecting mundane but essential items, notably the cargo shipping needed to supply the far-flung garrisons which their ambitious strategy had scattered about the western Pacific. American air and submarine attacks accounted for a million tons of Japanese shipping in 1942, while their own losses were insignificant in comparison. Japan's dream of being able to establish and defend the Greater East Asia Co-Prosperity Sphere, relying on the oil resources of the Dutch East Indies, foundered on the combination of lack of tankers to move it and the demands of the armed forces to defend the huge area they had conquered.

The decision to hang on to New Guinea and the western Solomons cost them dear, as MacArthur and Halsey developed their combined offensive in October 1943. Bougainville was held by 40,000 soldiers and 20,000 marines but, when the Americans landed on the far side of the island, the Japanese sent only 15,000 men against them, half of whom became casualties in vainly attacking the Americans, who, having lost only 300 men, withdrew and left the Japanese to wither on the vine. Meanwhile, in a series of amphibious moves, MacArthur isolated the five divisions of the 18th Army centred on Wewak on the north coast of New Guinea.

While MacArthur and Halsey had been isolating the outer defences of the ring in the south, Nimitz was pressing them back from the east, thus issuing a challenge to Yamamoto's successor, Admiral Toyoda. His first attack was directed against Makin and Tarawa in the Gilberts, 7,000 men being sent against the former and 18,000 against the latter, supported by 1,000 aircraft operating from six fleet and five light fleet carriers, with five new battleships providing massive naval fire support. The defenders of Tarawa inflicted heavy casualties on the US marines before they immolated themselves in a succession of suicidal counterattacks. The Marshalls were Nimitz's next target, but the attack was less direct. By-passing the eastern islands, where the Japanese defences were strongest, he aimed his assault at Kwajalein and Eniwetok, where his casualties were much lighter in spite of the suicidal tactics of the 8,000-strong garrison of the former. Nimitz's carriers successfully prevented the Japanese fleet from interfering and, with the rapid construction of airfields on the captured islands, Truk, 700 miles away, was threatened with land-based air attack. It had already been attacked during the operations by carrier-borne aircraft, which sank thirty-two ships and destroyed 250 aircraft in addition to the 150 destroyed over the sea.

Far away to the north-west, in Burma, the Japanese army's string of victories on the mainland of South-East Asia had met with a severe rebuff. Its conquest of Burma served three purposes: exerting a strangle-hold on supplies to Chiang Kai-shek; a defensive bastion between India and the Greater East Asia Co-Prosperity Sphere; and exploitation of the resources of Burma itself. At the end of 1943 General Kawabe, commanding Burma Area Army from Rangoon, had two subordinate armies, General Mutaguchi's 15th in Northern Burma and General Sakurai's 28th in Arakan. Signs that the British, under Mountbatten's newly created South-East Asia Command, were planning an offensive in 1944 prompted Mutaguchi to propose to pre-empt it by an offensive of his own into the Manipur hills, his ambitions extending to a break-through into Assam and the establishment of Subhas Chandra Bose and his Indian National Army as a puppet regime in India.

His proposal met with considerable opposition both among his subordinates and the staffs of his superiors, who regarded it as unrealistic, especially on logistic grounds. After intensive in-fighting between the staffs of the generals concerned, both Kawabe and Terauchi appearing almost indifferent, Tojo gave his final approval, while having his bath, to the order which stated: 'For the defence of Burma, the Commander-in-Chief Southern Army [Terauchi] shall destroy the enemy on that front at the appropriate juncture and occupy and secure a strategic zone in North East India in the area of Imphal.' Kawabe's plan to implement this was for Sakurai's 28th Army to launch an offensive against Lieutenant-General Sir Philip Christison's 15th Corps in Arakan in order to tie it down while Mutaguchi's 15th Army, with three divisions, crossed the Chindwin and eliminated Lieutenant-General Sir Geoffrey Scoones's 4th Corps. A new Army, the 33rd under General Honda, was formed to deal with Stilwell's Northern Combat Area Command threatening Myitkyina in the upper reaches of the Irawaddy and in the valley of the Salween.

Sakurai's offensive, launched at the beginning of February 1944, met with an unexpectedly stubborn resistance, as the beleaguered 7th Indian Division was kept supplied by air. Air supremacy, which the Japanese had enjoyed in 1942, had passed decisively to the British and Americans. When Mutaguchi launched his offensive from the Chindwin a month later, Sakurai's threat to the Arakan had failed, and Mountbatten, using American aircraft diverted from flying supplies to Chiang Kai-shek, was able to transfer a whole division by air from Arakan to block Mutaguchi's thrust to Kohima, threatening Slim's base at Dimapur, while also supplying Scoones at Imphal, where he had concentrated his three divisions in the nick of time. Mutaguchi had counted on a swift encirclement to prevent this, and on the capture of Slim's supplies to solve his logistic problems. He nearly succeeded in the former but, having failed in that, inevitably failed also in the latter. Resistance at both Imphal and Kohima, covered by overwhelming air superiority, was stubborn and effective, and by mid April all his divisions had suffered heavy casualties and had almost totally exhausted their supplies of food and ammunition. Nevertheless, in face of strong protests from his divisional commanders, Major-Generals Sato, Yamauchi and Yanagida, he persevered, backed by his superiors, reluctant to admit defeat. By the end of June, however, with Stilwell threatening Myitkyina, towards which the Ledo road had been driven through the mountains from India, Kawabe, a sick man, decided to call a halt. Terauchi and Tojo agreed. Sato, an old enemy of Mutaguchi since the *Kodo-ha* struggle with the *Tosei-ha*, had refused to continue, leading his division away from the front in search of food. On 11 July Kawabe at last prevailed on Mutaguchi to stop counterattacking and withdraw to the Chindwin, his exhausted army disintegrating as it did so. Three weeks later, having been besieged by Stilwell's almost equally exhausted troops for a month, the Japanese garrison of Myitkyina withdrew, its commander, Major-General Mizukami, committing suicide. Of the 156,000 men under Mutaguchi's command in March, 60,643 had become casualties. Slim's casualties over the same period totalled 16,667. The balance in Arakan was different, Christison losing 7,951 to Sakurai's 5,335. It was the severest defeat that the Japanese army ever suffered, and it was not surprising that Kawabe, Mutaguchi and all his divisional commanders were replaced, Kawabe by General Kimura, with the formidable Major-General Tanaka as his chief of staff, and Mutaguchi by General Katamura.

Japan's 'Zone of Absolute National Defence' was now severely threatened, the shipping crisis forcing Tojo to tell Terauchi that his forces in South-East Asia were to exist on the resources of the area they occupied, their only offensive task being to try to interrupt the movement of supplies to Chiang Kai-shek from India. The situation in the Pacific was now critical. The American success in the Marshalls was followed by a major assault on the Marianas, where the airfields on Saipan, Tinian and Guam were of major importance, as American possession of them would bring Tokyo, at 1,350 miles, within range of their Flying Fortresses. All these islands were strongly held by garrisons of 32,000, 9,000 and 8,000 men respectively in intricately constructed defences. To meet the threat of the US Admiral Spruance's 5th Fleet of fifteen carriers, carrying 950 aircraft, escorting one army and three marine divisions, supported by seven battleships and twenty-one cruisers, Admiral Ozawa deployed five fleet and three light carriers with five battleships and a large number of cruisers and destroyers, formed into three separate forces, the concept being to entrap Spruance between the carrier-borne aircraft and those based in the islands. Spruance began his assault on 15 June 1944, Ozawa intervening with his carrier-borne aircraft four days later, with results disastrous to himself, three of his fleet carriers being sunk and 480 aircraft lost in one day. Saipan fell on 5 July, the other two islands a month later, opening the way to an American assault on the Philippines.

The Japanese defeat in this battle, known to them as that of the Marianas and to the Americans as the Battle of the Philippine Sea, could not be concealed from the public, as many of the previous setbacks had been. For some time the *Jushin*, the conference of ex-prime ministers which had succeeded the *genro* as the body of elder statesmen, had been intriguing against Tojo, and now decisively moved against him. Tojo retaliated, inviting three of them to join his cabinet, replacing Admiral Shimada, who since February had doubled the post of Chief of the Naval General Staff and Navy Minister,* by Admiral Nomura, recalled from Berlin, and handing over the post of Army Minister to General Umezu from the Kwantung Army. But his colleagues deserted him and he resigned. The *Jushin* wanted Admiral Yonai to replace him, but he refused and, Tojo having refused to agree to Marshal Terauchi, Supreme Commander of the southern theatre of war, General Koiso, Governor-General of Korea, was appointed. Tojo tried to hang on to the Army Ministry, but had to hand it over to Sugiyama.

By this time the navy and the civilians realized that the war was irretrievably lost and were intent on bringing it to an end, but the army, still in occupation of almost all its conquests, and at a strength of 171 divisions, of which 55 were in Japan itself, refused to accept it. Konoye was prominent among the civilians who feared that further attempts to resist would bring about an internal crisis from which only the communists could benefit, the result of which would be worse than capitulation to the Americans. The cabinet was prepared to make concessions to the Soviet Union in Manchuria and the Kuriles, in return for Soviet help either in establishing peace with Germany or in the event of a German collapse. It came as a severe shock when Stalin, after secretly agreeing with his allies in October that he would join them in the war against Japan after Germany had been defeated, publicly branded her as an aggressor state in November 1944.

By that time MacArthur's troops had landed in the Philippines and the Japanese navy had again suffered severe losses – 306,000 tons of shipping – in the Battle of Leyte Gulf, against the Americans' only 37,000 tons. It was the real end of the Japanese fleet as a fighting force. Thereafter they relied on *kamikaze* suicide missions by land-based aircraft to inflict damage on the American fleet and ships transporting troops to the Philippines, to Iwo Jima, which was attacked in February 1945, and Okinawa on 1 April. Japanese resistance on both was, if anything, even fiercer and more suicidal than in previous operations. At Iwo Jima both sides had lost about 25,000 men when the few remnants of the garrison were finally rounded up after six weeks'

---

* Tojo had at the same time assumed the post of Chief of the Army General Staff in addition to his portfolios as Prime Minister and Army Minister.

intense fighting at the end of March. The battle of Okinawa lasted until 22 June, all but 7,400 of the garrison of over 100,000 having fallen, inflicting nearly 50,000 casualties on the Americans. By then 1,450 *kamikaze* pilots as well as several hundred others had lost their lives and the great battleship *Yamato* had sallied out on an equally suicidal misson.

The news from further west was no better. Slim's 14th Army in Burma had turned to the offensive at the beginning of December 1944, as soon as the monsoon ended. Kimura, whose instructions were to defend southern Burma, had already decided not to fight for the plain north of Mandalay, but to base his defence on that city and the Irawaddy south of it. He was outmanœuvred by Slim, who slipped his 4th Corps, in command of which Lieutenant-General Sir Frank Messervy had replaced Scoones, with a brigade of tanks, down the Myittha valley to cross the Irawaddy 100 miles south-west of Mandalay, south of its junction with the Chindwin, on 14 February 1945. Messervy drove east against little opposition to Meiktila, cutting off both the 15th and 33rd Armies, Meiktila falling on 5 March and Mandalay two weeks later. Kimura did his best to extricate what was left of his armies, but the defence of Burma collapsed as the monsoon started, Mountbatten's forces landing at Rangoon on 4 May.

While Slim had been enjoying the satisfaction of returning and taking his revenge in Burma, MacArthur's obsession with the need to do the same in the Philippines had cost both his forces and the Japanese dear. He had landed on the main island of Luzon in January and finally eliminated the last pocket of resistance on 4 March. The Japanese army was now in the situation that, even if it had wanted to evacuate its forces, it could not do so.

The only area in which there was any relief from the depressing tale of defeat was in China. Reacting to the American deployment of B-29 bombers in the area of China occupied by Chiang Kai-shek, the Japanese took the offensive in south-west China in the summer of 1944, forcing the evacuation of several airfields and causing Chiang Kai-shek anxiety about the safety of Chungking, as a result of which Chinese divisions were withdrawn from Stilwell's command in Burma and American advisers despatched to train sixteen divisions to a standard which might make them effective against the Japanese. The latter, however, had overstretched their logistic resources and remained content to secure their rail link with Indo-China, although they did attempt a further limited offensive in Central China in the spring of 1945, which the new Chinese divisions successfully parried.

In an attempt to secure better co-ordination of the higher direction of the war and greater authority for it, Koiso introduced a Supreme Council for the Direction of the War, which consisted of the former Liaison Conference with the difference that it met in the presence of the Emperor, who could actively intervene if he wished to, and not merely receive a report, as had been the rule of a full Imperial Conference. The secretariat, however, was still provided by the Bureaux of Military Affairs of the army and the navy, so that its deliberations were immediately known to more junior officers. Koiso wished to take over also the Army Ministry but, as a retired officer, he was prevented from doing so by the 1936 law. Frustrated in this and discredited by further defeats, he resigned on 4 April, three days after the assault on Okinawa had been launched.

In spite of opposition by Tojo, who favoured continuation of the war, the eighty-year-old Admiral Suzuki was appointed to succeed him. Although the new Army Minister, General Anami, was opposed to capitulation, he and the Chiefs of the Army and Navy General Staffs, Umezu and Toyoda, were not fanatically opposed to seeking peace, which was favoured by the Foreign Minister Togo and Navy Minister Yonai, as well as by the Emperor himself. One of the first measures was to exclude the Military Affairs Bureaux from the Supreme War Council. The day the new government was formed, Sato, the ambassador in Moscow, was informed by Molotov that the Soviet Union did not intend to extend the 1941 Neutrality Pact when it expired in April 1946, at the same time making reassuring statements which concealed Stalin's decision, announced at the Yalta conference in February, to attack them in return for a secret guarantee that he would receive south Sakhalin, the Kuriles and concessions in Manchuria.

The surrender of Germany on 8 May appeared to the Japanese government to improve their chances of negotiating peace. As the final fall of Okinawa became certain in mid June, the Supreme War Council agreed to seek peace through the offices of the USSR, the Emperor summoning them on 22 June to tell them to waste no time about it. The news that the leaders of the USA, the USSR and Britain were going to meet at Potsdam, and that Chiang Kai-shek's Foreign Minister T.V.Soong, had seen Stalin in Moscow, prompted the Emperor to summon Suzuki and emphasize the urgency. As a result, it was decided to send Konoye to Moscow as a special Imperial envoy, offering cooperation with the Soviet Union, but making clear that unconditional surrender would not be accepted. Stalin stalled, while discussing with the Americans the terms of a Soviet-Chinese agreement and what should be the pretext for a Soviet attack on Japan. The Americans obliged with a reference to the 1943 Moscow Declaration and the draft of the UN Charter, which referred to 'joint action on behalf of the community of nations to maintain peace and security'.

Having assured himself that there would be no American objection to his seizing what he was after in the Far East, Stalin was happy to acquiesce in the American proposal to bring the war to an end with the use of the atomic bomb. The War Council met on 27 July to consider the Potsdam Declaration, which seemed to them to offer the possibility of negotiation, once an armistice had been obtained, rather than unconditional surrender. Togo stressed that there would be serious consequences if it were not accepted, and the Emperor agreed. But the Army and Navy Ministers and their Chiefs of Staff insisted on rejection and on concealing from the public the more lenient of its terms. The Council decided to wait and see what resulted from their approach to the Soviet Union, and meanwhile to make no reply to the Declaration, giving orders to the press to play it down. Unfortunately the octogenarian Suzuki, either in error or under pressure from the army, told a press conference on 30 July that the government intended to ignore the Declaration, which was nothing more than a rehash of the one made after the Cairo conference in December 1943, and that it would make no difference to Japan's determination to fight to the end.

On 6 August the first atomic bomb fell on Hiroshima, and next day Molotov told Sato that he would see him at 5 p.m. Moscow time on 8 August. When he did he informed him that their countries would be at war on the morning of 9 August, Japan time, which was almost immediately. On that day a second atomic bomb was dropped on Nagasaki, and the combination of the two events persuaded the War Council to accept the Potsdam Declaration, provided that the Imperial House was preserved; but the Army Minister and the two Chiefs of Staff insisted on three more conditions: that there should be no occupation of the Japanese homeland; that they should withdraw, disarm and demobilize their soldiers and sailors themselves; and that the Japanese Government would be responsible for trying those accused of war crimes. No agreement could be reached in the War Council, in the Cabinet or in a full Imperial Conference. Against all precedent, the Emperor was asked to make the decision, a procedure for which the peace party had prepared him, taking the hard-liners by surprise. He agreed with Togo that the Potsdam Declaration should be accepted with the sole proviso about the Imperial House. 'I think that now is the time to bear the unbearable', he declared, recalling the Emperor Meiji's reluctant acceptance of the Triple Intervention in 1895, and a message to that effect was despatched through Switzerland and Sweden.

The American reply, broadcast on 12 August, which allowed no conditions to surrender, precipitated further argument between the civilians and the military, Suzuki and Yonai wavering between the two. The military were genuinely concerned that unconditional surrender would provoke a mutiny among their subordinates which they would be unable to control. All hung on the authority of the Emperor, who was supported by the Princes of the Blood. A full Cabinet meeting on 13 August having failed to reach agreement, an Imperial Conference was summoned. It met in the morning of 14 August, the Emperor giving a firm order that the Allied reply must be accepted. He made a recording of the Imperial surrender rescript which was to be broadcast to the whole nation and overseas.

Some of the Imperial Guard, at the instigation of officers of the army General Staff, broke into the palace and tried to destroy it. They were foiled by loyal troops, and on the morning of 15 August a message accepting the Allied terms was sent through the Swiss government. Suzuki resigned and, on the recommendation of Kido, Lord Keeper of the Privy Seal, whose influence throughout had been on the side of reason and peace, Prince Higashikuni, the Emperor's uncle-in-law and a general on the active list, took his place. The Emperor issued an edict ordering the army and navy to surrender and other royal princes were sent to Saigon, Singapore, China and Manchuria to convey it to the forces there, another being charged with the same task to the forces defending Japan itself.

It was the authority of the Emperor which in the last resort persuaded them to face the dishonour of surrender and saved Japan, and the USA also, from further casualties. What would have happened if the Imperial forces had known that the USA had no more atomic bombs to deliver remains one of the great enigmas. It is not easy to determine the moral account, balancing the casualties and suffering, immediate and long term, of Nagasaki and Hiroshima, against that which would have resulted, on both sides, from a continuation of the war into the main islands of Japan itself. Into the balance must also be set the demonstration to the whole world of the horrifying effect of the nuclear weapon.

Japan's total war casualties, including civilian dead, are put at about two and a half million. Her losses at sea were enormous. Of the warships in commission at the time of the attack on Pearl Harbor, nine battleships (two converted to carriers), ten aircraft-carriers, thirty-four cruisers and 100 destroyers were sunk; of those completed during the war, two battleships, eleven aircraft-carriers, six cruisers and 935 destroyers were lost. The fleet which surrendered at the end consisted of one battleship, four aircraft-carriers, four cruisers and thirty-nine destroyers. If Japan had played her cards more skilfully, the war might never have taken place, or, once it had begun, might not have ended so disastrously for her. The first major error was to become involved in China beyond Manchuria. If she had limited her ambitions on the mainland to that province there is no reason why Japan should have come into conflict with the USA. It was her struggle with the Kuomintang and her penetration into China which was the basic cause of that. Her next chance came when Germany invaded the USSR. She had a choice then between two potentially favourable strategies: either to join Germany in her attack on the Soviet Union, at the same time combining with her Axis partners in the naval war against Britain, especially in the Indian Ocean; or to use Germany's attack on the USSR as a pretext for abandoning the Axis Pact, and to seek an agreement with the USA which recognized her position in Manchuria and perhaps also, at least until the end of the war in which Germany was involved, in French Indo-China. The key element in the latter would have been acceptance of peace with the Kuomintang on terms which did not insist on the retention of a Japanese garrison in Central China. In the north, Chiang Kai-shek and the Americans might have conceded a Japanese military presence as a counter to Mao Tse-tung's communists. The major error, once the war had started, was to extend her conquests south of the Philippines, but even if that had not been done, Japan could never have challenged the potential power of the USA at sea and in the air with any hope of victory.

Unconditional surrender was one thing which the military would not accept, not only because it went against all their traditions, but also because it appeared to involve their extinction. The navy had almost reached that state by the end, but the army had not. Japan's last chance of saving something from the wreck was after the loss of the Philippines, but it would have meant accepting the terms of the Cairo Declaration of December 1943 which insisted on Japan handing back to China not only Manchuria, but the islands acquired by treaty in 1895, Formosa and the Pescadores; on her being and remaining disarmed, and subjected to occupation by Allied forces. With such terms, her opponents did not seem interested in seeking peace. Their anti-Japanese feeling arose not only from desire for revenge for their humiliation, but also from anger at the brutality with which the Japanese army had treated prisoners of war and the civilian population that had fallen into their hands. When the day of reckoning came, the Imperial Japanese army had much to answer for.

The people of Japan recognized this. They acquiesced, and in many quarters welcomed, Article 9 of the constitution designed by General MacArthur, which forbade the creation of military forces after the surrender. To many of them, including the Prime Minister, it came as a shock when, at the time of the outbreak of the Korean War, the USA reversed its attitude and began encouraging the Japanese government to raise forces both for internal security and for its defence, under the guise initially of a National Police Reserve, and then of Self-Defence Forces. The original ban on any ex-regular member of the army or navy joining these forces was soon abandoned when insufficient volunteers came forward.

In 1954 a mutual defence agreement with the USA was signed, as a result of which a Defence Agency was established, a civilian department controlling Ground, Maritime and Air Self-Defence Forces, the purpose of which was defined as the defence of the homeland against both direct and indirect aggression. Four years later the last American ground troops left Japan, except for Okinawa in the Ryukyus, but US air force units, supporting US forces in Korea, remained there. Recruitment to the Self-Defence Forces is voluntary, the initial engagement being for only two years, after which less than half re-engage. There is no reserve liability, although there is a voluntary officers' reserve, which involves an annual refresher course of five days and is generously rewarded. Most of its members are not ex-regulars, but civilians with a military bent. By 1986 the Self-Defence Forces numbered 245,000 men, the army forming thirteen so-called divisions, one of them armoured, which other armies (except the French) would classify as brigades, the navy manning sixty-three ships, none of them large, with sixty-four helicopters and eighty-four land-based aircraft, the air force operating 270 aircraft and three surface-to-air missile groups, paid for by less than one per cent of Japan's flourishing gross national product.

The USA, short-sightedly perhaps, continues to press Japan to spend more than the one per cent of GNP which she does on defence, partly on the grounds that Japan should contribute more to the security of the Pacific-Indian Ocean area, particularly of its sea routes, from which she is a principal beneficiary, but also, no doubt, because Japan is an economic competitor and it is felt to be unfair that her economy should benefit from such a low burden of defence expenditure. The increase in Soviet strength based in the Far East is cited as justification. Japan, keenly aware of the lessons of the past, realizes that her security and her current prosperity depend upon her maintaining peaceful relations with both China and the Soviet Union, and that no country in the Far East would welcome a resurgence of her military power. A marginal increase in her strength would have no effect on Soviet policy, and a major increase could raise ancient fears among all her neighbours.

The existence, organization and attitudes of the Self-Defence Force are the subject of some political controversy in Japan itself. The Socialist Party, among others, regards their very existence as unconstitutional, and has criticized the attempts by the Force to propagate a favourable image of itself. When General Kurisu stated publicly that it was absurd that the Force had to await a political decision before it could engage enemy forces involved in aggression, he was dismissed. The allegiance of the Force to the Emperor is another controversial issue, the more senior officers favouring it, but many of their juniors emphasizing allegiance to the nation rather than to the Emperor. Fortunately for all concerned, the Force has not yet been called upon to take action, other than in disaster relief operations at home. However, both action in aid of the civil power at home and action to protect Japanese interests abroad, such as securing threatened oil supplies, could provoke a political crisis, raising all the old issues once again.

8

CHINA

1900–31

The Boxer 'rebellion' in 1900 marked the climax of a period which saw the disintegration of the power of the Manchu Ch'ing dynasty that had ruled China since 1644. The inability of the imperial regime to resist the attempts of the 'barbarian' European nations and the USA to penetrate China with their trade and to remove from Chinese suzerainty outlying tributary areas, such as Tibet, Mongolia, Burma, Annam and Laos, fomented general discontent among the Chinese against their Manchurian rulers, whose authority declined sharply during the reign of the weak-witted Emperor Tao-kuang, who died in 1850. As his son Hsien-feng succeeded him a rebellion broke out in southern China, led by a young Hakka peasant who had failed his examinations in the Confucian classics, the essential method of entry into the bureaucracy. Known as T'aip'ing t'ien-kuo, 'Heavenly Kingdom of Great Peace', Hung Hsui-ch'uan, influenced by a Chinese convert to Protestant Christianity, believed himself to be appointed by God to overthrow the Manchu dynasty. The movement quickly spread northwards from Canton, exploiting all the discontents both of the peasantry and of those who resented the activities of foreigners of all kinds. European merchants were initially inclined to take a favourable view of a movement which had a Christian origin and might shake the inefficient and obstructive imperial regime out of its medieval torpor, but as the Taiping rebellion spread, causing chaos in the countryside and threatening foreign trading activities, they turned against it and, after the expedition to Peking in 1860, which forced the dissolute Emperor Hsien-feng to accept foreign legations, they gave active help to the imperial regime in suppressing the rebels, who had set up a rival capital in Nanking, and were threatening the approaches to Shanghai.

In previous times the Emperor had relied on two forces to support his authority over the huge and varied mass of people and territory that comprised the Chinese Empire, the Manchu Banners, recruited in Manchuria, and the Green Standards, recruited locally. Both theoretically owed allegiance to the Emperor, but as his authority weakened, especially at a distance from Peking, the forces in each province tended to become private armies of the Governor, until it was difficult to determine whether the Governor had an army by virtue of his appointment, or was appointed Governor because he owned an army. The inability of the imperial regime to provide an effective administration led to each governor exacting taxes and generally oppressing the populace in order to maintain his own power, which he tended to do through his soldiers. To suppress the Taiping rebellion the Emperor was dependent on the provincial armies of these governors, principally those of Tseng Kuo-fan in Hunan and Li Hung-chang in Ahnwei. It was with the latter that the British General Gordon's Ever Victorious Army, raised to defend Shanghai, co-operated.

The rebellion was finally crushed in 1864, three years after the death of Hsien-feng, leaving a five-year-old son whose mother, the concubine Yehonola, had established herself the *de facto* ruler by the mid 1870s, and was known as the Dowager Empress Tzu-hsi. Both for the maintenance of her authority in China itself and for dealing with external threats, of which the Japanese became the most immediate, she depended on the provincial armies that had defeated the Taipings. Li Hung-chang held the important post of Viceroy of Chihli, the provinces surrounding Peking, his protégé, Yuan Shih-kai, also wielding considerable influence, as did Tseng Kuo-fan, who was both a better general and less inclined to feather his own nest than the other two. The argument about whether to resist or adopt Western influence, which had resulted in the Meiji restoration in Japan, became a major issue in the closing years of the century, particularly after China's defeat by Japan in 1895. Li, Yuan, and to a lesser extent Tseng, favoured modernizing their armed forces as the only method of keeping barbarian influence out of the country. The Empress and the Mandarinate were opposed to any Westernization as the thin end of a wedge which would abolish them both.

Her son had died of smallpox in 1875, two years after assuming the throne, and had been succeeded by her three-year-old nephew, Kuang Hsu, restoring her to the regency. When the latter came of age and assumed the throne in the last decade of the century, he fell under the influence of the reforming scholar K'ang Yu-wei and wished to emulate the Japanese Emperor as a modernizing monarch. In 1898, after the death of the elder statesman Prince Kung, he dismissed Li Hung-chang, who had misappropriated funds intended for modernizing the navy, and announced a series of reforms, covering every aspect of government activity including the armed forces and the educational system. He called on Yuan Shih-kai to support him against the furious conservative reaction, led by the Dowager Empress, but Yuan betrayed him. Tzu-hsi moved swiftly against her nephew, forcing him to abdi-

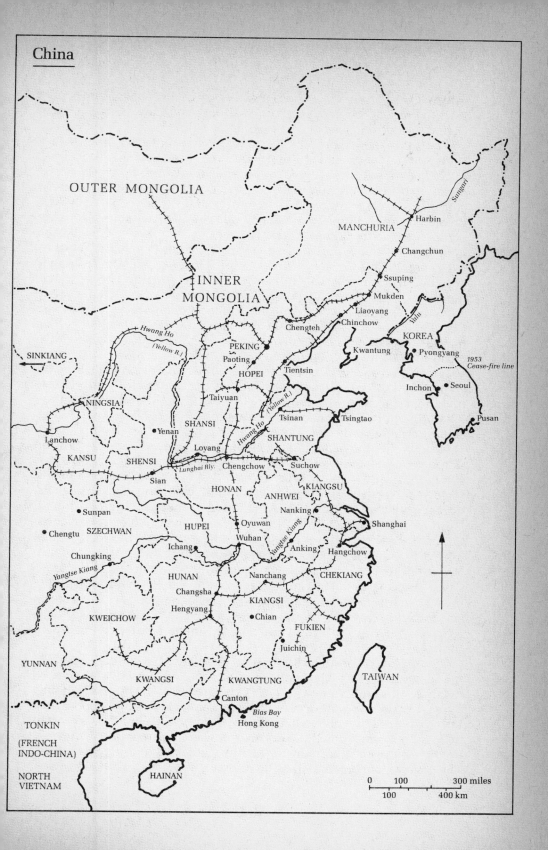

China

OUTER MONGOLIA

MANCHURIA

Sungari

Harbin

Changchun

INNER
MONGOLIA

Ssuping

Mukden

Liaoyang

Chengteh

Chinchow

Hwang Ho

(Yellow R.)

PEKING

Paoting

Kwantung

KOREA

Pyongyang

SINKIANG

HOPEI

Tientsin

Yalu

1953
Cease-fire line

Taiyuan

Inchon

Seoul

NINGSIA

SHANSI

(Yellow R.)

Tsinan

Tsingtao

Lanchow

● Yenan

Loyang

Hwang Ho

SHANTUNG

Pusan

KANSU

SHENSI

Lunghai Riv.

Chengchow

Suchow

Sian

HONAN

KIANGSU

ANHWEI

Nanking

● Sunpan

HUPEI

Oyuwan

● Chengtu

SZECHWAN

Wuhan

Yangtse Kiang

Anking

Shanghai

Chungking

Ichang

Hangchow

Yangtse Kiang

HUNAN

Nanchang

CHEKIANG

Changsha

KIANGSI

Hengyang

● Chian

KWEICHOW

FUKIEN

Juichin

YUNNAN

TAIWAN

KWANGSI

KWANGTUNG

Canton

TONKIN

Bias Bay

(FRENCH
INDO-CHINA)

Hong Kong

NORTH
VIETNAM

HAINAN

| 0 | 100 | | 300 miles |
| 100 | | 400 km | |

cate in favour of a cousin, nominated heir-apparent, son of Prince Tuan who shared the Empress's hatred of all foreign innovations. Li Hung-chang was reinstated as Viceroy of Chihli.

When, two years later, the Boxer rebellion erupted in the province of Shantung, Tuan and the Empress secretly encouraged it. Forced by German pressure to remove the provincial governor, who was merely transferred to another province, they appointed Yuan Shih-kai in his place. However, realizing that support of the Boxers could only lead to foreign intervention, he sat firmly on the sidelines as his fears were borne out and the Empress and her retinue fled from Peking. When Li died in 1901, and the Dowager Empress and her retinue had returned to Peking, Yuan succeeded him as Chihli Viceroy, and a limited policy of modernization was put in hand. Only one of Kuang Hsu's innova-tions, the University of Peking, had been allowed to remain, but in 1905 the classical Confucian examination system was abolished, and the Japanese defeat of Russia in that year stimulated the flow, which had already started, of students going to Japan: 500 went in 1902, 8,000 in 1905 and 13,000 in 1906. Yuan, who, as well as being Chihli Viceroy, held the post of Peiyang (North China) Great Official, with responsibility for trade, military and political affairs for North China, undertook the modernization of the army, Hsu Shih-ch'ang acting as his Chief of Staff. Together they created the Peiyang Army of six divi-sions, the commanders of which were their personal nominees.

Meanwhile a movement in favour of modernization had been created in the south, antagonistic to the imperial regime and led by Sun Yat-sen. Born near Macao in 1866, he had first organized a revolutionary society among the Chinese in Honolulu and was involved in an abortive uprising in Canton in 1895. After being kidnapped by the Chinese legation in London and freed through the intervention of an ex-Gover-nor of Hong Kong, he went to Japan, where, in 1905, he amalgamated his China Renaissance Society with two others to form the Alliance Society (T'ung Meng Hui), the aims of which were to expel the Man-chus, restore a national Chinese state as a republic, and redistribute the ownership of land. In response to protests from Peking, Sun Yat-sen was forced to leave Japan and moved to French Indo-China, while his movement gained adherents, especially among students in southern China.

On 14 November 1908 Kuang Hsu died, Tzu-hsi following him to the grave next day. She had arranged that his successor should be his three-year-old nephew P'u Yi, to be known as the Emperor Hsuan T'ung, whose father, Prince Ch'un, was nominated regent. Disregarding Kuang Hsu's dying command that Yuan Shih-kai should be executed for his treachery, he nevertheless 'allowed him to retire' with foot trouble; but he did not have to wait long before returning to power. The government faced increasing difficulties, having promised constitutional reforms, which the conservative elements strongly opposed, and sinking deeper into debt to foreign countries which were forcing ambitious railway construction projects on them. In addition they faced rebellion in the provinces of the west and south, while Japan and Russia carved up Manchuria and Korea between them.

Matters came to a head in October 1911 with rebellions in Szechwan and the important middle Yangtse trio of towns at Wuhan. The only effective military force available to the government was the Peiyang Army, and it owed allegiance primarily to Yuan, who was appointed Hukuang Viceroy and Imperial Commissioner of all Chinese Armed Forces. In December Prince Ch'un resigned as regent and was succeeded by the Dowager Empress Lung Yu, P'u Yi's mother, who granted Yuan full powers to settle with the rebels, with whom he set about negotiating, not to save the dynasty but to satisfy his own ambition to wield supreme power. In two days they agreed that a republican form of government should be established; that the Emperor and his family should be favourably treated; and that whoever overthrew the dynasty should become President, agreement also being reached that Sun Yatsen, who was in America, would provisionally occupy the post. He promised Yuan that, once the Manchus had abdicated and Yuan publicly declared his support of a republic, he would resign. Yuan had no difficulty in persuading the Regent to sign an abdication document on behalf of her son and to grant him full powers to 'organize a provisional republican government and confer with the Republican Army as to methods of union, thus assuring peace to the people and tranquillity to the Empire, and forming one Great Republic of China by the union as heretofore of the five peoples, namely, Manchus, Chinese, Mongols, Tibetans and Mohammedans, together with their territory in its integrity'.

Yuan had promised the rebels, whose leaders were members of Sun Yat-sen's T'ung Meng Hui, that the republic's capital would be at Nanking, but he staged disorders which made this appear imprudent, and he was proclaimed provisional President in succession to Sun Yatsen in Peking on 10 March 1912. He soon found himself at loggerheads with the T'ung Meng Hui, who had reorganized themselves to form a political party as the Kuomintang, the *de facto* leader of which was Sung Chiao-jen. Sun Yat-sen was not greatly interested in party politics and concentrated his activity on railway construction, while Yuan attempted to negotiate a reorganization loan with foreign bankers. In elections held early in 1913 the Kuomintang won a clear majority of the seats in both houses of parliament. They were opposed to Yuan's loan, just as he regarded their three concepts of nationalism, democracy and 'the people's livelihood' as dangerously radical. He therefore had Sung murdered in March.

When Parliament declared the signature of the loan agreement by

Yuan and two of his ministers illegal, he was faced with a military revolt by the Kuomintang governors of Kiangsi and Kwantung provinces, whom he promptly dismissed. Li Lieh-chun, the dismissed Governor of Kiangsi, joined forces with Huang Hsing at Nanking and Ch'en Ch'i-mei at Shanghai to oppose the three armies Yuan despatched to the Yangtse, but the combination of Huang's poor performance as overall commander of the Kuomintang forces and the availability of funds from the reorganization loan, with which Yuan could bribe other potentially rebellious governors, led to the collapse of the rebellion in two months, the Kuomintang losing control of six provinces, including Kwantung and Hunan. In October the cowed parliament formally elected Yuan as President, almost universal foreign recognition following that of the USA, which had been given in May, after the reorganization loan had been agreed. Yuan had demonstrated that power grew not only out of 'the barrel of the gun' as Mao Tse-tung, who had taken part in the revolt as a twenty-year-old soldier in Hunan, was later to write, but also out of the money-bags, a lesson which the Kuomintang took to heart.

Yuan's problems were not over. The reorganization loan ran out in April 1914 and he was forced to negotiate a new one on even more unfavourable terms, before the First World War broke out in August, leading to Japan's seizure of German concessions, followed by her Twenty-One Demands in January 1915. Yuan felt unable to reject them altogether; he knew that his forces were no match for the Japanese; he needed money and, with the European nations engaged in war, Japan was an important potential source. He was also plotting to have himself acknowledged as Emperor, for which he needed support from the war-lords of the north who, although nominally his supporters, were also his rivals. Two of the most important, who were opposed to his plans, were Feng Kuo-chang, governor of Kiangsu province, and Tuan Ch'i-jui, Minister of War, a graduate of the Peiyang Military Academy at Paoting, founded by Li Hung-chang, as was also Chiang Kai-shek, from the province of Chekiang, south of Shanghai. The latter went to pursue his studies at the Imperial Military Academy at Tokyo and returned from there in 1911, at the age of twenty-four, to join the revolutionary forces in Shanghai.

Foreign disapproval of Yuan's imperial ambitions encouraged all those who were opposed to it, from his rivals in the Peiyang clique to the genuine republicans of the Kuomintang and the southern and western war-lords who resented control from Peking. Revolts in Yunnan and Szechwan spread to Kweichow, Kwantung, Hunan and finally to Kwangsi. With the loss of all the southern provinces, Yuan's Peiyang generals, Tuan, Feng and Hsu Shih-chang, forced him not only to abandon the idea of becoming Emperor, but to retire from the presidency. Already a sick man, this brought on his death on 6 June 1916. Li Yuan-hung, the Vice-President, stepped into his shoes. Li, a native of Hupeh province and a supporter of the Kuomintang, had started life in the navy, his ship having been sunk in the Sino-Japanese War. Although he had fought as a brigade commander in 1911, he was fundamentally a weak character, and found himself at the mercy of the Peiyang clique, headed by Tuan, while Sun Yat-sen, from Shanghai, began to act as if he were President and Li merely his representative in Peking. At the same time the individual provincial governors took little notice of either and acted as if they were independent. Yuan had signally failed in discharging the duty imposed on him by the last Dowager Empress of 'forming one Great Republic of China'.

Rivalries within the different factions in Peking intensified. Whether or not China should join in the conflict on what, by 1917 with American entry into the war, was clearly going to be the winning side, was the principal political issue, although the real one was a struggle for power between the war-lords of the Peiyang clique. When eventually Tuan triumphed over his rivals, after an abortive attempt by Chang Hsun in July 1917 to reinstate the eleven-year-old P'u Yi as Emperor, Li handed the presidency over to Feng, and in August Tuan declared war on the Central Powers. Sun Yat-sen, who was suspected of taking money from the Germans, opposed the rupture of diplomatic relations in March 1917 and on the declaration of war sailed to Canton. There, with the precarious support of southern war-lords, he proclaimed a rival Chinese National Military Government, set up an alternative parliament, supported by a Movement for the Protection of the Constitution, and assembled a Constitution Protection Army which set off in September to march towards Peking. It was opposed by Feng's forces in Hunan and Szechwan, which succeeded in expelling it from those two provinces. However, Tuan could not restore Peking's authority over the provinces further south, and faced revolt also in the north, where Yang Tseng-hsin exercised independent power in Sinkiang, Chang Tso-lin in Manchuria and Yen Hsi-shan in Shansi. Tuan, however, succeeded in maintaining his ascendancy over his Peiyang rivals, Feng Kuo-chang and Ts'ao K'un, partly by yielding the presidency to Hsu Shih-chang and himself holding the post of Premier. Sun Yat-sen's movement was also beset by rivalries. His two deputies in the Southern Military Government were Lu Jung-t'ing, a former governor of Kwangsi province, who also controlled Kwantung, and T'ang Chi-yao, the virtually independent ruler of Yunnan. In May 1918 the Emergency Parliament replaced Sun as Generalissimo by a board of seven, of which only he and two others were Kuomintang men: the others, including Lu and T'ang, were war-lords, and one of them, Ts'en Ch'un-hsuan, from Kwangsi, was nominated chairman. In a huff, Sun left for Japan and moved from there to Shanghai. China was divided between different sets of war-lords.

As the First World War neared its end, Tuan's hopes rested on China being able to profit from its declaration of war, from its provision of a labour corps to work in France, and from its support of the Allied force in Eastern Siberia, to obtain the restoration of outlying territories, abolition of 'unequal' treaties and also financial help. In 1915 an agreement had been made with Russia by which the latter recognized Chinese theoretical suzerainty in Outer Mongolia, while Russia exercised *de facto* control over its so-called government. In 1919 Tuan sent his former Chief of Staff, Hsu Shu-cheng, with a brigade to take over the province. This annoyed Chang Tso-lin, the war-lord of Manchuria, who, in co-operation with the Japanese, encouraged a White Russian Cossack force under the 'mad' Baron Ungern-Sternberg to intervene. That in turn provoked the Mongolian princes and the deposed Chinese Resident Commissioner to appeal to the Soviet Red Army, which occupied the country and converted it into the Outer Mongolian Soviet Socialist Republic, which it has remained ever since.

Disappointment at the failure of China to achieve any of its expectations from the Versailles Conference stimulated xenophobic feeling, including student riots at Peking University, and provoked the May Fourth Movement. The European powers, the USA and Japan had all proved broken reeds and many of those who wished to see both a new order in China and a removal of the humiliation she had suffered at foreign hands for so long, looked to revolutionary Russia both as a model and as a source of support. It stimulated both a flow of students and the establishment of Marxist groups in China, although most young men going abroad to study still went to Japan. Among those who went to France and imbibed vintage Marxism on the Left Bank were Chou En-lai and Teng Hsaio-p'ing. It was in this period that Mao Tse-tung established a Marxist cell at Changsha in Hunan. Shanghai was the centre from which communist influence was directed, and from which contact was maintained with the Soviet Comintern.

In 1919 an attempt was made to negotiate peace between north and south. Since Sun's departure in 1917 T'ang Chi-yao, the war-lord of Yunnan, had ruled the roost in the south, Sun's principal military supporter being Ch'en Chiung-ming, who extended his sway from Kwangtung into Fukien province. After arguing from February until August, the negotiations, led by T'ang Shao-i on behalf of the south, broke down. In the spring of 1920 Sun tried to convert the Military Government in the south into a constitutional one, but was opposed by the powerful Kwangsi element, led by Lu Jung-t'ing and Ts'en Ch'un-hsuan. Sun ordered Ch'en to eliminate the control they exercised, but Ch'en hesitated as a major clash broke out between rival war-lords in the north: the Chihli clique, led by Ts'ao kun and Wu P'ei-fu supported by Chang Tso-lin, the virtual ruler of Manchuria, against the Anfu clique, supporters of Tuan and Hsu Shu-cheng. The Chihli clique won, Wu supplanting Tuan as the real source of power, while Hsu Shih-chang remained President. Ch'en promptly moved in the south, occupying Canton and driving Lu and his Kwangsi forces back to their native province, as Lu and Ts'en announced the dissolution of the Military Government.

In November Sun arrived in Canton and, at the end of the year, after negotiations with Ch'en and T'ang Chi-yao, they announced the resurrection of the Military Government, Sun to be Minister of the Interior and Ch'en War Minister as well as Governor and C-in-C of Kwantung and High Commissioner for Kwantung and Kwangsi. In January the old parliament met, electing Sun provisional President. When he formally assumed the post on 5 May 1921 he announced a programme of local autonomy, an Open Door for trade, development of industry and peaceful national unification, but his first step was far from peaceful: an advance by Kwantung forces to occupy Kwangsi, successfully concluded in July. Nor was his next proposed step towards unification peaceful: a combination of the south's forces with those of the Anfu clique and Chang Tso-lin from Manchuria to overthrow the Chihli clique in Peking. Ch'en, who was opposed to this, prevaricated and indulged in a series of measures to obstruct it.

Meanwhile, in February 1922, T'ang was ousted from power in Yunnan, leaving Ch'en as the most powerful military figure in the south, his position strengthened in March by the murder in Canton, probably at Ch'en's instigation, of Sun's trusted Chief of Staff and commander of the Kwangtung First Army, Teng K'eng. Ch'en resigned as Minister of War, moved his forces out of Canton and began negotiations with the Chihli clique. In co-operation with the Kwangsi General Yeh, he outmanœuvred the troops loyal to Sun, who was forced to withdraw in August to a British gunboat in the Pearl River, where he was joined by one of Ch'en's former officers, the thirty-five-year-old Chiang Kai-shek. Since 1912 Chiang had been involved in the shady world of politico-financial dealings in Shanghai before joining Ch'en. Together they sailed from Hong Kong in a British ship, *The Empress of Russia*, to Shanghai. At the same time the Chihli clique defeated Chang Tso-lin, who had moved his forces south of the Great Wall in April, announcing that he was joining forces with the south and proclaiming Manchuria independent of Peking. Wu and Ts'ao Kun, clear victors over the Anfu clique, forced Hsu Shih-chang to hand over the Presidency to the reluctant Li Yuan-hung, who accepted on condition that all provinces surrendered control of their military forces. All the principal military figures of the north agreed, having not the slightest intention of doing so.

The situation appeared to the Soviet Union to be ripe for exploitation. In August 1922 their representative, Adolph Joffe, had arrived in Peking to try to settle outstanding border issues and control of the Chinese Eastern railway in Manchuria before diplomatic relations were established. In January 1923 he went to Shanghai to see Sun Yat-sen. Contact had already been established with both the Kuomintang and the Chinese Communist Party by a Dutch member of the Comintern, H.Sneevliet, using the pseudonym of Maring, who did his best to persuade the two bodies to co-operate.

Events in 1923 revived Sun's hopes. Further dissension among the northern war-lords, offering the possibility of co-operation with Chang Tso-lin, and a revolt by Yunnan and Kwangsi forces, in co-operation with some of those of Kwantung, forced Ch'en out of Canton, to which Sun returned in February, proclaiming himself Generalissimo. He now realized that, if his political aspirations were to be fulfilled, he must have his own army as well as the co-operation of the communists. In August he sent a mission to Moscow, headed by Chiang Kai-shek, accompanied by Chang T'ai-lei, head of the Communist Youth Corps. In return Moscow sent Mikhail Borodin* as personal and political adviser to Sun. At the same time a mission was maintained with Peking. After Chiang's return from the Soviet Union at the end of the year, a strong Russian military mission, headed by General Blyukher, using the pseudonym Galen, was sent to Canton to help form an army on the Soviet model. The most significant step was the establishment of the Whampoa Military Academy to train young officers in military skills and indoctrinate them politically. Chiang was appointed its military commandant and Blyukher its chief instructor, directing a staff of Russians and Chinese. The American-born Liao Chung-kai, a loyal follower of Sun Yat-sen since 1905, was appointed acting head of the preparatory committee, the communist Chou En-lai being one of his subordinates. Both Blyukher's instructors from Moscow and Sun's opening address to the Academy stressed that the function of the academy was to create 'a revolutionary army', the struggle of the party having so far failed for lack of one; but it was not long before disagreement arose between the outlook of the military instructors, products like Chiang of the Paoting and Yunnan military academies, and the communist element both among the political staff and the Soviet military. Chiang clashed with them before the academy opened its doors.

The cadets did not have to wait long before putting their instruction into practice. In October 1924 a Soviet ship brought 8,000 rifles and 40,000 rounds of ammunition to Canton, and a month later the 490 graduates of the first class of the academy led 3,000 men, calling themselves 'The Party Army' (tang-chün) in successful attacks to clear Ch'en's forces further away from Canton.

---

* A Russian Jew whose real name was Gruzenberg, he had emigrated to the USA where he was a Comintern agent. He was expelled from Britain in 1922. He died after arrest in Stalin's purge of the 1930s.

While this development had taken place in the south, the northern war-lords were again at each others' throats, one of the causes being Chang Tso-lin's dislike of Peking's negotiations with the Russians for control of the railway which ran through his territory. Taking advantage of fighting between the war-lords of Kiangsu and Chekiang provinces, he advanced towards Peking in September 1924, while Sun Yat-sen announced that he would lead an expedition from the south to co-operate with Chang in overthrowing the regime of Ts'ao, who had appointed himself President in place of Li Yuan-hung in June 1923. As the latter's General Wu P'ei-fu faced Chang Tso-lin on the Great Wall, two of his army commanders, Feng Yü-hsiang and Hu Ching-yi, changed sides and marched on Peking, ousting Ts'ao and Wu, and bringing Tuan back as President. The latter, in co-operation with Chang and Feng, then invited Sun Yat-sen to come to Peking to negotiate a political settlement. Travelling via Shanghai and Japan, Sun arrived there on the last day of the year and remained engaged in fruitless discussions until his death from cancer on 12 March 1925.

In the ten weeks he had spent there, Chiang Kai-shek had expanded his military control to cover almost the whole of Kwangtung province and in the process had increased his army to a strength of 5,000. When Sun died, T'ang Chi-yao, the former war-lord of Yunnan, whom Sun had appointed his deputy as Generalissimo, tried to succeed him, relying on the Yunnan troops of Yang Hsi-min and the Kwangsi forces of Liu Chen-huan; but the First Kwangsi Army, loyal to Sun and supported by Chiang Kai-shek, under Li Tsung-jen, Pai Ch'ung-hsi and Huang Shao-hung, defeated him.

In May 1925, while this struggle for Kwangsi was in train, Chiang Kai-shek was appointed commander of the Kuomintang Party Army. On 1 July the Military Government at Canton was restyled the Nationalist Government of China, all thought of arriving at a political settlement with Peking having been discarded. In that month the Nationalist cause received a strong fillip from the anti-foreign feelings aroused by the shooting by British police in Shanghai of twelve rioting students. The two chief contenders for Sun's position as leader of the Kuomintang, Hu Han-min and Wang Ching-wei, became Chairman and Minister of Communications, Liao Chung-k'ai was appointed Minister of Finance, and Hsu Ch'ung-chih War Minister. Their unity did not last long. Liao was assassinated in August, Hu going abroad as his brother was involved. His departure was followed by that of Hsu, leaving the left-wing Wang Ching-wei in uneasy partnership with Chiang, a large proportion of their right-wing supporters also leaving for Shanghai, denouncing the growing influence of the communist and Soviet advisers. During the year Ch'en acknowledged defeat and departed for Hong Kong, and in January 1926 agreement was reached with Li Tsung-jen, Pai and Huang to incorporate their Kwangsi forces under

Li's command as the Seventh Army of the National Revolutionary Army. In March Chiang, without consulting Wang, declared military law and arrested a number of communists, whom he claimed were plotting against him, Wang himself moving to North China and thence abroad. Borodin then threatened to cut off Soviet aid and a compromise was reached by which Chiang promised to curb the right wing of the Kuomintang, if communist activities were restricted and Borodin withdrew his opposition to an expedition against the north.

The prospects for success of the latter had been improved by the internecine struggles for power between the northern war-lords. In the summer of 1925 Chang Tso-lin's son, 'the Young Marshal' Chang Hsueh-liang, led 2,000 men as far south as Shanghai, provoking a successful counteroffensive by the war-lord of Chekiang, Sun Ch'uan-fang, who acquired control of the forces of Kiangsu, Ahnwei, Fukien and Kiangsi. Chang's withdrawal in the face of Sun's forces was brought about by the intervention of Fen Yü-hsiang, who had built up his efficient Kuominchun army to a strength of 275,000. After he had captured Tientsin, he persuaded one of Chang Tso-lin's army commanders, Kuo Sung-lin, to join him. Chang was saved by Japanese support in Manchuria and by joining hands with Wu Péi-fu in North China. When the war-lord of Shansi, Yen Hsi-shan, threatened to join them, Feng went abroad, his subordinates withdrawing to the Great Wall, only fifty miles from Peking. Tuan, who had compromised himself with Feng, took refuge in the Japanese concession of Tientsin, while Wu established his power in Central China, acquiring control of Szechwan, Honan, Hupeh and the northern part of Hunan. One of the latter's generals, T'ang Sheng-chih, thereupon joined the Nationalists, his army becoming the Eighth Revolutionary Army. Chiang Kai-shek could now muster some 100,000 men to oppose five times that number, if the forces of Chang, Wu and Sun Ch'uan-fang combined against his; but, calculating that his opponents could not rely on the loyalty of either their subordinates or each other, and that popular feeling, fomented by the communists, would support his forces, he launched the long awaited northern expedition on 9 July 1926.

The northward thrust through Hunan to the Yangtse at Wuhan, led by the communist-sympathizing 4th (Kuomintang) Army under Li Chi-shen and the 7th (Kwangsi) under Li Tsung-jen, joining hands with T'ang Sheng-chih's 8th (Hunan), proved remarkably successful, the trio of cities, with their important arsenal, being secured in October. Wu, deserted by the Szechwan army to the west and given no help by Sun in the east, was forced north of the Yangtse, uncertain of any support from Chang, who controlled the Chihli area to the north, or from Feng, returned from Moscow, whose formidable Kuominchun army had concentrated in Shensi, to the north-west. At the last moment Sun thrust westward through Kiangsi, in an attempt to cut off the southern forces at Wuhan, which brought him into head-on conflict with Chiang's 1st Army, soon joined by Li Tsung-jen's 7th, switched east to support him after Wuhan had been secured. In a series of bloody battles, in which losses on both sides were heavy, Chiang defeated Sun, whose casualties are said to have amounted to 100,000, and occupied in succession Nanchang, Shanghai and Nanking, the last on 24 March 1927.

By that time a deep split had developed in the Nationalist camp. The political element, dominated by the left wing, had transferred the Nationalist Government to Wuhan, where, supported by the communist-dominated 4th Army, it instigated a campaign of communizing the province of Hunan, persecuting landlords and industrial employers to the consternation of the Kuomintang right wing, including Chiang himself. His demand that the seat of government be transferred to Nanchang, when he captured it, was ignored by the left-wing clique in Wuhan. Chiang's links with rich Chinese in Shanghai reinforced his position, although his capture of the city owed much to the activities of Chou En-lai in fomenting support for him against Sun. Matters came to a head in the spring of 1927. On 1 April Wang Ching-wei returned from exile in Europe and joined the Wuhan faction. On the 18th Chiang proclaimed a separate Nationalist government in Nanking, twelve days after a raid by Chang Tso-lin's police on the Soviet embassy in Peking, in which they seized documents revealing the extent of Soviet support to the Nationalists and Communists. At this stage Wu, lacking support from either Chang or Feng, withdrew into Western Honan, pursued by the 4th and 8th Nationalist Armies, which then came face to face with Chang's forces in the north, suffering heavy casualties. Feng then intervened with an advance from Shensi down the line of the Lunghai railway and the valley of the Yellow River, threatening the Nationalist left flank. Having taken Chengchow, he opened negotiations with both Nationalist factions, first with the Wuhan element, who were accompanied by Blyukher. They agreed to withdraw from hard-won Honan, acknowledging Feng as Chairman of the Honan Provincial Government Commission, but received nothing in return. Feng went on to meet Chiang Kai-shek on 20 June at Hsuchow, which the latter had recently captured, where they agreed to co-operate against the Communists, who had recently received a directive from Stalin to intensify Communist political activity, including the dispossession of landlords, making an exception in the case of those who were Nationalist officers. The activities of the left-wing elements at Wuhan provoked a general purge of communists, forcing Borodin, Blyukher and the Soviet advisers to leave the country and dispersing the Communist element within the Nationalist movement, many of them being killed.

With the defeat of Wu, the defection of Feng, and Yen in Shansi sitting on the fence, Chang Tso-lin's only ally, with whom to face the Nationalists, was Sun Ch'uan-fang. Some of his supporters argued for a withdrawal to Manchuria, but his more resolute chief of staff, Yang Yü-ting, won the day for a policy of accepting a division of the country, leaving the Nationalists in control south of the Yangtse. The first step was to exploit the differences in their camp and drive them south of the river, which they rapidly proceeded to do in the eastern part of the plains in August. Sun's success caused a split in Chiang's ranks, the Kwangsi element calling for a reconciliation with the Wuhan faction. Chiang resigned and, after political talks in Shanghai, went off to Japan. Sun then overplayed his hand by crossing the Yangtse with 70,000 men, who were promptly cut off by Li's forces, only 20,000 escaping north of the river.

Attempts to patch up differences between Wuhan and Nanking made little progress and finally broke down when a Communist-inspired uprising in Canton in December provoked a strong Nationalist reaction. Chiang had already returned to Shanghai and used the incident to undermine the position of his left-wing rival Wang Ching-wei, who once again left China, this time for a rest-cure in France. In that month also Feng and Ho Ying-ch'in, the latter commanding Chiang's 1st Army, recaptured Hsuchow and drove Chang's forces back into the Shantung peninsula, provoking the Japanese to advise Chang to withdraw to Manchuria. Chiang, reinforced by the financial support of rich friends in Shanghai, reassumed his posts of C-in-C of the Nationalist Army and Chairman of the Kuomintang Executive Committee in January 1928, adding to them that of Chairman of the Military Affairs Commission. The Kwangsi element having renewed their co-operation with him, the northward thrust was renewed in April, provoking the Japanese to land troops in northern Shantung and warn Chang that, unless he withdrew to Manchuria and secured hold of his native province, they might be forced to take it over themselves. After his forces had lost Paoting, eighty miles south of Peking, Chang finally took their advice, abandoning the capital on 2 June. As his private train approached Mukden two days later, he was killed as it was blown up at the instigation of an extremist clique of officers of the Japanese Kwantung Army.

On 10 October 1928, the Nationalist Government of China was proclaimed, with its seat at Nanking. Dominated by Chiang Kai-shek and the right wing of the Kuomintang, its 'Organic Law' provided for a one-party government, closely resembling its predecessors in Peking. Chiang had done his best to restore a Chinese national state, but Inner Mongolia, Tibet, Sinkiang and Manchuria were at that stage not under his control. Authority over other areas rested on agreement with Li of Kwangsi; Yen, known as 'the Model Governor' in Shansi; Feng, 'the Christian General', in Inner Mongolia and Shensi; and with 'the Young

Marshal', Chang Hsueh-liang in Manchuria. He paid little attention to the principle of democracy, and no more than lip-service, if that, to 'the People's Livelihood'. The Young Marshal rejected pressure from Japan and from two of his father's most powerful generals, Yang Yü-ting and Ch'ang Yiu-huai, to declare Manchuria independent, and nailed his flag to the Nationalist mast, being appointed Governor of Manchuria in return and having the two generals murdered.

The Nationalist forces had reached a total of probably about 1,600,000, imposing a heavy burden on the new state. At a meeting in July 1928, Chiang, Feng, Yen and Pai, on behalf of Li, had agreed that the armed forces should be reduced to between fifty and sixty divisions and a gendarmerie of 200,000. Nothing had been achieved by January 1929, when another meeting was held to discuss the issue. At that time Chiang controlled directly 420,000, Li and Pai 230,000, Feng 220,000 and Yen 200,000. If Chiang had been prepared to accept a high degree of devolution, perpetuating control of their own forces by provincial governors, a split might have been avoided; but it was clear that, as demobilization proceeded, he intended to concentrate control of armed forces under the Kuomintang and therefore in his own hands. There is no doubt that, thanks to Whampoa Academy, they were the most efficient.

The first to challenge Chiang's authority were the Kwangsi forces but, with the support of those in Kwangtung, Chiang, with 150,000 men, defeated their 120,000 at Wuhan in April, and a large proportion of their 7th Army swung to Chiang's side. Next to revolt was Feng. Chiang had gone back on his word to give him control of Shantung province, and had bribed one of his generals to switch his allegiance to Chiang in October 1929. Feng rallied the support both of Yen and of the defeated Kwangsi forces, and launched an offensive against Chiang in May 1930, Wang Ching-wei having returned from abroad to join them. Both sides appealed for support to the Young Marshal, Feng's coalition having been weakened by a successful attack on Kwangsi from Kwantung, while Yen and Feng were heavily engaged with Chiang's army in Ahnwei and Shantung. At this stage Chang Hsueh-liang was offered a bribe of ten million yuan and the prospect of governing the whole of China north of the Yellow River. As a result, in September 1930, he declared his support of the Central Government. Feng disappeared from the power game and Yen took a back seat, while Wang Ching-wei and the Kwangsi element, having made peace with Kwangtung, proclaimed a separate government in Canton in May 1931.

Rich Nationalist war-lords and Wang's separatist government were not the only challenges to Chiang's authority. Since their expulsion from the Nationalist ranks and the purge of 1927, the communists had been gradually organizing themselves into soviets, with their own armed forces, in different provinces. The largest was the Oyuwan

Soviet in Hopei and Ahnwei, its 1st Red Army of some 6,000 men in 1929 being led by Hsu Hsiang-chien, known as 'the Ironside', who had been involved in the abortive Canton commune rising in December 1927. But the most significant militarily was the 4th Army, formed by an amalgamation of the forces Mao Tse-tung had assembled at Ching-kang-shan, an inaccessible mountain lair on the borders of Kiangsi and Hunan, with those which Chu Teh had gathered together in Kiangsi, after an abortive attempt to capture the port of Swatow.

Chu Teh, born in Szechwan in 1886, was a product of the Yunnan Military Academy who had served in the old Imperial army and commanded a brigade of the Yunnan forces which fought against Yuan Shih-kai. In 1921 he abandoned his dissolute life-style and went to Europe to study. There he met Chou En-lai who persuaded him to join the ranks of the communists, which he did on returning to China in 1926. An able, experienced and popular commander, he remained for long the principal military figure attached to Mao. By May 1928 they had assembled 10,000 men, although they did not have weapons for that number, and had evolved a partnership, in which Mao acted as political director and policy maker and Chu Teh organized, trained and commanded the troops. They agreed on a strategy, based on four 'Golden Rules', each expressed by only four Chinese characters: 'when the enemy advances, we retreat; when the enemy halts, we harass; when the enemy avoids battle, we attack; when the enemy retreats, we follow'. The strategy was supported by 'Three Cardinal Rules of Discipline': 'obey all orders; take nothing from the peasants; pool all confiscated goods'.

Before 1928 their principal enemy had been the local forces of the landlords, the *Ming Tuan*, but, from the middle of that year onwards, Nationalist forces closed round them, one of their brigade commanders, P'eng Teh-huai, a graduate of Whampoa, defecting with most of his brigade to join Mao. Meanwhile Liu Chih-tan, also an ex-Whampoa cadet, had established a soviet in Shensi which dominated most of the countryside, confining the war-lord's troops to the towns. In January 1929, leaving P'eng to hold the base, Chu and Mao with 4,000 men broke out of the ring and marched into Kiangsi. In their absence P'eng was forced temporarily to evacuate the base, while Chu and Mao moved their small band across Kiangsi to Fukien, attempting, with varying success, to organize soviets and forces to support them. Meanwhile Liu tried to exploit the famine in Shensi by raising a revolt, which was suppressed by Feng, 'the Christian General', whom Liu then joined, earning himself the nickname of 'the Turncoat'. In June 1930 the communist forces were organized into armies, the 1st under Chu Teh in Kiangsi, the 2nd in Hunan and Hupei under Ho Lung, the 3rd in Hunan-Kiangsi under P'eng and the 4th in the Oyuwan Soviet area under Hsu 'the Ironside'. Total numbers for whom there were weapons probably amounted to no more than 20,000. There were many more partially trained men without arms, as well as partisans, who also demanded rifles, but the Central Executive Committee, headed by Li Li-san, decided that all arms should go to the 'regular' armies. After Chu Teh had successfully defeated an attack by Yunnanese Nationalist forces in April, the Committee, against the advice of Mao, decided on a policy of attacking cities. Mao had also pressed for some arms to go to the partisans. The Committee's policy was influenced by the fact that it was dominated by figures who looked to the urban masses for support, while Mao and Chu believed that the peasantry should be their power base. The Committee's policy appeared to be justified when P'eng seized Changsha in Hunan. However he only held it for ten days, the masses failing to give him support. An attempt by Chu Teh to take Nanchang at that time also threatened from the north, failed with heavy casualties, as did a second attempt by Chu and P'eng to regain Changsha in September. They succeeded in capturing Chian, but could not hold it, after which Chu insisted that without artillery they must revert to guerrilla warfare. Li's policy was discredited, and the Comintern intervened, condemning his 'revolutionary adventurism' and removing him to Moscow.

Having dealt with his non-communist Nationalist rivals, Chiang Kai-shek gave priority to sorting out the communists. When the Soviet advisers were dismissed in 1927, Chiang turned to the Germans, of whom, since his days in Tokyo, he had been an admirer; but not initially to the German government. In December 1927 Colonel Bauer, a former associate of Ludendorff, met Chiang Kai-shek in Shanghai and took a Chinese military mission back to Germany. He returned a year later with forty-five instructors of various ranks and set about reorganizing and training Chiang's army on German lines. He also gave personal advice on the conduct of operations, which, when followed, usually led to success. He died in May 1929 from smallpox contracted while engaged in the successful battle for Wuhan. He was replaced by Lieutenant-Colonel Kriebel, who, on taking his leave at the armistice negotiations in 1918, had said to the Allied representatives: 'See you again in twenty years!' He was succeeded by Lieutenant-General Wetzell, who, in 1933, persuaded Seeckt himself to come on a tour of inspection.* After three months in China, Seeckt's recommendations, not surprisingly, reflected what he had achieved with the Germany army: that quality should be preferred to quantity; that emphasis should be given to officer training; that there should be centralized control of personnel and finance matters, and that, when operations were undertaken, a C-in-C in the field should be appointed,

---

* In 1922 the Nationalists had hoped to persuade Falkenhayn to come, but, when he died, first approached Seeckt in 1923.

separate from the political head, and principal organizer of the forces, and, finally, that a period of external peace was required while the army was being reorganized. Wetzell, more in tune with realities in China, disagreed, and was replaced by Seeckt himself in the spring of 1934, with the full agreement of the German government, by then headed by Hitler. One of the principal motives of all the German advisers, including Seeckt, was to use China as an outlet for German industry, which would contribute to the establishment of a Chinese war industry. Seeckt brought with him as his Chief of Staff General Falkenhausen, who succeeded him when ill-health forced his return to Germany in 1935. Their plan was the creation of a force of sixty fully modernized and equipped divisions, backed by an equal number of semi-modernized divisions. By 1935 the number of German military advisers had risen to seventy.

1931–45

In this period of reorganization, Chiang had to face a clear choice of priorities after the Mukden incident in 1931, between the growing threat from Japan and that from the communists. Believing that the latter could be disposed of before he had to face the full force of the former, he adopted the policy: 'First pacify the country, then oppose the foreigner'. In the event it proved to be a fatally wrong choice. At a time of grave national discontent and hardship caused by flooding of the great rivers, he might have been able to unite all factions in opposition to Japanese encroachment. Instead he gave first priority to 'bandit-extermination', as his campaigns against the communists were called, appointing Lu Ti-p'ing, governor of Kwangsi, as the overall commander of anti-communist forces.

The first move was made against Chu Teh in Kiangsi and Fukien by Chang Hui-tsan with eight divisions, totalling 100,000 men. Chu Teh had 40,000, only half of whom were armed, and for several months, starting in December 1930, he led the Nationalists a dance, exhausting them and risking only one general engagement, at Lungkang, when Chu's men surprised one of their divisions, capturing its commander and most of his men. The campaign died down in February 1931 with no conclusive result. Chu Teh's 1st Red Army had been reduced to 20,000 men, but had gained invaluable experience and was still intact. When Chang withdrew it still controlled most of Kiangsi and Fukien. In May Chiang gave the task of renewing the offensive to Ho Ying-ch'in, his faithful subordinate from the earliest days of the 1st Kuomintang Army, who was given 150,000 men in fifteen divisions, while another 50,000 were allotted to other generals to support local forces in dealing with the other soviets and their Red armies. While Ho advanced cautiously into the hills where the communists either avoided or harassed them, Chu dealt a series of blows at the forces of other generals, in Fukien and Kiangsi. After six weeks Ho's campaign fizzled out and Chiang himself decided to take the field with 130,000 troops in July.

Although Chu's tactics again proved successful, he was almost in dire straits when Chiang's effort was diverted by preoccupation with the consequences of the Mukden incident in September 1931. In the pause which followed, Mao strengthened his position, establishing a 'Central Soviet Government' in Juichin, on the borders of Hunan and Kiangsi, parallel to the Central Executive Committee in the Oyuwan Soviet, which now controlled an area containing three million people and had suffered little from the extermination campaigns. It moved to join the Kiangsi Soviet in this period. The armed forces of the soviets had been renamed, those of the Kiangsi Soviet, the 1st Front Army, numbering 30,000 under Lin Paio, Chu Teh exercising a more general command of all the soviets' military affairs; of the Hunan, the 2nd, numbering 10,000 under Ho Lung; and those of the Oyuwan, the 4th, under Hsu 'the Ironside', numbering 20,000. P'eng continued to command the 3rd. Mao sent a message to Chiang suggesting co-operation against Japan but, having received no answer, himself formally declared war on Japan in April 1932, after Chiang's troops had clashed with Japanese on the outskirts of Shanghai and Japan declared Manchuria independent as Manchukuo, P'u Yi having been resurrected from obscurity as the Regent (later Emperor) Hsuan Tung. By then 'the Young Marshal', on Chiang's advice, had withdrawn south of the Great Wall.

During 1932 the Nationalists gradually tightened the noose round the two principal soviets. Advised by Seeckt, Chiang launched his 4th Extermination (known to the communists as Encirclement) Campaign in June against both the Oyuwan and the Central (Kiangsi) Soviets, employing some 600,000 troops, supported by 100 aircraft, under his own command. He moved first against the Oyuwan Soviet and, having dispersed its forces by December while fighting off attempts by Lin Piao's First Front Army to disrupt his offensive, he switched the main Nationalist effort to Kiangsi. Skilful operations by Lin Piao and P'eng brought this to a halt by the end of April 1933, both sides having suffered heavy casualties.

During the next five months both sides reorganized their forces, Chu Teh and Chou En-lai playing important parts in converting the communist forces to a more regular pattern. On German advice, Chiang decided to adopt a methodical strategy of encirclement of the Kiangsi Soviet, based on the establishment of blockhouses, combined with a scorched earth policy and a strict ban on movement to deprive the communists of supplies. In October 1933 he launched his 5th Extermination Campaign, deploying some 800,000 troops, 500,000 in his North Route

Army and 150,000 each in the West and South Route Armies, against Chu Teh's 150,000, whom he attacked from all directions.

In April 1934, with the loss of Kuang Ch'ang, the communists were forced to consider seriously the evacuation of their forces from Kiangsi. The final decision to do so and embark on the Long March was made sometime between July and September.

The Long March started in October 1934, by which time the different armies and soviets could communicate with each other by radio. Hsu 'The Ironside's' 4th Army of some 50,000 men had already left the Oyuwan Soviet area after Chiang had attacked it in the autumn of 1932, meeting little opposition as he moved west through southern Honan towards Sian in southern Shensi, as Chiang was preoccupied with his operations in Kiangsi. The war-lords through whose territory Hsu passed prudently refrained from attacking so large a force, while assuring Chiang that they were doing so. Most of Hsu's losses arose from Nationalist air attacks and from the bitter winter weather. Early in 1933 Hsu halted south of Sian, while argument raged about his final destination. Reports from his namesake in the mountains of northwest Shensi round Yenan were discouraging. Most of his soldiers were from the low plains of Central China, and were reluctant to move into the rugged, bare western mountains; but Feng, Yen (who had returned and been reinstated as Governor of Shansi) and the Young Marshal were concentrating their forces to threaten him. Chang Kuotao, chairman of the Oyuwan Soviet, took charge, with Hsu as his military commander, and in the spring of 1935 ordered a move westward into Szechwan, where he could wait to join hands with Mao and Chu Teh and hope to persuade them to stay there until they were strong enough to move back down the Yangtse valley. Reduced in strength they reached the high plateau of Sunpan in western Szechwan in June, stopping there to wait for the 1st Army.

Mao and Chu Teh faced a much fiercer struggle to break out. Chu Teh's 1st Front Army contained 90,000 men in five corps, accompanied by 10,000 political and other personnel. Leaving behind Chen Yi with some 13,000 partisans both to keep communism alive in the area and to distract the attention of the surrounding Nationalist troops, they fought a fierce battle to break out of the first ring on 21 October, in two columns, the 3rd Corps led by P'eng and the 1st by Lin Piao, another ex-Whampoa cadet: command alternated between them. Harassed throughout by Nationalist air attacks, they succeeded in breaking through three more Nationalist lines before, reduced to 60,000, they reached the comparative safety of Kweichow province in January 1935, where the army, out-manœuvring the pursuit, spent four months moving about the province, recruiting, collecting supplies and making converts. Chu Teh may also have hoped that Ho Lung's 2nd Army from northern Hunan would join him there, but Ho Lung, whose forces had risen to 40,000 when he was joined by a local war-lord, preferred to stay put in his secure area west of the lakes. Chiang correctly appreciated that at some stage Chu Teh would try to cross the Yangtse into Szechwan and transferred twenty divisions from Central China to guard the crossings in that province, at the same time ordering his Yunnan forces to attack Chu Teh's south of the river. To distract the latter, Chu Teh sent one corps southwards in April, while the rest of the 1st Army moved west to where the Yangtse changes direction in the mountains of northern Yunnan. By a combination of rapid movement, deception, skill and courage, Chu Teh's leading troops forced a crossing on 4 May at Leng Kai, the whole army moving across in nine days.

Chiang then prepared to entrap them as they crossed the Tatu, 200 miles further north; but by moving through a forested mountain area inhabited by the Lolos, normally hostile to Chinese, Chu Teh forestalled the Nationalists and forced a crossing of the Tatu against the local war-lord's forces at An Shung Chang three weeks later. A critical situation arose as the flooded river rose and Nationalist aircraft appeared. Only Lin Piao's 1st Corps of 8,000 had crossed: the rest, headed by P'eng, had to move upstream on the south bank, where they captured a partially demolished chain suspension bridge over the gorge at Lin Ting, by which the rest of the army crossed. Over the next two months the army struggled over high mountains, harassed by the local tribesmen, until, reduced to 45,000, they joined the 4th Front Army in June in the north-west corner of Szechwan.

Differences between Mao and Chang soon raised their heads. Mao proposed to establish a base in Kansu, from which to operate against the Japanese, who by then had moved their forces into Inner Mongolia and up to the Great Wall, occupying Jehol. He saw that the communists could gain popular support from being anti-Japanese at a time when the Nationalists appeared to be appeasing them. Chang, however, stuck to his concept of building up strength, preferably in the mountains of Sikang, also known as Chamdo, west of Szechwan, in anticipation of a return down the Yangtse valley to restore soviets in Central China. In August they went their separate ways, Chu Teh, with 15,000 men, joining Chang, leaving only 30,000 with Mao, accompanied by Lin Piao and P'eng. Orders were sent to Ho Lung to move out of Hunan and join Chang. Entering Kansu in October, Mao's men defeated the Nationalist 19th Route Army and Muslim cavalry from whom they captured some much needed horses and, reduced to 20,000, joined forces with those of Hsu 'the Peasant' and the North Shensi Soviet, moving into the town of Pao An. Their Long March had covered more than 6,000 miles and taken a year.

Ho Lung, who set off in the spring of 1936, had to fight almost every mile of the way against Nationalist and war-lord troops as he crossed Kweichow into Yunnan. When he eventually reached Sikang and joined Chang and Chu Teh in July, his original strength of 45,000 had been whittled down to 20,000. His arrival swung the balance against Chang, and he and Chu Teh persuaded Chang that they should move their 80,000 men to join Mao in North Shensi. They set off in August and arrived, reduced to 50,000, in December, just two years after the break-out from Oyuwan and Kiangsi. Their arrival brought the strength of the Red armies in the North Shensi Soviet to 95,000, including 10,000 who had deserted from the Young Marshal's Manchurian army and from Chinese Muslims.

In the year Mao had spent there he had taken practical measures to train and organize the 1st Army, P'eng being placed in charge of operations and Lin Piao of training. A start had been made on organizing the area as a base for local supply of all the army's needs, including armaments. At the same time P'eng had had to fend off the half-hearted efforts made by Feng, Yen and the Young Marshal, on Chiang's orders, to crush his forces. All three were more concerned to oppose Japanese encroachment into the area, and Chou En-lai succeeded in establishing an undercover relationship with the Young Marshal, lending him some instructors for his military school.

Chiang's attempt to force Feng, Yen and the Young Marshal to undertake a '6th Extermination Campaign' led to a curious incident at Sian, capital of Shensi, in December 1936. The Manchurian generals supported the Young Marshal in resisting a new offensive and wished to call off the campaign against the communists. In an attempt to put pressure on Chiang, they kidnapped him and held him for two weeks, during which P'eng extended his activities over much of Shensi. They then released him, and the Young Marshal, as a token of good faith, accompanied him back to Nanking, where Chiang clapped him into prison. He was later transferred to Taiwan where, in 1961, he was released to house arrest, in which he apparently remains. Meanwhile P'eng reoccupied Yenan, from which Manchurian troops had ousted him, and Mao moved there, declaring it the capital of 'The Chinese Workers' and Peasants' Soviet Republic', which it remained until 1947. Chiang sent Hu Tsung-nan, one of his most virulently anti-communist generals, to break up the Manchurian army and replace it with his own divisions. He rejected an appeal by Mao, conveyed to Nanking personally by Chou En-lai, to place his forces under Nationalist overall command to form a united front against Japan. In a further effort to accommodate anti-communist opposition to this, Mao changed the name of his 'government' to 'Special Area Government' two months before the Lukouchiao incident on 7 July 1937 sparked off war with Japan.

From China's population of 450 million, Chiang Kai-shek controlled a standing army of 1,700,000 regulars, backed by 518,400 reservists.

He was to conscript another fourteen million between 1937 and 1945, the peak strength of his army in 1941 rising to 5.7 million, of whom three million were front-line soldiers. In theory all able-bodied men between eighteen and forty-five were liable for conscription, but only a proportion were actually selected, on average one man from each group of about a dozen families. Many of them in fact never reached their units, getting 'lost' in one way or another during their walk of a hundred miles or more from the recruiting centre to their training unit. General Stilwell estimated in 1943 that over 50 per cent were lost in that way. In 1937 this large army provided only forty divisions, manned by 300,000 men, based on the reorganization carried out on the German pattern, and only 80,000 of those had modern weapons. Fighting with vigour, they lost heavily in the unsuccessful attempts to hold Nanking and Shanghai, the loss being particularly severe among junior officers, 10,000 of whom became casualties, equivalent to a three-year output from the Military Academy, which had been moved from Wampoa to Nanking. The loss represented one-ninth of Chiang Kai-shek's properly trained officer strength.

The first Japanese moves to consolidate their positions round Peking brought them up against inferior forces of local war-lords, whom they pushed aside, occupying Suiyuan, penetrating into the north of Shansi and setting up a puppet regime in Inner Mongolia. Meanwhile fighting had also started at Shanghai. In spite of Germany's close relationship with Japan, sealed in the 1936 anti-Comintern pact, Falkenhausen remained with Chiang Kai-shek, as the German government hoped to mediate between the two belligerents. The mission was reduced to thirty by the time it was removed altogether in 1938 as a result of Japanese pressure on the German government.

Direct help, however, was promptly supplied by the Soviet Union, which on 21 August concluded a non-aggression agreement and sent six squadrons, totalling 400 aircraft, to support the Nationalists, as well as a mission headed by General Cherepanov. Military supplies were also promised, to be sent overland through Sinkiang, where Moscow maintained close relations with its virtually independent ruler Sheng Shih-ts'ai. Both Chiang's slender stock of aircraft and the Russian squadrons were concentrated to defend the area of the Yangtse valley.

In this critical situation, Chiang responded more favourably to a renewed offer by Mao to form a United Front. After Chou En-lai and Chu Teh had visited Nanking, which was soon under direct threat, he agreed that the latter's troops in North West Shensi should be named the 8th Route Army. In return Mao renamed his 'government' the North-west Border Region, and agreed to accept Chiang's overall direction and refrain from creating new soviets. Chiang refused to recognize other soviets. Chu Teh's 8th Route Army of 45,000 men was formed into three divisions, the 115th, led by Lin Piao, consisting of the 1st Front Army that had stayed with Mao, the 120th, commanded by Ho Lung, formed from the 2nd Army, and the 129th, by Liu Po-cheng, 'the One-Eyed General', formed from Hsu 'the Ironside's' 4th and that part of the 1st which had split off and moved with Liu and Chu Teh to accompany it. Hsu was his second-in-command. P'eng remained in command of the 50–60,000 soldiers remaining in Northwest Shensi and acted as Mao's military adviser. Chu Teh moved his army into Shansi where he theoretically came under the command of Yen's 18th Army Group and 2nd War Zone. Lin Piao registered an early success against a Japanese mechanized brigade which he ambushed in the P'inghsing Ruan pass.

Having secured their position westward from Peking by mid October, the Japanese began a two-pronged drive south, through Shansi towards Yen's capital at Taiyuan and down the railway from Peking towards Chengchow against Sung Che Yuan's 29th Route Army at Paoting, which fell on 24 September. In early November the Japanese General Matsui landed a force in Hangchow Bay, captured Shanghai and advanced to Nanking, Chiang Kai-shek moving to Wuhan and sending his government offices to Chungking, a further 600 miles up the Yangtse in Szechwan. Nanking fell in mid December; the Japanese rape of the city recalled the mayhem that had followed the Nationalist capture of it ten years earlier, and aroused widespread international indignation, especially in the USA.

This left Han Fu-chü with 70,000 men in Shantung threatened with encirclement. When the Japanese began a two-pronged offensive, from north and south-east, at the end of the month, Han offered little resistance, and Li Tsung-jen, sent to defend the vital railway junction at Hsuchow, arrested him and sent him to Wuhan, where he and nine other senior officers were executed. With thirty divisions Li, inflicting significant reverses on Terauchi's Northern Japanese Army, notably at Taierhchuang, held out until 19 May 1938, by which time General Hata's force from the south-east had almost surrounded him, a large proportion of his forces escaping into Honan.

Sugiyama, who had replaced Terauchi, now planned to move west up the Lunghai railway south of Shansi, before striking south to Wuhan. He had begun to do so, relieving General Doihara, isolated at Lanfeng, on the way, when Chiang ordered the flooding of the Yellow River, bringing operations in Central China to a halt and devastating the countryside. The main Japanese effort switched to the Yangtse valley, the Wuhan trio of cities being captured at the end of October, and Changsha, capital of Hunan, in mid November. This, combined with the landing of a force of 30,000 men under General Fususko at Bias Bay, south-east of Canton, brought the campaign to a halt for the time being. Chiang Kai-shek had withdrawn to Chungking; his forces, according to Japanese sources, had suffered 800,000 battle casualties, while theirs numbered 50,000.

Chiang rejected offers from Japan of an agreement by which they would co-operate against the communists in return for Japanese concessions in China, equivalent to the 1915 Twenty-One Demands, which Chiang rightly foresaw as reducing China to tributary status. He calculated that Japan's ambitions would bring her into conflict with either the Soviet Union or the USA, her clash with the former on the borders of Manchuria in the month that Stalin signed his non-aggression pact with Hitler lending support to his thesis. Meanwhile he would exploit the anti-Japanese feeling of both powers in seeking aid, which he would use to strengthen his own internal position against the communists, whom he would encourage to turn their attention towards the Japanese, who might then do some of his dirty work for him. His appreciation was basically correct: his mistake was to pay more attention to the exploitation of anti-Japanese sentiment abroad, leaving the field free for Mao to exploit it internally, and to think of strengthening his own internal position in terms of military strength and politically in terms of those who had in the past wielded political power.

Mao appreciated the politico-military strategic situation in parallel, but sought to exploit anti-Japanese feeling internally among the masses and to strengthen his own position by a combination of military guerrilla action and political activity both in the area his forces occupied and behind the Japanese lines. Meanwhile he would try to remove the threat from the Nationalists and hope to gain both Soviet and American support by appearing to co-operate with and, perhaps, surpass Chiang in opposition to the Japanese occupation. Chou En-lai was left with Chiang at Chungking as the agent of this policy. Mao did not succeed either in removing the Nationalist threat or in obtaining American or Soviet support, both of which were provided to Chungking, and none passed on to Yenan.

Even before the Second Sino-Japanese War started in 1937 Chiang Kai-shek had been cultivating the strong anti-Japanese and pro-Chinese lobby in the USA, incongruously supported by both industrial and financial tycoons and missionaries, mostly Protestant. His marriage, after discarding a previous wife, in 1927 to Mayling, sister of Sun Yat-sen's second wife and of T.V.Soong, scion of a rich Christian financial family in Shanghai with strong American links, fortified his ability to play that card. To strengthen it he adopted his wife's religion. In 1937 he employed a USAAF captain, retired for deafness, Claire Chennault, to train his air force, replacing an Italian team. Chennault, a fighter ace, was a passionate devotee of Mitchell's belief in the self-sufficiency of air power, but had no opportunity to demonstrate it before the Nationalist air force, equipped with a miscellany of obsolete foreign aircraft, was swept out of the sky by the Japanese.

Not only was Chiang reluctant to share any support he might obtain with the communists, but he decided, while the Japanese threat was quiescent, to move against them and restict their ability to sovietize the countryside. In December 1939 he ordered Hu to drive them out of southern Shensi. The area controlled by 'The Border Region' was severely reduced and Mao's ability to move guerrillas and political cadres out of it restricted by a tight ring imposed by Hu, although an easier way out remained on the eastern flank, controlled by Yen. After the Japanese had thrust further up the Yangtse valley, capturing the 'gateway' at I-chang in June 1940 and frequently bombing Chung-king, Mao in August launched an offensive in the north, known as the Hundred Regiments campaign. He claimed that 500,000 men were employed in widespread attacks on railways, roads and isolated Japa-nese installations which, it was said, inflicted 20,000 casualties on the Japanese, 30,000 on Chinese fighting with them, and destroyed 300 miles of railway and 900 of roads. It was some time before the Japanese reacted. When they did they inflicted heavy losses on the 8th Route Army, killing perhaps 100,000.

The 8th was not the only Communist army theoretically forming part of the United Front. The partisans and soldiers left behind with Chen Yi in Kiangsi had been formed into the 4th Route Army and moved to the south bank of the Yangtse. Chiang, suspecting that they were proselytizing the area between them and Shanghai, ordered Chen Yi to move, using a route north of the river. After much argument he began to do so, and when three out of his four divisions were over, the Nationalists attacked the remainder, killing or capturing nine out of ten thousand, including the commander Yeh T'ing, wounded and arrested, and his deputy Hsiang Ying killed. Chen Yi then established a new soviet on the north bank. After this and a Nationalist-inspired attack on the Border Region in April 1941 by Chinese Muslim forces in Kansu and Ninghsia, co-operation of any sort between Mao and Chiang ceased, although Chou En-lai remained in Chungking for the sake of appearances and in the hope of gleaning some crumbs from the aid provided by the USA. That from the Soviet Union dried up when Germany attacked her in June.

The prospects for US military aid increased dramatically with the passage of the Lend-Lease Act in March 1941. For some time before that Chiang had been making strenuous efforts to ensure that his claim for help was not prejudiced by Roosevelt's desire to help Britain. In November 1940 he sent T.V.Soong to Washington to plead for an American-manned force of 500 aircraft, including B-17 bombers, and an Anglo-American loan of $200–300 million. He appealed both to the guilty conscience of Americans who felt that, ever since the US had acquiesced in Japanese retention of German concessions in Shan-tung after 1918, they had failed to support China, even when it was led by a Christian who professed to exercise democracy. Soong had little success with the Roosevelt administration, but took the opportunity to prepare the background for the organization which he set up as soon as Lend-Lease was approved – China Defence Supplies Inc, controlled by himself and his other brother-in-law, H.H.Kung, who was Chiang's Finance Minister. This handled all US aid, material and financial, and proved a highly profitable business to Chiang and his family.

With the advent of Lend-Lease, Chennault's plan was accepted, the first step being the delivery of 100 fighters and the recruitment of American volunteer pilots for them, paid high salaries and a bonus of $500 for every Japanese aircraft destroyed. Officially known as the American Volunteer Group, unofficially as Chennault's Flying Tigers, they were to exercise a significant influence on American policy towards Nationalist China. Chennault's exaggerated claims for the effect that air forces based in China could have on Japan appealed to those, in both the USA and in China, who hoped to defeat her without having to fight bloody battles. In addition to provision of the AVG, support was promised to organize, train and equip thirty of Chiang's 300 divisions to a modern standard, which should have clear internal political implications. To supervise the programme an American Military Mission was sent, headed by Brigadier General Magruder, who had been military attaché in Peking on two occasions.

After Pearl Harbor in December 1941, Chiang expected and argued for more support. The only route by which material help could reach him was the Burma Road, which the British had closed, under Japanese pressure, from July to October 1940, ostensibly in the hope of bringing about an agreement between China and Japan. Supplies piled up at Rangoon and all along the route. The British were suggesting that it would be better to employ them in defence of Burma than to transport them along the rapidly deteriorating road up to Chungking. Chiang also feared that priority would be given to the defence of American and British bases in the Far East. He made great play of the danger that, if he were not supported, he could not contain Japanese forces in China, which would then be available for use against the British and Americans. He also resented that he was not represented at the ARCADIA conference in Washington at the end of December, at which the ABDA command was established, with Wavell as its head. Chiang was fobbed off with being appointed C-in-C of a separate Allied China Theatre. In that capacity he was to be provided with an American Chief of Staff, who would also command all US army, including USAAF, forces in China, Burma and India.

The man selected was the fifty-eight-year-old Joseph W.Stilwell, promoted Lieutenant-General for the post. He had an unrivalled knowledge of China, having been attached to the US embassy in Peking as an intelligence officer from 1920 to 1923, commanded the US battalion at Tientsin from 1926 to 1929, and been military attaché from 1935 to 1939. He was also given responsibility for the distribution of Lend-Lease Aid to the Chinese armed forces, a role which brought him into conflict with both Chiang and Chennault. He did not reach Chungking until 6 March 1942, the day before the Japanese captured Rangoon. When Wavell, as British C-in-C India, had visited Chungking before the ARCADIA conference, Chiang offered two armies, the 5th, one of his best, and the 6th, of lower grade, for the defence of Burma. Wavell had declined, asking only for a reinforced division. His reaction annoyed Chiang. It was based partly on Wavell's belief that a large Chinese force could cause more trouble than it would be worth, particularly if it demanded help with logistic support, partly on a suspicion that Chiang had an ulterior motive in detaching Burma from India. Later, when the threat to Burma became more direct as the Japanese advanced down Malaya, Wavell changed his mind.

By the time Stilwell arrived the 6th Chinese Army was moving into the Shan states in north-eastern Burma. The 5th had not moved although, on 1 March, Chiang had released one of its divisions to join the 6th. After Stilwell had taken command of the Chinese forces in Burma, theoretically accepting the overall command of the British General Alexander, he persuaded Chiang to send the rest of the 5th. A further cause of dissension arose over the AVG. It was based in northern Burma and had provided most of the air support to the British and Chinese forces as they withdrew into northern Burma. Stilwell regarded them as American forces under his command, while Chennault, backed by Chiang, insisted that they formed part of the Chinese Air Force. Stilwell eventually won the argument, Chennault being given command of all USAAF in China, designated the 14th Air Force, USAAF in India forming the 10th. At the disastrous end of the 1942 campaign in Burma, Stilwell himself with remnants of two divisions, the 22nd and the 38th, some 8,000 men, reached India, while others struggled back into China.

Stilwell's return to Chungking in June 1942 set off a long-running dispute between himself, Chiang Kai-shek and Chennault. Stilwell's aim was to turn the large Nationalist army into an effective force to fight the Japanese, which would involve a major reduction in its size and a revolution in the way in which it was commanded, trained and administered – much the same line as Seeckt had taken. US aid should be tailored to that end. With the airlift from India over the 'hump' the only route, Chennault demanded that priority be given to the build-up of the force of 500 aircraft, which would be a more effective form of US aid than a major programme for the army. Chiang tended to back Chennault, while at the same time demanding unrealistic amounts of supplies for his existing army; and that US divisions should be sent to India to help in the recapture of Burma in order to open the road, while a new one, the Ledo road, was being constructed through the mountains from India. Chiang's resistance to Stilwell's persistent demand for concentration of effort in modernizing thirty divisions was primarily motivated by the fear that some other general might gain control of them and oust him from power. The more Stilwell sought out generals who favoured his concept, the more Chiang suspected him of plotting against him. Stilwell's undiplomatic manner aggravated the differences between them.

Meanwhile neither the Nationalists nor the Communists were taking any significant offensive action against the Japanese, who made their main effort in China against Chu Teh's 8th Route Army, reducing it to a strength of 300,000 and confining it to North West Shensi, while they took ruthless action against Communists all over the area they occupied, in the campaign known as 'The Three Alls' – Kill All, Burn All, Loot All. In spite of this Chen Yi, north of the Yangtse, gradually increased his 4th Route Army until, in 1943, it numbered some 80,000. Although organized into divisions, the Red armies at this time engaged only in guerrilla operations and in parapolitical activity, including help to the peasantry, as well as growing their own food and providing their own supplies. The 'regular' forces were supplemented by the *Ming Ping*, a militia, originally called Home Guards or Self-Defence Corps, which acted both as a local part-time force and as a reserve force from which replacements for the regular force were recruited. In theory all were volunteers, but in practice, as the need for replacements grew, considerable pressure, virtually equivalent to conscription, was applied in areas controlled by local soviets. The militia, although trained by the 'regulars', remained under local control. Not all the *Ming Ping* could be armed, although captured Japanese weapons were distributed to them. By 1944 their total strength rose to two million.

In 1943 the Japanese launched major offensives on the northern Shansi border and Hopeh against the Communists and in Hupeh and Hunan against the Nationalists. While the strengthened forces of both struggled to contain them, Stilwell tried to persuade Chiang to give priority to the recapture of Burma, to which twenty well-trained Chinese divisions should be allotted, holding out the enticing prospect of thus restoring a flow of supplies to his armies. The British should co-operate by capturing the Andaman islands as a step towards a landing at Rangoon. After much argument, in the course of which Chiang renewed his attempts to wrest control of the use of US aid away from Stilwell, he agreed, on condition that the British participated both on land and at sea and that the operations were supported by an adequate air force. His demands were always backed by the threat that, if they were not met, he could not contain the Japanese, with a veiled hint that he might even come to terms with them. British refusal to embark on a campaign to regain Burma in 1943 caused Chiang to back out also, to the chagrin of Stilwell.

High-level American visits to Chungking resulted in promises of more aircraft being devoted to the airlift over the hump and of major increases in the strength of the forces allotted to Chennault. The whole question of Allied strategy in the area was tackled at the QUADRANT conference in Quebec in August 1943, at which the British South East Asia Command, with Mountbatten as Supremo, was established, Stilwell being appointed his deputy, while continuing to hold his other posts. Almost immediately a division between them arose over strategy. To Stilwell and the Americans, first priority should be given to recapturing Northern Burma in order to improve communications with China, while Mountbatten, principally concerned with restoring the British position in the Far East, thought in terms of amphibious operations. The issue was decided by the Japanese who, launching an offensive towards India in March 1944, made themselves vulnerable to a counteroffensive, which was launched in November, the month after Stilwell was replaced by Wedemeyer, and which led almost without a pause to the complete clearance of Burma by May 1945.

Stilwell's replacement was primarily caused by a struggle for power in Chiang's immediate entourage, in which Soong's hopes of replacing Chiang were opposed by his sisters; but it was aggravated by Stilwell's constant pressure for reduction in the size of the Nationalist army in favour of organizing more effective divisions, Stilwell proposing plans for a second thirty division programme. It was intensified by the tug of war between them over the allocation of forces to Burma, where Stilwell had been trying to capture Myitkyina in co-operation with Wingate's Chindits, and by the threat posed by a Japanese offensive, launched in April in reaction to the deployment of USAF B-29 bombers, capable of reaching Japan, with Chennault's 14th Air Force. Chiang also disapproved of Stilwell's pressure for greater co-operation with the Communists.

Stilwell was the unfortunate pawn in a game between Chiang and Roosevelt. The former relied on the USA, with some contribution from Britain, to defeat Japan, while he kept his powder dry to deal with his real enemy, the Communists. Roosevelt wished to use China both to distract Japan on land, while the US forces fought her at sea, and to act as a permanent balance to Japanese power after the war. Stilwell himself was seized with a passionate desire to prove that the Chinese peasant, for whom he had formed a great affection, if decently led, fed, clothed, equipped and trained, tough and inured to hardship as he was, could prove the equal of the Japanese. In the end none of them achieved their aims.

The first Japanese thrust came down the Peking-Hankow railway in Honan, dispersing the thirty-four divisions Chiang maintained in that province, and allowing the Communists to filter in from Shansi to take their place. In June they drove south from Wuhan, threatening the 14th Air Force's bases in Hunan and continuing through Kwangsi until the whole line of the railway to Hanoi was securely in their hands. In addition they pushed further up the Yangtse towards Chungking, causing Chiang to threaten withdrawal of his forces from Burma, just before the offensive there was due to start, provoking a sharp reaction from Washington, demanding that Stilwell be placed in command of *all* forces in China. Chiang's response was to demand Stilwell's recall, and that responsibility for handling Lend-Lease should be transferred to himself. He immediately told his own Central Executive Committee that he had done so and that to accept an American commander of all his forces would be a new form of imperialism: if he agreed, they would be nothing but puppets and might as well all go and join Wang Ching-wei. President Roosevelt gave way and Stilwell was replaced by Lieutenant-General Wedemeyer, at the time deputy chief of staff to Mountbatten. He resolved the crisis in Chungking by providing transport aircraft to reinforce its defences from the forces deployed against the Communists in the north-west. By this time the American enthusiasm for supporting the Nationalists had waned, as their concept of landing their forces on the mainland of China had been cancelled and they could bomb Japan from the islands they had captured.

1945–53

In the early months of 1945 it was clear to all concerned that it could not be long before the war came to an end, and both Chiang and Mao were concerned about their respective positions if and when Japan began to withdraw. Mao doubled the ranks of the 8th Route Army by reinforcing it with militia, until its total numbers are said to have reached 800,000, while Chen Yi increased his New 4th Army to 110,000. The total strength of political cadres rose above a million, and Mao ordered the expansion of soviets in every region, the number of large ones increasing from fourteen to nineteen. He claimed to control 300,000 square miles, inhabited by ninety-five million Chinese.

In June 1945, as the Japanese began to withdraw down the Yangtse valley and later, in North China, towards the coast, Hu Tsung-nan launched an offensive against Chu Teh, but the latter held him, although he had to cede some ground in the north-west. When the atom bombs fell on Japan and the Soviet Union joined in the war against her, Chiang's and Mao's forces were watching each other even more suspiciously than usual. In the eight-year-long Second Sino-Japanese War the Nationalists suffered some three million casualties out of fourteen million men recruited. The Communists admit to only 500,000, but the true figure was probably nearer two million. It is not possible to say how many of the casualties of both were incurred fighting each other. The Communists claimed to have captured 500,000 men serving in forces operating under Wang's puppet government, 3,500 Japanese and 34,000 Nationalists, with over 200,000 rifles, 3,000 machine-guns and 150 artillery pieces, their principal source of arms.

On 11 August 1945 Chiang gave an order to Chu Teh not to move any of his forces or to accept the surrender of any Japanese forces. Chu Teh firmly rejected it, telling his subordinates to accept surrenders and take over control of areas occupied by the Japanese and Wang's puppet regime. At the same time he sent Lin Piao with 100,000 men, mostly unarmed, into Manchuria, uncertain of how they would be received by the Russians who, in agreement with their allies, were to accept the surrender of Japanese forces there. However, they connived in Lin Piao obtaining a substantial arsenal of equipment off the Japanese, including seventy tanks complete with their Japanese crews, and in his recruiting Manchurians to bring his strength up to 200,000 by the end of the year. Meanwhile Chu Teh's and Chen Yi's forces were busy sabotaging the rail system to prevent the transfer of Nationalist forces to the north. The latter, eventually carried out by air and sea with American help, was delayed by Chiang's insistence that only troops from the thirty-nine modernized divisions of his Central Army, deployed in the extreme south, could be employed. He feared that, if he used other forces, such as the Kwangsi men under Li Tsung-jen and Pai Chung-hsi, or the armies of Fu Tso-yi and Sun Lien-chung, they might seize power from him, perhaps coming to some accommodation with Chu Teh.

Seeds of distrust within the Nationalist ranks, which had never been absent, were germinated by Chiang's attitude at this time. As the Communists established themselves more and more firmly in the north and east, Chiang, whom MacArthur had nominated as the sole authority to receive Japanese surrenders in China, gave orders to the Japanese not to surrender to the Reds but to Chinese forces who had served Wang, until Nationalist forces arrived. At the same time four offensives against the Red armies were to be launched: by Li's forces against Chen Yi's 4th Route Army in the Lower Yangtse valley, by his own Central Army northwards from Wuhan, by Yen against Ho Lung in Shansi and by Fu against Nieh Jung-chen's Soviet round Peking. Only the first met with any success, Chen Yi being forced north towards Shantung, destroying the communications as he went.

Attempts by the American ambassador, General Hurley, to bring about a reconciliation between Chiang and Mao met with little success, until some half a million Nationalist troops had been transported in American aircraft and ships to the north in late October, preceded by 53,000 US marines who landed to secure the Tientsin-Peking area. The prestigious figure of General Marshall was sent at the end of

November to try to halt the internecine war. By then Mao and Chu Teh had been forced to recognize that their ill-equipped forces could not challenge the modernized Nationalist troops in a straight fight. A cease-fire between them came into effect on 13 January 1946.

Chiang, who had asked for a US military mission to help him, had already set about reorganizing his army. All the thirty-nine American-trained divisions of his own Central Army would be retained, as would 180 others of the provincial and central armies, led by the more trusted generals, the reorganization to be complete by June 1946. As a first step one and a half million men out of his total of five and a half million were to be discharged. This wholesale demobilization, which by the end of the year included 200,000 officers, affected most the forces other than those of Chiang's personal Central Army, and caused grave discontent not only among those discharged, many of whom soon joined the Reds, but in the higher ranks, who saw it as confirmation that Chiang was using American support to strengthen his own personal supremacy. Marshall's mediation resulted in a Political Consultative Conference in Chungking in January 1946, at which agreement was reached on a programme for national reconstruction and a democratic political structure. It was followed by an agreement that the Nationalist and Communist forces would be merged into one army, which by mid 1947 would have been reduced to sixty divisions, ten of which would originate from the Communists. Neither side had much, if any, intention of implementing any of the political or military agreements, both using them as a method of trying to obtain American support.

A breakdown in the truce between them soon developed over the arrangements for handover by the Russians of authority in Manchuria, where they had been busy removing everything of material value. In spite of tacit support to Lin Piao's presence and strengthening of his forces, they adhered to their recognition of Chiang's Nationalist Government and, after several delays, handed over to his forces at Changchun and Mukden, Chiang having rejected Mao's demand for a share in the administration of the province. Fighting between Lin Piao's 'North-eastern Army' and Sun Li-jen's First Nationalist Army broke out in April, the Communists capturing Changchun only five days after the Russians had handed it over to Nationalist troops, who were airlanded there. This was followed by a month-long battle at Ssuping (also known as Szeping (kai)), sixty miles further south on the railway, in which seven Nationalist divisions, totalling 70,000 men, successfully defended it against over 100,000 Communists, who then evacuated Changchun. Having occupied it at the end of May, the Nationalists advanced up the railway, driving Lin's forces over the Sungari river to Harbin before another truce came into force on 7 June. While the Nationalists, under the overall command of the incompetent but politically reliable Tu Yu-ming, had scored successes in that area, they lost control of important areas in southern Manchuria, where a Nationalist division from Yunnan defected to the Communists. Outside Manchuria, the Nationalists had cleared a large area of the Central Plains, driving Communist forces westward into the hills, although Chen Yi continued to hold the key town of Tsinan in Shantung. For three weeks General Marshall shuttled to and fro between the two sides, but Mao rejected Chiang's demands that he move his forces out of certain key areas, and fighting resumed in July, as the internecine struggle broke into full-scale war.

At that time Chiang had some three million men under arms and continued to receive substantial military and financial help from the USA, to which the sale at knock-down prices of surplus military equipment, originally procured at a cost of $900 million, made a significant contribution. Mao's forces, receiving virtually no external aid of any kind, probably numbered about 1.4 million, backed by at least as many in the militia. In July he declared them 'The People's Liberation Army', of which Chu Teh was C-in-C. It was subdivided into the Yenan army, commanded by P'eng, who was also Chu's deputy, the North Eastern Liberation Army under Lin Piao in Manchuria, and the East China Liberation Army under Chen Yi. During the rest of the year all of them suffered reverses at the hands of the Nationalists, the area controlled by the Communists being reduced by about a quarter from that which they had held in January.

Chiang announced a cease-fire in November, when he convened the National Assembly at Nanking, to which his government had returned. It was boycotted by both the Communists and the Democratic League, and its value as proof of the Kuomintang's adherence to democracy did not impress the Americans, who, having withdrawn the marines and reduced their military presence in China from 113,000 to 12,000, withdrew also from the mediation arrangements which General Marshall had established, although the US navy continued to maintain a naval training group at Tsingtao and the military mission remained at Nanking.

The Nationalist position was not as strong as it looked. Their forces had advanced and occupied a number of towns, but they did not control the countryside around them, where the Communists were active, their ranks swelled by further Nationalist demobilization, including those who had served in Wang's puppet forces. Dissension within the Nationalist ranks was spreading, as the Whampoa clique extended their control of the higher appointments and the Kwangsi element and other outsiders were thrust aside and given appointments which might appear prestigious but in which they wielded no real power or influence. 1947 saw a renewal of Nationalist offensives. In January Li Tsung-jen's Kwangsi Army took the field again and by March had driven Chen Yi back into the hills. As he did so, Hu Tsung-wan launched an offensive against P'eng, forcing Mao to abandon Yenan, and in Manchuria Sun Li-jen* inflicted a severe defeat on a force of 60,000 which Lin Piao had sent south of the Sungari, killing a third of them. Chu Teh and Mao, forced to move eastward from Yenan, found themselves threatened by Fu Tso-yi's powerful forces in the north, while P'eng also had to fend off attacks by Muslim cavalry under Ma Pu-fang in the west.

At this low ebb in his fortunes, Mao reconsidered his strategy. His attempt to seize control of Manchuria and North China had failed, and his forces had suffered heavy casualties in direct assault with massed forces against the better armed Nationalists. He decided to concentrate on guerrilla activity, limited offensives over a wide area being authorized to distract the Nationalists while Chu Teh prepared a winter offensive in Manchuria, to be followed by concentration against other Nationalist armies, one at a time. Liu, 'the One-Eyed General', moved south across the Lunghai railway, joining forces with those of Ch'en Keng in the Tapieh mountains in western Honan. Together they formed the Central Plains Liberation Army under Liu's command. Chen Yi, recovered from the battering he had taken from the Kwangsi Army, returned to the offensive from Shantung, disrupting the Lunghai railway and claiming to have inflicted 60,000 casualties on the Nationalists defending it. Lin Piao in Manchuria, having extended his control over most of the countryside, had almost isolated the Nationalist garrisons in the major cities, forcing Chiang to supply them by air. Chu's forces still relied almost entirely on infantry. He had about 6,000 artillery pieces, most with Lin Piao, the rest with Chen Yi, which had been formed into artillery divisions, often manned by Nationalist deserters. Lin Piao also had a small force of tanks, captured from the Japanese or the Nationalists and manned by deserters from the latter, although some Japanese instructors and drivers still served with him.

In December 1947 Lin launched an attack to isolate Mukden before he began his winter offensive in February 1948. In that month he secured the important steel manufacturing town of Anshan, south of Mukden, and in March took Ssuping on the railway between Mukden and Changchun. The Nationalists evacuated Kirin, reducing their presence in Manchuria to the garrisons of Changshun, Mukden, and Chinchow, which Lin was unable to reduce before his offensive exhausted itself at the end of the month. Meanwhile Chiang had been trying to deal with Liu, deploying 300,000 men against him in the mountains

---

* Like General Marshall a graduate of the Virginia Military Institute in the USA, he had distinguished himself in command of the 38th Division in Burma, which was built up again in India.

bordering the Central Plains. He failed disastrously, losing over half his strength and allowing Liu to capture Loyang on the Lunghai railway in March, lose it and take it again in April. By then P'eng had attacked Hu and driven his forces away from Yenan, allowing Chu Teh to return there. In Shantung Chen Yi's activity forced the Nationalists to withdraw from all the coastal ports except Yentai, and in September, when his East China Liberation Army had reached a strength of 145,000, he sent 60,000 under Su Yu to attack Tsinan. After an eight-day battle he took it, capturing a large quantity of American supplies. Meanwhile Lin Piao, whose North China Liberation Army numbered over 300,000, had failed in attempting, with nearly half that number, to take Chinchow on the railway between Mukden and Tientsin.

By then the dissension in high places in the Nationalist camp had come to a head over elections by the constitutional assembly to the Presidency and Vice-Presidency, the Kwangsi element leading the opposition to domination by Chiang's Whampoa clique. The election of one Kwangsi general, Li Tsung-jen, to the Vice-Presidency against Chiang's wishes led to the dismissal of another one, Pai Chung-hai, from the post of Defence Minister. Several of the principal generals in the field, Fu Tso-yi, Hsueh Yueh and Sun Lien-chung, were kept short of troops and refused permission to raise local militia for fear that they would establish rival power-bases.

In October 1948 Mao felt sufficiently confident of his position to declare that he had formed a North China People's Government, and he reorganized the PLA, now at a strength of 2.6 million, into five armies designated by area, but later by numbers. They were the 1st (North West) under P'eng at Yenan; the 2nd (Central Plains) under Liu, mostly in Honan; the 3rd (East China) under Chen Yi in Shantung; the 4th (North East) under Lin Piao in Manchuria; and the 5th (North China) under Nieh Jung-chen in Shansi, Hopei and Suiyuan. Chu Teh remained C-in-C, with P'eng as his deputy, a well-tried and experienced team. The five armies were organized into small divisions of about 6,000 men, almost all infantry. Artillery, when available, was organized into divisions, which came directly under the army commander, as did any tank units. A few Nationalist aircraft had been captured, but at that stage had not been used in action.

Although most of the PLA's attempts to assault the defences of major towns had failed, Chu Teh decided that the key ones must now be tackled. Chunchow, defended by 70,000 Nationalist troops, was attacked on 14 October and taken within twenty-four hours. Meanwhile the Nationalist commander at Changchun, Tseng Tse-sheng, was in secret communication with the Communists, and on 17 October brought his Yunnanese 60th Corps over to their side. Mukden was now isolated, and on 1 November surrendered, Chengteh (Jehol) in Suiyuan having already fallen.

The whole of Manchuria, in which the Nationalists had lost 300,000 men in seventeen divisions, was now in Red hands and the position of Fu Tso-yi, in the area around Peking and Tientsin, seriously threatened.

Before dealing with him, Chu Teh decided to concentrate against the fifty Nationalist divisions, organized into six army groups round Hsuchow, Chen Yi's 3rd Army attacking from the north, while Liu's 2nd came in from the west. A complicated series of battles ensued, lasting sixty-five days and known as the Hwai-Hai campaign, in which the lack of an effective overall command on the Nationalist side proved fatal. Rather than recall the obvious candidate, Pai,* Chiang appointed the ineffective Liu Chih, with the equally feeble Tu Yü-ming as his deputy, and then interfered himself at crucial moments of the campaign. By the end of the year the Nationalists, having lost some 300,000 men, had withdrawn south to within thirty miles of Nanking. As they did so, Lin Piao, farther north, began to close the ring round Fu in the 'Peking pocket'.

Chiang now gave up hope and began preparing his bolt-hole in Formosa, or Taiwan as he had renamed it. He handed over the Presidency to Li Tsung-jen and left him to make the best arrangements he could with Mao. Fu promptly surrendered his 250,000 men in the Peking pocket to Lin Piao. When Chu Teh demanded free passage across the Yangtse, Li refused, and an uneasy period ensued in which inconclusive negotiations were conducted, while the Communists increased their strength and extended their control in the north and Chiang removed the bulk of his original Central Army and his air force to Taiwan. When negotiations finally broke down on 20 April, Liu's 2nd Field Army was poised to cross the Lower Yangtse between Wuhu and Anking, while Chen Yi's 3rd attacked Nanking itself, heading for Shanghai. Lin Piao's 4th was on its way south from Peking, heading for Wuhan. They met with little resistance.

Meanwhile Nieh's 5th cleared up Shansi, the seventy-three-year-old Yen escaping from the province of which for so long he had been 'the Model Governor'. The rabidly anti-Red General Hu was not so easily dealt with. As P'eng began to drive him out of Shensi into Szechwan, Hu, co-operating with Muslim cavalry from the west, delivered a sharp counterattack, from which P'eng recovered when Nieh moved into Shensi to join him, forcing Hu to withdraw into Szechwan. By then the Nationalist Cabinet, having transferred its seat of government to Canton, resigned, Yen taking over the reins and moving successively to Chungking, Chengtu in Szechwan, and thence by air to join Chiang in Taiwan.

* One source suggests that at the last moment Pai was offered the job, but turned it down.

Pai made a brief stand at Hengyang in southern Hunan and Hu held out in Szechwan until October, by which time the PLA had occupied the whole of North and Central China and the coast down to Canton. Kiangsu, Ninghsia and Shensi were all then cleared. The PLA was now over four million strong, 1.7 million Nationalist troops having joined it since the Hwai-Hai campaign. On 1 October 1949 Mao declared the formation of the 'People's Republic of China' in Peking, and there was no force left on the mainland to challenge his authority.

Chiang's failure was to have neglected the need to foster public support, to have relied increasingly on military and financial power to support his party, which became increasingly a personal political power-base, not a political movement. He had reverted to the old-fashioned Chinese warlordism, which it had been the principal aim of Sun Yat-sen's *T'ung Meng Hui* movement to eradicate. He compounded his error by failing to fight the Japanese and by becoming too obviously dependent on foreign help, allowing the Communists to exploit the natural xenophobia of the people, as well as the sufferings they experienced as a result of the country being constantly in a state of war, internal or external or both. The modernization of his armed forces, which he at first resisted and then became too dependent on, provided him with illusory victories in the opening stages of the Civil War, but in the end proved counterproductive. Had he entrusted the security of the Nationalist cause to a greater degree to the war-lords who did not belong to his Whampoa clique, he might have fared better; but his determination to create a centralized control of China's armed forces, and to concentrate that in his own hands, proved his undoing. Mao was to be more successful in that aim, to a large extent due to his being prepared to trust his principal military subordinates, who remained at the head of their armies through thick and thin.

One of Mao's first acts on proclaiming the new People's Republic of China was to approach the Soviet Union in order to discuss recognition and the possibility of aid. Recognition was granted almost immediately, but practical measures to implement it and support Mao's regime were not forthcoming until Mao went to Moscow at the end of the year and spent two and a half months there engaged in tough negotiations, the outcome of which undoubtedly disappointed him. Stalin agreed to surrender the old Japanese concessions in Manchuria, including Port Arthur, and the Chinese Eastern Railway (which Russia had *sold* to Japan), but not for another three years: Sinkiang was recognized as part of China, but the Soviet Union's interest was perpetuated in the form of joint stock companies. A loan equivalent to $300 million could not be regarded as generous, and Mao had to abandon any idea of regaining Chinese suzerainty over Outer Mongolia. A Soviet military mission would be sent to China to help modernize the PLA, but it did not become significant until the Korean War started.

Modernization was not entirely what the PLA needed at that time. Admittedly, from being a revolutionary organ it had to assume the responsibilities both of protecting the state against external threats and of suppressing internal ones, but it was only for the former that modernization was needed and the only one seemed to be from Chiang's forces in Taiwan. The PLA's principal tasks were to extend Peking's authority over areas where it did not yet run – Sinkiang, Tibet, Hainan and, eventually, Taiwan; to impose order, which included the suppression of 'bandits' of all kinds, including pockets of Nationalist sympathizers, almost all south of the Yangtse; to restore and improve the country's infrastructure – its roads, railways and waterworks of all kinds; to participate directly and indirectly in restoring and developing industrial and agricultural production, gravely affected by the war years and natural disasters; and to support the political cadres in spreading the Communist gospel and implanting the control of the Communist state over every aspect of life and society. At the same time the financial burden of maintaining armed forces, which had almost reached the five million mark, had to be reduced, elements of doubtful political reliability discarded and formations returned to the area from which they originated, in order that all these tasks, and demobilization, could be carried out in sympathy with the local population, whose support Mao, and his principal political supporter, Liu Shao-ch'i, had always recognized as being cardinal to the achievement of their aims.

The basic organization of the PLA, as it had developed in the long years of war, was retained for the time being, the Field Armies becoming responsible for the regions in which they had been operating. P'eng's 1st Army in the North West extended control over Kansu, Ninghsia and Sinkiang, absorbing the Muslim cavalry and the remnants of the anti-Communist General Hu's army. The One-Eyed Liu's 2nd Army took over the south and soon became involved in indirect support of Ho Chi Minh and Giap in fighting the French in Indo-China. Chen Yi's 3rd was charged with preparing an amphibious assault on Taiwan. A first step, to capture the island of Quemoy in October 1949, proved disastrous; 13,000 of his men, crowded in unsuitable craft, were killed or drowned and 7,000 captured.

The other attempts to extend the government's control to outlying areas were more successful. P'eng, after a remarkable march across the Gobi desert at the end of 1949, established control of Sinkiang, and in April 1950, after one rebuff, Hainan island was secured. In January 1950 Ho Lung was put in charge of the campaign to take over Tibet, the force of 35,000, five divisions from the 2nd and two from the 1st Army, being commanded by Chang Kuo-hua. Difficulty in making their way through the mountains delayed his entry into Tibet, defended by 8,000 of the Dalai Lama's primitive paramilitary forces, until October. After a skirmish at Chamdo, the Tibetan forces offered no further resistance, but the Kham tribesmen did, and accounted for a division of 3,000 men who never returned from an expedition into their area. In addition to them, Chang's force lost 2,000 killed, 2,000 frozen to death and 3,000 dead of sickness. When the Dalai Lama fled to India a settlement with the leading Tibetans was agreed in May 1951.

The hopes of the Central People's Government Council and the Peoples' Republic's Military Committee, of both of which Mao was Chairman and Liu Shao-ch'i, formerly head of the political department of the 4th Route Army, was Vice-Chairman, that the PLA could be significantly reduced in strength and devote most of its effort to production and political work, were dashed by the outbreak of the Korean War in June 1950 and America's intervention to rescue South Korea. Liu's propagation of a policy of support for an armed struggle against all forms of colonialism, backed by a People's Liberation Army, certainly gave theoretical support to the North Korean action. But the prompt US reaction, in spite of the January 1950 declaration by their Secretary of State, Dean Acheson, which seemed to exclude Korea and Taiwan from the area they regarded themselves as committed to defend, came as a surprise. It applied not only to Korea, but also to Taiwan, where the US Seventh Fleet interposed itself between the island and the mainland, putting paid to the prospects of Chen Yi's expedition. After MacArthur's landing at Inchon in September, Chou En-lai gave a clear warning that any advance north of the 38th parallel would involve direct Chinese entry into the war on the side of Kim Il-sung. This had been agreed in August at a conference with Soviet and North Korean representatives, on condition that the former provided military equipment and support to the Chinese forces involved, which would be based on Lin Piao's 4th Field Army, whose strength, which had fallen to 120,000, was brought up to 850,000 by November from other armies.

When MacArthur's forces reached the Yalu river in November, Lin Piao's 'People's Volunteer Army', lined up on the far side, was 300,000 strong, organized in twenty so-called armies (actually corps), of three infantry divisions, each about 10,000 strong, backed by only one artillery battalion of twelve guns. They were supported by a small number of Soviet Mig-15 aircraft, manned by hastily-trained Chinese pilots. On 26 November Lin Piao attacked with fourteen divisions, surprising MacArthur and by-passing his forces by tramping over the broken hilly country while the Americans and South Koreans clung to the few roads. He pushed MacArthur back to and beyond the 38th parallel, capturing Seoul and Inchon on 4 January 1951. When he tried to assault the line south of that, on which Ridgway had organized his defence, his troops suffered heavy casualties, their mass infantry assaults repulsed by concentrated artillery fire. When Ridgway counterattacked in March, their casualties were equally high and Lin Piao was forced to withdraw behind the 38th parallel, surrendering Seoul, effective command being assumed by P'eng. The PVA then numbered about 500,000 in forty-five divisions, but twenty of them had been badly knocked about, so that his effective strength was no more than thirty.

Reinforced with artillery, provided by Russia, he counterattacked in April and a series of battles ensued until the front was stabilized in June, by which time most of his divisions had lost at least half their men. From then until July 1953, when the armistice agreement was signed, there was no major fighting, the two sides facing each other in conditions reminiscent of France in 1915—18. P'eng's forces, including the North Korean Army, were kept at a strength of over a million, providing sixty divisions, of which just over a quarter were held in the front line in seven Chinese armies and two North Korean corps. They were backed by artillery divisions of thirty-six guns, in addition to the twelve in every division, by T-34 tanks, generally used as mobile pill-boxes, and by about 1,500 aircraft, manned by Chinese and North Korean pilots. After the armistice, the PVA remained in Korea, but was gradually reduced until, 350,000 strong, it finally left Korea in March 1958.

1953—86

The Korean War had a number of significant effects on the PLA. The process of demobilization and concentration on help to the civilian field was reversed. Something like 1,750,000 men were recruited to swell the ranks and replace the 300,000 killed. At the time of the armistice in 1953, the Field Armies numbered about three million and the Garrison Armies half that number. The former were mostly Communist veterans of the Civil War, while the latter owed their origin to Nationalist or Wang's puppet formations. Both were backed by a militia which increased in strength from five million in 1950 to nearly thirteen million, and from which recruits were drawn for the Field Armies, from which they might 'volunteer' for service in Korea. The Korean War, which turned the USA into Public Enemy Number One, led to the formation of a whole series of campaigns to raise public enthusiasm in support of production and of political activity. One appeal, to which considerable attention was given, was the organization of help to dependants of soldiers, particularly those serving, or who had become casualties, in Korea. The local community was encouraged to help dependants by a productive effort to replace that of the absent, incapacitated or defunct soldier.

Emphasis on the need for extra popular effort in production was stimulated by disappointment at the scale and form of the help provided by the Soviet Union. In no field was this greater than in that of aviation. The PLA's air force dated from the defection of Captain Liu Shan-pan from the Nationalist Air Force, who flew his B-24 bomber from Chengtu to Yenan in June 1946. A few aircraft were acquired from those left behind by the Japanese in Manchuria and 200 young men were sent to the Soviet Union to be trained, a flying school being established in Manchuria in 1948. After liberation at the end of 1949, the People's Air Force was officially created, headed by Liu Ya-lou, then Chief of Staff to Lin Piao, with a strength of 100 aircraft. At the start of the Korean War it probably had about 500 aircraft, the majority propeller-driven fighters. By the end of 1951 the force supporting the PVA in Korea totalled 1,200–1,500, half of which were Mig-15 jets, the total by 1953 rising to 2,000, half of which were jets and only a hundred IL-28 bombers. Throughout the Russians were reluctant to provide the PLA with a strong offensive capability. After 1953 the main effort of the air force was redeployed to the south, covering the airspace between the mainland and Taiwan, including that over the offshore islands of Quemoy and Matsu, still occupied by the Nationalists and protected by the US Seventh Fleet. In 1955 the first Mig-17s were received, and manufacture under licence started in Manchuria. When Soviet aid was removed in 1960, the PAF's strength stood at 2,500–3,000 aircraft, almost all jets and 80 per cent of them fighters; the rest were IL-28 bombers, although there may also have been one squadron of long-range TU-4s.

The war and the US defence of Taiwan and the offshore islands also stimulated some interest in the navy, which originated with the craft captured by the PLA when it crossed the Yangtse and took Shanghai in 1949, but until recently it has never been more than a coastal force. When the Russians left Port Arthur in 1954 they handed over eighty obsolescent craft, including two fast destroyers of 1,600 tons and five submarines, two of which were ocean-going. Thereafter Soviet 'W' class submarines were built under licence at Kianguan.

With the end of the Korean War, the problem of organizing the PLA on a permanent basis was tackled, conscription being introduced by law in 1954. All able-bodied men on reaching the age of eighteen had to register for military service and could be called up at any time within the next five years, the local authority selecting them to fill the quota they were given. Students and only sons were exempt. Service was for three years in the army, four in the air force and five in the navy, but in 1965 this was increased to five years for all except the infantry, where it was four, and service with a sea-going element of the navy, where it was six. After conscript service a reserve liability, which was shared by those who were not called up, continued to the age of forty.

The reorganization also officially established the existence of officers, granted them ranks and laid down maximum ages, except for officers above the rank of major-general. Commissions were granted by the Ministry of Defence to graduates of the military schools and academies, based on their scholastic performance, although entry from the ranks was possible for soldiers who had shown exceptional ability and been given special courses. Promotion was by time served, provided certain standards were met, and could be approved at the lower levels by local commanders, but above the rank of colonel it was reserved to the State Council. In 1955 Mao Tse-tung created the ten leading military figures Marshals of the People's Republic of China: they were Chu Teh, P'eng, Lin Piao, Liu 'the One-Eyed', Ho Lung, Chen Yi, Nieh, Hsu 'the Ironside', Yeh Chien-ying, who had been Chief of Staff to the 8th Route Army, and Lo Jung-kuan, a commissar who was the then head of the PLA's political department.

These reforms were the subject of considerable argument, highlighting the division, which was to continue, between those who, with the experience of Korea behind them, wished to modernize the PLA, and those who clung to its revolutionary traditions and wished to place emphasis on the contribution it should make to completing the revolution in every field and building up the People's Republic into a strong, enthusiastic popular state. They feared that imitation of the western, or indeed the Soviet Union's, pattern of armed forces would lead in the direction of 'formalism' or, worse, 'bureaucratism and warlordism', the road which the Kuomintang had followed. The same argument against reliance on 'technical' solutions, instead of on the masses, as had raged in the Soviet Union in the 1920s, raised its head in the 1950s in China, as did the struggle between the military commanders and the commissars of the political department of the PLA. The latter was represented at every level of command, down to platoon, and undertook most of what in other armies would have been regarded as personnel and administrative duties, especially welfare, education and public relations. The political officers, instructors and fighters took a leading part in the Party Committees which also existed at every level, the highest being the Military Affairs Committee of the Central Committee. In the early years the senior army commanders had often been also the major political figures, and after liberation were also the heads of regional governments. The reorganization of 1954 was intended to be a step away from this.

A Ministry of National Defence was created, headed by P'eng, who also appears to have succeeded Chu Teh as C-in-C of the PLA with his own GHQ, served by the General Staff, the general political department and the general rear services department. Through them he exercised command over thirteen military regions, nine garrison commands and thirteen arms directorates, including the navy, air force and militia. Reaction to this modernization grew and had become serious by 1956. At the Party Congress in September of that year P'eng himself joined in the chorus of criticism of the army for neglecting the tradition of unity of officers and men and unity of the higher levels with the lower levels: neglecting democracy: leadership characterized by stress of administrative orders, neglect of ideological work and departure from the mass line: doctrinairism and formalism divorced from reality: estrangement of the army from the people and local Party and Government organs. This criticism coincided with the 'Let a Hundred Flowers Bloom' movement which Mao had intended to produce an outburst of energy and enthusiasm to replace that drummed up in the Korean War. However, many of the blooms which flourished were not to the Party's liking, the criticism of the government which erupted being described as 'poisonous weeds', so a rectification campaign was instituted. The political department's campaign within the army was part of it, and led to a greater emphasis on the army's participation in production and political activity, which the professional military complained interfered with the training which was essential if the forces were to be modernized. In 1956–57, the army's productive effort was concentrated on agriculture; in 1958–59 on industry, and in 1960–61 on flood relief.

The Great Leap Forward took place in 1958, and the PLA was expected to play a major part in it. The rectification campaign had already led to an 'Officers to the Ranks' campaign, by which officers, including senior ones, spent a period serving in the ranks as soldiers. Young political officers were also made to serve six months in the ranks before taking up their political work. In 1957 the People's Militia had been merged with the PLA's reserve, absorbing both ex-conscripts and those who had not been called up. Mao's 'Everyone a Soldier' campaign was a reaction against this step towards 'formalism'. He not only believed that the way to deal with an invader was to allow his forces to be swamped by a sea of guerrilla fighters and that China's dispersed masses could absorb modern attack – 'a sea formed by several hundred million militiamen is something that no modern weapon can destroy' – but he criticized 'some comrades [who] take a purely military view of the militia organizations and overlook the part played by militia organizations in promoting socialist construction; or else take the view that the war for national defence and against aggression is the business of the army, not of the whole people'.

The climax of the argument between Mao and the professional military was reached in September 1959 when P'eng was dismissed and replaced as Defence Minister by Lin Piao. The difference between them lay primarily over the use and organization of the PLA and the militia, but also over relations with the Soviet Union. P'eng disapproved not only of the economic naïvety of Mao's Great Leap Forward, but also of his indifference to the state of relations between the Soviet Union and the USA. Mao thought that the former should adopt a more aggressive stance in furthering the cause of international Communism and should not be so concerned about America's superiority in modern weapons. He believed that imperialism was a 'paper tiger'; that the 'east wind of socialism and national revolution was already prevailing over imperialism and its running dogs', and that 'men, not materials are the determining factor in war'. P'eng argued for closer links with the Soviet Union, at the expense, if necessary, of accepting some restrictions on freedom of action. Mao, on the other hand, believed in self-sufficiency and that it could be achieved through the Great Leap Forward. The crunch came when, on P'eng's visit to Moscow in 1959, the Soviet authorities refused any support to China's development of her own nuclear weapons, although it was claimed that, in a general scientific agreement in 1957, they had promised it.

P'eng's replacement by Lin Piao was followed by the publication of Mao's *Imperialists and all Reactionaries are Paper Tigers*, which included his dissertation *On Protracted War*, and two years later his *Selected Works*, which Lin Piao applauded. Mao's dismissal of P'eng was followed in 1960 by the departure of all the Soviet advisers and the cessation of all forms of economic and military aid, a sharp blow to the modernization of the forces, notably of the air force. The relations between the two countries deteriorated over the next few years, China's short war in 1962 with India, then enjoying close relations with Russia, and Mao's criticism of Khrushchev over the Cuba crisis at the end of that year, made them worse. The war with India arose out of disputes over the frontiers of Tibet, which China suspected India of attempting to subvert. With little difficulty the Chinese forces drove the Indians back on the frontiers both of Ladakh and of Assam, withdrawing to the McMahon line in the latter, when they had achieved their aim of ensuring that the Indians did not attempt to challenge their position north of the Karakoram mountain range, where they had built a road to link Lhasa with Kashgar in southwestern Sinkiang.

By 1963 it was clear that China was challenging the Soviet Union's leadership of international Communism, particularly in colonial or ex-colonial countries in which liberation struggles were in progress, notably in South East Asia, where the USA was involved in Vietnam. China had already begun to develop her own nuclear weapons and saw the Soviet Union's agreement with the USA and Britain on a Limited Test Ban Treaty in 1963 as directed against her. On 16 October 1964 she exploded her first nuclear device in the test area she had developed in Sinkiang. The extension of the Vietnam War, brought about by the American bombing of the North, led to arguments within the hierarchy both about whether China should herself become directly involved, which Lin Piao firmly opposed, and about whether or not reconciliation with the Soviet Union should be sought in order to present a united front against American imperialism. Professionalism in the PLA was also a subject of debate. The hard-liners won; Chou En-lai stated publicly that peaceful coexistence with American imperialism was 'totally excluded' and criticized the Soviet Union by inference as being 'modern revisionists', not true Marxist-Leninists. Meanwhile the swing to the left internally continued, the PLA being employed increasingly on non-military tasks at the expense of professional training and modernization, and the officers' ranks, introduced in 1954, being abolished.

These trends culminated in the inauguration, in August 1966, of the Great Proletarian Cultural Revolution and the establishment of the Red Guards, controlled by the Party independently of the formal organs of government, to the annoyance of both Chou En-lai and Chen Yi, the Foreign Minister. Red Guards attacked foreign embassies and instigated 'popular movements' in Macao and Hong Kong, as well as on the Manchurian border with the Soviet Union. Mao made no secret of the fact that he supported them and was opposed to all forms of 'formalism' and 'bureaucratism'. Lin Piao, in spite of this challenge to the authority of the PLA, came out publicly on Mao's side but nevertheless made it clear that 'the main component of the state is the army'. The Cultural Revolution was accompanied by a major purge which started at the top – twelve Politburo members, seven deputy premiers and forty-two Cabinet Ministers or Vice-Ministers – and extended into every city and village, where anybody who did not toe the Red Guards' party line was accused of being bourgeois or some other Communist deviation and, if not eliminated, deprived of any means of earning his or her livelihood. The PLA did not altogether escape its clutches, although, in the early phases, it managed to remain generally aloof.

Reaction against the chaos caused by the excesses of the Cultural Revolution began to develop momentum. Although Lin Piao continued to give full vocal support to Mao's policy of preferring 'the masses' to either the official Party apparatus or the organs of government and the PLA, the leaders of the latter were increasingly disturbed at the growing power and influence of the Red Guards and at the economic, social and political chaos into which the country was disintegrating. In October 1968 the Party's Central Committee called for an end to the turmoil, at the same time emphasizing the need for Party control over the Army, and convened a National Party Congress, which had not been held for some years, in April 1969, in an attempt to heal the deep divisions within it, of which the Cultural Revolution was both a symptom and a cause. At that conference the decision was taken to use the PLA to restore order. Lin Piao gave extravagant praise to Mao, accusing his old associate Liu Shao-ch'i of having plotted a return to capitalism. In return he was nominated as Mao's successor.

A new constitution concentrated more power in the hands of the Chairman at the expense of the Central Committee, which was enlarged to a figure of 279, only 84 of whom had been members of its predecessor. Of these, 80 per cent had been Party members before the Long Marches, 112 of them were military and, significantly, there was no representation of the Red Guards. The downgrading of the Party in favour of the Chairman was signalled by the abolition of the Party secretariat, headed by Teng Hsiao-ping.

Serious differences at the top emerged at the meeting of the full Central Committee at Lushan in August 1970, when Lin, supported by Ch'en Po-ta and several senior military figures, proposed to maintain the post of State Chairman in the draft of a new constitution, in addition to that of Party Chairman, a measure Mao was known to oppose, fearing a threat to his own position. A year later this rift came to a dramatic end when Lin, his wife and son, were killed in a Chinese military aircraft which crashed in Outer Mongolia on 13 September 1971.* Immediately afterwards Ch'en Po-ta and the Chief of the General Staff, the chief of the general rear services department, the commander of the air force, and the chief political commissar of the navy were relieved of their posts and arrested.

Chou En-lai was the beneficiary of these developments and he presided over the 10th Party Congress in 1973, which Mao attended, but without speaking. Although Chou condemned Lin and his associates for planning an armed uprising, he frankly admitted that 'For a long time to come, there will be two-line struggles within the Party' and that 'Lin Piaos will appear again, and so will persons like Liu Shao-ch'i, P'eng Teh-hua and Kao Kang'. The struggle did indeed continue but, during the few remaining years left to Chou and Mao, the pendulum swung back to the right, clear evidence of this being given in 1975 by the appointment of the nearly eighty-year-old veteran Marshal Yeh Chien-ying to the post, left vacant since Lin Piao's death, of Minister of Defence and the re-emergence of Teng Hsiao-ping as Vice-Premier to Chou. At the same time a new article of the constitution stated that 'The Chairman of the Central Committee of the Communist Party commands the country's armed forces'. Teng presided over Chou's memorial service in January 1976, only to find himself thrust aside and insulted by the clique surrounding Mao, who were manipulating the Politburo in favour of the comparatively obscure Hua Kuo-Feng. This reversal of fortune was the precursor of significant events: the death of old Chu Teh in July, serious earthquakes in North China, Szechwan and Yunnan and, on 9 September, the death of Mao himself, who had survived Chiang Kai-shek by a year.

---

* There have since been suggestions that they were not in the aircraft, but were shot at a dinner-party in Peking.

Reaction against the clique which had surrounded Mao was swift. The 'Gang of Four' – Mao's widow, the former actress Chiang Ch'ing, Wang Hung-wen, the young Vice-Chairman of the Party; the Shanghai journalist, Yao Wen-yuan; and Vice-Premier Chang Ch'un-ch'iao, who expected to become Prime Minister, were arrested, and Hua Kuo-Feng was declared successor to the Great Helmsman. It was the influence of the PLA, with the venerated Marshal Yeh at its head, that swung the balance decisively in favour of the pragmatists against the revolutionaries. They prepared the way for the rehabilitation of Teng Hsaio-ping who, in July 1977, was reinstated as Chief of Staff of the PLA, in addition to being a Vice-Premier, Vice-Chairman of the Party's Military Commission, Vice-Chairman of the Central Committee and a member of the Politburo. He was backed not only by Yeh but also by Wang Tung-hsing, who controlled the army unit responsible for guarding the area of Peking in which government leaders lived. Wang had commanded Mao's bodyguard on the Long March and, as Vice-Chairman of Public Security, had effected the arrest of the Gang of Four. Indicative of the resurgence of army influence was the appointment to the Central Committee and of its Military Commission of Lo Jui-ch'ing, a former Chief of the General Staff and associate of Lin Piao, whose suicide, after denunciation by Red Guards, had been announced in 1967. In the new Politburo eleven out of the twenty-six members had a military background.

The previous eight years had seen a revolution in China's attitude to the USA, of which her participation in the Paris talks about the Vietnam War had been the first sign. The motives appear to have been mixed: fear of the potential effects of the war: anxiety about Soviet aggressiveness and military build-up on her borders: the détente between the USSR and the USA, which meant that China could not expect 'to use the barbarians to fight the barbarians'; and her own parlous economic plight. Kissinger's visit to Peking in 1971, followed by Nixon's in 1972, confirmed the adoption of a policy of peaceful coexistence between the two nations. Lin Piao's opposition to this trend, arguing for a rapprochement with the Soviet Union, may have been one of the causes of his eclipse. With the re-emergence of Teng Hsaio-ping after Mao's death, the pragmatic line, exploiting peaceful coexistence with the West, has prevailed. At the time of writing it appears that it may extend to the Soviet Union also.

From 1977 onwards the argument between those who favoured professionalism and those who harked back to the revolutionary principles of 'the people's war' has been won by the former, covered by the formula 'people's war under modern conditions'. The need for modernization was clearly demonstrated by the poor performance of the armed forces employed in the punitive expedition against Vietnam in 1979. It revealed grave deficiencies in the co-ordination of all arms on the battlefield, in the arrangements for command, control and communications and for logistic support in the field. Since then, although the defence budget has been reduced, a considerable effort has been devoted to modernization in every field, particularly to training in the use and co-ordination of more modern weapon systems. This has involved a strong emphasis on professionalism, producing forces, in their organization and rank structure, more similar to those of other major military powers, especially the Soviet Union. The ambitious plans of the aged Marshal Yeh, who died in 1986, were scaled down, a firm decision being taken that general modernization and development of the economy must take priority over that of the armed forces. The size of the army and of its reserves has been significantly reduced, and that of the navy and the air force is planned.

In 1986 the strength of the standing army was 2,110,000, of whom about one million were conscripts, providing thirteen armoured, 118 infantry, seventeen field and sixteen anti-aircraft artillery divisions in thirty-five armies. The armies, which in 1986 were in the process of reorganization, correspond to corps in European armies. They consist of up to four divisions and are of an average strength of 46,300 men. In peacetime they are grouped into Military Regions, of which some would become Fronts (zhanxian) in war. The number of these Regions, which are further subdivided into Military Districts, has been reduced from eleven to seven. These forces man some 11,450 tanks, 12,800 guns and 2,800 armoured personnel carriers. The great majority of the equipment is produced in China, most of it copies of Soviet models, although some original Chinese designs are appearing and a small amount of foreign equipment is purchased. The degree of modernization varies from region to region, priority being given to forces based in the north, which would have to face the principal threat, that from the Soviet Union.

China's general strategy is no longer based on relying primarily on guerrilla warfare to absorb the shock of an invasion, but on a combination of modern defence, mobility, guerrilla warfare and popular resistance, backed by nuclear weapons. With her great size, long frontiers and poor internal communications, it is clearly not possible for China either to defend her frontiers everywhere against a concentrated attack, or to move her forces rapidly from one part of the country to another. Military regions are expected to be strategically largely independent and are allotted forces, both land and air, by the General Staff to be so. Their requirements obviously differ widely. In addition to regular armies, they also command local reserve and border forces totalling some seventy divisions, at present in process of reorganization. These are raised from the four million strong Armed Militia, formed from those who have completed their regular or conscript service, which is three years for the army and the marines, four for the air force and five for the navy. In addition there is an unarmed Ordinary Militia of six million, as well as a number of paramilitary armed police and border security forces, totalling perhaps two million. The army's strategy is expected to take the form of holding defensive positions on the major invasion routes, while operating against the enemy's lines of communication with both mobile regular and local forces of all kinds.

The air force is regarded as closely linked to the support of the army, air armies being assigned to the support of Military Regions or Fronts. In peacetime it is organized into seven Military Air Regions, and has a total strength of 490,000 men of whom 160,000 are conscripts. It suffered most from, first, the withdrawal of Soviet support in 1960, and then from the disruption caused by the Cultural Revolution. The result is that its stock of aircraft consists almost entirely of home-produced copies of out-of-date Soviet types, the majority being fighters.* A considerable effort is being made to develop the aircraft industry, but both in design and production capability it has a long way to go to catch up with established aircraft industries elsewhere.

The navy, which has 350,000 men of whom 115,000 are conscripts, has profited most from modernization. Although still primarily a coastal defence force, it has developed an ocean-going capability and a significant submarine arm in response to several factors. One is Chinese determination to reduce, and as far as possible abolish, her former great dependence on foreign shipping for her overseas, and even her coastal, trade. Another is her anxiety at the increasing activity of the Soviet Far East Fleet, and particularly its use of the Cam Ranh Bay base in Vietnam. She is sensitive to the threat of Soviet support of Vietnam and of Soviet naval and amphibious operations being combined with a land invasion. Finally, China's claim to islands in the South China Sea, including the Paracels, which significantly extend the Economic Exploitation Zone off her coast, is disputed by both Vietnam and the Philippines. The navy now has two ballistic missile nuclear submarines, with possibly four more on order, three nuclear-powered attack and 113 diesel submarines, fifteen guided weapon destroyers and thirty-one frigates, of which twenty-six carry guided weapons. In addition it mans a very large number of smaller coastal and river craft. Its air force, all shore-based, has 800 aircraft, three-quarters of them fighters, and it has its own land force of 5,500 marines.

After exploding her first nuclear device in 1964, China formed the PLA's 'Second Artillery', equipped with a 700-mile-range ballistic missile, delivering a twenty kiloton warhead. Since then it has been joined by others of longer range, the arsenal in 1986 consisting of fifty DF-2

---

* Including 400 J-5 (Mig-17), 3,000 J-6 (Mig-19), 200 J-7 (Mig-21) and 30 J-8 (Mig-23) fighters; 120 H-6 (Tu-16) and 500 H-5 (IL-28) bombers; some 550 transport aircraft, mostly Y-5 (An-2) and 400 helicopters, most of them Z-5 and 6 (Mig-4 and 8).

(or CSS-1) of the same characteristics, sixty DF-3 (CSS-2), delivering a two-megaton warhead to a range of 1,700 miles, four DF-4 of 4,375-mile range and a three-megaton warhead, and two DF-5 (both CSS-3) of 9,000-mile range and a five-megaton warhead. It is possible that a copy of the Soviet FROG-7 short-range ballistic missile is under development and that some aircraft carry nuclear weapons. The navy's submarine is equipped with a modified DF-3. China's nuclear force appears to be intended as a deterrent, based not on retaliation against Soviet cities, but both to ensure that the Soviet Union would pay a heavy price, if it were to attempt an invasion, and as a hedge against the threat of intimidation in a limited conflict. No doubt it also serves, as do the British and French nuclear forces, as a status symbol.

The development of these modernized forces has imposed considerable strain on the internal composition and structure of the PLA. In all its ranks it had been strongly based on the peasantry; but the requirements of modern forces demand higher standards of education and technical skill, which were to be found mainly in the urban areas. Modernization is drawing the armed forces further and further away from the original pattern of the PLA. At the same time it has been gradually extricated from the control and influence it exerted in the political and administrative field before and during the Cultural Revolution, and in restoring order after it. The Party and its governmental machinery have progressively asserted their control, extending it to the defence industry and certain fields of logistics. The PLA is still expected to make its contribution to the development of civilian industry and to other aspects of the economy and the country's infrastructure. At the time of writing, with Teng Hsiao-ping apparently exercising firm control, the PLA does not appear to be an active rival to the Party.

The Chinese armed forces are large, but not excessively so for a country with such a large population and vast territory. They have very limited offensive capability, particularly to operate at any distance beyond her frontiers. China poses little threat to her neighbours, although it is conceivable that, in circumstances that were favourable, she might attempt to reassert the very loose form of suzerainty which she exercised historically over places like Burma, Laos, Vietnam and Korea. But the fear of Soviet aggression on her vulnerable northern frontier would always act as a restraining factor in any major commitment to southern adventures. It does now appear that control by the central government has been firmly established, but the traditional 'warlordism' might raise its head if serious divisions arose within it. The restoration of Taiwan remains a firm aim, but the present regime seems content to approach that awkward hurdle delicately, perhaps hoping that regaining Hong Kong may prove an example that whoever is in control of Taiwan then, whether Chiang Kai-shek's descendants or local Taiwanese, might be prepared to follow. Much will depend on internal developments in China between now and then.

China's defence policy is fundamentally defensive, and her forces, although much inferior in quality of equipment and in the sophistication of the command organization to operate it, are probably adequate to make her confident in her ability to maintain her independence. The only serious threat could be from the Soviet Union, and that is more likely to take a limited form than an attempt to occupy the central area of the country. In the distant future, however, it is possible that a significant development of China's military strength could provoke a Soviet reaction on her borders.

9

CONCLUSION

The preceding chapters have shown how different have been the influences affecting the development of the armed forces of the major military nations in this century. Some have been influenced more by the domestic political situation of their nation than by external threats, the British and the Americans less than others. All have been affected by the legacy of nineteenth-century imperialism; securing its acquisitions, continuing to pursue its policies, or reacting against it. The essence of that imperialism was the concept that, in order to trade with or within an area, it was necessary to exercise some form of empire over it, either to establish satisfactory conditions of order in which trade could flourish, or in order to exclude rivals.

The development of the US armed forces, particularly that of the US navy, was a reaction to this: to ensure that imperialism was excluded from the whole American continent and that American trade was not excluded from other areas. The development of Japan's armed forces was a direct reaction to its application to China, and led to her emulation of European imperialism. The history of China's various armies is solely concerned with that ancient country's struggle to expel the imperialist barbarians, whether European or Japanese. Britain's forces throughout, even after the Indian sub-continent had cast off the imperial yoke, were consistently inclined to be more concerned with the security of far-flung dependencies than with maintaining the balance of power in Europe. Those of France, particularly after 1945, were profoundly affected by that interest. Germany's desire not to be excluded from imperial competition, or to compensate for it by extending her dominion within Europe, lay behind the two great convulsions which changed the face of Europe, disturbed the balance of power and ended with the dissolution of Bismarck's achievement in uniting the German-speaking people.

But the two World Wars, which involved all these major nations (although Japan and China played little part in the First), were not initially provoked by nor fought for these imperialist issues. Each was primarily concerned with who should exercise power on the European continent between the Urals and the Atlantic. That remains the central issue today, in the definition of which a vast array of weapons is amassed; the North Atlantic Treaty Alliance, resting primarily on the military power of the USA, facing in the centre of Europe the Warsaw Pact, which rests even more essentially on the military power of the USSR. In 1900, at the start of the period reviewed, that could not have been foreseen.

The only true imperialist power today, that is one which believes that in order to trade one must exercise authority over the area in which one wishes to do so, is the Soviet Union. The difficulty in persuading that vast country to accept what the other ex-imperialist powers, China, Britain, France, Japan and Germany, have accepted – the Open Door policy advocated by the USA – is that it would undermine the political and social concept on which the whole state is organized. The principal purpose of the huge armed forces of the Soviet Union, therefore, is to preserve the fruits of the Revolution, as it was originally of the US armed forces after 1776, of the Napoleonic armies in France, and throughout has been of the People's Liberation Army in China. That of the German army had been to preserve and extend the politico-military achievements of Bismarck in uniting Germany against the animosities it aroused in her neighbours. One has to retrace one's steps to the restoration of the English monarchy in 1660 to be able to regard the British army in the same light: as the guardians of the unity of the nation. In varying degrees, armed forces have been designed as much to meet the internal needs of the nation and to support its political structure as they have been tailored to meet a specific external military threat.

Nevertheless, although created perhaps initially to meet an internal purpose, they have acquired external ambitions, partly because their development has created anxieties among neighbours and rivals. The development of the German and Russian armed forces before 1914, and of the former after the advent to power of Hitler in 1933, are examples, as are that of the Japanese up to 1945 and that of the Soviet Union's forces since then. But it does not appear that it was the development of the armed forces themselves – the so-called arms race – which caused any of the wars which have marked the century. The pre-1914 naval arms race was certainly not a major cause of the First World War.

All the wars described in previous chapters were the result of the policies which the nations chose to pursue. None was a direct result of competition in an arms race, although fear that a rival might develop a greater military superiority, if it were not pre-empted, was undoubtedly an influence in the decision of Germany to support Austria-Hungary in 1914, of Britain and France finally to challenge Germany in 1939, and Japan the USA in 1941. Today this fear fuels the nuclear arms race between the USA and the USSR.

The principal cause of war has been the belief of one of the parties to it, normally the aggressor, that he can achieve his aim without great loss in a short period of time, as Bismarck did. In the twentieth century it has generally been a false hope which, in the case of the two World Wars, proved to be a disastrous miscalculation, although Germany's initial victories, particularly in the Second World War, appeared to confirm the judgement of her leaders. Japan's judgement in challenging Russia in 1904 proved well-founded, as it had against China in 1895. If she had limited her acquisition to Manchuria in the 1930s, the whole history of that part of the world could have been different, as would that of Europe if Hitler had not attacked the Soviet Union in 1941. The prospect of rapid victory, to which the accumulation of strong enough armed forces to make it appear possible makes a notable contribution, tempts the leader who wishes to change the status quo to his own advantage. The nation that wishes to preserve it has then to decide how to meet that challenge. If it is not adequately prepared, it has to fall back on methods which will exhaust the aggressor, and it is likely to come near to exhaustion itself in the process.

The nature of the armed forces themselves has been a factor which has made it difficult for either contestant to bring a war between major military nations to an end before the point has been reached at which both sides lose more than they could gain from its continuation. The size of the forces, raised normally by universal conscription, and the provision of the equipment they need to fight on land, at sea and in the air, has involved the total mobilization of the nation's resources, human and material. To maintain popular support for that, governments have had to represent the war as a combination of crusade and struggle for survival. To bring it to an end, before its declared aims have been achieved, is seen as a betrayal of those who have made considerable sacrifice, perhaps that of their lives. When wars were fought by smaller professional forces, the support of which did not affect the nation so closely, it was easier for statesmen to relate the military operations more rationally to the importance of the issues at stake. The problem has acquired an entirely new dimension with the advent of the nuclear weapon.

The tendency for the armed forces themselves to influence policies which might lead to war has varied between the extremes of Japan and China at one end of the scale and Britain and the USA at the other. The armed forces of Germany cannot themselves be accused of favouring war in 1939, although many of them looked forward to regaining territory removed by the Treaty of Versailles at some time by a short, sharp war or the threat of it. Neither the French nor the Soviet armed forces could be accused of favouring war, but both they, as well as the Germans, exerted a strong political influence at different times, the French in their defensive attitude in the 1930s, in seeking an armistice in 1940, and over the affairs of Indo-China and Algeria after 1945. De Gaulle himself was the extreme example.

The direct influence of the US armed forces on policy has been much less, General MacArthur's dismissal by President Truman illustrating the firm domination of the armed forces by the civil executive. Eisenhower's presidency did not increase their influence; if anything, the reverse, as he knew how to keep them in order. Their major influence on policy has been through the influence on industry of armament orders of all kinds, from nuclear weapons and their delivery systems to a myriad of items of equipment. The vested interest of the military/industrial complex in the USA is now a major political factor, as it is in the Soviet Union.

The extent of the influence of the Soviet military on policy is difficult to judge. Although it is a firm tenet of Soviet doctrine that the armed forces are subservient to the Communist Party, there can be no doubt that their influence is great, particularly if allied to one of the other two major sources of power and influence, the KGB or the Party. On the whole, it appears that their influence is conservative and cautious, resistant to any weakening of the strength of their forces.

The political influence of the British armed forces has been minimal, Ireland on more than one occasion having been a potential scene for it. Preservation of what was the Empire and then became the Commonwealth, and the weakness of Britain's armed forces in relation to all the commitments it involved, influenced British military leaders throughout the period to view with dismay the prospect of having to fight a major war in Europe. At no time, at any rate since the Boer War, could the British armed forces be accused of war-mongering.

None of the nations considered can be said to have evolved an entirely satisfactory organization for the political control of their armed forces and, allied to it, their higher direction in war. An organization designed to ensure strict political and financial control in peacetime has generally proved unsuited to either the development of a sound politico-military strategy or the effective higher command of operations. The problem has been complicated by the need to act in alliance with other powers. In the early years of the century the problem was complicated, as far as Germany, Russia and, to a slight degree, Japan, were concerned, by the position of their monarchs. The co-ordination of armies and navies proved difficult enough. When aircraft added a third dimension, a further complication was added, whether or not the air force was recognized as a separate service.

Hitler's assumption of Supreme Command himself, although providing a unified overall higher direction of strategy, proved in the end fatal, partly because he would not restrict his command function to that, but insisted on interfering in operational command at lower levels. Stalin did better. Of all the systems of command in the Second World War the Soviet one seems to have functioned the best, combining a firm general direction with a fairly high degree of freedom in execution at lower levels. It had the advantage over the American and British of not having to consider co-ordination of operations with its allies.

The British system of command by inter-service committee was a patent failure, redeemed only in London by the position of Churchill who, as Prime Minister and Minister of Defence, acted also as Supreme Commander. The American insistence on the appointment of a Supreme Commander in the theatre of operations improved on the British system, but was not applied in their major theatre of war, the Pacific, and was limited, as far as control of air forces were concerned, in Europe. Unified higher direction of strategy in Washington was not easily achieved and was almost entirely dependent on agreement between General Marshall and Admiral King. Neither they nor General Arnold attempted to direct operations, as Churchill and the British Chiefs of Staff tended to do. The American Chiefs of Staff left that to the individual Commanders-in-Chief – MacArthur, Nimitz, Eisenhower, Doolittle and Spaatz. The Japanese faced the same problem of balancing command between army and navy, but were saved the complication of having a powerful separate air force. Neither Chiang Kai-shek nor Mao Tse-tung were troubled by the problem, having virtually no navies or air forces. The French navy played such a small part in both World Wars and in France's colonial struggles that co-ordination of naval with army and air force strategies and operations did not raise major problems.

Since 1945 the creation of NATO on the one side and the Warsaw Pact on the other would appear to have greatly complicated the problem. It certainly has in theory, but the practicality is that the former accepts overall higher direction and command by the command organization of the United States, and the latter, even more firmly, that of the Soviet Union. Doubts about the US organization arise from the independent strength of the three separate services and the weakness of the central direction, which is itself divided between the Defense Department, itself often divided, the National Security Council and the President's own office. The organization's ability to develop sound strategies or to exercise effective command of operations does not arouse great confidence in many of its allies.

How well suited have the forces themselves been to fulfil the purposes for which they were maintained? There have been some remarkable successes and some notable failures. Japan's achievement in developing a modern and efficient army and navy, capable of defeating a major power, less than forty years after she had emerged from a state bordering on the medieval, was truly remarkable. The sophistication, strength and efficiency of her armed forces in 1941, particularly the use of aircraft in support of both the navy and the army, were also notable. Her eventual defeat was due to the pursuit of an unrealistic strategy, not to any inferiority in the quality or the operation of her armed forces.

In spite of the example of the Russo-Japanese War, none of the participants in the First World War were well prepared for the sort of conflict which it turned out to be, either at sea or on land. Given the geographic relationship between Germany and Britain, it was unrealistic of the former's navy to imagine that its battle fleet could dominate the North Sea, or even break out of it, and of the latter's that it could force a challenge by operating close to the German coast. It was the stalemate between the battle fleets, and between the armies on land, which turned what both sides expected would be a short war into a prolonged one of attrition that made the submarine guerre de course against merchant shipping the principal naval activity, for which the German submarine fleet was better prepared than was the British navy to counter it.

It was by no means inevitable that the war should have been a long one. Germany might have achieved a victory in France in 1914, as a result of which the war on both the western and eastern fronts might have been brought to an end in that year. Having failed to achieve it, she could have sought an end to the war on terms which might have been acceptable to the other powers. In default of either of those, the ability to move by rail huge conscript armies and their supplies to the front, combined with the strength of the defence provided by artillery, machine-guns and earthworks, protected by barbed wire, restricted the ability of the generals, at least on the Western Front, to execute any decisive manoeuvre. Cavalry, which had made little effective contribution in the Russo-Japanese War, although the terrain and conditions were well suited to horsemen, was rendered impotent, but all the participants were reluctant to abandon the arme blanche. The British invention of the tank appeared to have broken the deadlock, but its use cannot be said to have been decisive in bringing victory in 1918, nor was that favourite British application of maritime warfare – blockade. The decision was obtained by the application of superior resources, imposing losses on the opposing alliance which exhausted it, psychologically and physically, even though one of the principal allies, Russia, had fallen out as a result of just that sort of exhaustion.

Artillery and machine-guns had dominated the battlefield, in attempting to advance against which the infantry of all armies had suffered terrible losses. In these conditions there was little opportunity for superior generalship to affect the issue, and no commander of that war stands out as achieving results by the use of superior imagination, forethought, skill or inspiration. Some, such as Brusilov, did better than others, but it was as general managers of huge military machines that men like Haig, Pétain, Foch, Falkenhayn and Ludendorff excelled, rather than as strategists or by skill in the 'operational art'. Much the same could be said of the admirals. Apart from a few brief encounters in the North Sea, in which neither side could boast of showing the Nelson touch, the submarine campaign was a matter of general management, at which the German admirals could lay claim to a slight edge over the British. The admirals of other nations had little to do.

Post-war reaction to the experience took different forms. The French, recognizing the strength of modern defence and with bitter experience of the cost of their pre-war devotion to the doctrine of the offensive, placed their trust in a static defensive strategy. Britain returned to her preoccupation with the security of her Empire, hoping to avoid any continental entanglement, or at least to limit it. The USA, having refused to accept any commitment to Europe, kept a watchful eye on Japan, clearly a matter primarily for the navy. Britain's hopes of limiting her commitment rested, to a certain degree, on the exploitation of two of the war's innovations, the aircraft and the tank, the successful operation of which depended on a third, the development of radio communications, or wireless. The air enthusiasts held out a hope: that the enemy's will to resist could be rapidly undermined by air attack on his capital, and perhaps other cities, possibly using chemical weapons, thus rendering large armies and long costly campaigns unnecessary, and that his navy could also be rapidly and cheaply eliminated by land-based aircraft.

The proponents of army mechanization, including those who wished to concentrate it on tanks, also saw their target as the enemy's will, linked to his ability to control operations. They believed that restoration of the art of manœuvre, designed to paralyse the opponent's military nerve system, rather than the Clausewitzian aim of the destruction of his forces, could make smaller, professional and fully mechanized armies, working closely with air forces, more effective than huge conscript forces, relying on massed infantry and artillery. In all the armies considered, this gave rise to an argument, which continues to this day, between those who extol the virtues of infantry and press for the maximum number of men, and those who favour a higher proportion of effort to be devoted to firepower: to weapons and the mobile platforms from which they can be delivered, aircraft, ships and automotive vehicles. The argument takes different forms, the crudest being that between 'the masses' and 'the technical factor', which raged in both the Soviet and the Chinese communist armies.

At the other end of the ideological scale, this has coloured the arguments in Britain between a continental and a maritime strategy;

between the supporters of mechanization in the 1920s and 1930s and their opponents; between those who gave first priority to imperial security and those who advocated its application to the maintenance of the European balance of power; and today between the advocates of high technology and those of 'alternative strategies'. In France de Gaulle's *Vers l'Armée de Métier* put the same radical cat among the conservative pigeons, and it was a source of dissension in the Japanese army in the same period. In the USA, which has throughout tended to seek a firepower rather than a manpower solution to battlefield problems, it was at the heart of Maxwell Taylor's opposition in the 1950s to undue reliance on nuclear weapons, and today it is reflected in the criticism by an important element of the US defence academic community of the Pentagon for excessive reliance on technological solutions and neglect of the human factor.

In Germany, it was not until Hitler came to power in 1933, determined to overturn the terms of the Versailles Treaty, including the restrictions imposed on the German armed forces, that they began to give priority to tanks and aircraft, the latter being given greater priority. Although in 1934 they were not as fully prepared for modern mechanized war as they had planned to be when the time came for Hitler to put matters to the test, the German army and air force showed themselves far better prepared to wage it than their opponents, who had not suffered from any constraints in developing the equipment needed for it, other than self-imposed ones: financial, political and conservative. But, as in the First World War, Germany's leaders underestimated both the resolution and the ability of their opponents, in the case of Britain and the Soviet Union, to muster the resources necessary to wage another long war of attrition. Their calculation was correct about France, but fatally wrong about the Soviet Union. The latter's ability, after what by any standard would be regarded as crippling initial losses, to resurrect huge armies, including masses of tanks and aircraft, provided by industries which had had to be displaced by hundreds of miles, was a fantastic feat, which the Germans cannot be blamed for discounting. Nobody else thought it possible.

The clash of arms between the enormous German and Soviet armies over a vast tract of Russian territory, the greatest conflict in history, offered opportunities for combining manœuvre with firepower in which both sides displayed high qualities both of generalship and of human courage and endurance, the former marred in the case of the Germans by the vagaries of Hitler's intervention in the direction of operations. No less remarkable was the speed and efficiency with which, after the Japanese attack on Pearl Harbor, the USA raised, trained, equipped and moved across the oceans fleets, air forces and armies which, while inflicting a resounding defeat on the Japanese in the Pacific, assumed the main burden, after that of the Soviet Union, in the defeat of Germany in Europe. In these operations, the most notable developments were in the use of aircraft and of amphibious forces. There was nothing particularly novel about their land operations, although Patton's virtually unopposed tank drive across France in the summer of 1944 was hailed by some as a masterpiece of armoured warfare. The development of the technique of amphibious landings, which owed much to the US Marine Corps, exploited by Mountbatten in Britain, made possible not only the defeat of Japan, but also the deployment of Anglo-American forces onto the continent of Europe, without which Germany could not be defeated or prevented from making a separate peace with the Soviet Union.

It was, however, in the application of air power that the Anglo-American alliance differed most both from its ally and from its opponents. For neither the Soviet Union nor Germany was naval aviation an important arm, as it was for Japan. The latter both developed and employed its air forces, naval and army, with efficiency and skill, but had no reason to develop a separate air arm, offensive or defensive. The Chinese Communists had no air force until they had defeated Chiang Kai-shek's forces, whose air arm was never a significant factor, although its harassment of the Long March caused them significant losses. The Soviet air force was closely tied to the army, and had no good reason to devote effort to strategic bombing of Germany, even if its military leaders had believed in its value, as the British and Americans were already applying considerable resources to it.

Some air enthusiasts can be found to argue that, if the Soviet Union had devoted a similar effort to strategic bombing, the overall effect on German industry and civilian morale could have been decisive, but co-ordination of operations would have posed major difficulties. Although Germany's *Luftwaffe* was an independent force, its operations were closely tied to those of the army, and the navy had good reason to complain of the failure to devote sufficient air effort to its support, particularly of the anti-shipping campaign which alone held out hope of a decisive effect on the British war effort. The bombing campaign against Britain was its only independent strategic effort, and failed to have any decisive effect, even when the weapons of the future, cruise and ballistic missiles, were used in 1944. Had they been available earlier, when their launching sites were not in danger of being overrun by the Anglo-American armies, they might have produced more significant results.

The British, followed by the Americans, applied considerable effort to independent strategic air operations. Their effectiveness remains a matter of argument. Until the advent of the atom bomb their effect certainly fell far short of the claims made for them by the protagonists of independent air power. A good case can be made for arguing that the human and industrial effort devoted to them would have been better used in more direct support of the British and American armies or navies, although the US navy seldom lacked what it needed, nor in the final stages of the war did the Anglo-American armies. What is not in doubt is that the Anglo-American bombing campaign against Germany forced the latter to devote an ever-increasing effort to air defence at the expense of her capability to launch offensive operations. In general, the development of air forces had a greater effect on maritime than on land warfare. The ability of the German army to continue to provide an effective defence, when the Anglo-American air forces totally dominated the air over Italy and North West Europe, was proof of that; as was the defeat of the Japanese navy by air attack, while it still retained a significant surface fleet.

One of the developments of the air war which was to have considerable impact on the future was electronic warfare. The development of radar as a means of detecting aircraft and subsequent developments designed to overcome it, as well as for navigation, led to a war of electronic measures, countermeasures and counter-countermeasures. Linked to it was the radio war of interception and code-breaking, using primitive forms of computer, a field in which the Anglo-American alliance gained a significant advantage over its opponents.

Navalist claims that victory showed the superiority of a maritime over a continental strategy have little validity. The combined military effort of the USA and Britain could not have prevailed over that of a Germany in control of all Europe west of Russia. It was the immense war effort of the latter which was the decisive factor in Germany's defeat, although it could probably not have prevailed without the added pressure of the Anglo-American forces. Japan was defeated by a maritime strategy because, being an island maritime power, she, like Britain, was particularly vulnerable to such a strategy. Although the Second World War, on land and sea, and less obviously in the air, being generally one of manœuvre, offered greater opportunity than the First for the exercise of general/admiral-ship, its final outcome cannot be said to have been greatly affected by the respective skills of either side in that field. Like the First, it was finally decided by the combined pressure of superior resources of all kinds, applied on land, at sea and in the air, by the British Commonwealth, the USA and the USSR. There was no short cut to victory, no magic formula – maritime strategy, independent air power or the indirect approach – which made it possible to evade the need to defeat, indeed almost totally destroy, the enemy's armed forces in battle. Not until then, in the case both of Germany and of Japan, was the object of strategy to be attained: the imposition of the victor's will on the vanquished.

Nor can it be said that victory was due to any significant qualitative superiority of the human material, the soldiers, sailors and airmen of all ranks. As in previous conflicts, all the forces involved displayed

shining examples of courage, determination, endurance of great hardship and loyalty to their country and their comrades, even, in many
cases, when ill-led, ill-fed, ill-clothed, ill-equipped and ill-trained.
Some forces went through bad patches. The French and British in
the early stages certainly did so, but other forces were not exempt
at certain times. The performance of Slim's British-Indian army in
1944 and 1945 proved that there was nothing wrong with the human
quality of the British and Indian soldiers, whose performance two years
earlier had been so disappointing. Well led, equipped, trained and
supported, in operations that were strategically soundly based, they
proved themselves more than a match for the formidable Japanese,
as their American allies had already done.

After 1945 two new forms of warfare appeared: revolutionary war and
war conducted with, or under the threat of, nuclear weapons. Neither
affected Japan, which ceased to be a military power, and the former
did not affect Germany. Initially military leaders assumed that nothing
had basically changed: that nationalist movements in colonial territories could be suppressed by the methods previously used, and that
the atomic bomb was merely a logical extension of the armament of
strategic air forces. Plans continued to be made to meet the possibility
of major campaigns of the Second World War type in Europe, the
Middle and the Far East. NATO was formed to prepare, and possibly
execute, the first; the Central Treaty Organization (CENTO) the second;
and the South-East Asia Treaty Organization (SEATO) the third. The
Korean War appeared to demonstrate the need for such preparations.

   But the threat took a different form and, although receiving general
support from the Communist bloc, came primarily from a different
source: that of nationalism. The British faced it in a less virulent form
and on a smaller scale than the French, largely because they were
prepared to give way to pressure, if standing up to it seemed too difficult, as in Palestine and India. Elsewhere they developed a method
of warfare which combined concentrating force against the hard core
with making political concessions to the majority in order to win its
support, the consequence of which was to grant independence to a
non-Communist régime which might be expected to be favourable to
Britain. The policy succeeded, except in Aden.

   The French and Americans in Indo-China, the French alone in
Algeria, and the Portuguese in their colonies in Africa, failed to find
a military solution to revolutionary warfare conducted on a large scale
on the general lines advocated and to a certain extent practised by
Mao Tse-tung, just as Chiang Kai-shek had failed before them. Superior
fire-power and mechanical mobility, conferred by aircraft of all kinds
as well as land vehicles, were not sufficient to overcome not only
superior numbers of men on their feet, supported, whether willingly

or under threat, by the majority of the populace, both in rural and urban areas, but also the determination of the nationalist forces, communist-led in Indo-China, to go to any limit, of time or of human effort and loss, to achieve their aims. The French and American electorates were not prepared to support an unlimited effort, and that expended by Portugal precipitated a political revolution at home, as it almost did in France over Algeria. Even the Soviet Union, which does not have the same problem with its electorate, appears to have set some limits on the effort it is prepared to devote to its similar campaign in Afghanistan.

Theories developed about the conduct of limited war were disproved. Both sides have to accept limits, before war itself can be limited. Minor examples of that were seen in the British campaign against Indonesia in Borneo in the 1960s and in the affair of the Falkland Islands twenty years later. Other wars also took place which were limited in one way or another. Those between India and Pakistan and between Israel and her Arab neighbours were limited in that neither bombed the cities of the other for fear both of retaliation and of offending international opinion. Both were also affected by their dependence on other powers for armament supply. The rate of loss of major equipment, such as aircraft, tanks and anti-tank missiles, set a limit to the length and intensity of wars before further supply was needed. The sources of supply, one or other of the major powers, could therefore exert pressure on their clients; when the suppliers of both sides wished hostilities to cease they could bring the war to an end, as they did on more than one occasion in the case both of Indo-Pakistani and Arab-Israeli wars. Unfortunately, that has not yet been effective in the case of the war between Iran and Iraq.

One very significant way in which war has been limited has been that, in spite of the extraordinary proliferation in the numbers and types of nuclear weapons and their means of delivery, until, in the 1980s, the total number of warheads has risen to some 50,000, the great majority almost equally divided between the USA and the USSR, no nuclear weapons have been used. That is very largely due to the fact that the possibility of their being used has itself acted as a strong deterrent against any of the powers which possess them from becoming engaged in armed conflict with another such power which had the capability to reply in kind. Also, the international and domestic political and moral odium would be aroused by their use against an opponent who did not have them. It has not inhibited nuclear powers from engaging in conventional operations against non-nuclear ones, as the British and French did against Egypt over Suez in 1956, the Americans against North Vietnam, and the British in their liberation of the Falkland Islands from the Argentine invasion of 1982.

The principal difficulty that faces military planners and those responsible for organizing and training the armed forces to execute, if necessary, those plans, as the twentieth century nears its end, is to reconcile the paradox at the heart of this policy of deterrence. If one wishes to deter one's potential opponent from embarking on a military adventure by the threat of the possible use of nuclear weapons against his forces or his country, one must appear to be prepared to use them in certain circumstances. But if one actually does so, and is answered back in kind, one is likely to finish up very much worse off than if one had not initiated their use, even if the enemy restricts their use to the same level as one has oneself; and there is no way by which one can confidently assume that.

Many theories and possible scenarios of limited nuclear war have been evolved; but they all fail on three counts: first, that there is no way of ensuring that the enemy's response is limited; second, that at the heart of deterrent theory lies the threat of escalation to attack on the enemy's heartland, that is, that the exchange could develop into something unlimited; and third, that even the most limited use, and even if observed by both sides, would have such terrible consequences, principally on the civilian population of the theatre of war, that the end result could not conceivably be called defence. Attempts to limit the enemy's response have been made by the development of counterforce strategies, to be executed principally by land-based missiles with multiple warheads, designed to destroy the opponent's similar systems; but they have merely led to the proliferation both of these systems and of others, delivered from submarines, surface ships and aircraft. It is the vain hope of establishing a superiority, which could limit the enemy's response, that is primarily responsible for the huge arsenals of nuclear weapons in the hands of the superpowers. The attempt to develop a strategic defensive system is the latest move in this field, which seems likely to prove as vain as the offensive counterforce strategies that have preceded it.

While military planners, arms control officials and politicians grapple with this abstruse field of so-called defence, they also have to face two others which are affected by it: how to plan and execute operations with non-nuclear weapons, at sea, on land and in the air, against an opponent who has a nuclear capability at all levels. The problems of fighting the enemy in all three elements under the threat that nuclear weapons could be used at any moment, and of conducting operations while they are being used by both sides, has never had to be faced in reality. Plans for doing so generally have such an air of unreality that they raise the question of whether it is possible in those circumstances to conduct operations which could meet Clausewitz's definition of war as the continuation of state policy by other means. That raises the further issue of whether armed forces can now be considered as a means of furthering policy. Sensibly, the countries of Western Europe, which for centuries have fought each other in the furtherance of what they considered to be their national interests, have clearly decided that they cannot, as far as any clash of interest between each other, and between them and most other countries in the world, including the USA, Japan and China are concerned. It is to be hoped that this welcome change of attitude about conducting relations with other nations will be further extended, especially to the relations between the superpowers, representing rival political, social and economic systems.

Several question marks remain. The first is how to ensure that the Soviet Union does not extend the imperium she exerts over Eastern Europe and Afghanistan into other countries in Europe and Asia. Secondly, how to counter movements elsewhere in the world which, combining nationalistic sentiment with revolutionary Marxism, might exclude them from the democratic and capitalist Western politico-economic system and join them to the Soviet imperium. And finally, how to ensure that nationalism, antagonism between countries, such as the Arabs and Israel, Iran and Iraq, or internal upheavals, do not seriously affect the interests of individual countries of the Western system or the system as a whole. The part that military power can and should play in that context, and how, if called upon to do so, it should be applied, is not easy to judge.

Meanwhile technical innovation, the electronic age already appearing to be about to give way to the photonic age, proceeds at a fantastic rate, both provoked by and provoking a technical arms race. Weapon systems and all equipment associated with them become out of date almost before they have entered service, causing an inflation in the cost of keeping armed forces up to date which forces a reduction in their size, if defence budgets are to be kept under control. That factor sharpens the old debate between the relative value of manpower-intensive forces, equipped with comparatively unsophisticated weapon systems and platforms for them, and of firepower-intensive forces, relying on the products of high technology to provide information about the enemy, including 'target acquisition', and to deliver and direct the warhead of the weapon to the target. The latter involves an ever-increasing effort devoted to electronic (in future also photonic) measures to defeat the enemy's electronically dependent systems. The ultimate result may be to make electronic countermeasures a more effective method of frustrating the enemy's operations than hitting him, or his weapon platform, physically with a kinetic or chemical energy projectile.

The ability to detect the enemy's forces and other military targets by means other than the human eye or a camera from an aircraft, and to deliver an effective warhead to the target by remotely guided missiles, fired from the land or sea surface, or under the surface at sea, combined with the increased vulnerability of the aircraft to detection and destruction by anti-aircraft systems of all kinds, calls into question the future of combat aircraft for both reconnaissance and strike missions in land warfare. At sea, on the other hand, their ability to act as platforms for long-range reconnaissance and delivery of anti-submarine weapons, in areas where they are out of range of ship-based anti-aircraft systems, will probably provide aircraft with an even more important role in maritime warfare than they have played in the past, important as that has been. The probable replacement of the piloted aircraft in many combat roles by the surface-launched missile, ballistic or cruise, and by pilotless aircraft raises a further question: whether an independent air force, as a third armed force, is the best solution for the employment of aircraft in military operations, and whether or not a reversion to the organization by which air forces formed an arm respectively of the army and the navy would not be both a more economical and a more effective use of resources. The alternative could be to integrate the three services into one, but, for most countries, the likelihood of having to execute operations in which all three services were more or less equally balanced is small. Where a nation's maritime operations are likely to be confined to waters which are well within the range of land-based air forces, in the Mediterranean and Black Sea, for example, an integrated solution might, however, be appropriate.

The technical possibilities for change in the structure and methods of employment of the armed forces of the major military nations in the twenty-first century appear almost limitless, but history shows us that fundamental changes seldom come about except under the stress of actual war itself. It has also shown that futuristic prognostications are seldom borne out by the actuality in the early stages of a war, although they often prove remarkably farsighted about developments which come in the later stages or afterwards. The fanciful concepts of Jules Verne and H.G.Wells in the early years of the century have been realized in its second half.

One thing is certain: that those responsible for planning the allotment of the nation's resources to defence, and the future organization and training of the armed forces to implement it, will be faced with exceptionally complex decisions. Examples taken from the past may often usefully act as warnings of pitfalls which should be avoided; but, apart from general principles of warfare, in which Clausewitz remains still a remarkably sound mentor, they are seldom reliable as a guide for the future. Short of actual experience, which is clearly to be avoided as far as possible, those who plan for the military future have to achieve a judicious balance between imagination and down-to-earth realism, the past being the only guide to the latter. It is not a waste of time for those concerned with the military problems of today and tomorrow to improve their understanding of those of yesterday.

Select Bibliography

GENERAL

Bond, Brian, *War and Society in Europe 1870–1970*, London 1984
Carver, Michael (ed.), *The War Lords*, London 1976
    War Since 1945, London 1980
Gooch, John, *Armies in Europe*, London 1980
Hart, Basil Liddell, *History of the First World War*, London 1930
    History of the Second World War, London 1970
Howard, Michael, *The Causes of War*, London 1983
Hunt, Barry, *War Aims and Strategic Policy in the Great War*, London 1967
Joll, James, *Europe Since 1870*, London 1973
Jordan, Gerald (ed.), *Naval Warfare in the Twentieth Century*, London 1977
Kennedy, Paul, *Strategy and Diplomacy 1870–1945*, London 1983
Paret, Peter (ed), *Makers of Modern Strategy*, Princeton 1986
Stone, Norman, *Europe Transformed 1879–1919*, London 1983
Wilkinson, Samuel, *The Politics of Grand Strategy*, Harvard 1969

BRITAIN

Barnett, Correlli, *Britain and Her Army 1509–1970*, London 1976
Bond, Brian, *British Military Policy between the Two World Wars*,
    Oxford 1980
Carew, Tim, *The Vanished Army*, London 1964
Carver, Michael, *The Seven Ages of the British Army*, London 1984
Darby, Phillip, *British Defence Policy East of Suez 1947–68*, London 1973
Dean, Maurice, *The Royal Air Force and Two World Wars*, London 1979
Dunlop, John K., *The Development of the British Army 1899–1914*, London
    1938
Fraser, David, *Alanbrooke*, London 1982
    And We Shall Shock Them, London 1983
Gooch, John, *The Plans of War. The General Staff and British Military Strategy
    1900–1916*, London 1974
Howard, Michael, *The Continental Commitment*, London 1962
Kennedy, Paul, *The Rise and Fall of British Naval Mastery*, London 1976
Luvaas, Jay, *The Education of an Army, 1815–1940*, London 1964

Roskill, Stephen, *Naval Policy between the Two World Wars*, vol. I, London
  1968; vol. II, London 1976
Sims, Charles, *The Royal Air Force: The First 50 Years*, London 1968
Terraine, John, *The Right of The Line. The Royal Air Force in the European
  War 1939–1945*, London 1985
Tyler,J.E., *The British Army and The Continent 1904–1914*, London 1938

FRANCE

Cook, Don, *Charles de Gaulle*, London 1984
De Gaulle, Charles, *Vers l'armée de métier* (Towards a Professional Army),
  Paris 1934
Gorce, Paul-Marie de la, *The French Army*, London 1963
Griffiths, Richard, *Marshal Pétain*, London 1970
Hammer, Ellen J., *The Struggle for Indo-China*, Stanford 1960
Haute, André van, *The French Air Force*, London 1974 and 1975
Horne, Alistair, *The Price of Glory*, London 1962
  To Lose A Battle, London 1969
  A Savage War of Peace, London 1977
  The French Army and Politics, London 1984
Jenkins,E.H., *A History of the French Navy*, London 1973
Porch, Douglas, 'The French Army and The Spirit of the Offensive', in Bond,B.,
  and Roy,I. (eds), *War and Society*, London 1975
  The March to the Marne. The French Army 1877–1914, London 1981
  The Conquest of the Sahara, London 1986
Rose, François de, *La France et la défense de l'Europe*, Paris 1976

GERMANY

Carsten,F.L., *The Reichswehr and Politics 1918–1933*, Oxford 1966
Cooper, Matthew, *The German Army 1933–45*, London 1975
  The German Air Force 1933–45, London 1981
Craig, Gordon, *The Politics of the Prussian Army*, London 1964
Fischer, Fritz, *Germany's Aims in the First World War*, New York 1962
Guderian, Heinz, *Panzer Leader*, London 1952
Herwig, Holger H., *Luxury Fleet. The Imperial German Navy 1888–1918*,
  London 1980
Killen, John, *The Luftwaffe – A History*, London 1967
Kitchen, Martin, *The German Officer Corps 1890–1914*, Oxford 1968
  A Military History of Germany, London 1975
Klee, Dr Karl, *The Decisive Battles of World War II. The German View*, London
  1965
Lambi, Ivo Nikolai, *The Navy and German Power Politics 1882–1914*, London
  1985
Lee, Asher, *The German Air Force*, London 1946
Mason,H.M., *The Rise of the Luftwaffe*, London 1973
Raeder, Admiral Erich, *Struggle for the Sea*, London 1959
Ritter, Gerhard, *The Sword and the Sceptre*, vol. 2, 1890–1914, London 1973

Rosinski, Herbert, *The German Army*, London 1939
Seaton, Albert, *The German Army 1933–45*, London 1982
    The Russo-German War 1941–45, London 1971
Taylor, Telford, *Sword and Swastika*, New York 1952
Tuleja, Thaddaeus V., *Eclipse of the German Navy*, London 1958
Wheeler-Bennett, J.W., *The Nemesis of Power. The German Army in Politics
    1918–1945*, London 1967

RUSSIA AND THE SOVIET UNION

Bonds, Roy (ed.), *The Soviet War Machine*, London 1977
Erickson, John, *The Soviet High Command 1918–1941*, London 1961
    The Road to Stalingrad, London 1975
    The Road to Berlin, London 1983
Garden, Michael, *A History of the Soviet Army*, London 1966
Golovin, N.N., *The Russian Army in the World War*, New Haven 1931
Hart, Basil Liddell, *The Soviet Army*, London 1956
Higham, Robin, and Kipp, Jacob W., *Soviet Aviation and Air Power*, London
    1978
Lee, Asher, *The Soviet Air Force*, London 1956
Mackintosh, Malcolm, *Juggernaut*, London 1962
Mitchell, Donald, *A History of Russian and Soviet Sea Power*, New York 1974
Stone, Norman, *The Eastern Front 1914–1917*, Oxford 1975
Walder, David, *The Short Victorious War: the Russo-Japanese Conflict*, London
    1973

UNITED STATES OF AMERICA

Carrison, Daniel, *The U.S. Navy*, New York 1968
Hagan, Kenneth J. (ed.), *In Peace and War*, Westport 1978
MacCluskey, Monro, *The U.S. Air Force*, New York 1967
Macintyre, Donald, *Sea Power in the Pacific*, London 1972
Morison, Samuel, *The Two-Ocean War*, Boston 1963
Taylor, Maxwell D., *Swords and Plowshares*, New York 1972
Weigley, Russell F., *History of the U.S. Army*, London 1968
Westmoreland, William C., *A Soldier Reports*, New York 1976

JAPAN

Allen, Louis, *Burma. The Longest War 1941–45*, London 1984
Committee of Imperial Defence, *The Official History of the Russo-Japanese
    War*, London 1909
Dull, Paul S., *The Imperial Japanese Navy. A Battle History (1941–1945)*,
    Cambridge 1978
Jones, F.C., *Japan's New Order in East Asia. Its Rise and Fall 1937–45*, Oxford
    1954
Lehmann, Jean-Pierre, *The Roots of Modern Japan*, London 1982
Mayer, S.L. (ed.), *The Japanese War Machine*, Feltham 1976
Presseisen, Ernst L., *Before Aggression*, Tucson 1965

Storry, Richard, *A History of Modern Japan*, London 1982
    Double Patriots, London 1957
Walder, David, *The Short Victorious War; the Russo-Japanese Conflict*, London
    1973
War Office, *Handbook of the Japanese Army*, London 1928
Warner, Denis and Peggy, *The Tide at Sunrise*, London 1974

CHINA

Bonds, Roy (ed.), *The Chinese War Machine*, London 1979
Clubb, O.Edmund, *Twentieth Century China*, New York 1978
Gittings, John, *The Role of the Chinese Army*, London 1967
Griffith, Samuel B., *The Chinese People's Liberation Army*, London 1968
Hibbert, Christopher, *The Dragon Awakes*, London 1970
Liu,F.F., *A Military History of Modern China*, Princeton 1956
Nelsen, Harvey W., *The Chinese Military System*, Boulder 1977
O'Ballance, Edgar, *The Red Army of China*, London 1962
Segal, Gerald, and Tow, William T., *Chinese Defence Policy*, London 1984
Whitson, William W., with Chen-Hsia Huang, *The Chinese High Command*,
    London 1973
Wilson, Dick, *When Tigers Fight*, London 1982

Notes

1: BRITAIN (pages 3–39)

1 Paul Kennedy, *The Rise and Fall of British Naval Mastery*, London 1976, p. 209

2 Sir Thomas Sanderson, PUS at the Foreign Office 1907, quoted in Michael Howard, *The Continental Commitment*, London 1962, p. 11

3 Samuel Wilkinson, *The Politics of Grand Strategy*, Harvard 1969, p. 37

4 Barbara Tuchman, *The Guns of August*, New York 1962, p. 195

5 Michael Howard, *The Continental Commitment*, London 1962, pp. 54–5

6 *Ibid.*, p. 20

7 CID paper, 1366-B

8 DP (P) 22

9 Brian Bond, *British Military Policy between the Two World Wars*, Oxford 1980, p. 307

10 John Terraine, *The Right of the Line*, London 1985, pp. 219–20

11 *Ibid.*, p. 261

2: FRANCE (pages 40–89)

1 Michael Carver (ed.), *The War Lords*, London 1976, p. 124

2 Quoted in Alistair Horne, *The French Army and Politics*, London 1984, p. 46

3 Alistair Horne, *To Lose a Battle*, London 1969, p. 515

3: GERMANY (pages 90–170)

1 Alistair Horne, *The Price of Glory*, p. 36

4: RUSSIA (pages 171–206)

1 N.N.Golovin, *The Russian Army in the World War*, New Haven 1931, pp. 82–92

5: THE SOVIET UNION (pages 207–59)

1 John Erickson, *The Soviet High Command 1918–1941*, London 1961, p. 506

7: JAPAN (pages 331–78)

1 Ernst L.Presseisen, *Before Aggression*, Tucson 1965, p. 186
2 Denis and Peggy Warner, *The Tide at Sunrise*, London 1974, pp. 301, 433, 448

Index

Abe, General, 357
Abrams, General Creighton W., 324
Aden, 35, 36
Adenauer, Konrad, 58, 169
Adlerangriff, Operation (German/air attack on Britain, 1940), 140, 149
Agadir incident, 10, 44, 95
Ainsworth, General Frederick, 269
Airborne forces, 30, 83, 152, 216
Aircraft, naval, 21–3, 36, 115, 121
Airships, 109, 114–15, 118
Alamein, El, Battle of, 31–2, 148, 153, 367
Alanbrooke, Field Marshal Viscount, 294, 301–2, 309–10
Albert, King of the Belgians, 108
Alexander, Field Marshal Earl, of Tunis, 75, 301, 305, 308, 310, 406
Alexeyev, Admiral E.I., 175, 177–8, 180, 182
Alexeyev, General M.V., 193, 196–202, 208
Algeria, 35, 42, 73, 79–86, 367, 434, 442
Alksnis, General Ya.I., 216, 221–2
Allenby, Field Marshal Viscount, 18
Alsace-Lorraine, 20, 40, 45–6, 57, 72, 91, 93, 108, 110
Anderson, Lieutenant-General Sir Kenneth, 299
André, General, 41
Antonescu, Marshal Ion, 241–2
Antonov, General A.I., 243, 246–7, 250

Anvil, Operation (Planned Anglo-US landing in southern France 1944), 303–5
Arakan, 370–1
Araki, General, 350, 356
Arcadia Conference (Washington, December 1941), 294, 405
Ardant du Picq, Colonel Charles, 42–3
Ardennes, 66, 135, 165–6, 244–5, 309
Argenlieu, Admiral Thierry d', 77
Arnhem, Battle of, 32
Arnold, General Henry H., 289, 294, 297, 300, 302
Arnold-Forster, H.O., 8
Artillery,
  anti-aircraft, 24, 150, 254
  anti-tank, 30, 127
  coast-defence, 177, 268
  field, 16, 43, 46, 96, 105, 143, 155, 167, 188–9, 235, 247, 268, 395, 436–7
Asquith, Earl of Oxford and, 11–12, 16–17
Atlantic, Battle of the, 29, 302
Atomic bombs, *see* Nuclear weapons
Aube, Admiral, 47
Auchinleck, Field Marshal Sir Claude, 30–1
Australian forces, 14, 18–9, 30, 58, 323
Austria (also Austro-Hungarian Empire), 14, 48, 91, 93–4, 97, 100, 109–10, 125, 166, 190, 201, 432

Badoglio, Marshal Pietro, 156, 301

Bagramyan, General I.Kh., 236, 239, 242–3, 246

Baker, Newton, 270–1, 274

Balfour, Arthur, 7

Balkans, 40, 91, 97, 155, 171, 301, 303

Balloons, 21, 187

Baranov, General P.I., 216

Barbarossa, Operation (German invasion of Russia 1941), 142, 156

Barratt, Air Marshal Sir Arthur, 68

Baruch, Bernard, 278, 299

Bauer, Colonel, 395

Beatty, Admiral of the Fleet Earl, 13, 16, 105–6, 279

Beaufre, General André, 43, 81

Beck, General (Austrian, World War I), 97

Beck, General Ludwig (German, World War II), 119, 122–6

Bedell-Smith, General Walter, 309

Belgium, 9. 11–12, 20, 27, 60, 64, 66, 68, 95, 99, 100, 104, 127–9, 135, 139, 168, 224

Benson, Admiral, US Navy, 279

Beria, Laurenti, 227

Berlin, Battle of, 165–6, 245–9
Congress of, 171

Berzin, Yan, 220

Bethmann-Hollweg, Theodore von, 104, 108, 112

Bidault, Georges, 83

Billotte, General Pierre, 68

Bismarck, Otto, Prince von, 91–2, 97, 112, 431–2

Blanchard, General, 66

Blaskowitz, General I., 162

Bliss, General Tasker, 277

Blomberg, Field Marshall Werner von, 120, 122–5

Blum, Léon, 59, 63

Blyukher, Marshal Vasilii, 210, 212, 218, 220–1, 388, 391

Bock, Field Marshal Fedor von, 128–9, 134–6, 143–4, 146

Boer War, 6, 44, 339, 343, 434

Bolero, Operation (Build-up of US forces in Britain 1942–4), 297

Bolsheviks, 111, 184, 203–5, 207–8, 347

Bor-Komorowski, General Tadeusz, 238–9

Borodin, Mikhail (Gruzenberg), 218, 388, 390–1

Boulanger, General Georges, 40

Boxer rebellion, 175, 331, 338, 379, 382

Bradley, General of the Army Omar N., 250, 289, 301, 303–4, 306–9, 315

Brauchitsch, Field Marshal Walter von, 125–6, 128–9, 136, 142, 145

Brest-Litovsk, Armistice and Treaty of, 52, 203, 207–8

Bret, Major-General, USAAF, 289

Brezhnev, Leonid, 255–6

Briand, Aristide, 51, 53

Britain, Battle of, 28, 140–1

British (Royal) Air Force,
casualties, 28, 33, 151, 158, 160
doctrine, 21–2, 25
equipment, 25, 35–6, 140–1
organization, 21–2, 25, 35
strength, 21, 28, 33, 37, 65, 140

British armed forces,
finance for, 22, 24
higher command, 7, 9, 26, 39

British Army,
casualties, 16–7, 19, 33–4, 49, 51–2, 69
doctrine, 10, 21, 26
organization, 3–6, 9, 24, 26
strength, 5, 12, 14, 18, 26, 33–4

British (Royal) Navy,
casualties, 29, 33, 107
doctrine, 10, 23–4, 37–8
equipment, 36–8
organization, 7, 8, 23–4, 37–8
strength, 5, 11, 13, 18–19, 22–3, 29, 33, 37

Brodrick, St John, 6

Brooke, Field Marshal Sir Alan, *see* Alanbrooke

Brüning, Heinrich, 119
Brusilov, General Alexei, 107, 193, 197, 199–202, 437
Brussels Treaty, 168
Buckner, General Simon Bolivar, 311
Budenny, Marshal Semen, 210–11, 215, 220–2, 227, 231
Bugeaud, Marshal Thomas-Robert, 42–3, 79
Bukharin, Nikolai, 220
Bulganin, Nikolai, 227, 254
Bulgaria, 108, 142, 162, 190, 198, 241
Bullard, Major-General Robert L., 277
Bülow, Bernhard, Prince von, 90, 96–7
Bülow, General Karl von, 100
Burma, 31, 294–5, 300, 303–4, 366, 368, 370, 373, 405–8
Busch, Field Marshal E., 143, 155, 162
Byng, Field Marshal Viscount, 18

Cambon, Paul, 9
Cambrai, Battle of, 18, 111
Cameron, Major-General George H., 277
Campbell-Bannerman, Sir Henry, 9
Canadian forces, 18–19, 33, 58, 306, 310
Carter, President Jimmy, 327–8
Casablanca conference (Symbol), 73–4, 300
Castelnau, General Noel de, 50–1
Catroux, General Georges, 42
Cavalry, 16, 24, 179, 184, 190, 215, 436
Chaban-Delmas, Jacques, 82
Chaffee, Major Adna E., 287
Challe, General Maurice, 83–5
Chamberlain, Neville, 26–7
Chang Hsueh-liang (The Young Marshal), 351, 390, 393, 398, 400
Chang Hui-tsan, 396
Chang Kuo-hua, 417–8
Chang Kuo-tao, 398–400
Chang Tso-lin, 351, 385–90, 392
Chemical warfare, 15, 27, 51, 214
Chemin des Dames, Battle of, 51, 275

Ch'en Chiung-ming, 386–8
Chennault, Major-General Claire L., 403–6, 408
Ch'en Po-ta, 425
Chen Yi, 398, 404, 407, 409–10, 412–15, 417, 421, 424
Cherepanov, General, 401
Cherevichenko, General Ya.T., 229
Chernyakhovsky, General I.D., 236–7, 239, 244–6
Cherrière, General Paul, 79
Chiang Kai-shek, Generalissimo, 218–9, 281, 302–3, 311, 354–60, 362, 367, 370–1, 373–4, 377, 384, 387–93, 395–7, 399–404, 406–16, 425, 429, 439, 441
Chinese Communist forces, 250
    air forces, 418, 420, 428
    casualties, 396, 409–10, 417, 419
    doctrine, 394, 396, 421, 423, 427
    equipment, 410, 412, 420, 427–8
    high command, 421, 425
    naval forces, 420, 428
    organization, 393–5, 397, 401–2, 404, 407, 411–12, 414, 417, 419–20, 426–7, 429
    political attitude, 421–2, 424–5
    strength, 394–6, 397–8, 400–2, 404, 407, 409–10, 414, 416, 418–19, 427
Chinese Nationalist forces,
    air forces, 398, 401, 403–4
    casualties, 391, 403, 406, 409, 415
    equipment, 401, 403–5, 412
    organization, 389, 393, 401, 411
    political attitude, 391, 393, 414
    strength, 388–90, 393, 396-8, 400–1, 411
    training, 388, 395–6
Chou En-lai, 386, 391, 397, 400–1, 403–4, 418, 424–5
Christie, J. Walter, 287
Christison, General Sir Philip, 370–1
Chuikov, General V.I., 231–3, 237–8, 245, 249
Churchill, Sir Winston, 10–11, 14, 16, 30, 70, 231, 235, 241, 244–5, 249, 294, 297, 301, 303–4, 310, 435

Chu Teh, 394–402, 407, 409–12, 414–15, 421, 425

Citadel, Operation (German attack at Kursk July 1943), 155

Clark, General Mark W., 75, 289, 299, 301–2, 304, 308, 310

Clausewitz, Major-General Karl-Maria von, 1, 42–3, 90–1, 113, 307, 437, 443–4

Clemenceau, Georges, 53–4, 57–8

Committee of Imperial Defence (Britain), 7, 16

Coningham, Air Marshal Sir Arthur, 313

Conrad von Hötzendorf, Field Marshal Franz, 97, 101, 103–4, 107–8, 193–5, 197–8, 201

Coolidge, President Calvin, 281

Cot, Pierre, 65–6

Coty, President René, 82–3

Craid, General Malin, 287, 289

Crerar, General Henry, 306

Crimea, 3, 4, 42, 112, 143, 145–6, 155, 173–4, 205, 212, 253

Cuba, 261–3, 267, 320, 423

Curragh incident, 12

Custer, Colonel George, 265

Czechoslovakia, 26, 125–6, 166, 239–40, 242, 249–50, 258, 282

Dakar, 70–1

Daladier, Edouard, 61, 63

Danilov, General, 190, 192–3

Darlan, Admiral François, 69–71, 73

Davis, Dwight F., 287

Debré, Michel, 84

Delbeque, Léon, 82

Dempsey, General Sir Miles, 306–7, 309

Denikin, General Anton, 208, 211

Denmark, 95, 134–5, 224

Dentz, General Fernand, 71

Devers, General Jacob L., 304, 306, 308

Dick, Congressman, 268

Dickman, Major-General Joseph T., 276

Dien Bien Phu, 78

Dill, Field Marshal Sir John, 294, 302, 310

Dog, Plan (US general strategic plan December 1940), 284–5

Donetz, basin and river, 112, 142–3, 145, 147, 154, 212, 227–8, 233

Dönitz, Admiral Karl, 110, 120, 122, 139, 148–9, 166, 295, 310

Doolittle, General James H., 295, 312–13, 435

Douhet, General Giulio, 129

Doumenc, General Joseph, 62, 68

Doumergue, Gaston, 61

Dowding, Air Chief Marshal Lord, 28, 140

Dragoon, Operation (Franco-American landing in southern France 1944), 75, 305–6, 308

Dreadnought class of battleships, 8

Dreyfus affair, 41

Drum, Major-General Hugh A., 288

Dukhonin, General, 208

Dunkirk, 27–9, 140

Eaker, General Ira C., 312

Edward VII, King, 44

Egypt, 7–10, 81, 151–2, 442

Eisenhower, General of the Army Dwight D., 29, 31–2, 73, 75, 148, 164, 166, 246, 249–50, 287, 289, 299–301, 303, 305–7, 309–10, 313, 315, 317, 319, 325, 435

Ely, General Paul, 82, 85

Emmons, Lieutenant-General, USAAF, 289

Engineers, 24

Esher, Lord, 1

Evert, General, 193, 197, 199, 202

Falkenhausen, General, 396, 401

Falkenhayn, General Erich von, 14, 50, 101, 103–5, 107–8, 196, 198–9, 201, 395, 437

Falkenhorst, General Nikolaus von, 133

Falkland Island, 36, 38, 442
Feng Kuo-chang, 384–5
Feng Yü-hsiang (The Christian
    General), 219, 389–90, 392–4,
    398, 400
Ferdinand, Archduke Franz, of
    Austria, 99
Finland, 64, 112, 134, 222–4, 236,
    240, 251
Fisher, Admiral of the Fleet Sir John,
    8, 13, 16, 37
Foch, Marshal Ferdinand, 19, 42–3,
    49, 53–4, 56–7, 274–77, 437
Fock, Major-General, 178, 183
Foelkerzam, Admiral, 186
Ford, President Gerald, 327
Forrestal, James, 315
Franchet d'Esperey, Marshal Louis,
    42
François, General Hermann von, 101,
    192
Fredendall, Major-General Lloyd R.,
    299
Free French, 71, 73–5
Freikorps, 116
French Air Force,
    casualties, 68
    doctrine, 65, 66, 86
    equipment, 47, 55, 65, 88
    organization, 47, 54–5, 65, 88
    plans, 65, 68
    strength, 47, 63, 65, 88
French armed forces,
    finance for, 60, 63, 76
    high command, 44, 53, 59–61, 77,
        87
    political attitude, 41, 59, 69, 72,
        76, 78–9, 81, 87, 89
French Army,
    casualties, 17, 48–9, 51, 56, 69, 78,
        105
    doctrine, 42–4, 48, 59, 86
    equipment, 41, 46, 61, 86, 88
    organization, 40–1, 60–1, 75, 87–8
    plans, 40, 45–6, 61–3, 66, 80, 86
    strength, 41, 44, 46, 60, 62, 64–5,
        75, 88

French Navy,
    casualties, 56, 70
    doctrine, 47
    equipment, 48, 56, 70, 88
    organization, 56, 69–70, 88
    plans, 48
    strength, 48, 56, 70, 88
French, Field Marshal Sir John (Earl,
    of Ypres), 12, 14, 49
Freycinet, Charles, 41, 47
Fritsch, General Werner, Freiherr
    von, 122–5, 128
Fromm, General F., 164
Frunze, General Mikhail, 210–14
Fuller, Major-General J.F.C., 27, 30
Fu Tso-yi, 410, 413–15

Gaillard, Felix, 82
Galicia, 14, 101, 107, 190, 193, 196,
    199, 203, 222
Gallieni, General Joseph-Simon, 42,
    50
Gallipoli, 14, 16, 56
Gamelin, General Maurice, 59, 61,
    63–5, 67–8
Garrison, Lindley, 270
Gaulle, Brigadier-General Charles de,
    59, 61, 66–7, 69, 71–7, 79, 82–6,
    88, 434, 438
Gaza, Battle of, 18
Geddes, Sir Eric, 279
Gelb, Operation (German invasion of
    France, Belgium and Holland
    1940), 134
Gensoul, Admiral Marcel, 70–1
Georges, General Joseph, 66, 68
German Air Force (Luftwaffe),
    casualties, 115–6, 133, 139–41,
        150–1, 158–9
    doctrine, 115, 131
    equipment, 114–16, 118, 130–2,
        140, 149–51, 157–8, 160–1
    organization, 114–15, 129–23
    plans, 132, 140, 149
    strength, 28, 65, 114–16, 130, 132–
        3, 140–1, 150–1, 153, 157–9, 168

German armed forces (*Wehrmacht*),
high command, 92, 95, 122–6, 129,
145, 153, 170
political attitude, 92, 116, 119,
123, 127
German Army,
casualties, 19, 51, 54, 69, 105, 137,
145–7, 164, 166, 193, 202, 237,
241–3, 245, 249
doctrine, 93
equipment, 46, 127, 154
organization, 90, 124, 127, 142,
154, 169
plans, 93–4, 96–8, 125–8, 134, 143
strength, 58, 64, 92–4, 100, 113,
117, 124, 126–8, 135, 142–3, 154,
165, 167–9, 191, 232, 235, 243
German Navy,
casualties, 107, 114, 148–9
doctrine, 120
equipment, 95–6, 99, 118, 120–2,
149
organization, 94, 98, 120–1
plans, 95–6, 98–9, 106, 120–2, 133,
137–8
strength, 5, 11, 95, 114, 120–2, 137,
148
Ghormley, Admiral Robert L., 296
Giap, Vo Nguyen, 78, 323–4, 417
Giraud, General Henri, 42, 66, 72–4
Gneisenau, General Augustus, Graf
von, 90
Goering, Reichsmarschall Hermann,
28, 115, 117, 121, 124–5, 129,
136, 140–1, 151, 156–7, 160
Goethals, Major-General, 278
Golikov, General F.I., 230–3
Goltz, Field Marshal Colmar von der,
91, 208
Gordov, General V.N., 231
Gorshkov, Admiral S.G., 255
Gort, Field Marshal Viscount, 28, 65,
68
Gough, General Sir Hubert, 18, 19
Gouraud, General Henri, 42, 277
Govorov, Marshal L.A., 229, 235, 240,
246

Grandmaison, Colonel Louis de, 43
Grant, General Ulysses S., 265–6
Grechko, Marshal A.A., 255
Greece, 30, 142, 146, 151, 198, 241
Grey, Edward, Viscount, of Fallodon,
9. 11, 99, 109
Grippenberg, General D.K., 184
Gröner, General Wilhelm, 116, 119
Guadalcanal, 296, 367, 368
Guderian, General Heinz, 66–7, 124,
128, 135–6, 143–5, 153–4, 161,
165, 227
Guépratte, Admiral, 56
Guingand, Major-General Sir Francis
de, 309
Gymnast, Operation (First plan for
Anglo-American landing in
French North Africa 1942), 297,
299

Haig, Field Marshal Earl, of
Bemersyde, 12, 16–20, 22, 51,
53–4, 107, 274, 276, 437
Haldane, Lord, 9, 12
Halder, General Franz, 125–6, 129,
144–7, 164
Halsey, Admiral William F., 295–6,
300–1, 368–9
Hamaguchi, Yuko, 352
Hamilton, General Sir Ian, 12, 178
Hammerstein-Equord, General K.,
Freiherr von, 119, 122
Hankey, Lord, 7, 9
Harbord, General, us Army, 286
Harding, President Warren, 280–1
Harkins, General Paul D., 322
Harpe, General Josef, 244–5
Harris, Air Chief Marshal Sir Arthur,
28, 151, 305, 312
Hart, Sir Basil Liddell, 20, 24, 26, 30,
32, 113
Hata, General, 357–8, 362, 402
Hay, Congressman, James, 270
Hayashi Senjuro, General, 350, 354
Hayashi Tadasa, 339
Henderson, Air Chief Marshal Sir
David, 22

Heydrich, Reinhard, 221

Himmler, Heinrich, 167, 246

Hindenburg, Field Marshal Paul von, 101, 103–4, 108–9, 111–13, 116, 118–20, 122–3, 201, 203

Hines, Major-General John L., 277

Hipper, Admiral Franz von, 105–6, 113

Hiramuma Kiichiro, 357

Hirohito, Emperor, 348, 351, 355–6, 360, 362, 374–6, 378

Hiroshima, 180, 319, 376

Hirota Koki, 354, 356

Hitler, Adolf, 26, 33, 59, 60–1, 64, 72, 79, 117, 119–20, 122–3, 125–7, 135–6, 139, 141–8, 151–3, 155, 157, 160–7, 222, 227, 230, 235, 243, 248–9, 289, 291, 308, 349, 358, 396, 432–4, 438

Ho Chi Minh, 77, 322, 417

Hodges, General Courtney H., 304, 306, 308–9

Holland (The Netherlands), 32, 64, 66, 95, 127, 129, 139, 160, 168, 224, 282–3, 358, 360

Ho Lung (the Ex-bandit), 394, 397–400, 402, 410, 417, 421

Honda, General, 370

Höppner, General (German, World War I), 114

Höppner, General Erich (World War II), 143, 145

Hoover, President Herbert, 281

Hoth, General H., 143, 146–7, 155, 232–3

Ho Ying-ch'in, 392, 396–7

Hsien-feng, Emperor, 379

Hsu Ch'ung-chih, 389

Hsu Hsiang-chien (The Ironside), 394, 397–8, 402, 421

Hsu Shih-ch'ang, 382, 385, 387, 398

Hsu Shu-cheng, 386

Hua Kuo-feng, 425, 426

Huang Shuo-hung, 389

Hu Ching-yi, 389

Hu Han-min, 389

Hull, Cordell, 290, 360, 362

Hungary, 108, 162, 241–2, 258

Huntziger, General Charles, 69

Husky, Operation (Anglo-American landing in Sicily 1943), 301

Hu Tsung-nan, 400, 404, 409, 413–15, 417

Ikki Kita, 349–50

India, 218, 302, 370–1, 406–8, 423, 442

Indian Army, 1, 15, 19, 30–1, 310

Infantry, 24, 436, 437

Ingenohl, Admiral, 105

Inouye Kaoru, 334, 336, 339

Ireland, 35

Inukai, Ki, 350

Italy, 8, 15, 26, 32, 34, 48, 107, 117, 126, 152, 156, 282–3, 301, 368, 440

Ito Hirobumi, 334, 336, 339–40, 345–6

Ivanov, General, 193–6, 199, 202

Iwo Jima, 311, 372

Japanese Army,
  air force, 365
  casualties, 178, 181, 183, 185, 311, 367–9, 371–3, 376, 403, 410
  doctrine, 341–3
  equipment, 334, 341, 364
  high command, 180, 372–4
  organization, 341–2, 350
  political activity, 337, 345, 348–51, 353–4, 356–7, 375
  strength, 185, 331, 333, 340–1, 348, 357, 363, 366–7, 369, 371
  training, 178, 187, 333, 335–8

Japanese Navy,
  air arm, 365
  casualties, 187, 367–9, 371–3, 376
  doctrine, and plans, 177, 352, 363
  equipment, 341–2, 352, 363
  high command, 372
  organization, 334
  political activity, 337, 345, 352–3, 357, 375
  strength, 341–2, 352–3, 363, 366, 368–9, 371, 376
  training, 176, 187, 334, 337, 363

Japanese Self-Defence Forces, 377–8
Jellicoe, Admiral of the Fleet Earl,
    13–14, 16–17, 99, 105–7
Jeschonnek, General H., 129, 133, 149,
    151, 156, 159
Jodl, General Alfred, 126, 128–9, 135,
    143, 153–4, 164, 166, 310
Joffe, Adolph, 387–8
Joffre, Marshal Joseph, 12, 14, 17, 42,
    45, 49, 51, 100
Johnson, President Andrew, 264
Johnson, General Harold K., 324
Johnson, President Lyndon B., 323–4
Juin, Marshal Alphonse, 42, 72, 75–6
Jutland, Battle of, 13, 16, 107

Kalinin, K.A., 216
Kamagushi Minseito, 350
Kamenev, L.B., 213, 220
Kamenev, General S.S., 210, 212–13
Kamimura, Vice-Admiral, 344
Kammhüber, General Josef, 150
Katamura, General, 371
Kato Komei, 346
Katsura, General, 338–40, 345–6
Kawabe, General, 370–1
Kawamura, General, 184–5
Keitel, General Bodewin, 125, 145
Keitel, Field Marshal Wilhelm, 124–
    6, 128, 145–7, 164, 166
Kenly, Major-General, 273
Kennedy, President John F., 320,
    322–3
Kenny, General, USAAF, 313
Kerensky, Alexander, 203
Kesselring, Field Marshal Albert, 129,
    133, 139, 140–1, 152–3, 156
Khalepsky, General I., 221
Khruschev, Nikita, 227, 231, 254–6,
    423
Kiel canal, 95, 169
Kim Il-sung, 418
Kimmel, Admiral Husband E., 284,295
Kimura, General, 371, 373
King, Admiral Ernest J., 282, 291,
    294–7, 300–2, 304, 311–12, 435
Kirponos, General M.P., 225

Kissinger, Henry, 324, 426
Kitchener, Field Marshal Earl, 6, 12,
    14, 16
Kleist, Field Marshal Paul von, 66–7,
    135–6, 146–7, 230
Kluck, General Alexander von, 49, 97,
    100–1
Kluge, Field Marshal Gunther von,
    128–9, 143, 145, 154–5, 163–4,
    228, 308
Kodama, General, 340, 342–3
Kodo-ha, 349–51, 356, 371
Koenig, General Marie-Pierre, 74
Koiso, General, 372–4
Kolchak, Admiral Alexander, 204–5,
    208, 210, 217, 347
Komura, Minister, 339–40
Kondo, Admiral, 366
Koniev, Marshal Ivan, 228–9, 235–40,
    244–5, 247–50
Königgratz, Battle of (also known as
    Sadowa), 91, 93, 100
Konoye, Prince Fumimaro, 291, 354–
    60, 362, 372, 374
Korea, 38, 174–5, 177, 185, 316–19,
    332, 334–5, 340, 345, 349, 377,
    383, 418–19
Kostenko, General F.Ya., 229
Kosygin, Alexei, 255
Kozlov, General K., 230
Kriebel, Lieutenant-Colonel, 395
Krylenko, Nikolai, 208
Kuang Hsu, Emperor, 380, 382
Küchler, Field Marshal G., 128, 143
Kuomintang, 218, 219, 352–5, 376–7,
    383–5, 389–91, 393, 412
Kuo Sung-ling, 390
Kuroki, Major-General, 176, 178, 181,
    184–5, 342
Kuropatkin, General A.N., 174, 177,
    179–82, 185, 199, 202, 340
Kursk, Battle of, 155
Kurusu Saburo, 362
Kuznetsov, General F.I., 225
Kuznetsov, Admiral N.G., 254–5
Kwantung army, 219, 250, 349, 351,
    353–4, 356

Lacoste, Robert, 81–2
Lagaillarde, Pierre, 82
Lansdowne, Lord, 9, 20
Lattre de Tassigny, Marshal Jean de, 72, 75–8
Laval, Pierre, 61, 72
Leahy, Admiral William D., 282, 294
Leclerc, Marshal Jacques, Vicomte de Hautecloque, 74, 77
Leeb, Field Marshal Wilhelm, Ritter von, 135, 143–5, 228
Leigh-Mallory, Air Marshal Sir Trafford, 305
Le May, General Curtis, 313
Lenin, Vladimir Ilyich (Ulyanov), 104, 203, 205, 207–9, 212–13, 218
Leningrad (St Petersburg, Petrograd), 104, 142–5, 155, 162, 171, 184, 202, 204, 208, 233, 227–8, 235
Liaotung peninsula, 174, 336, 346, 348
Libya, 30, 71, 152, 294
Liggett, Major-General Hunter, 275–6
Li Hung-chang, 335, 339, 380, 382, 384, 390
Li Li-san, 395
Lin Piao, 317, 397–400, 402, 410–15, 418–19, 421–6
List, Field Marshal Wilhelm, 142, 146–7, 231
Li Tsung-jen, 389–92, 402, 410, 412, 414–15
Liu Chih-tan (The Turncoat), 394
Liu Po-cheng (The One-eyed General), 402, 413–14, 417, 421
Liu Shao-ch'i, 417–18, 424–5
Lii Yuan-hung, 385, 387, 389
Lloyd-George, David, Earl, of Dwyfor, 17, 18, 20, 54, 57, 109
Löhr, General A., 133, 151, 241
Lo Jung-kuan, 421
Long March, 398–400, 439
Lorillot, General Henri, 81
Lovett, Robert A., 289
Ludendorff, General Erich von, 17–19, 51, 53, 101, 103–4, 107–13, 115–16, 193, 195, 197, 203, 395, 437

Lu Jung-t'ing, 385–7
Luftwaffe, see German Air Force
Lu Ti-p'ing, 396
Lüttwitz, General W. Freiherr von, 117
Luxemburg, 100, 168
Lyautey, Marshal Louis Hubert, 42, 59, 79, 82

MacArthur, General of the Army Douglas C., 286–7, 290, 295–6, 301, 304, 311, 316–17, 326, 368–9, 372–3, 377, 410, 418, 435
McCreery, General Sir Richard, 310
Machine-guns, 43, 46, 96, 179, 188, 436
Mackensen, Field Marshal August von, 104, 108, 192, 195–6, 198
McKinley, President William, 267
Macmillan, Sir Harold (Earl of Stockton), 34, 36–7
McNair, General Lesley J., 290, 297, 299, 313
McNamara, Robert S., 320, 322–3
Maginot, André, 60
Maginot Line, 60, 62, 64–5, 129, 135–6
Mahan, Captain Alfred Thayer, 8, 47, 262–3, 279
Mahan, Dennis Hart, 265
Makarov, Admiral S.O., 177
Malaya, 29, 31, 35, 295, 304, 322, 359, 363, 366, 368
Malenkov, Georgi, 227, 254
Malinovsky, Marshal R.Ya., 229, 232, 235–6, 240–2, 245, 248–50, 255
Malta, 6, 7, 11, 29, 30, 151–2, 245
Manchu dynasty, 174, 379, 382
Manchuria (also Manchukuo), 174–5, 177, 180, 185, 187, 218, 223, 241, 336, 338–41, 346, 349, 351, 353, 355, 372, 374, 376–7, 379, 383, 386, 388, 390, 392, 397, 403, 410, 413, 415–16, 433
Mandel, Georges, 69
Mangin, General Charles, 51, 57

Mannerheim, Field Marshal Carl
Gustaf, Baron von, 208, 233, 240
Manstein, Field Marshal Erich von,
134–5, 147, 154, 161–3
Manteuffel, General Hasso von, 247,
249
Mao Tse-tung, 34, 218–19, 250, 354,
384, 386, 394–5, 397–401, 403–4,
409–14, 416, 418, 421–6, 441
March, General Peyton C., 278, 285–6
Marco Polo Bridge (Lukouchiao)
incident, 354, 400
Marita, Operation (German invasion
of Greece 1941), 142, 146, 151
Marshall, General of the Army George
C., 268, 276, 284–5, 289–90, 294–
6, 299–303, 312, 315–16, 410–13,
435
Massu, General Jacques, 81, 83–4, 87
Matsui, General, 355, 402
Matsuoka Yosuke, 358
Maurin, General Joseph-Léon-Marie,
62
Meiji (Mutsuhito), Emperor, 331–4,
336, 345
Mekhlis, L. Z., 220
Mendès-France, Pierre, 79
Meretskov, General K.A., 223, 225,
229, 235, 250
Mers-el-Kebir, 70, 74
Messervy, General Sir Frank, 373
Mexico, 264, 269–70, 273
Midway Islands, Battle of, 295–6,
367–8
Mikolajczyk, Stanislaw, 238–9, 241
Mikoyan, A.I., 216, 251, 256
Mil, M.A., 216, 256
Milch, Field Marshal Erhard, 129,
139–40, 151, 156–8, 160
Miles, General Nelson, A., 267–8
Milyutin, General Count D.A., 174,
250
Missiles,
anti-aircraft, 325
anti-tank, 442
ballistic, 160, 254–7, 319–20, 439
cruise, 159–60, 319–20, 439

Mitchell, Brigadier-General William
(Billy), USAAF, 273, 275, 281,
287–8, 403
Model, Field Marshal Walter, 155,
165–6, 239
Mollet, Guy, 80, 86
Molotov, Vyacheslav, 223, 227
Moltke, General Helmuth Johann von
(the Younger), 49, 96–101
Moltke, Field Marshal Helmuth Karl,
Graf von (the Elder), 90, 91, 93,
112
Montgomery, Field Marshal
Viscount, of Alamein, 22, 29, 31–
2, 74, 148, 247, 300–310, 313
Morocco, 9, 59, 73, 79, 96, 98, 299,
301
Mountbatten, Admiral of the Fleet
Earl, of Burma, 31–2, 37–9, 303,
370, 373, 408–9, 439
Mukden, 175, 179, 181, 185, 338–9,
343, 351–2, 355, 392, 396–7, 411,
413–14
Mussolini, Benito, 79, 152, 156, 301
Mutaguchi, General, 370
Muto, General, 349

Nagano, Admiral, 361, 363
Nagasaki, 319, 376
Nagumo, Admiral, 363, 365
Nasser, Colonel Gamal Abdul, 81
Nazi Party (NSDAP – National
Socialist German Workers Party),
119, 120
Nebogatov, Admiral N.I., 185–7
Neurath, Freiherr von, 125
New Zealand forces, 14, 18, 19, 30, 58,
310, 323
Nicholas II, Tsar of Russia, 103–4,
171, 175, 179, 183–4, 188, 192,
199, 202–3, 339
Nicholas, Grand Duke, 101, 189–90,
192, 196–7
Nieh Jung-chen, 410, 414–15, 421
Nimitz, Fleet Admiral Chester W.,
295–6, 300, 311–12, 367, 369, 435

Nivelle, General Robert, 17–18, 50–2, 274

Nixon, President Richard W., 324, 426

Nogi Maresuke, General, 180–1, 183–5, 334, 342, 346

Nomura, Admiral, 290, 357, 359, 361–2, 372

Normandy, 22, 31–2, 164, 305–8

North Atlantic Treaty Organization (NATO), 76–7, 80, 84, 86–7, 89, 168–70, 316, 319–20, 432, 435, 441

Norway, 27, 64, 133, 135, 166, 224

Novikov, Marshal A.A., 251

Nozu, General, 180–1, 342

Nuclear weapons, 34–6, 86–7, 234, 253–5, 257, 312, 317, 319–20, 326–7, 374, 376, 423, 428, 433, 439, 442, 443

Octagon Conference (Quebec, September 1944), 311

Oikawa, Admiral, 360

Okada, Admiral, 351

Okinawa, 311, 372–4, 378

Oktyabrsky, Admiral F.E., 252

Oku, General, 178, 181, 184, 342

Okuma Shigenoku, 346–8

Omura Masujiro, 332–3

Orange plan (US naval plan of operations in the Pacific 1907–17), 264, 281–3

Orlov, Admiral V.M., 217, 222

Ortiz, Joseph, 85

Ottoman Empire, 23, 110, 171, 190

Overlord, Operation (Anglo-US landing in Normandy June 1944), 301–3

Oyama Iwao, General, 180–2, 184–5, 332, 334, 336, 338, 342–3

Ozawa, Admiral, 371

Pai Ch'ung-hsi, 389, 393, 410, 414–16

Painlevé, Paul, 53

Palestine, 18, 34

Palitsyn, General F.F., 189

Palmer, Brigadier-General John M., 285, 316

Panay incident, 282, 355

Papen, Franz von, 119

Patton, General George S., 32, 249–50, 287, 299, 301, 304, 307, 309, 439

Paulus, Field Marshal Friedrich, 146–7, 157, 230–3

Pavlov, General D.G., 225, 227

Pearl Harbor, 31, 146, 264, 280, 282–4, 291, 294, 297, 311, 362, 366, 376, 405, 438

P'eng Teh-huai, 394–5, 398–400, 413–15, 417, 419, 422–3, 425

Perry, Commodore Matthew, 174

Pershing, General of the Army John J., 19, 271, 273–8, 285–7, 289

Pétain, Marshal Philippe, 18, 43, 50–54, 56–7, 59, 60, 63, 65, 68–9, 72–3, 274–6, 437

Petrov, General I., 236, 241–2, 248

Philippine Islands, 262, 264, 267–8, 281, 283, 290–1, 294–5, 304, 311, 359, 363, 366–7, 372–3, 377

Pilsudski, Marshal Joseph, 59, 211–2

Pleven, René, 76, 168

Plumer, Field Marshal Viscount, 18

Poincaré, Raymond, 46

Poland, 27, 59, 61, 64, 103–4, 111–2, 118, 126–8, 166, 190–1, 193–4, 197, 211–12, 222–3, 225, 238–9, 241, 289

Polikarpov, N.N., 216

Polish forces, 30, 33, 308, 310

Popov, General M.M., 225

Portal, Marshal of the Royal Air Force Viscount, of Hungerford, 28

Port Arthur, 174–5, 177, 180–2, 184, 187, 250, 336, 338–9, 343, 416, 420

Pound, Admiral of the Fleet Sir Dudley, 70

Prittwitz, General von, und Gaffron, 98, 101, 191–2

Purkayev, General M.A., 250

P'u Yi, Prince (Emperor Hsuan Túng), 382–3, 385, 397

Quadrant conference (Quebec 1943), 301–3, 408

Radzievsky, General A.I., 238
Raeder, Grand Admiral Erich, 120, 124–5, 133, 137–9, 148
Rainbow plan (US operational plans in the Pacific 1939–41), 283–4
Ramsay, Admiral Sir Bertram, 305
Reagan, President Ronald, 328
Reichenau, Field Marshal Walter von, 120, 122–4, 129, 145
Reiter, General M.A., 233
Rennenkampf, General Pavel K., 179, 184, 191–3
Reynaud, Paul, 57, 69
Rhineland, 58, 60, 62–3
Richardson, Admiral, US Navy, 283–4
Richthofen, Captain Manfred, Freiherr von, 115–16
Richthofen, General W., Freiherr von, 129, 142, 157
Ridgway, General Matthew B., 289, 317, 320, 418–9
Ritchie, General Sir Neil, 30, 148
Roberts, Field Marshal Earl, 6, 12
Robertson, Field Marshal Sir William, 16, 18, 20
Rokossovsky, Marshal Konstantin, 210, 232–4, 236–9, 244–9
Rolling Thunder, Operation (US bombing of North Vietnam), 323
Romania, 27, 61, 108, 111, 162, 201, 224–5, 236, 240–1
Rommel, Field Marshal Erwin, 30, 66, 146, 148, 151–2, 156, 161–2, 297
Roosevelt, President Franklin D., 235, 248, 281–2, 284, 288–91, 295, 297, 299, 302–4, 310–11, 360, 404, 409
Roosevelt, President Theodore, 187, 263, 267, 270
Root, Elihu, 267–70

Rostov-on-Don, 112, 145, 147, 229, 231–2
Round-Up, Operation (Anglo-US plan for landing in Europe 1943), 296–7, 300–1
Rozhdestvensky, Admiral Z.P., 179, 182, 184–6, 344
Rundstedt, Field Marshal Gerd von, 126–8, 134–6, 143, 145, 161–3, 165, 227
Rupprecht, Crown Prince of Bavaria, 100
Russian Air Force (pre-1918), 205–6
Russian armed forces (pre-1918), finance for, 188
    high command, 174, 189, 190, 196, 198
    political attitude, 204
Russian Imperial Army,
    casualties, 178, 183, 185, 193, 197
    doctrine, 187
    equipment, 187
    organization, 172–3, 189, 196, 198
    strength, 112, 171–2, 175, 185, 189, 191–2, 198, 200, 203
Russian Imperial Navy
    casualties, 177, 180–1, 183, 187, 194
    equipment, 171, 188, 204
    organization, 174, 188, 203
    strength, 171, 188, 191
Ruzski, General, 193–6, 199, 203

S.A. (Sturm Abteilung), 118, 123
Saarland, 58, 63
Saigo Takamuri, 332, 334
Saint Mihiel, Battle of, 276, 287
Saint Petersburg see Leningrad
Saionji, Prince, 345, 348, 350
Saito Makota, Admiral, 350–2, 362
Sakharov, General, 177–200
Sakurai, General, 370
Salan, General Raoul, 78, 81–3, 85
Salerno, 31, 302
Salisbury, Lord, 7
Salonika, 50, 198

Samsonov, General Alexander V.,
179, 191–3
Sarrail, General Maurice, 49
Scapa Flow, 13, 114, 133
Scharnhorst, General Gerhard von,
90–1
Scheer, Admiral Reinhard, 13–14,
106–7, 110, 113
Schleicher, General Kurt von, 118–
20, 123
Schlieffen, General Alfred, Graf von,
44, 92–3, 95–6, 99–100
Schmundt, General R., 126, 145
Schörner, Field Marshal Ferdinand,
243, 245, 249
Schweppenburg, General L., Freiherr
Gehr von, 163
Scoones, General Sir Geoffrey, 370,
373
Scott, General Hugh, 277
Sedan, 66, 91, 135, 275
Sealion, Operation (German plan for
invasion of Englans 1940), 137,
141
Seeckt, General Hans von, 117–8,
196, 395–7, 406
Serbia, 20, 97–100, 104, 190, 193, 198,
242
Sevastopol, 188
Sextant conference (Cairo, November
1943), 303, 375
Shafter, Major-General, 267
Shaposhnikov, Marshal Boris, 209,
214–15, 221, 225, 227–30
Sharp, Admiral U.S. Grant, 325
Sherman, General William T., 265–6
Shidehara Kijuro, 351
Shimada, Admiral, 361, 372
Shimonoseki, 174, 332, 336
Shipping, merchant, 13, 16, 17, 29,
33, 106, 110, 148–9, 152, 372
Sicily, 32, 155, 300
Sikorsky, Igor, 205
Simpson, General William H., 310
Sims, Admiral, us Navy, 279
Singapore, 22–3, 31, 36, 185, 280,
283–4, 295, 304, 347, 376

Sledgehammer, Operation (Anglo-US
plan for landing in Europe 1942),
296–7
Slim, Field Marshal Viscount, 31,
370–1, 373, 441
Smuts, Field Marshal Jan, 22
Sokolovsky, Marshal Vasili, 234–5,
248, 255
Somervell, General Brehan B., 298–9
Somerville, Admiral of the Fleet Sir
James, 70–1
Soong, T.V., 374, 403–5, 408
South African forces, 30
Soviet Air Force,
doctrine, 251
equipment, 216, 225, 234, 250–1,
253, 255
organization, 216, 256
strength, 216, 219, 225, 234, 237,
248
Soviet armed forces,
high command, 213–14, 225, 227,
252–5
political control of, 213–14, 221,
243, 253–5, 258
purge of, 214–17, 220–2
Soviet Army,
casualties, 167, 224, 227, 229, 249
doctrine, 210–11, 214–15, 236,
252–5, 257–9
equipment, 155, 215–16, 225, 234,
250, 253, 256–7, 259
organization, 208–11, 225, 229,
235, 255, 257, 259
strength, 155, 165, 208–9, 213, 219,
223–5, 228, 230, 232, 234–5, 237–
8, 240, 244, 253
Soviet Navy,
equipment, 256
organization, 217
strength, 217, 252
Spaatz, General Carl A., 305, 312–3
Spain, 29, 117, 261, 263, 282
Spanish Civil War, 63, 217
Spanish-American War, 261–2, 267
Spee, Admiral Maximilian, Graf von,
106

Speidel, General Hans, 170

Sperrle, Field Marshal H., 129, 139, 141

Spruance, Admiral Raymond A., 296, 312, 371

S.S. (*Schutz Staffel*), 123, 154, 163

Stackelberg, General G.K. von, 178, 181–2, 184

Stalin, Josef V. (Djugashvili), 209–10, 212–13, 215–16, 218–19, 221–3, 225, 227–8, 230–5, 237–9, 241, 243–8, 250, 253–4, 303, 310, 317, 358, 372, 374, 403, 416, 435

Stanhope, Edward, 3

Stark, Admiral Harold R., 283–4, 290, 294

Stauffenberg, Colonel Graf von, 163

Stilwell, General Joseph W., 370–1, 373, 401, 405–9

Stimson, Henry, 269–70, 289

Stoessel, Lieutenant-General A.M., 175, 179, 183

Strasser, Gregor, 120, 123

Strauss, General A., 143

Student, General Kurt, 129, 151

Submarines, 13, 14, 16, 29, 36, 47, 107, 110, 148–9, 217, 253, 278, 319, 352–3, 436

Suez Canal, 10, 34, 56, 81, 183, 185, 442

Sugiyama, General, 355–6, 361, 372, 402

Sukhomlinov, General V.A., 189–90, 192

Summerall, Major-General Charles P., 277, 287

Sun Ch'uan-fang, 390, 392

Sung Che Yuan, 402

Sun Li-jen, 411, 413

Sun Yat-sen, 218, 382–3, 385–6, 388, 389, 403, 400

Suzuki, Admiral, 351, 375–6

Sweden, 117, 134, 203

Symbol conference *see* Casablanca conference

Syngman Rhee, 316

Taft, President William H., 269

T'aip'ing rebellion, 379–80

Taiwan (Formosa), 174, 335, 377, 415, 417, 418, 429

Tanaka, General, 351

T'ang Chi-yao, 385–7, 389

T'ang Sheng-chih, 390

Tanks, 16, 18, 24, 30–1, 51, 61–2, 64–5, 127, 135, 142–3, 146, 153–5, 167, 214–16, 227, 234–5, 243, 250, 257, 275, 287, 364, 437–8, 442

Tannenberg, Battle of, 101, 103, 192–3

Taylor, General Maxwell D., 320, 322, 438

Tedder, Marshal of the Royal Air Force Lord, 306, 313

Teng Hsaio-p'ing, 386, 425–6, 429

Teng K'eng, 387

Terauchi Seiki, Marshal, 340, 345, 347, 354, 366, 370–2, 402

Tibet, 190, 383, 392, 417–18, 423

Timoshenko, Marshal Semen, 144, 210, 215, 222–5, 227–30, 234–5, 240

Tirpitz, Grand Admiral Alfred von, 6, 10, 13, 94–5, 98–9, 106

Togo Heihachiro, Admiral, 175–6, 178–80, 181, 186–7, 332, 334–6, 343

Tojo Hideki, General, 291, 343, 350, 354, 356, 358, 360–2, 368, 370–2, 374

Tolbukhin, Marshal F.I., 235–6, 240–2, 244–5, 248

Torch, Operation (Anglo-US landing in French North Africa November 1942), 299, 313

Tosei-ha, 350–1, 356, 371

Toyoda, Admiral, 359–60, 369, 374

Trenchard, Marshal of the Royal Air Force, Viscount, 21–2, 25, 28, 275, 287

Tributs, Admiral V.F., 252

Trotsky, Lev Davidovich, 111, 207–13, 215

Truman President Harry S., 248, 310, 312, 317, 434
Ts'ao kun, 386–7, 389
Ts'en Ch'un-hsuan, 385–7
Tseng Kuo-fan, 380
Tseng Tse-sheng, 414
Tsushima, Battle of, 8, 185–7, 263, 343–4, 352
Tuan Ch'i-jui, 384–7, 389–90
Tukhachevsky, Marshal Mikhail, 205, 210–2, 214–6, 220–1
Tupolev, A.N., 216, 251, 256
Turkey, 7, 10, 14, 18, 23, 49, 104, 108, 171, 174, 190, 300
Tyulenev, General I.V., 225
Tzu-hsi, Dowager Empress (Yehanola), 380, 382

Udet, Ernst, 115
Uehara, Marshal, 345, 349–50
Ugaki, General, 349–50, 354, 356
Umezu, General, 372, 374
United States Air Force (also US Army Air Force),
    casualties, 158, 306, 312
    doctrine, 288, 312–13
    equipment, 272–3, 289, 312, 314, 319–21, 326–7, 329
    organization, 272–3, 287–9, 312–13, 320–1, 329
    strength, 273, 275, 287–9, 312, 321, 325, 329
United States armed forces,
    high command, 277–90, 294–5, 299, 303, 315–16, 325–6, 330
United States Army,
    casualties, 277, 311, 317, 373
    doctrine, 275
    equipment, 266, 272, 286–7, 298–9, 319, 328
    organization, 264–71, 273, 285–6, 290, 297–99, 314–17, 320–1, 328
    strength, 266–7, 270–5, 277, 285–6, 289–90, 297, 302, 314–17, 320–4, 328
United States Marine Corps, 32, 274–5, 311, 313–16, 322, 324, 369, 439

United States Navy,
    casualties, 278, 291, 314
    doctrine, 261–2, 278, 326
    equipment, 278–82, 296, 319–20, 322, 326, 328–9
    organization, 261, 263, 279
    strength, 260, 278–84, 312, 314, 320, 322, 328–9
Upton, General Emory, 265, 268, 285

Van Fleet, General James, 317
Vasilevsky, Marshal Alexander, 230–3, 236, 243–4, 246, 248, 253
Vatsetis, General I.I., 209–10
Vatutin, General N.F., 232, 235–6
Verdun, Battle of, 50, 52, 54, 105, 199
Versailles, Treaty of, 21, 58, 71, 116–18, 120, 124, 129, 212, 280, 347, 386
Vichy, French Government of, 71–4, 299
Vietinghoff, General Heinrich von, 156
Viet Minh, 77, 79
Vietnam, 77, 79, 322–8, 417, 423, 426, 428, 442
Vitheft, Admiral, 179–81
Vlasov, General A.A., 230
Voronov, Marshal N.N., 233, 252
Voroshilov, Marshal Kliment, 210, 212, 214, 218–19, 223–4, 227–8
Vuillemin, General, 66–8

Wagner, Colonel Arthur, 265
Wakatsuki Reijero, 351, 353
Waldersee, Field Marshal Alfred, Graf von, 91–2, 96, 99, 338
Walker, General Walton, 316
Wang Ching-wei, 357–8, 389–93, 409–10, 412, 419
Washington Naval Conference, 1922, 23, 69, 280–1, 347
Wavell, Field Marshall Earl, 28, 146, 294, 405
Wedemeyer, General Alfred C., 409
Weichs, Field Marshal Maximilian, Freiherr von, 146–7

Westmoreland, General William C., 322–5
Wetzell, Lieutenant-General, 395–6
Weygand, General Maxime, 57, 59, 65, 68–9, 71
Whampoa Military Academy, 388, 393–4, 398, 401, 412, 414, 416
Wheeler, General Earle G., 325
Wilhelm II, Kaiser, 6, 9, 19, 91–2, 97, 99, 103–6, 108–9, 113, 182, 196, 339
Wilhelm, Crown Prince of Prussia, 100, 105
Wilson, Admiral Sir Arthur, 9, 14
Wilson, Field Marshal Sir Henry, 9, 10, 12, 20, 46
Wilson, Field Marshal Lord (Henry Maitland), 304, 310
Wilson, President Woodrow, 20, 57, 109, 270, 278–80
Wingate, Major-General Orde, 31
Wood, General Leonard, 267, 269–70
Wrangel, General Peter, Baron von, 205, 211–12
Wu P'ei-fu, 386–7, 389–91

Yalta conference (*Argonaut*, February 1945), 245–6, 249, 310–11, 374
Yalu, River, 175, 177, 317, 335–6, 340, 418
Yamagata Aritomo, Marshal, 180, 332–4, 336–7, 339–40, 345–8
Yamamoto Gonnohyoe, Admiral, 340, 344–5
Yamamoto Isoroku, Admiral, 291, 344, 363, 365–8

Yang Tseng-hsen, 385
Yang Yü-ting, 392–3
Yegorov, General A.I., 211–12, 214, 221
Yeh Chien-ying, 421, 425–7
Yeh Ting, 404
Yen Hsi-shan (the Model Governor), 385, 390–3, 398, 400, 402, 404, 410, 415
Yeremenko, Marshal Andrei, 227, 231, 234, 240, 242–3, 248–9
Yezhov, Nikolai, 221, 227
Yonai, Admiral, 357, 374–5
Yoshida, Admiral, 357–8
Yuan Shih-kai, 335, 346–7, 380, 382–5, 394
Yudenich, General N.N., 208

Zakharov, General G.F., 236–7, 244
Zasulich, Lieutenant-General M.I., 175, 177–8
Zeitzler, General Kurt, 147, 153–5, 161
Zeller, General Marie-André, 85
Zhdanov, Andrei, 227
Zhigarev, Marshal P.F., 251
Zhilinski, General, 192–3, 195, 198
Zhukov, Marshal Georgi, 166, 210, 215, 222–3, 225, 227–9, 232–3, 236, 243–50, 253–5, 357
Zhukovski, N.E., 216
Zinoviev, Grigori, 213, 220
Zof, V.I., 217
Zuckerman, Lord, 306